Introduction to Computational Science and Mathematics

The Jones and Bartlett Series in Computational Science and Applied Mathematics

Richard A. Tapia, Consulting Editor

Introduction to Computational Science and Mathematics
Charles F. Van Loan

Introduction to Computational Science and Mathematics

Charles F. Van Loan
Department of Computer Science
Cornell University

Jones and Bartlett Publishers
Sudbury, Massachusetts
Boston London Singapore

Dedicated to my parents
Ted and Elizabeth Van Loan

Editorial, Sales, and Customer Service Offices

Jones and Bartlett Publishers
40 Tall Pine Drive
Sudbury, MA 01776
508-443-5000
800-832-0034

Jones and Bartlett Publishers International
Barb House, Barb Mews
London W6 7PA
England

Library of Congress Cataloging-in-Publication Data

Van Loan, Charles F.
An Introduction to computational science and mathematics / Charles Van Loan.
p. cm.
Includes bibliographical references and index.
ISBN 0-86720-473-7
1. Computer science—Mathematics. 2. Electronic digital computers—Programming. I. Title.
QA76.9.M35V37 1996
004'.01'51--dc20 95-40892
CIP

Printed in the United States of America

00 99 98 97 96 10 9 8 7 6 5 4 3 2 1

Contents

Acknowledgments

For the last several years the National Science Foundation has helped support the redesign of our introductory computer science courses here at Cornell. I have benefited from this support and from the lively discussions that took place in the context of the various course renovations.

The programs in the text are written in *Think Pascal*, a product of the Symantec Corporation. I find that the Think Pascal environment is very effective when teaching the concepts in this book to undergraduates.

Stephen Vavasis taught from a preliminary version of the text and made many useful suggestions. He and Professor Norman Kretzmann (philospher, colleague, neighbor) put me in touch with the two greatest "problem-solving" texts that I have ever read:

H. Dorrie, *100 Great Problems of Elementary Mathematics* (Translated by D. Antin), Dover Publications Inc., New York, 1958.

H. Steinhaus, *Mathematical Snapshots*, Oxford University Press, New York, 1951.

I break out in a mathematical fever every time I look over these wonderful books and I will consider my manuscript a success if it can engender the same kind of enthusiasm for computer problem solving.

Cindy Robinson of Cornell University was my administrative assistant during the four years it took to develop this book. The completion of the manuscript would have been impossible without her contributions.

Software

The software used in this book is available via the following URL:

ftp://ftp.cs.cornell.edu/cv/bookfiles.sit

This includes all the example programs and all the programs specified in Appendix 1.

Mathematical Facts

The level of "mathematical maturity" required to navigate the text may be anticipated by browsing through the following summary of "facts." Judge yourself by the number of these facts that you would be comfortable in using and *not* by the number of facts that you can reconstruct or derive.

Numbers

A real number is an *integer* if it has no fraction part.

If there is no remainder when an integer a is divided by an integer b, then b is a *divisor* of a.

A positive integer is *prime* if its only positive divisors are one and itself.

The *greatest common divisor* of two positive integers a and b is the largest integer that is a divisor of them both.

A real number is *rational* if it is the quotient of two integers.

Complex numbers have the form $a + bi$ where a and b are real and $i = \sqrt{-1}$.

Important Functions

Exponential: The function $f(x) = e^x$ where $e = 2.718..$ is called the *exponential.*

Logarithm: The base-b logarithm of a positive number y is the solution of $b^x = y$. $\ln(x)$ denotes the natural, base e logarithm.

Sine and Cosine: The line segment from (0,0) to $(\cos(\theta), \sin(\theta))$ makes a counterclockwise angle θ with the positive x-axis. The sine and cosine functions have period 2π meaning that $\cos(\theta + 2\pi) = \cos(\theta)$, $\sin(\theta + 2\pi) = \sin(\theta)$.

Tangent: The tangent of an angle θ is given by $\tan(\theta) = \sin(\theta)/\cos(\theta)$ with the understanding that the tangent is infinite if $\cos(\theta) = 0$.

Arctangent: The arctangent of a real number y is the solution x to $\tan(x) = y$ that satisfies $-\pi/2 < x < \pi/2$. It is an example of an inverse trigonometric function.

Polynomials: A polynomial of degree d has the form $a_0 + a_1x + \cdots + a_dx^d$ where $a_d \neq 0$ and has d roots.

Rationals: A rational function is the quotient of two polynomials.

Angles

2π radians = 360 degrees, one degree = sixty minutes, one minute = sixty seconds.

Not being fussy about where the x and y axes fall, we say that

$$(x, y) \text{ is in the } \left\{ \begin{array}{c} \text{First} \\ \text{Second} \\ \text{Third} \\ \text{Fourth} \end{array} \right\} \text{ quadrant if } \left\{ \begin{array}{c} x \geq 0,\ y \geq 0 \\ x \leq 0,\ y \geq 0 \\ x \leq 0,\ y \leq 0 \\ x \geq 0,\ y \leq 0 \end{array} \right\}.$$

Notation

Dot-Dot-Dot: $a_1, \ldots, a_n$ means a sequence of n numbers a_1, a_2 "on up to" a_n.

Summation: $\sum_{k=1}^{n} a_k = a_1 + \cdots + a_n$.

Product: $\prod_{k=1}^{n} a_k = a_1 a_2 \cdots a_n$.

Positional: $(d_{t-1} \cdots d_1 d_0)_{10}$ represents the value $d_0 + d_1 10^1 + d_2 10^2 + \cdots + d_{t-1} 10^{t-1}$.

Absolute Value: $|x| = \left\{ \begin{array}{ll} x & x \geq 0 \\ -x & x < 0 \end{array} \right.$.

Polygons

A polygon's *edges* meet at its *vertices.* (We assume that the edges do not cross.) If all the interior angles are equal and all the sides are equal in length, then the polygon is *regular.*

Polygons are known by the number of their sides. The *triangle* (3), *quadrilateral* (4), *pentagon* (5), *hexagon* (6), and *octagon* (8) are important special cases.

Triangles with two equal sides are *isosceles.* Triangles with three equal sides are *equilateral.*

A quadrilateral with one pair of parallel sides is a *trapezoid* and with two pairs of parallel sides is a *parallelogram.* A parallelogram with four equal sides (angles) is a *rhombus* (*rectangle*). A rectangle that is also a rhombus is a *square.*

If every line segment that connects two points on a polygon is inside the polygon, then the polygon is *convex.* (This means that it has no inward pointing corners.)

Distance, Area, Volume

Distance between points (x_1, y_1) and (x_2, y_2): $d = \sqrt{(x_1 - x_2)^2 + (y_1 - y_2)^2}$.

Area of a triangle with sides a, b, and c: $A = \sqrt{h(h-a)(h-b)(h-c)}$, $h = (a + b + c)/2$.

Area of a triangle with sides a and b and angle θ in between: $A = \frac{1}{2} ab \sin(\theta)$.

The area of a polygon with n vertices equally spaced around a circle of radius r:

$$A = \frac{nr^2}{2} \sin\left(\frac{2\pi}{n}\right).$$

The surface area of a radius r sphere: $A = 4\pi r^2$.

The volume of a radius r sphere: $V = \frac{4}{3}\pi r^3$.

The volume of a radius r, height h cone: $V = \frac{1}{3}\pi r^2 h$.

More Formulas

The average of $x_1, \ldots, x_n$:

$$A = (x_1 + \cdots + x_n)/n$$

The roots of a quadratic equation $ax^2 + bx + c = 0$:

$$\frac{-b \pm \sqrt{b^2 - 4ac}}{2a}, \qquad a \neq 0.$$

Powers:

$$x^n x^m = x^{n+m}, \qquad x^{-n} = 1/x^n, \qquad a^x = e^{\ln(a)x}$$

The line through the points (x_1, y_1) and (x_2, y_2):

$$y = y_1 + \frac{x - x_1}{x_2 - x_1}(y_2 - y_1) \qquad x_1 \neq x_2.$$

The circle with center (x_c, y_c) and radius r:

$$(x - x_c)^2 + (y - y_c)^2 = r^2.$$

The ellipse with center (x_c, y_c) and semiaxes r_x and r_y:

$$\left(\frac{x - x_c}{r_x}\right)^2 + \left(\frac{y - y_c}{r_y}\right)^2 = 1.$$

Parametric Equations

A line through the distinct points (x_0, y_0) and (x_1, y_1):

$$\begin{aligned} x(t) &= x_0 + t(x_1 - x_0) \\ y(t) &= y_0 + t(y_1 - y_0) \end{aligned} \qquad -\infty < t < \infty$$

A circle with center (x_c, y_c) and radius r:

$$\begin{aligned} x(t) &= x_c + r\cos(t) \\ y(t) &= y_c + r\sin(t) \end{aligned} \qquad 0 \leq t \leq 2\pi$$

An ellipse with center (x_c, y_c) and semiaxes r_x and r_y:

$$\begin{aligned} x(t) &= x_c + r_x \cos(t) \\ y(t) &= y_c + r_y \sin(t) \end{aligned} \qquad 0 \le t \le 2\pi$$

Combinations

The number of ways that n objects can be arranged is given by n *factorial*:

$$n! = 1 \cdot 2 \cdot 3 \cdots n.$$

The number of ways that k objects can be selected from n objects is given by the *binomial coefficient* n-choose-k:

$$\binom{n}{k} = \frac{n!}{k!(n-k)!}.$$

Trigonometric Identities

$$\begin{aligned} \cos(\theta/2) &= \sqrt{\frac{1+\cos(\theta)}{2}} && 0 \le \theta \le \pi/2 \\ \sin(\theta/2) &= \sqrt{\frac{1-\cos(\theta)}{2}} && 0 \le \theta \le \pi/2 \\ \cos(\theta+\phi) &= \cos(\theta)\cos(\phi) - \sin(\theta)\sin(\phi) \\ \sin(\theta+\phi) &= \sin(\theta)\cos(\phi) + \cos(\theta)\sin(\phi) \end{aligned}$$

Sets

If A is a set, then $x \in A$ means that x belongs to A.

$\mathbb{R}$ denotes the set of real numbers.

Types of Intervals:

$$\begin{aligned} [a,b] &= \{x \in \mathbb{R} : a \le x \le b\} \\ (a,b) &= \{x \in \mathbb{R} : a < x < b\} \\ (a,b] &= \{x \in \mathbb{R} : a < x \le b\} \\ [a,b) &= \{x \in \mathbb{R} : a \le x < b\} \end{aligned}$$

The first of these reads as "the set of all real x that are greater than or equal to a and less than or equal to b.

The *union* of two sets A and B: $A \cup B = \{x : x \in A \text{ or } x \in B\}$.

The *intersection* of two sets A and B: $A \cap B = \{x : x \in A \text{ and } x \in B\}$.

Differentiation

If $f(x)$ is differentiable at x_0, then $f'(x_0)$ is the slope of f at x_0 and is a limit:

$$f'(x_0) = \lim_{h \to 0} \frac{f(x_0 + h) - f(x_0)}{h}$$

The values of x where $f'(x) = 0$ are called *critical points*. The second derivative $f''(x)$ determines the nature or a critical point:

$$\begin{aligned} f'(x_*) = 0, f''(x_*) > 0 \quad &\Rightarrow \quad x_* \text{ is a local minimum} \\ f'(x_*) = 0, f''(x_*) < 0 \quad &\Rightarrow \quad x_* \text{ is a local maximum} \end{aligned}$$

To determine the maximum (minimum) of a function $f(x)$ on an interval $[a, b]$, check the value of f at the endpoints and at all local maxima (minima) in the interval.

The Mean Value Theorem:

$$\frac{f(b) - f(a)}{b - a} = f'(x_*) \qquad \text{for some } x \in [a, b].$$

Some well-known derivatives:

$f(x)$	$f'(x)$
$\sin(x)$	$\cos(x)$
$\cos(x)$	$-\sin(x)$
e^x	e^x
$a_0 + a_1 x + \cdots + a_d x^d$	$a_1 + 2a_2 x + \cdots + d a_d x^{d-1}$
$g(x)h(x)$	$g(x)h'(x) + g'(x)h(x)$
$\frac{g(x)}{h(x)}$	$\frac{h(x)g'(x) - g(x)h'(x)}{h(x)^2}$

Integration

The integral of a function $f(x)$ from a to b is denoted by

$$I = \int_a^b f(x)dx$$

and is the signed area under the graph of the integrand.

The fundamental theorem of calculus:

$$\int_a^b F'(x)dx = F(b) - F(a).$$

Preface

I wrote this book to help students learn something about computational science while they are developing their basic programming skills. The focus is on building intuition in certain key areas by studying well-chosen examples. The following article clarifies the philosophy that supports this approach. It first appeared in the September 1995 and October 1995 issues of SIAM News.

During the late 1970s I was asked to develop a "Programming for Poets" course for the non-technically oriented undergraduate. One approach would have been to water down the "serious" introductory programming course we offer to engineers and other quantitative types. But just at that time, I had the good fortune to read Donald Knuth's 1974 Turing Award lecture, "Computer Programming as an Art." Knuth points out that computer science is directly concerned with several of the seven original liberal arts,[1] an observation that started me thinking the right way about what we could achieve in "Poets." I became convinced that at the introductory level we should do more than just teach skills and I became intrigued by the following question: Given the importance of computer science, what should liberal arts students know about it?

Thinking about this question turned a potentially low-level, mundane teaching assignment into an exciting educational adventure, forcing me to think about what my colleagues and I do for a living and how it fits into the big picture.

A course with a double agenda unfolded. Yes, there was a practical-skills component. The students learned a programming language, refined some of their quantitative abilities, and solved a few interesting problems. But, at another level, through history and timely examples, they came to appreciate the notion of an algorithm and the culture of computing.

Liberal education is all about the different ways that human beings can express what they know, and algorithms are right up there with painting, the novel, and other revered forms of expression. No single form of expression is by definition any better than the others. The mere possession of a computer simulation doesn't imply that a university-trained ecologist knows more about acid rain than a Native American whose livelihood is threatened by the phenomenon. Programming for Poets gave students who came from every imaginable discipline – including history, dance, and landscape architecture – the critical skills necessary to deal effectively with the self-proclaimed computer experts.

More recently I have been thinking about how we teach computing to technically oriented students. The increasingly high profile of "computational science" in the research community led me to dwell upon a second classroom question: Given the importance of computational science, what part should it play in the freshman scientist/engineer's experience?

Thinking about the second question again led to a course with a double agenda, one that is best described in terms of the twin ideals set forth by the co-founders of Cornell University, Ezra Cornell and Andrew D. White. Ezra Cornell espoused the ideal of practical education. He wanted a university where the sons and daughters of farmers could acquire the scientific skills necessary to improve the state of agriculture. When this objective is mapped into the 21st century, it means

[1] Logic, grammar, geometry, arithmetic, rhetoric, music, and astronomy.

that computer science should provide undergraduate instruction in practical programming that enables students to return to their "computational farms" and make useful contributions. This defines the first agenda, which is to teach programming, i.e., computer problem solving.

Andrew White was the 19th-century embodiment of the ideal of liberal education and the perfect counterpart to the practically minded Cornell. To White, it was not enough for the fledgling university to provide skills-based instruction in the technical arts; history and literature were as critical to the uplifting of rural life as genetics and chemistry. It is unfortunate that these liberal education ideals are not appreciated with the same vigor today. This is partly due to the pressure to specialize, a pressure that is crowding the role that mathematics and computer science should play in the education of the computational scientist. That is why we should teach introductory programming with a second agenda: to trigger a life-long interest in how these two subjects fit into the technical culture.

What Is Computational Science?

In one way, "computational science" is the ultimate triumph of terminology, coupling the respectability of science with whatever is computational. But instead of regarding the term as a marketing ploy, think of computational science as a point of view. Science can be seen as a triangle with theory, experiment, and computation at its vertices (see Figure 1). Each vertex represents a style of research and provides a window through which we can look at science in the large. The vibrancy of what we see inside the triangle depends upon the ideas that flow around the edges of the triangle. A good theory couched in the language of mathematics may be realized in the form of a computer program, perhaps just to affirm its correctness. Running the program results in a simulation that may suggest a physical experiment. The experiment in turn may reveal a missed parameter in the underlying mathematical model, and around we go again.

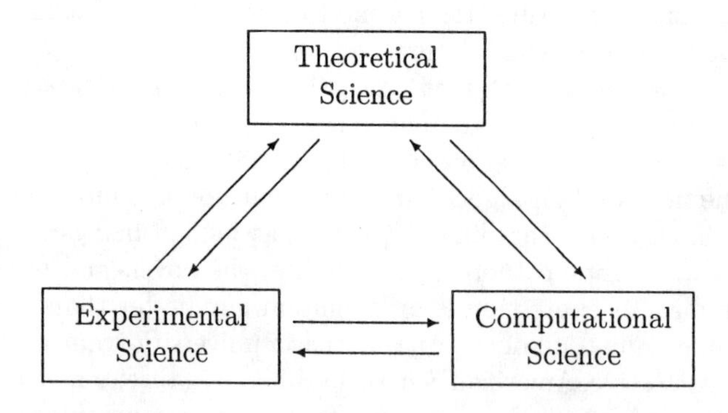

FIGURE 1. *The Research Triangle*

But interesting ideas can flow in the other direction as well. A physical experiment may be restricted in scope for reasons of budget or safety, so the scene shifts to computer simulation. The act of writing the program to perform the simulation will most likely have a clarifying influence, prompting some new mathematical pursuit. Innovative models are discovered, leading to a modification of the initial set of experiments, and so forth.

A parable will serve to clarify these interactions. Three scientists show up at the Leaning Tower of Pisa. E is experimental, T is theoretical, and C is computational. E arrives with a 16-pound shot-put, climbs the tower, releases the weight, and times the free fall. A paper is written and appears in a journal of experimental physics. T thinks about the experiment and then develops a mathematical model that captures the essence of the phenomenon. A paper is written and appears in a journal of theoretical physics. E looks over the equations and wonders what would happen if w were set to 1600 pounds. For practical reasons, the obvious experiment cannot be performed. However, C responds to the lament of E by transforming T's mathematics into a program. The program is run with $w = 1600$, and it is discovered that the hundredfold increase in weight does not have much of an effect upon the time of free fall. A paper is written and appears in a journal of computational physics. Moral: The interactions between mathematics, physical experimentation, and computer simulation are crucial to the scientific enterprise.

If we fail to communicate the dynamics illustrated by the parable, then the technical student will graduate with the anemic, weakly connected view of how things work in science and engineering shown in Figure 2. The time to start building the required appreciation is during the freshman year, and introductory programming is an appropriate vehicle. Think of the freshman computing experience as a trip along Route 66 from Chicago to Los Angeles, with computational science being the view outside the window of the car. If the trip is a success, then the freshman arriving in L.A. cannot stop talking about the landscape: "I got my kicks on Route 66 and want to revisit all those great sights." If it is a failure, then the student arrives in L.A. and cannot stop complaining about the car problems sustained along the way: "I got kicked on Route 66 and never want to go back." We must avoid the latter situation, where the syntactic side of computing and the anomalies of the system dominate the student's experience at the expense of what is truly important.

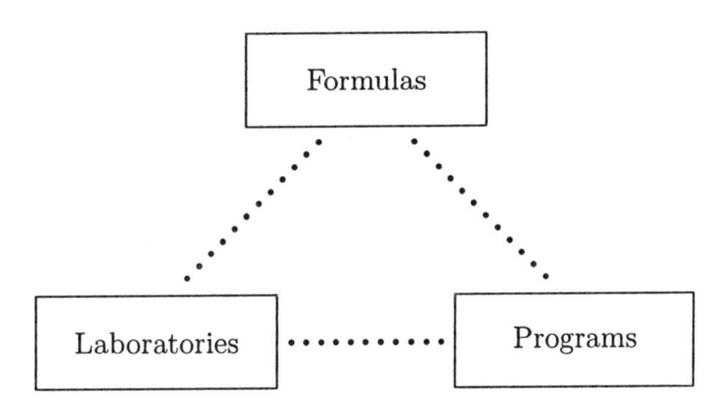

FIGURE 2. *The Research Triangle (Anemic Version)*

The key is to paint the landscape in colors so brilliant that the student will be absolutely entranced with the view. Is this possible with freshman-level mathematical maturity, typically defined by a semester or two of calculus? If our ambition is to build "computational intuition," the answer is yes.

The Five Computational Senses

Much has been written by philosophers and psychologists on the subject of intuition, and just about every scientist I know has something to say about it. Instead of deliberating on the topic, let us just assume that intuition is a sense of direction, essentially no different from the sense of direction that enables you to find your way around an old childhood neighborhood without a map. *The key is that you've been there before.*

If intuition is a sense of direction, then computational intuition is a sense of computational direction. Those who have it can find their way around the computational side of engineering and science. Success requires five keen senses, so we ask the freshman:

1. Do you have eyes for the geometric? The ability to visualize is very important to the computational scientist. Of course computer graphics plays a tremendous role here, but the visualization tools that it offers do not obviate the need to reason in geometric terms. The student must be totally at home with sines and cosines, polygons and polyhedra, metrics and proximity, and so forth.

2. Do you have ears that can hear the "combinatoric explosion"? Many design and optimization problems involve huge search spaces with an exponential number of possibilities. It is important to be able to anticipate this complexity and to have the wherewithal to handle it with intelligent heuristics.

3. Do you have a taste for the random? Many important processes have a random component. Having a sense of probability and the ability to gather and interpret statistics with the computer is vital.

4. Do you have a nose for dimension? Simulation is much more computationally intensive in three dimensions than in two dimensions – a hard fact of life that is staring many computational scientists right in the face. An accurate impression of how computers assist in the understanding of the physical world requires an appreciation of this point. Moreover, being able to think at the array level is essential for effective, high-performance computing.

5. Do you have a touch for what is finite, inexact, and approximate? Rounding errors attend floating-point arithmetic, terminal screens are granular, analytic derivatives are approximated with divided differences, a polynomial is used in lieu of the sine function, and the data acquired in a lab are correct to only three significant digits. Life in computational science is like this, and you just can't fall apart in the presence of such uncertainty. A steady balance is required along the fence that separates the continuous from the discrete.

We teach our children to use their physical senses, and we can teach our freshman to use their computational senses. In both arenas, experience with examples is critical.

Intuition and Rigor

Computational intuition is built by hanging around the right set of examples, a process that should begin during the freshman year. The first year of college is not the time to stress mathematical rigor or formal program correctness proofs. Interest in computational science is jeopardized if these things are pushed before the student is ready. However, the freshman year can be used to set the stage for the precision that mathematics has to offer if the connection between intuition and formality is understood:

$$\begin{aligned} \text{Formalism First} &= \text{Rigor Mortis} \\ \text{Intuition First} &= \text{Rigor's Mortise} \end{aligned}$$

Experience must precede abstraction. Formal methods that support the development and verification of programs are more easily learned by students who have spent a semester programming at a more informal level. Likewise, a mathematical concept that supports an interesting computation is more easily mastered if the student has quite literally played with the concept beforehand on the computer.

Critics of this philosophy tend to equate intuitive problem solving with ad hoc problem solving. This is unfortunate because it denigrates the role of intuition. There is a lesson to be learned here from the school of phonetic spelling. Many first-grade teachers find it easier to foster creative writing skills by accepting phonetic spelling. The "axioms" of correct spelling are gradually enforced in a way that does not stifle the child's originality. Correct spelling *is* important, but it is co-developed with other writing skills, not in advance of them. Likewise, the student's facility with mathematical rigor must be co-developed with other computational skills.

To illustrate these points further, I offer two examples. First, consider the path from freshman computing to complexity theory, a branch of computer science that deals rigorously with what makes certain computational problems hard to solve. A beginning student may start by benchmarking (timing) a number of programs that all solve the same problem. The sorting problem is a favorite because there are so many different methods from which to choose. After benchmarking, say, the methods of bubble sort and merge sort over a range of input lengths, the student discovers that they behave differently. The intuition acquired from the computational experience sets the stage for a discussion of running-time classifications. Merge sort turns out to be an $O(n \log n)$ algorithm, while bubble sort is $O(n^2)$. Here, as n (the number of items to be sorted) increases, the ratio of the bubble sort running time to the merge sort running time grows like $n/ \log n$. By observing this experimentally, the student develops intuition about running time.

Later, in follow-up courses, the student is brought into contact with other running-time classifications: $O(\log n)$, $O(n)$, $O(n^2)$, and so forth. This prompts a curiosity about lower bounds for running time and naturally leads to questions about inherent problem-solving difficulties. In this way the student is brought gracefully to the house of complexity theory and is ready to enter.

Another instructive example involves the path from elementary array manipulation to advanced scientific computation. For most students, the presentation of arrays in introductory programming is the first time they see n things portrayed as a single object. Matrix-vector multiplication is a classic example of an array computation that can and should be introduced at this level. However, the words "matrix" and "vector" cannot be mentioned because the student has yet to study linear algebra. But this situation doesn't last long, because, in one or two semesters, the student is engaged in a formal presentation of linear algebra by the mathematicians on campus. At this point, the prior experience in programming with arrays has built a facility with subscripts and an intuition about array-level operations that free the student to think about such central mathematical concepts as basis, independence, and rank.

The matrix/vector intuition acquired "over in math" then sets the stage for a return visit to elementary numerical analysis "over in CS." But the presentation of Gaussian elimination and its analysis in CS requires the use of matrix norms, a topic not fully covered in that first linear algebra course. So back over to math the student goes, perhaps stepping into a first course in functional analysis. With that expertise the student is ready for yet another return visit to CS and a course in the numerical solution of partial differential equations. And so it goes.

In this particular shuttle view of CS/math interactions, the student is more than just a Ping-Pong ball going back and forth between the two departments. The intuition acquired in one course sets the stage for the rigors of the next. Our job is to ensure a proper tread-to-riser ratio as the student climbs this curriculum staircase. Intuition is the tread, the firm, nonslip surface that sets the stage for the next level of abstraction. Rigor is the riser, and it should be negotiated only when the firmly planted foot says "ready."

Examples

The time has come to talk about examples and I would like to revolve the discussion around my motto-level knowledge of Latin:

1. *exempli gratis* (by way of example);
2. *E pluribus unum* (out of many, one);
3. *caveat emptor* (let the buyer beware);
4. *semper fidelis* (always faithful).

We can foster the development of the five computational senses defined above by teaching computer programming *exempli gratis*.

In teaching, writing, and research, there is no greater clarifier than a well-chosen example. Above and beyond clarification, however, examples generate interest. There seems to be a positive correlation between student enthusiasm and the use of examples. In a single lecture, examples illustrate and enliven when sprinkled among the theorems and abstractions. Throughout a semester, detailed, aptly positioned examples can make a particular course memorable. During a four-year undergraduate experience dominated by the weekly problem set, it is the "project course," with its focus on a single, large example, that often best prepares the student for the outside world. No matter what the educational time scale, examples play an important role. They engage the student.

The Frisking of Examples

There is good reason to invite examples into the computational science classroom. But just because an example shows up at the door doesn't mean that you should let it enter. *Caveat emptor*. When freshman are involved, all examples need to be frisked and interrogated with the following questions:
(a) Do you complement or uplift the student's mathematical expertise?
(b) Do you symbolize what goes on in computational science?
(c) Do you engender a burning curiosity for mathematics and/or computer science?
(d) Do you squelch naive views about computer problem solving?

Let us look at some examples that, when scrutinized in this fashion, get by the security guards.

Reinforcing Examples

Consider the problem of determining whether a real number x is inside an interval $[a, b]$. One way to do this is to see if x is to the right of a and to the left of b:

```
In:=(a<=x) and (x<=b)
```

Another way is to make sure that x is neither to the left of a nor to the right of b:

```
In:=not ((x<a) or (b<x))
```

A discussion of these two alternatives amounts to a "90% proof" of De Morgan's law,

$$A \wedge B \equiv \neg(\neg A \vee \neg B),$$

a well-known theorem in propositional logic that relates the "and" and "or" operations. The formal proof of this law may be presented in a discrete mathematics course, but until then the student is well served by this proof by example.

Just as a computing problem can set the stage for future mathematical work, it can also serve as an occasion to use a bit of mathematics already learned. Consider this illustration of nested conditionals:

```
if a<=b then
    if a<=c then
       min:=a
    else
       min:=c
else
    if b<=c then
       min:=b
    else
       min:=c
```

The illustration happens to assign the smallest of a, b, and c to min, and, as such, it is a fine example of nested conditionals. But if the student knows first-semester calculus, to use this illustration is to squander an opportunity: A better illustration is to have the student compute the minimum value of a quadratic $x^2 + bx + c$ on the interval $[L, R]$:

```
L_Slope:=2*L+b;
R_Slope:=2*R+b;
if L_Slope>=0 then
    min:=L*L+b*L+c
else
    if R_Slope<=0 then
       min:=R*R+b*R+c
    else
       min:=c-b*b/4
```

This, too, illustrates nested conditionals, but it also serves to reinforce the student's calculus expertise. It makes use of the fact that the minimum of a continuously differentiable function on an interval occurs either at the endpoints or at a point where the derivative is zero.

Symbolic Examples

The symbolic example is a snapshot that illustrates some particular concern or worry of the computational scientist. Out of many such snapshots should emerge one view of the discipline. *E pluribus unum.*

For example, the Earth's surface area in square kilometers can be computed from the formula where $A = 4\pi r^2$ is the radius in kilometers:

```
c:=4.0*3.14159;
r:=6378;
A:=c*r*r;
```

This looks harmless enough as a vehicle for talking about expressions and assignment. If we poke around, however, we find that there is a lot more to the example than meets the eye. Is the Earth a sphere or an oblate spheroid? (A model error question.) Is the radius exactly 6378 kilometers? (A data-error question.) Does π equal 3.14159? (A mathematical error question.) Does the computer multiply 4 and 3.14159 exactly to get 12.56636? (A roundoff-error question.) All kinds of error beset the computational scientist, and this simple example is symbolic of these difficulties.

When are three points collinear? This seemingly simple problem can be used to introduce a basic programming skill such as the writing of Boolean-valued functions. But in a world filled with fuzzy data and inexact arithmetic, we quickly discover that life isn't so Boolean after all. To handle collinearity problems in practice, we need to develop suitable measures of near-collinearity, a surprisingly complicated problem that is guaranteed to make any curious student reflect upon the descriptive power of mathematics. Accepting that nothing is simple whenever computers are involved is the first step on the road to becoming an enlightened computational scientist. The collinearity problem dramatizes this point.

What is the design process, and how do computers fit in? "Industrial-strength" design problems are typically very complex, often involving thousands of parameters and requiring sophisticated computer environments for their solution. Still, the key ideas can be communicated with a small symbolic example. Consider the design of a mountain bike that requires the selection of three pedal sprockets and seven wheel sprockets. The problem is to choose the ten sprockets so that the 21 gear ratios are uniformly spread across the real interval [1,4] subject to constraints. One constraint, imposed by the marketing experts, could be that the lowest and highest ratios should be 1 and 4, respectively. Another constraint, imposed by the wholesale buyer, could be that the sprockets come in a limited number of sizes.

Although this example is small, it is rich enough to support the discussion of key issues, such as the notion of parameter-space dimension, the combinatoric explosion of possibilities, the choice of objective function, the role of heuristics, and the implications of constraints. A simple interactive environment can be set up to illustrate how the computer can accelerate the search for the optimal design. The 10-parameter mountain bike design problem is symbolic of all design problems.

Seductive Examples

No matter what course of study the freshman ultimately pursues, our goal should be that he or she maintain a faith in mathematics and computer science. *Semper fidelis.* Well-chosen, seductive examples have the effect of making the student hungry for computer science and mathematics by pointing to their role in the modern conduct of science and engineering.

Consider the problem of finding the smallest ellipse (in area) that encloses n given points in the plane. A simple interactive environment can be set up that facilitates the "clicking in" of trial ellipses. The student may in fact help build the environment as a programming exercise by writing, for example, a function that can test for inclusion of the point set. However, after enough experimentation, the student will begin to wonder if the search for the optimum ellipse can be automated. The tumblers will fall into place, and the example will have unlocked an interest in computational geometry.

Sobering Examples

Computational science has its share of belligerent know-it-all types who bully their way around by underplaying the realities of computing. One such reality is the chasm between formula and production software. The computer belligerent thinks that the mere transcription into code of a math book formula like

$$x_{n+1} = x_n - \frac{f(x_n)}{f'(x_n)}$$

is all that is required to produce usable software. But a freshman-level excursion into Newton's method will expose the fallacies of this wishful thinking and have a sobering effect.

Another type of computational belligerence has to do with the finiteness of computer arithmetic. Lack of understanding in this area sets the stage for an exaggerated view of computational accuracy and makes possible tragedies like the Patriot missile disaster of the Gulf War. But the freshman who hangs around the right set of sobering examples will not be prone to pitfalls like this.

A final kind of sobering example has a counterintuitive, less-is-more theme. If a spherical Earth ($r = 6378$ km) is glazed with a one-micron layer of gold, what would the increase in surface area be? The exact increase, ΔA, is given by:

$$\Delta A \approx 4\pi((r + \Delta r)^2 - r^2)$$

and gives rise to the computation

```
r:=6378;
Delta_r:=0.000000001;
Delta_A:=4*pi*(sqr(r+Delta_r) - sqr(r)).
```

An approximation to the increase (derived using the calculus) is given by:

$$\Delta A \approx 8\pi r \Delta r$$

and leads to:

```
r:=6378;
Delta_r:=0.000000001;
Delta_A:=8*pi*r*Delta_r.
```

On many computers, however, the exact-formula method gives zero while the approximate-formula method gives an answer that is much closer to the true value. Exact, closed-form recipes are a breeding ground for unjustified confidence and set the stage for computational belligerence.

The nice thing about sobering examples is how easy they are to discover. Almost all freshman-level concepts in mathematics have subtleties when you play with them on the computer. The right combination of sobering examples can send a powerful message to the student.

Conclusion

The examples used when we teach freshman computing act as a set of "basis vectors"; everything that the student learns is in their span. Our job as professors is to choose that basis carefully to ensure the development of computational intuition.

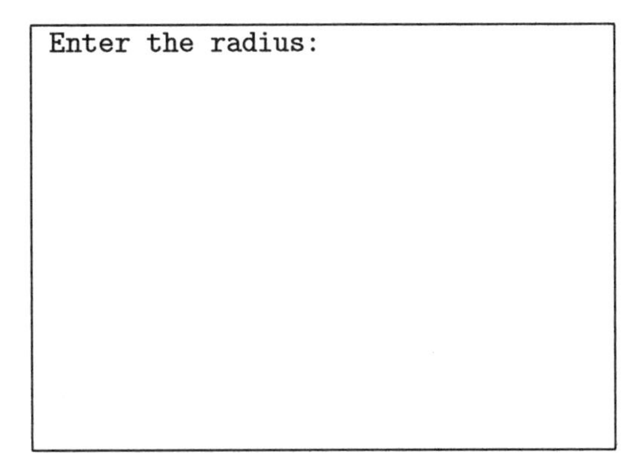

FIGURE 1.2 *Output In Text Window*

variable definitions can be given at judicious points in the program body. This is be illustrated in §1.2.

Notice how colons separate the variable name and the designated type. A semicolon comes after each declaration. Right now we just have the `real` type. Other types are described later. The following rules always apply:

- Every variable used in the program must be declared exactly once.
- The order in which variables are declared does not matter.

We now consider the program body of `Example1_1`. It consists of five statements that are separated from each other by semicolons:

```
ShowText;
writeln('Enter the radius:');
readln(r);
A := 4*3.14159*r*r;
writeln('The surface area = ', A:10:6);
```

The statement

```
ShowText;
```

says that program output will be displayed on the screen in a designated area called *TextWindow.*

After the `ShowText` statement, the action flows rather naturally. The statement

```
writeln('Enter the radius:');
```

prints the message "Enter the radius" in the TextWindow. See FIGURE 1.2. All `writeln` statements have the form

`writeln(`⟨*Description about what is to be printed.*⟩`)`

The first `writeln` prints a message. The message is enclosed in single quotes. This message is referred to as a *prompt* because its sole mission is to prompt the typist to enter some data, in this case, the radius.

The processing of the program user's response is handled by the next statement in the program:

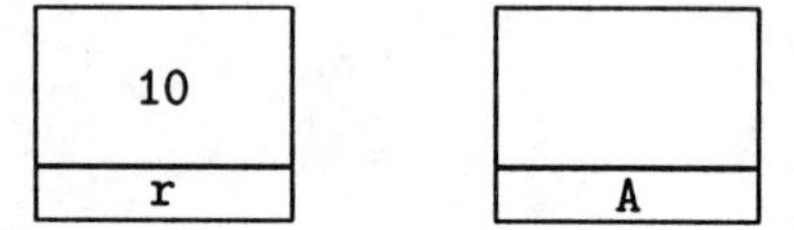

FIGURE 1.3 *Assignment to* r

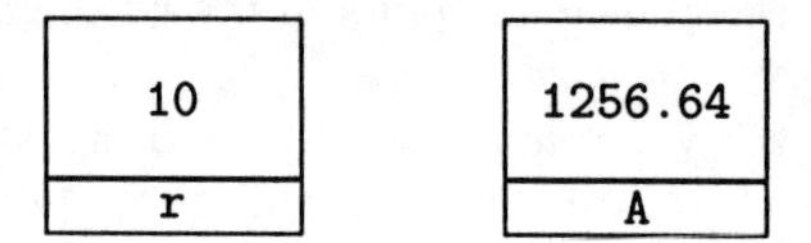

FIGURE 1.4 *Assignment to* A

```
readln(r);
```

This takes the entered number and stores it in the variable `r`. For example, if we respond to the prompt by typing "10" and then striking the $< \mathit{return} >$ button, then the value of 10 is placed in the variable `r` as shown in FIGURE 1.3. Once the data for the problem is in the computer's memory, we are (at long last) ready to invoke the $A = 4\pi r^2$ formula:

```
A := 4.0*3.14159*r*r;
```

This is an *assignment statement.* The righthand side is a recipe for a numerical value. The left hand side names the variable (the box) where the computed value is to be stored. FIGURE 1.4 shows the status of `r` and `A` after the assignment. The assignment of a value to a real variable has the following form

$\langle$*real variable name*$\rangle$:= $\langle$*real arithmetic expression*$\rangle$;

Notice the punctuation. The *assignment operator* ":=" separates the recipe that produces the value from the name of the variable that is to receive the value. The statement ends with a semicolon. Although it is tempting to read the assignment as "A equals $4\pi r^2$", it is much more accurate to say that "A is *assigned* the value of $4\pi r^2$". This is because the assignment is an action, not a statement about equality. This point is clarified later.

When literal real values like "4" and "3.14159" are included in an arithmetic expression, some care must be exercised. It is not necessary to include a decimal point. Thus,

```
A := 4*3.14159*r*r;
```

is equivalent to `A := 4.0*3.14159*r*r`. A scientific notation is also available:

```
Enter the radius:
10
The surface area = 1256.64
```

FIGURE 1.5 *Output In Text Window*

```
A := 0.04E+2*3.14159*r*r;
```

although in this case, it is not a convenience. The one point of inflexibility is that any decimal point must be preceded by a digit. Thus,

```
A := .4*31.4159*r*r;
```

is illegal.

The assignment statement under consideration has a particularly simple arithmetic expression: `4*3.14159*r*r`. Asterisks are used to designate multiplication. Because the arithmetic expression is a recipe for a value, it is essential that (a) all the ingredients be present and (b) the variable used to store the result has been declared. Rule (a) is violated if we try to assign the surface area to `A` before acquisition of the radius `r`, e.g.,

```
A := 4*3.14159*r*r;
readln(r);
```

The order of statements in a program body is very important. Likewise, it is essential that `A` be declared. Otherwise there would be no "place" to put the value prescribed by `4*3.14159*r*r`.

The last action to be performed is the printing of the result and for that purpose we have a second `writeln` statement:

```
writeln('The surface area = ', A:7:2);
```

The line printed has two parts, a message and a number. See FIG.1.5 To specify the number that is to be printed and its appearance, we name the variable that contains the number and use a pair of integers to specify the format. `A:7:2` says "print the value of `A` allocating 2 digits for the fraction part and a minimum of 7 "spaces" for the number overall.

Choosing a format so that the output looks nice requires some care. It is typically a detail that should not be addressed until after a preliminary version of the program is working, for then you know the size of the numbers involved and can apply sensible formats. Another consideration is the precision of the computer's arithmetic. This is discussed in the next section. Italics are used to indicate user-responses to prompts from the program.

We give another example program to clarify the ThinkPascal constructs just learned. `Example1_2` solicits a radius r (assumed to be in miles) and a modification Δr (assumed to be in inches) and prints the increase in spherical surface area (in square miles) when the radius is increased from r to $r + \Delta r$: First we see that in a declaration, it is possible to declare several variables "at once." Thus,

```
program Example1_1;
{Computes the surface area of a sphere.}
var
   r:real; {The radius of the sphere.}
   A:real; {The surface area of the sphere.}
begin
   ShowText;
   writeln('Enter the radius:');
   readln(r);
   A := 4.0*3.14159*r*r;
   writeln('The surface area = ', A:7:2);
end.
```

Sample output:

```
Enter the radius:
10
The surface area = 1256.64
```

```
program Example1_2;
{Explores how the surface area of a sphere changes with changes}
{in the radius.}
var
   r,newr:real; {The two radii in miles.}
   delta:real; {The increase in radius in inches}
   Area,NewArea:real; {Area = A(r), NewArea = A(newr)}
   increase:real; {The increase in surface area.}
begin
   ShowText;
   writeln('Enter r (miles) and delta (inches):');
   readln(r,delta);
   newr := r + ((delta/5280.0)/12.0);
   Area := 4.0*pi*r*r;
   NewArea := 4.0*pi*newr*newr;
   increase := NewArea - Area;
   writeln('Increase in area (square miles) = ', increase:10:6)
end.
```

Sample output:

```
Enter r (miles) and delta (inches):
4000      10
Increase in area (square miles) = 16.000000
```

```
r,newr:real;
```

sets up real variables r and newr. This is often convenient when the simultaneously declared variables are related, for then a single unifying comment can establish their connection.

A new feature of Example1_2 is the use of the *built-in constant* pi. The number π is so important that it is incorporated in the ThinkPascal language. When pi appears in an arithmetic expression, its value as defined by ThinkPascal is substituted.

Example1_2 also shows that it is possible to acquire more than one number with a single readln. Suppose the typist replies to the prompt by entering two separate numbers on a line like this:

4000   100.5

The effect of readln(r,delta) is to assign the real value 4000.0 to r and the real value 100.5 to delta.

Notice the use of parentheses in the assignment

```
newr := r + ((deltar/5280.0)/12.0);
```

It turns out that

```
newr := r + deltar/5280.0/12.0;
```

renders the same value because of the *rules of precedence.* These are rules that determine the order of operations in an arithmetic expression. Unless overridden by parentheses, multiplicative operations (*, /) are performed before all additive operations (+,−). A succession of multiplicative operations or a succession of additive operations are performed left to right. Thus, a := 1/2/3/4 is equivalent to a := ((1/2)/3)/4 and different from a := 1/(2/(3/4)). Always use parentheses in ambiguous situations.

Even though the above programs are short and simple, they reveal a number of very important aspects about errors in the computational process. Suppose that the program is used to compute the surface area of the Earth. To begin with, there is a *model error* associated with the assumption that the Earth is a perfect sphere. In fact, the shape of the Earth is better modeled by an oblate spheroid, i.e., an ellipse of revolution. Second, if we use the radius value $r = 3960$ and if this value is determined experimentally, say by a satellite measurement, then there is undoubtedly a *measurement error.* Perhaps the satellite instruments are sensitive to four significant digits and that the "real" Earth has a radius of 3960.2 miles. Third, there is the *mathematical error* associated with the assumption that $\pi = 3.14159$. A better, but still inexact approximation of this constant is 3.1415892653589. Finally, there is the *roundoff error* associated with the actual computation A := 4*3.14159*r*r. Computers do not do exact real arithmetic. Just as the division of 1 by 3 on your calculator produces something like .333333 and not 1/3 exactly, so may we expect the computer to make a small error every time an arithmetic operation is performed. This is discussed in §2.3.

A great deal of computing experience is required before the interplay between these factors can be fully appreciated. One of our goals is to communicate these subtleties and to build your intuition about them.

PROBLEM 1.1. Assume that the Earth is a sphere with radius 4000 miles. By running `Example1_2` three times, determine the increase in surface area if the Earth is uniformly paved with 1, 5, and 10 inches of cement. The results suggest that the method of calculation is flawed for very small paving depths. The reason is related to the precision of the computer's arithmetic. Modify `Example1_2` so that it also computes the approximate surface area increase via the following formula:

$$\Delta A = A(r + \Delta r) - A(r) \approx 8\pi r \Delta r.$$

This follows the derivative of $A(r)$ can be approximated very well by a divided difference if Δr is very much smaller than r:

$$A'(r) \approx \frac{A(r + \Delta r) - A(r)}{\Delta r}.$$

Compare the two methods using the following choices for r and Δr:

r	Δr
4000	1
4000	5
4000	10
4000	1000

An explanation will follow in §2.3.

PROBLEM 1.2. The surface area of an oblate spheroid such as the Earth is given by $A = 4\pi r_1 r_2$ where r_1 is the equatorial radius and r_2 is the polar radius. Write a program that reads in these two radii and computes the difference between $4\pi r_1 r_2$ and $4\pi((r_1 + r_2)/2)^2$. Use the Earth data $r_1 = 3963$, $r_2 = 3957$.

1.2 Sines and Cosines

Let us continue the discussion of formulas and programs using as examples some well-known trigonometric identities. Our first calculations involve the *half-angle* formulae for cosine and sine:

$$\cos(\theta/2) = \sqrt{(1 + \cos(\theta))/2} \qquad 0 \le \theta \le \pi/2.$$
$$\sin(\theta/2) = \sqrt{(1 - \cos(\theta))/2} \qquad 0 \le \theta \le \pi/2.$$

These recipes can be used to compute sines and cosines of various angles from a sines and cosines of other angles. For example, since $\cos(\pi/4) = 1/\sqrt{2}$ we can use the half-angle formula for sine to compute $\sin(\pi/8)$:

```
a := 2;
c:= 1/sqrt(a);
s = sqrt((1 - c)/2);
writeln('sin(pi/8) =',s:10:6);
```

An excerpt from a program like this is called a *fragment.* We use fragments to communicate new programming ideas whenever the inclusion of the whole program burdens us with unnecessary detail. Right now, we are *not* interested in var declarations or program headings. The focus is on the assignment statement and the fragment illustrates some new features about this construct. We may infer that a, c, and s are declared variables of type `real`.

The fragment makes use of the `sqrt`, one of several *built-in* functions that are part of the ThinkPascal language. If a is a real variable that contains a nonnegative value, then `sqrt(a)` returns the value of its square root. The `sqrt` function can accept literals and so it is legal to replace the first two assignments with

```
c = 1/sqrt(2);
```

More generally, sqrt can be applied to any arithmetic expression that produces a nonnegative value, e.g.,

```
c := 1/sqrt(2);
s := sqrt((1-c)/2)
```

We can "collapse" the fragment even further:

```
s = sqrt((1-(1/sqrt(2)))/2)
```

But now the role of the half angle formulae is obscured by all the parentheses. Avoid "dense" one-liners like this. It is better to spread the computation of over a few lines thereby highlighting some of the important intermediate results that arise during the course of computation.

The little cosine/sine computation gives us the opportunity to clarify three different types of error. If we type

```
a := 2;
c:= 1/sqrt(a);
s := sqrt((1 - c/2);
writeln('sin(pi/8) =',s:10:6);
```

instead of (1.2.1), then a *syntactic* error results because there are unbalanced parentheses in the third assignment statement. Syntactic errors are detected at *compile time* because the ThinkPascal compiler has discovered a grammatical violation. Failing to declare a variable is another common syntactic error.

Even if a program compiles it may not run to completion because of a *run-time* error. For example,

```
a := -2;
c:= 1/sqrt(a);
s := sqrt((1 - c)/2);
writeln('sin(pi/8) =',s:10:6);
```

is syntactically correct, but sqrt breaks down because it is applied to a negative number. The fragment does not run to completion.

In addition to syntactic and run-time errors, there are *programmer errors*:

```
a := 2;
c:= 1/sqrt(a);                                   (1.2.5)
s := sqrt(1 - c/2);
writeln('sin(pi/8) =',s:10:6);
```

This compiles and runs to completion, but the desired output is not produced. The deleted parentheses in the assignment to s means that the program is computing $\sqrt{1-(c/2)}$ instead of $\sqrt{(1-c)/2}$. Programmer errors are often the hardest to detect because the editor and the compiler are not there to point out our mistakes. Moreover, we may be so excited that our program actually runs that we overlook its correctness!

The issue of program correctness is particularly complex, especially as programs get long. One of our goals is to develop problem-solving strategies that are organized in such a way that we can be confident about a program's correctness. [1]

Let us move on to a more complicated example. We can compute $\sin(\pi/16)$ by repeated application of the half-angle formulae:

[1]Just being *confident* in a program's correctness does not mean that we can *guarantee* its correctness. That's a very mathematical exercise and beyond the scope of this introductory text.

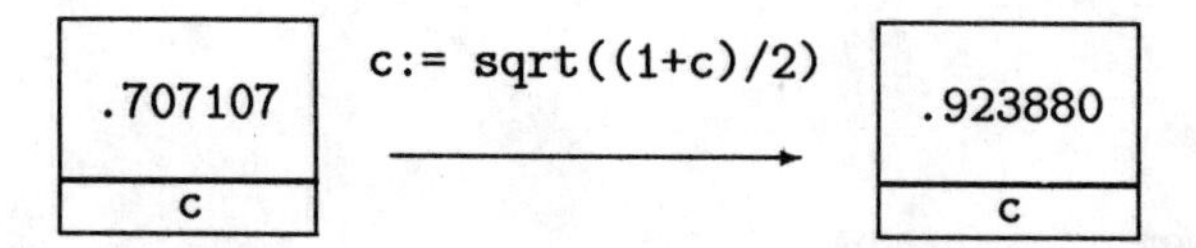

FIGURE 1.6 *Overwriting*

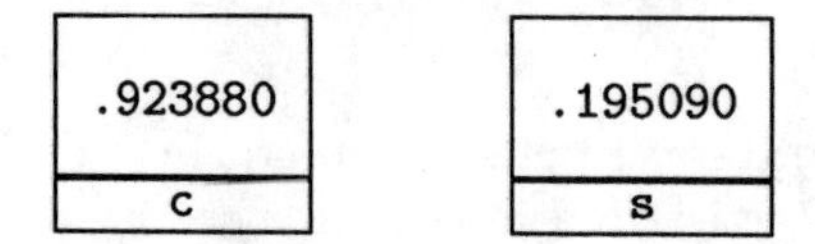

FIGURE 1.7 *The Final Cosine/Sine Values*

```
c = 1/sqrt(2); { c = cos(pi/4) }
c:= sqrt((1+c)/2); { c = cos(pi/8) }
s:= sqrt((1-c)/2); { s = sin(pi/16) }
writeln('sin(pi/16) = ',s:10:6);
```

The inclusion of comments helps us trace what happens as the computation unfolds. These cryptic, mathematical comments are called *assertions*. After the assignment of $1/\sqrt{2}$ to c we can *assert* that $c = \cos(\pi/4)$.

The next assignment computes $\cos(\pi/8)$ from $\cos(\pi/4)$ using the cosine half-angle formula. As usual, to the right of the ":=' is a recipe, or more precisely, a real-valued arithmetic expression. To the left of the assignment symbol is the name of the variable where the result is to be stored, in this case, c. Thus, even though c is involved in the expression, it is the "target" of the operation. FIGURE 1.6 shows how we should visualize the update of c's value. The term *overwriting* is often used in this context. The current contents of a variable can be overwritten with a new value. The old value is "erased" much as the current contents of a audio cassette is erased during recording.

Finally, the sine is computed and the values displayed in FIGURE 1.7 are displayed.

These two values could be computed directly by using the built-in functions `sin` and `cos`. These functions assume input values in radians. Thus, if `pi` houses the value of π, then

```
s:= sin(pi/16);
```

is mathematically equivalent to

```
c:= 1/sqrt(2);
s:= sqrt(1-c/2)
```

Not surprisingly, there is often more than one way to compute the same thing.

```
program Example1_3;
{Illustrates the double angle formulae for sine and cosine.}
var
   a,theta:real; {The input angle in degrees and radians.}
   c,s:real; {For cosines and sines.}
   cosError,sinError:  real; {Errors.}
begin
   ShowText;
   writeln('Enter angle (degrees):');
   readln(a);
   theta := a*(pi/180); {the angle in radians}
   c := cos(theta);
   s := sin(theta);
   ctemp := c;
   c := c*c - s*s; {cos(2*theta)}
   s := 2*ctemp*s; {sin(2*theta)}
   cosError := abs(c-cos(2*theta));
   sinError := abs(s-sin(2*theta));
   writeln('cos(2*theta) = ', c:8:6, ' cosError = ', cosError);
   writeln('sin(2*theta) = ', s:8:6, ' sinError = ', sinError);
end.
```

Sample output:

```
Enter angle (degrees):
30
cos(2*theta) = 0.500000       cosError = 4.6e-9
sin(2*theta) = 0.866025       sinError = 3.0e-8
```

Next we consider the double angle formulae:

$$\begin{aligned}\cos(2\theta) &= \cos^2(\theta) - \sin^2(\theta) \\ \sin(2\theta) &= 2\sin(\theta)\cos(\theta)\end{aligned}$$

The program Example1_3 solicits an angle and then computes the sine and cosine of the double angle. Note that the input angle is given in degrees and is then converted to radians[2]. This is necessary because sin and cos assume radian arguments.

After the conversion to radians, the $\cos(\theta)$ and $\sin(\theta)$ are computed and stored in c and s respectively. Using the double angle formulae, these values are replaced by $\cos(2\theta)$ and $\sin(2\theta)$. Note that the variable ctemp is necessary to hold $\cos(\theta)$. The fragment

```
c := cos(theta);
s := sin(theta);
c := c*c - s*s; {cos(2*theta)}
s := 2*c*s; {sin(2*theta)}
```

[2] 180 degrees = π radians

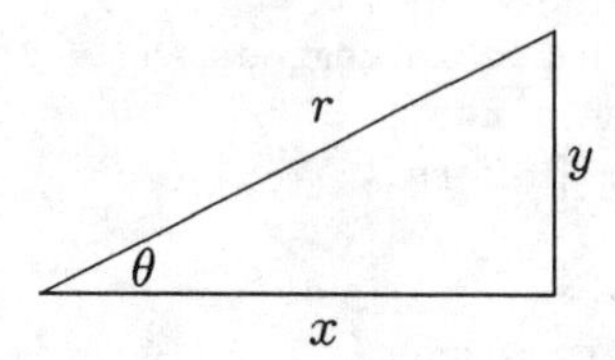

FIGURE 1.8 $\sin(\theta) = y/r$ *and* $\cos(\theta) = x/r$

does *not* assign $\sin(2\theta)$ to s since the value of c used in the last assignment is $\cos(2\theta)$ and not $\cos(\theta)$.

For both cosine and sine, the absolute value function abs is used to compute the discrepancy between the double angle method and the direct method. No formats are given in the writeln and as a result, the values are printed in "scientific notation."

As final example, suppose x and y house positive real values x and y and that θ is defined by $\tan(\theta) = y/x$. Our goal is to assign the values $\cos(\theta)$ and $\sin(\theta)$ to real variables c and s. One approach is to use the built-in function arctan.[3] Thus, if theta is a declared real variable, then

```
theta := arctan(y/x);
c:=cos(theta); s:=sin(theta)
```

makes the required assignments to c and s. Alternatively, from FIGURE 1.8 we also have

```
r:=sqrt(sqr(x) + sqr(y));
c:=x/r; s:=y/r;
```

This makes use of the built-in function sqr that can be used to square values. The assignment to r is equivalent to

```
r:=sqrt(x*x + y*y);
```

Again we see that there is more than one way to compute the same thing. This is especially true in the trigonometric area where there are so many identities to work with.

PROBLEM 1.3. Modify Example1_3 so that it applies the double angle formulae twice to produce the sine and cosine of 4θ where θ is the input angle. Print the discrepancies from the values sin(4*theta) and cos(4*theta).

PROBLEM 1.4. Given that $\cos(60^o) = 1/2$ and $\cos(72^o) = (\sqrt{5} - 1)/4$, write fragments that print the following values: **(a)** $\sin(18^o)$, **(b)**$\cos(3^o)$, and **(c)** $\sin(27^o)$. Do not make use of the built-in functions sin, cos, or arctan. Use the half-angle formulae and the identities

$$\begin{aligned}
\cos(\theta_1 + \theta_2) &= \cos(\theta_1)\cos(\theta_2) - \sin(\theta_1)\sin(\theta_2) \\
\sin(\theta_1 + \theta_2) &= \cos(\theta_1)\sin(\theta_2) + \sin(\theta_1)\cos(\theta_2) \\
\cos(-\theta) &= \cos(\theta) \\
\sin(-\theta) &= -\sin(\theta)
\end{aligned}$$

[3] Recall that if $u = \arctan(v)$, then $\tan(u) = v$.

1.3 Max's and Min's

Let us pose some questions about the behavior of the the quadratic function

$$q(x) = x^2 + bx + c$$

on the interval $[L, R]$:

> Q_1: Which is smaller, $q(L)$ or $q(R)$?
>
> Q_2: Does the derivative $q'(x)$ have a zero in the interval $[L, R]$?
>
> Q_3: What is the minimum value of $q(x)$ in $[L, R]$?

To answer each question it is necessary to make a comparison of values and then branch to an appropriate course of action. The `if` construct is required to handle this kind of situation.

Consider the first problem where we want to decide whether $q(x)$ is larger at $x = L$ or at $x = R$. Assume that `L`, `R`, `b`, and `c` are initialized real variables and that $L \leq R$. Here is a fragment that prints a message based upon the comparison of $q(L)$ and $q(R)$:

```
{Fragment A}
qL := sqr(L) + b*L + c;
qR := sqr(R) + b*R + c;
if qL < qR then                                        (1.3.1)
   writeln('q(L) < q(R)')
else
   writeln('q(L) >= q(R)');
```

The fragment begins by storing the value of $q(L)$ and $q(R)$ in the variables qL and qR respectively. Clearly, these variables should have type `real`. Once these values are stored, a comparison is made. If the value of qL is smaller than the value qR, then the message

```
q(L) < q(R)
```

is printed. Otherwise,

```
q(L) >= q(R)
```

is printed.

Alternatives like this are very common in computing and the `if-then-else` construct is designed to navigate the "fork in the road":

```
if ⟨Condition⟩ then
   ⟨Something to do if the condition is true.⟩
else
   ⟨Something to do if the condition is false.⟩
```

Three reserved words are associated with this statement: `if`, `then`, and `else`. The condition is called a *boolean expression*. Just as arithmetic expressions produce numbers, so do boolean expressions produce true-false values. Once qL and qR have been assigned values, the boolean expression `qL < qR` is either true or false. The symbol `<` stands for "less than" and is one of the several *relational operators* listed in FIGURE 1.9. There are often several ways to organize an `if-then-else`. For example, the fragment

Operator	Meaning
>	greater than
>=	greater than or equal to
=	equal to
<>	not equal to
<=	less than or equal to
<	less than

FIGURE 1.9 *Relational Operators*

```
{Fragment B}
qL := sqr(L) + b*L + c; qR := sqr(R) + b*R + c;
if qL >= qR then
   writeln('q(L) >= q(R)')
else
   writeln('q(L) < q(R)');
```

is equivalent to `Fragment B`. If we play with the underlying mathematics, then other possibilities unfold. For example, since

$$q(L) - q(R) = (L^2 + bL + c) - (R^2 + bR + c) = (L^2 - R^2) + b(L - R) = (L - R)(L + R + b)$$

we also have

```
{Fragment C}
if (L-R)*(L+R+b) > 0 then
   writeln('q(L) > q(R)')
else
   writeln('q(L) < q(R)')
```

Sometimes there is "more than one thing to do" as a result of a comparison. For example, we may want the `if-then-else` to handle the following situation:

```
if (L-R)*(L+R+b) > 0 then
   ⟨Print a message 'q(L) ≥ q(R)' and compute the slope of q at L.⟩
else
   ⟨Print a message 'q(L) < q(R)' and compute the slope of q at R.⟩
```

To solve this problem, the statements that are conditionally executed must be grouped with a `begin-end` pair:

```
if (L-R)*(L+R+b) > 0 then
   begin
      writeln('q(L) >= q(R)');
      slope := 2*L + b
   end
else
   begin
      writeln('q(L) < q(R)')
      slope := 2*R + b
   end
```

(Recall that the slope of q at x is given by $q'(x) = 2x + b$.) The begin-end pairs have the effect of "gluing" statements together. In so doing, they form what is called a *compound statement*. There is no need to place a semicolon after the statement that immediately precedes the end. A semicolon *must not* follow the end that precedes the else.

This illustrates the general structure of the if-then-else:

```
if ⟨Condition⟩ then
   begin
      ⟨Things to do if the condition is true.⟩
   end
else
   begin
      ⟨Things to do if the condition is false.⟩
   end
```

Sometimes there is "nothing to do" if the comparison in the if is false. In this case, just delete the "else" part. The fragment

```
qL := sqr(L) + b*L + c;
qR := sqr(R) + b*R + c;
if abs(qL-qR) <= 0.001 then
   begin
      writeln('q(L) is close to q(R)');
      ave := (qL+qR)/2;
   end
```

prints the message

```
q(L) is close to q(R)
```

if $|q(L) - q(R)| \leq .001$ and assigns the average value to ave. Nothing is done if this boolean expression is false. Here is the general structure of a simple if-then:

```
if ⟨Condition⟩ then
   ⟨Something to do if the condition is true.⟩
```

Now consider the second of the three questions posed above: does the derivative of $q(x) = x^2 + bx + c$ have a zero in the interval $[L, R]$? Since $q'(x) = 2x + b$, the zero is given by $x_c = -b$. Let us write a fragment that prints a message indicating whether or not $L \leq x_c \leq R$ is true. Two comparisons must be made. We must compare x_c with L and x_c with R. Only if the comparison $L \leq x_c$ is true *and* the comparison $x_c \leq R$ is true may we conclude that x_c is in the interval. Here is a fragment that performs these two checks and prints an appropriate message:

```
{Fragment D}
xc := -b;
if (L <= xc) and (xc <= R) then
   writeln('xc is in [L,R]')
else
   writeln('xc is not in [L,R]');
```

The expression

```
(L <= xc) and (xc <= R)
```

a	b	a and b
false	false	false
false	true	false
true	false	false
true	true	true

FIGURE 1.10 *The* and *Operation*

a	b	a or b
false	false	false
false	true	true
true	false	true
true	true	true

FIGURE 1.11 *The* or *Operation*

is a boolean expression and therefore has a value of true or false. The and operation has the form

⟨*Boolean Expression*⟩ and ⟨*Boolean Expression*⟩

It compares two boolean values and returns another boolean value according to the following table in FIGURE 1.10. Fragment D is equivalent to:

```
{Fragment E}
xc := -b;
if (L > xc) or (xc > R) then
   writeln('xc is not in [L,R]')
else
   writeln('xc is in [L,R]');
```

This illustrates the or operation. This logical operation is defined in FIGURE 1.11. A third logical operation called the not operation negates truth value. See FIGURE 1.12. Here is another rewrite of Fragment D that uses the not operation:

```
xc := -b;
if not((L <= xc) and (xc <= R) ) then
   writeln('xc is not in [L,R]')
else
   writeln('xc is in [L,R]');
```

a	not a
false	true
true	false

FIGURE 1.12 *The* not *Operation*

As a last example, let us compute the minimum value of the function $q(x) = x^2 + bx + c$ on the interval $[L, R]$. Calling this minimum m, here is the "formula":

$$m = \begin{cases} q(x_c) & \text{if } x_c \in [L, R] \\ \text{the smaller of } q(L) \text{ and } q(R) & \text{if } x_c \text{ is not in } [L, R] \end{cases}$$

This either/or situation calls for an if-then-else construct:

```
xc := -b;
if (L <= xc) and (xc <= R) then
   ⟨The minimum value is q(xc) = -b² + c.⟩
else
   ⟨The minimum value is the smaller of q(L) and q(R)⟩
```

If the condition is true, then `m := -sqr(b)+c`. If the condition is false, then the code necessary to identify the smaller of $q(L)$ and $q(R)$ involves another `if-then-else`:

```
if (L-R)*(L+R+b)<0 then
   m := L*L+b*L+c
else
   m := R*R+b*R+c;
```

Putting it all together we obtain the program `Example1_4`. The program shows how `if` statements can be *nested*. But notice that as we developed the program, we never dealt with more than one `if` at a time. The mission of the "outer" `if` is established first without regard to the details of the alternatives. We didn't have to think about the "inner" `if` until the issue of q's value at the endpoints surfaced. This is an example of *top-down* problem solving, a truly essential skill for the computational scientist. Examples of top-down problem solving permeate the rest of the text.

We mention in closing it takes a while to develop a facility with boolean expressions. Arithmetic expressions pose no comparable difficulty, because we have had years of schooling in mathematics. But through a carefully planned sequence of "boolean challenges", you will become as adept with `and`, `or`, and `not` as you are with "+", "-", "*", and "/".

PROBLEM 1.5. Modify `Example1_4` so that it handles quadratics with a variable quadratic coefficient, i.e., quadratics of the form $q(x) = ax^2 + bx + c$. Assume $a \neq 0$.

1.4 Formulas that Involve Integers

According to the rules of the Gregorian calendar, a year is a leap year if it is divisible by four with the exception of century years that are not divisible by 400. Thus, 1992 and 2000 are leap years while 1993 and 2100 are not.

The leap year "formula" involves integer arithmetic. Integer arithmetic is not quite the same as real arithmetic. For example, the division of one integer by another produces a *quotient* and a *remainder*. Using $\div$ to designate this operation, we see that $23 \div 8 = 2$ with remainder 5. Integer calculations in ThinkPascal should be performed using the integer type and should make use of the built-in functions `mod` and `div` that can be used for computing quotients and remainders. To illustrate these features, we solve a few "calendar problems." The program `Example1_5` indicates whether a given year is an ordinary year or a leap year. The program shows how to declare integer variables:

```
program Example1_4;
{Minimum of x^2 + bx + c on [L,R].}
var
   b,c:real; {Coefficients of the quadratic.}
   L,R:real; {The left and right endpoints of the interval.}
   m:real; { The minimum of the quadratic on [L,R].}
   xc:real; {The critical value of the quadratic.}
begin
   ShowText;
   writeln('Enter coefficients b and c:');
   readln(b,c);
   writeln('Enter L and R;');
   readln(L,R);
   xc := -b/2;
   if (L <= xc) and (xc <= R) then
      m := c-b*b/4
   else
      {xc is not in [L,R]. }
      if (L-R)*(L+R+b)<0 then
         m := L*L+b*L+c
      else
         m := R*R+b*R+c;
   writeln(m:10:6);
end.
```

Sample output:

```
Enter coefficients b and c:
-5      6
Enter L and R;
-10     10
min = -0.250000
```

```
program Example1_5;
{Leap year calculations.}
var
   year:integer; {The year to examine.}
begin
   ShowText;
   writeln('Enter year (integer):');
   readln(year);
   if (year mod 4 <> 0) then
      writeln(year:5,' is an ordinary year.')
   else
      if ((year mod 100) = 0) and (year mod 400 <> 0) then
         writeln(year:5,' is an ordinary year.')
      else
         writeln(year:5,' is a leap year.')
end.
```

Sample output:

```
Enter year (integer):
1900
1900 is an ordinary year.
```

```
year:integer;
```

As with the declaration of real variables, this sets up " a box" for holding numbers. However, the numbers in this case must be integral. The integer value must be in the range $[-(2^{15}-1), (2^{15}-1]$. The built-in constant `maxint` houses the value $2^{15} - 1 = 16383$.

In `Example1_5` a value is read into `year` and then a question is posed about its divisibility by 4. If the value in `year` is a multiple of 4, then `year mod 4` is zero. In general,

⟨*Integer-valued Expression*⟩ **mod** ⟨*Integer-valued Expression*⟩

is the remainder when the value of the left expression is divided by the value of the right expression.

Because integers have no fraction part, their formatting for output requires just a single number that stipulates the total number of spaces that are to be allocated. Thus, `year:5` indicates that 5 spaces are to be used for the printing of `year`.

It is instructive to "derive" `Example1_5` taking the "top-down" approach. Assume that `year` houses the value in question. We first examine `year` to see if it is divisible by 4:

```
if (year mod 4) <> 0 then
   writeln(year:5,' is an ordinary year')
else
   (The case when year is divisible by 4.)
```

If `year` is divisible by 4, then we must handle the century years special and the solution fragment expands to:

```
if (year mod 4) <> 0 then
   writeln(year:5,' is an ordinary year')
else
   if (year mod 100) = 0 then
      ⟨The century year is divisible by 100.⟩
   else
      writeln(year:5,' is a leap year')
```

This handling requires a check if the century year is divisible by 400:

```
if (year mod 4) <> 0 then
   writeln(year:5,' is an ordinary year')
else
   if (year mod 100) = 0 then
      if (year mod 400) <> 0 then
         writeln(year:5,' is an ordinary year')
      else
         writeln(year:5,' is a leap year')
   else
      writeln(year:5,' is a leap year')
```

Often, after a top-down development like this it is possible to simplify the derived code using and and or operations. With such a manipulation it is possible to remove the innermost `if` and that produces `Example1_5`.

The program `Example1_6` is concerned with the number of leap years that have occurred from 1900 to a specified year and introduces the `div` function. The `div` function is used for integer division. The value of 22 `div` 4 is 5. In general, the value of

⟨*Integer-valued Expression* ⟩ **div** ⟨*Integer-valued Expression* ⟩

is the whole number obtained by dividing the value of the right expression into the value of the left expression.

PROBLEM 1.6. Modify `Example1_6` so that it changes the "base date" from January 1, 1900 to January 1, 1600.

PROBLEM 1.7. Write a program that solicits a time period T in seconds and then prints its equivalent in units of hours, minutes, and seconds. Thus, if $T = 10000$, then

$$T = 2 \cdot 3600 + 46 \cdot 60 + 40$$

implying that 10000 seconds equals 2 hours, 46 minutes, and 40 seconds. Make sure that the value of`T` does not exceed `maxint`.

PROBLEM 1.8. Assume that `L` and `R` are integers whose values satisfy $0 \leq L \leq R$. Define m to be the largest value that the cosine function attains on the set $\{L^o, \dots, R^o\}$. Thus, if $L = 34$ and $R = 38$, m is the largest of the numbers $\cos(34^o)$, $\cos(35^o)$, $\cos(36^o)$, $\cos(37^o)$, and $\cos(38^o)$.

Many applications require computations that produce integers from reals and vice versa. A nice setting to practice these transitions deals with angles. We need a definition to get started. We say that an angle θ measured in degrees is normalized if $0 \leq \theta < 360$. *Any* angle d can be written in the form

$$d = 360w + \theta$$

```
program Example1_6;
{Computes the number of leap year days in between January 1, 1900}
{and December 31 of a prescribed year in the 21-st century.}
var
   year:integer; {The prescribed year.}
   n:integer; {The number of leap year days.}
begin
   ShowText;
   writeln('Enter the year (integer and in interval [1900,2099]):');
   readln(year);
   if (year<2000 then
      n := ((year-1900) div 4 );
   else
      n := ((year-1900) div 4 ) + 1;
   writeln('There are ',n:3,'leap year days during 1900,...,',year:4);
end.
```

Sample output:

```
Enter the year (integer and in interval [1900,2099]):
1994
Number of leap year days between Jan 1, 1900 and Dec 31,1994 = 23
```

where w is an integer called the *winding number* and θ is normalized. Let's consider the problem of computing the winding number and normalization of an angle that is specified in degrees. Assume that `d` is a real variable whose value is non-negative. We need to determine how many integral multiples of 360 are contained in d. This can be done using a built-in function called `trunc`. If `x` is real, then the value of `trunc(x)` is the integer obtained by throwing away the fraction part. Thus, `trunc(-4.7)` is the integer -4 while `trunc(2.1)` is the integer 2. The fragment

```
w = trunc(d/360);
theta := d - 360*w;
```

assigns the winding number to `w` and the normalization to `theta`.

While `trunc` takes a real and obtains an integer by "removing" the fraction part, `round` takes a real value and returns the value of the nearest integer. Thus, `round(-4.7)` is -5 while `round(2.8)` is 3. In case there are two equally distant nearest integers, then the one further away from 0 is selected. Thus, `round(2.5)` is 3. As an application, here is a statement that converts an angular measure in radians stored in `angle` to the nearest integral degree:

```
degree := round((angle/pi)*180)
```

PROBLEM 1.9. A sign on a taxi reads "5 dollars for the first eighth mile or fraction thereof and 2 dollars for each successive eighth mile or fraction thereof." Here is a small table that clarifies the method of charging:

Distance	Charge
0.10	5
0.20	7
0.99	19
1.0	19

Write a program that solicits the distance traveled and prints the charge for the trip.

PROBLEM 1.10. **(a)** Give a boolean expression that is true if the value in a real variable `x` is closer closer to an integer than to a real number whose fractional part equals one-half. **(b)** Assume that `a` and `b` are real with $a < b$. Write a fragment that prints the number of integers in the interval $[a, b]$. **(c)** Assume that `a`, `b`, and `c` are real variables with positive values and that a and b are not integral multiples of c. Write a fragment that prints the message "ok" if there is a real number strictly in between a and b that is an integral multiple of c. Indicate the type of any additional variables required by your solution. Assume that $a < b$.

PROBLEM 1.11. Assume that `L` and `R` are real with $L < R$. Write a program that reads L and R and prints the maximum value of $cos(x)$ on $[L, R]$.

PROBLEM 1.12. Suppose we have two rays which make make positive angles a and b with the positive x-axis. Write a Pascal program that reads the two angles (in radians), and determines whether or not the two rays make an acute angle. Examples: $a = \pi/6$ and $b = 23\pi/6$ do make an acute angle, while $a = \pi/6$ and $b = 3\pi$ do not.

PROBLEM 1.13. Write a program that reads in two nonnegative real numbers a and b and indicates whether or not the two rays that make positive angles a^o and b^o with the positive x-axis are in the same quadrant. For clarity we assume that if $0^o <= x < 360^o$ then

$$x \text{ is in the } \left\{ \begin{array}{c} \text{First} \\ \text{Second} \\ \text{Third} \\ \text{Fourth} \end{array} \right\} \text{ quadrant if } \left\{ \begin{array}{c} 0 \leq x < 90 \\ 90 \leq x < 180 \\ 180 \leq x < 270 \\ 270 \leq x < 360 \end{array} \right\}.$$

Chapter 2

Numerical Exploration

All work and no play does not a computational scientist make. It is essential to be able to *play* with computational idea before moving on to its formal codification and development. This is very much a comment about the role of intuition. A computational experiment can get our mind moving in a creative direction. In that sense, merely watching what a program does is no different then watching a chemistry experiment unfold: it gets us to think about concepts and relationships. It builds intuition.

The chapter begins with a small example to illustrate this point. The area of a circle is computed as a limit of regular polygon areas. We "discover" π by writing and running a sequence of programs.

Sometimes our understanding of an established result is solidified by experiments that confirm its correctness. In §2.2 we check out a theorem from number theory that says $3^{2k+1}+2^k$ is divisible by 7 for all positive integers k.

To set the stage for more involved "computational trips" into mathematics and science, we explore the landscape of floating point numbers. The terrain is *finite* and *dangerous.* Our aim is simply to build a working intuition for the limits of floating point arithmetic. Formal models are not developed. We're quite happy just to run a few well-chosen computational experiments that show the lay of the land and build an appreciation for the inexactitude of real arithmetic.

The design of effective problem-solving *environments* for the computational scientist is a research area of immense importance. The goal is to shorten the path from concept to computer program. We have much to say about this throughout the text, In §2.4 we develop the notion of an interactive framework that fosters the exploration of elementary computational ideas.

FIGURE 2.1 *Regular n-gons*

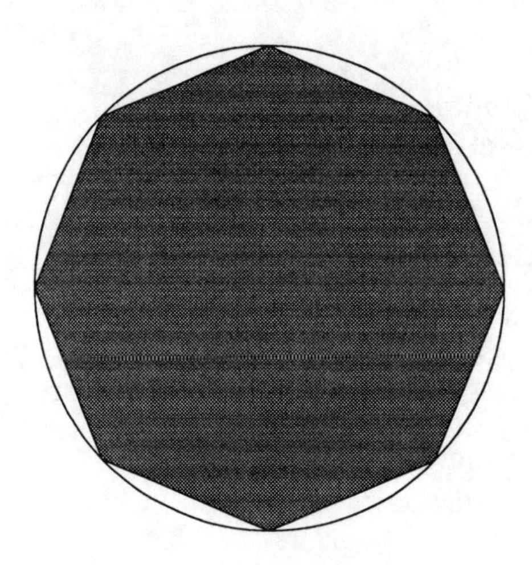

FIGURE 2.2 *Inscribed n-gon*

2.1 Limits

A polygon with n equal sides is called a *regular n-gon.* FIGURE 2.1 illustrates the first four interesting cases. Given n equally spaced points around a circle C, there are two ways to construct a regular n-gon. One is simply to connect the points in order. Each point is then a *vertex* of the n-gon which is said to be *inscribed* in C. See FIGURE 2.2. On the other hand, the tangent lines at each point define a regular n-gon that *circumscribes* C. See FIGURE 2.3. If C has radius one, then the areas of these two regular n-gons are given by

$$\begin{aligned} A_n &= (n/2)\sin(2\pi/n) \qquad \text{(Inscribed)} \\ B_n &= n\tan(\pi/n) \qquad \text{(Circumscribed)} \end{aligned}$$

These formulas can be derived by chopping the n-gon into n equal triangles and summing their areas. Now for any value of n that satisfies $n \geq 3$, the fragment

```
{Fragment A}
InnerA := (n/2)*sin(2*pi/n);
OuterA := n*sin(pi/n)/cos(pi/n);
writeln(n:3,InnerA:9:6,OuterA:9:6)
```

FIGURE 2.3 *Circumscribed n-gon*

prints n, A_n, and B_n. It follows that

```
n := 3;
InnerA := (n/2)*sin(2*pi/n);
OuterA := n*sin(pi/n)/cos(pi/n);
writeln(n:3,InnerA:9:6,OuterA:9:6)
n := 4;
InnerA := (n/2)*sin(2*pi/n);
OuterA := n*sin(pi/n)/cos(pi/n);
writeln(n:3,InnerA:9:6,OuterA:9:6);
n := 5;
InnerA := (n/2)*sin(2*pi/n);
OuterA := n*sin(pi/n)/cos(pi/n);
writeln(n:3,InnerA:9:6,OuterA:9:6);
n := 6;
InnerA := (n/2)*sin(2*pi/n);
OuterA := n*sin(pi/n)/cos(pi/n);
writeln(n:3,InnerA:9:6,OuterA:9:6);
```

produces a 4-line table that reports on the areas for n = 3,4,5 and 6:

```
  3  1.299038  5.196152
  4  2.000000  4.000000
  5  2.377641  3.632713
  6  2.598076  3.464102
```

This approach to table generation is tedious. It would be much handier if we could specify the repetition as follows:

⟨*Execute the following fragment for n= 3,4,5,and 6:*⟩

```
    a := (n/2)*sin(2*pi/n);
    A := n*sin(pi/n)/cos(pi/n);
    writeln(n:3,a:9:6,A:9:);
```

The `for`-loop is designed precisely for this kind of situation:

```
for n:=3 to 6 do
   begin
      InnerA := (n/2)*sin(2*pi/n);
      OuterA := n*sin(pi/n)/cos(pi/n);
      writeln(n:3,InnerA:9:6,OuterA:9:);
   end
```

Here is how it works. The value of "3" is assigned to the *count variable* n. Since the boolean expression 3 <= 6 is true, the *loop body*

```
InnerA := (n/2)*sin(2*pi/n);
OuterA := n*sin(pi/n)/cos(pi/n);
writeln(n:3,InnerA:9:6,OuterA:9:);
```

is executed. For this *pass* through the loop, the value of n is 3 and the areas of the inner and outer triangles are printed. After this, the value of n is increased by 1. Since 4 <= 6 is true, the loop body is executed again and the data for the inner and outer squares is printed. The pattern is repeated for $n = 5$ and $n = 6$. After the hexagon results are printed, n is incremented to 7. the boolean expression 7<=6 is false signaling the end of the iteration.

The beauty of this arrangement is that it is just as easy to produce a "3-to-100" table as it is to produce a "3-to-6" table. Indeed, the program Example2_1 enables us to let n range between any two integers of our choice. The example reveals the general form of the for-loop:

for ⟨*Count Variable*⟩ := ⟨*Lower Bound*⟩ **to** ⟨*Upper Bound*⟩ **do**
  begin
    ⟨*The fragment to be repeated goes here.*⟩
  end

The following rules must be obeyed:

- The count variable must be a declared variable of type integer.
- The lower and upper bounds must be integer-valued expressions.
- Within the loop body, the count variable can be referenced, but it should never appear to the left of the assignment operator.

Notice the judicious use of comments to assist in the understanding of the loop and the three variables n, InnerA, and OuterA. The definitions of InnerA and OuterA in terms of n appear at the "top" of the loop body. Definitions of this sort that are crucial to our understanding of the loop process should always appear in this place.

If $R < L$, then the body of the entire loop is skipped and execution continues with the terminating writeln statement.

Instead of soliciting the starting and stopping values of n, we may wish to have the user enter the starting value of n and the *length* of the table that is to be produced. Declaring m integer, we merely alter the beginning of the program body in Example2_1 as follows:

```
writeln('Enter starting value L and table length m:');
readln(L,m);
writeln;
writeln('n a(n) A(n)');
for n:=L to L+m-1 do
```

```
program Example2_1;
{Areas of regular polygons that are inscribed and circumscribed}
{in the unit circle.}
var
   L,R:integer; {The lower and upper bounds for n.}
   n:integer; {The number of sides.}
   InnerA,OuterA:real; {The inscribed and circumscribed areas.}
begin
   ShowText;
   writeln('Enter integers L and R:');
   readln(L,R);
   writeln;
   writeln('n A(n) B(n)');
   writeln('-------------------------------');
   for n:=L to R do
      {Compute areas of the inscribed and circumscribed n-gons.}
      begin
         InnerA := (n/2)*sin(2*pi/n);
         OuterA := n*sin(pi/n)/cos(pi/n);
         writeln(n:3,InnerA:9:6,OuterA:9:6);
      end;
   writeln('-----------------------------------');
end.
```

Sample output:

```
Enter integers L and R:
15 20

 n       A(n)          B(n)
----------------------------
15   3.050525      3.188349
16   3.061467      3.182598
17   3.070554      3.177851
18   3.078181      3.173886
19   3.084645      3.170539
20   3.090170      3.167689
----------------------------
```

This modification shows how integer-valued expressions can "show up" in the `for`-loop statement. To build confidence in the correctness of the loop bounds, it is often useful to check out a few cases "by hand:"

L	m	Desired Sequence	L+m-1
3	0	(None.)	2
3	1	3	3
3	4	3,4,5,6	6

`For`-loops are very regimented because the loop variable increases in steps of 1. However, it is not hard to carry out iterations that involve non-unit stepping. Consider the problem of producing a table of area values for $n = 5, 10, 15, \ldots, 100$. Repetition is certainly involved because the required table has 20 entries. Thus, a "1 to 20" loop seems appropriate:

```
for k:=1 to 20 do
    {Print the k-th line of the table}
```

Noting that the k-th line is to report the polygon areas for $n = 5k$, we obtain the program `Example2_2`.

PROBLEM 2.1. Write a program that reads in integers a, b, and m where $a \le b$ and $m > 0$ and prints a table of sine values. The table entries should range from a^o to b^o with spacing $(1/m)^o$. Thus, if $a = 1$, $b = 3$, and $m = 2$, then the table should report the sine of 1°, 1.5^o, 2°, 2.5^o, and 3°.

PROBLEM 2.2. Here are the *n-stars* of size 5 and 12:

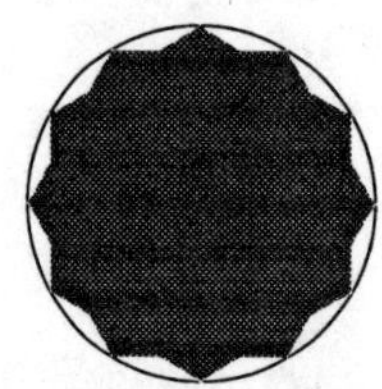

The precise definition of an n-star is not important. Suffice it to say that the area of an n-star that is inscribed in the unit circle is given by

$$A(n) = \begin{cases} n\dfrac{\cos(\pi/(2n)) - \cos(3\pi/(2n))}{2\sin(3\pi/(2n))} & \text{if } n \text{ is odd} \\[2ex] n\,\dfrac{1-\cos(2\pi/n)}{2\sin(2\pi/n)} & \text{if } n \text{ is even} \end{cases}$$

and its perimeter by

$$E(n) = \begin{cases} \dfrac{\sin(\pi/n)}{\sin(3\pi/(2n))} & \text{if } n \text{ is odd} \\[2ex] \dfrac{\sin(\pi/n)}{\sin(2\pi/n)} & \text{if } n \text{ is even} \end{cases}$$

Write a program that prints a table whose k-th line has the values n, $A(n)$, $E(n)$, $A(n+1)$, $E(n+1)$ where $n = 10k$. The value of k should range from 1 to 20.

As n increases, the regular inscribed and circumscribed n-gons converge to the circle. Since the area of the unit circle is π, we have

$$\lim_{n\to\infty} A_n = \pi \qquad \lim_{n\to\infty} B_n = \pi\,.$$

```
program Example2_2;
{Table of Areas for regular polygons inscribed and circumscribed }
{in the unit circle.}
var
   k:integer;
   n:integer; {The number of sides.}
   InnerA, OuterA:real; {Polygon areas.}
begin
   ShowText;
   writeln;
   writeln(' n A(n) B(n)');
   writeln('-----------------------------');
   for k:=1 to 20 do
      {InnerA is the area of a regular 5k-gon inscribed in}
      {the unit circle.}
      {OuterA is the area of a regular 5k-gon circumscribed around}
      {the unit circle.}
      begin
         n := 5*k;
         InnerA := (n/2)*sin(2*pi/n);
         OuterA := n*sin(pi/n)/cos(pi/n);
         writeln(n:3,InnerA:9:6,OuterA:9:6);
      end;
   writeln('----------------------------------');
end.
```

Output:

```
  n     A(n)          B(n)
----------------------------
  5   2.377641      3.632713
 10   2.938926      3.249197
 15   3.050525      3.188349
            ⋮
 90   3.139041      3.142869
 95   3.139303      3.142738
100   3.139526      3.142627
----------------------------
```

Moreover, for all $n >= 3$ we have

$$A_n < \pi < B_n \,.$$

This limiting behavior allows us to formulate iteration problems where the total number of steps is not known in advance. For example, suppose we want to print a "3-to-n table where n is the smallest integer such that

$$B_n - A_n \leq 0.0001.$$

We know that such an n exists because A_n and B_n each converge to π as n increases.

One solution would be to use a **for**-loop with a large, "safe" upper bound and an **if** inside the loop body to guard against unwanted printing:

```
for n:=3 to 10000 do
   begin
      InnerA := (n/2)*sin(2*pi/n);
      OuterA := n*sin(pi/n)/cos(pi/n);
      if OuterA - InnerA > 0.0001 then
         writeln(n:3,InnerA:9:6,OuterA:9:6);
   end
```

But this is a flawed problem-solving strategy for two reasons:

The guessing of a "safe" upper bound may be very difficult in practice. In our example, if the chosen upper bound is too small, then the table will be too short.

It is inefficient. In our example, once the last line is printed, all subsequent area computations are superfluous.

What we need is an ability to "jump out" of the iteration as soon as the condition

```
OuterA - InnerA > 0.001
```

is false. The **while**-loop is designed for this kind of situation and the program `Example2_3` highlights this point. Here is how the program works. Before the loop begins, the variables `n`, `InnerA`, and `OuterA` are assigned the values 3, A_3, and B_3 respectively. The condition in the **while** statement acts as a guard. If the inner and outer areas differ by more than .001, then the boolean expression is true and the loop body

```
writeln(n:3,InnerA:9:6,OuterA:9:6);
n = n+1;
InnerA := (n/2)*sin(2*pi/n);
OuterA := n*sin(pi/n)/cos(pi/n);
```

is executed. This prints a line in the table and updates the triplet of variables `n`, `InnerA`, and `OuterA`. Once again the boolean expression in the **while** statement is evaluated. If it is true, then the loop body is again executed. From what we know about the problem, eventually the inner and outer areas will get so close that the condition `OuterA - InnerA) >.0.001` is false. When this happens, the execution of the **while**-loop terminates and control passes to the final `writeln`.

The comments just after the **while** statement define the three variables `n`, `InnerA`, and `OuterA`. Although these variables change value during the course of the iteration, the definitions apply at the beginning and end of the loop body. **While**-loops should always include comments that state critical relationships of this sort.

In general, **while**-loops are structured as follows:

```
program Example2_3;
{Explores convergence of inner and outer areas.}
var
   n:integer;
   InnerA,OuterA:real;
begin
   ShowText;
   writeln(' n A(n) B(n)');
   writeln('----------------------------')
   n := 3;
   InnerA:= 3*sqrt(3)/4;
   OuterA:= 3*sqrt(3);
   while OuterA - InnerA > 0.001 do
      {InnerA is the area of the regular n-gon inscribed in }
      {the unit circle.}
      {OuterA is the area of the regular n-gon circumscribed}
      { outside the unit circle.}
      begin
         writeln(n:3,InnerA:9:6,OuterA:9:6);
         n := n+1;
         InnerA := (n/2)*sin(2*pi/n);
         OuterA := n*sin(pi/n)/cos(pi/n);
      end;
   writeln('----------------------------')
end.
```

Output:

```
  n     A(n)        B(n)
----------------------------
  3   1.299038    5.196152
  4   2.000000    4.000000
  5   2.377641    3.632713
  6   2.598076    3.464102
           :
173   3.140902    3.141938
174   3.140910    3.141934
175   3.140918    3.141930
176   3.140925    3.141926
----------------------------
```

```
⟨Initializations⟩
while ⟨Boolean Expression⟩ do
   begin
      ⟨Fragment to be repeated goes here.⟩
   end
```

Note that a `for`-loop can always be written as a `while`-loop. For example

```
for n:=3 to 6 do
   begin
      InnerA := (n/2)*sin(2*pi/n);
      OuterA := n*sin(pi/n)/cos(pi/n);
      writeln(n:3,InnerA:9:6,OuterA:9:6);
   end;
```

is equivalent to

```
n := 3;
while n<=6 do
   begin
      InnerA := (n/2)*sin(2*pi/n);
      OuterA := n*sin(pi/n)/cos(pi/n);
      writeln(n:3,InnerA:9:6,OuterA:9:6);
      n := n+1;
   end;
```

When confronted with an iterative computation, a `for`-loop is usually appropriate if the number of iterations is known in advance *and* the counting is "regular." Otherwise, the situation calls for a `while` loop.

To illustrate another type of counting, let's modify `Example2_3` so that n is repeatedly doubled instead of incremented:

```
n:=4; OuterA := 4; InnerA:= 2;
while OuterA - InnerA > 0.00001 do
   begin
      writeln(n:3,InnerA:9:6,OuterA:9:6);
      n = 2*n;
      InnerA := (n/2)*sin(2*pi/n);
      OuterA := n*sin(pi/n)/cos(pi/n);
   end
```

Notice how `n` houses the required powers of two which are obtained through repeated doubling process. The value of `n` starts out at 4. The first execution of `n:=2*n` replaces this value with 8. When `n:=2*n` is carried out during the next pass, the"8" becomes a "16", etc.

The repeated doubling also makes it possible to circumvent the explicit references to `sin`, `cos`, and `pi`. The idea is to use the half-angle formulae

$$\begin{aligned}\cos(\theta/2) &= \sqrt{(1+\cos(\theta))/2}\\ \sin(\theta/2) &= \sqrt{(1-\cos(\theta))/2}\end{aligned}$$

to produce the required sines and cosines. The angles that "show up" during the process are all repeated halvings of $\pi/4$:

n	Angle (Radians)
4	$\pi/4$
8	$\pi/8$
16	$\pi/16$
32	$\pi/32$
$\vdots$	$\vdots$

Since the sine and cosine of $\pi/4$ are both $1/\sqrt{2}$ we obtain `Example2_4`. Unlike `Example2_3` that uses π, `Example2_4` computes π.

PROBLEM 2.3. The volume of a pyramid whose base is an unit regular n-gon of radius r and whose height is one is given by

$$V(n) = \frac{nr^2}{6}\sin(2\pi/n).$$

The volume of a cone whose base is the unit circle and which has height one is given by

$$V_\infty = \frac{\pi}{3}.$$

Write a program that prints the smallest n so that $V_n/V_\infty > .99$.

PROBLEM 2.4. The area of a triangle whose sides have length a, b, and c is given by

$$E = \sqrt{\mu(\mu - a)(\mu - b)(\mu - c)}$$

where $\mu = (a + b + c)/2$. Thus, the area of an equilateral triangle with $a = b = c = 2$ is given by

$$E = \sqrt{3(3-2)(3-2)(3-2)} = \sqrt{3}.$$

Let T_k be the triangle with sides $a = 2$, $b = 2 + 1/2^k$, and $c = 2 - 1/2^k$. As k increases, T_k looks increasingly like an equilateral triangle and its area is increasingly close to $E = \sqrt{3}$. Write a program that prints the smallest value of k such that $|E_k - \sqrt{3}| \leq 0.0001$ where E_k is the area of T_k.

PROBLEM 2.5. Write a program that computes the smallest positive k such that $\text{ctn}(\pi/2^k) > 1000$. Make repeated use of the half-angle formulae. Here, $\text{ctn}(x)$ is the cotangent function: $\text{ctn}(x) = \cos(x)/\sin(x)$.

2.2 Confirming Conjectures

Number theory is a branch of mathematics that deals with the integers and their properties. Many number theoretic results can be couched in elementary terms and can be explored by programs that involve simple iterations. Let us consider the affirmation of the following fact:

If k is a nonnegative integer, then $3^{2k+1} + 2^{k+2}$ is divisible by 7.

This means that there is no remainder when we divide $3^{2k+1} + 2^{k+2}$ by 7. To acquire an understanding of *any* mathematical fact like this, it is best to begin with a few pencil-and-paper verifications:

	$k=0$	$k=1$	$k=2$	$k=3$
2^{k+2}	4	8	16	32
3^{2k+1}	3	27	243	2187
$2^{k+2} + 3^{2x+1}$	7	35	259	2219

```
program Example2_4;
{Computes pi.}
var
   n:  integer;
   OuterA, InnerA: real;
   c, s:  real;
begin
   ShowText;
   n := 4;
   OuterA := 4;
   InnerA := 2;
   c := 1 / sqrt(2);
   while OuterA - InnerA > 0.0001 do
      {InnerA = area of the regular n-gon inscribed in}
      {the unit circle.}
      {OuterA = area of the regular n-gon circumscribed around}
      {the unit circle.}
      {c = cos(pi/n).}
      begin
         writeln(n:5, InnerA:9:6, OuterA:9:6);
         n := 2*n;
         s := sqrt((1 - c) / 2); {s = sin(pi/n)}
         c := sqrt((1 + c) / 2);
         InnerA := n * s * c;
         OuterA := n * s / c;
      end
end.
```

Output:

n	A(n)	B(n)
4	2.000000	4.000000
8	2.828427	3.313709
16	3.061468	3.182599
32	3.121448	3.151727
64	3.136551	3.144121
128	3.140340	3.142232
256	3.141282	3.141755
512	3.141149	3.141267

It is easy to check that the sum of the indicated powers is indeed divisible by 7.

We now set out to write a program that verifies the conjecture for as many k "as possible." We must first figure out how to generate the necessary powers of two and three. Some programming languages have an exponentiation operator but Think Pascal does not. Powers have to be generated "by hand". But this just involves repeated multiplication. If `TwoPower` has type `integer`, then the fragment

```
TwoPower := 1;
for k:=1 to 10 do
   TwoPower := 2*TwoPower;
```

assigns to it the value $2^{10} = 1024$. If `n` is positive, then *in principle*

```
TwoPower := 1;
for k:=1 to n do
   TwoPower := 2*TwoPower;
```

assigns 2^n to `TwoPower`. However, integer variables can only store values up to $2^{15} - 1$ and so `n` must be less than 15 for the fragment to be successful.

To permit the manipulation of very large integers, we introduce the type `longint`[1]. Variables that have this type can store integers as large as $2^{31} - 1 = 2{,}147{,}483{,}647$. The built-in constant `maxlongint` equals this value. Recall that variables of type `integer` can hold values no larger than $2^{15} - 1 = 16{,}383$ and so the range of allowable integers is considerably extended by using `longint`. All the `mod`, `div`, and integer arithmetic discussion in §1.4 applies to `longint` variables.

Returning to the conjecture, we need to be able to compute 3^{2k+1} for $k = 1$, $k = 2$, etc. The value of this expression increases by nine if k is increased by one. Thus, the fragment

```
ThreePower:=3;
for k:=1 to 3 do
   begin
      ThreePower:=9*ThreePower;
      writeln(ThreePower);
   end
```

prints the numbers 27, 243, and 2187. Combining this with our power-of-two work we obtain `Example2_5` that checks the conjecture. In the lead comment we use the "hat" notation "^" to indicate powers.[2]

If the loop bound is increased to 10, then the value of `s` is too large for representation in a `longint` variable and *integer overflow* occurs. The program would keep running, but the value stored in `s` would not be the expected sum of `TwoPower` and `ThreePower`.

Note that the last column of output reveals the remainder of $2^{x+2} + 3^{2x+1} \div 7$ and that it is identically zero for the values $x = 0, \ldots, 9$. `Example2_5` does not *prove* anything. It merely confirms a few special cases cases of a general result. In a research context, a mathematician may choose to run an experiment like this before investing time in a rigorous analytical proof of a general result.

PROBLEM 2.6. The integer next above $(\sqrt{3} + 1)^{2n}$ is divisible by 2^{n+1}. Write a program that confirms this for $n = 1, \ldots, 6$.

---

[1] The `longint` type is not part of standard Pascal.

[2] We repeat. ThinkPascal does *not* have an exponentiation operator. In particular, `a := 2^(x+2)` *is not* a valid statement.

```
program Example2_5;
{ Confirms that 3^(2n+1) + 2^(n+2) is divisible by 7 for n=0,...,9}
var
   a,b,s:longint;
   n:integer;
begin
   Showtext;
   writeln(' n 3^(2n+1) + 2^(n+2) Remainder');
   writeln('------------------------------------------');
   writeln;
   a := 4;
   b := 3;
   writeln(0,a+b:15, (s mod 7):15);
   for n:=1 to 9 do
      {a = 2^(n+2)}
      {b = 3^(2n+1)}
      begin
         a := 2*a;
         b := 9*b;
         s := a+b;
         writeln(n,s:15, (s mod 7):15)
      end
end.
```

Output:

```
n        3^(2n+1) + 2^(n+2)     Remainder
------------------------------------------
0                         7             0
1                        35             0
2                       259             0
3                      2219             0
4                     19747             0
5                    177275             0
6                   1594579             0
7                  14349419             0
8                 129141187             0
9                1162263515             0
```

PROBLEM 2.7. The product of three consecutive whole numbers is exactly divisible by 504 if the middle one is the cube of a whole number. Verify this for the triplets $(n^3 - 1, n^3, n^3 + 1)$ with $n = 2, \ldots, 10$.

PROBLEM 2.8. Let a_n be the n-th non-perfect square among positive integers. Thus, $a_1 = 2$, $a_2 = 3$, $a_3 = 5$, etc. For $n = 1$ to 10000, confirm that $a_n = n + round(\sqrt{n})$.

PROBLEM 2.9. Let n be a positive integer and let $b(n)$ be the minimum value of $k + (n/k)$ as k is allowed to range through all positive integers. It can be shown that $b(n)$ and $\sqrt{4n+1}$ have the same integer part. Confirm this for all $n \leq 1000$. Hint: You can compute $b(n)$ without a loop.

PROBLEM 2.10. There are at least seven positive integers x that make $x(x + 180)$ the square of an integer. Write a program that confirms this conjecture.

PROBLEM 2.11. Write a program that reads in a positive integer q and prints a list of all powers of q that are less then or equal to `maxlongint`. Organize the computation so that integer overflow does not occur.

2.3 The Floating Point Terrain

We now turn our attention to the numerical landscape of real numbers. It turns out that real arithmetic is inexact. If we think of the computer as a telescope, then this effects its resolution. A good astronomer understands the limits of a telescope that is being used to observe the galaxies. A good computational scientist understands the limits of a computer that is being used to observe the "numerical versions" of those galaxies.

There are just a handful of things to know and the starting point is the representation of a numerical value in scientific notation:

$$x = +.123 \times 10^{+3} \qquad y = -.1000 \times 10^{-05} \qquad z = .00000 \times 10^{00}.$$

The fraction part is called the *mantissa.* It has a sign and length. The mantissas for x, y, and z have length 3, 4, and 5 respectively. The mantissa is always less than one in absolute value and for non-zero numbers, the most significant digit is non-zero. The representations $x = 1.23 \times 10^{+2}$ and $x = .0123 \times 10^4$, are not *normalized.*

The *exponent* part of the notation indicates the power to which the base is raised. In the above base-10 examples above, x, y, and z have exponents 3, -5 and 0 respectively. Like the mantissa, the exponent has a sign and a length.

The representation of numbers in `real` variables follows the above style. But there is an added wrinkle: *the amount of memory that is allocated for the mantissa and exponent is fixed and finite.* For example, a computer may permit 3-digit mantissas and 1-digit exponents.[3] Here is a list of some numbers and and their *floating point* representation:

$a = 12.3$	a =	+	1	2	3	+	2
$b = .000000123$	b =	+	1	2	3	−	6
$c = -12.3$	c =	−	1	2	3	+	2
$d = 0.0$	d =	+	0	0	0	+	0

[3]These are unrealistic parameters but they are good enough to communicate the main ideas. A real number in a computer would "typically" be represented in a base-2 floating point format. The mantissas would consist of (say) fifty-six binary digits and the exponents would consist of eight binary digits.

Note that with the limited mantissa length, some numbers can only be stored approximately:

$a = 12.34$	a =	+	1	2	3 ‖ +	2
$b = 12.37$	b =	+	1	2	4 ‖ +	2
$c = \pi$	c =	+	3	1	4 ‖ +	1

The reasonable thing to do if there isn't enough "room" to store the exact mantissa is to *round.* Since 12.37 is closer to $.124 \times 10^2$ than $.123 \times 10^2$, the former value is stored. In case of a tie, we assume that the computer rounds up.

To drive home the point that the set of floating point numbers is finite, we display the smallest and the largest positive floating point numbers:

$min = .0000000001$	min =	+	1	0	0 ‖ −	9
$max = 999000000$	max =	+	9	9	9 ‖ +	9

Some numbers just do not fit at all because the exponent length is too short. For example, if $x = 1234567890 = .1234567890 \times 10^{10}$, then a 2-digit exponent is required. Likewise, $x = .0000000000999 = .999 \times 10^{-10}$ cannot be represented because again, two digits are required to hold the exponent.

A reasonable way to model the addition, subtraction, multiplication or division of two floating point numbers is as follows:

1. Perform the operation exactly.

2. Put the answer in normalized scientific form.

3. Round the mantissa to the allotted number of mantissa digits.

Thus, the addition of x and y where

$x = 12.3$	x =	+	1	2	3 ‖ +	2
$y = 5.27$	y =	+	5	2	7 ‖ +	1

proceeds as follows:

- $12.3 + 5.27 = 17.57$.

- $17.57 = .1757 \times 10^2$

- `z:= x+y` z = | + | 1 | 7 | 6 ‖ + | 2 |

A consequence of this model of floating point arithmetic is that *rounding errors* usually attend every floating point arithmetic operation. This forces us to depart from the "exact arithmetic" mindset when developing programs that manipulate real numbers. This is a complicated issue and we will only be able to scratch the surface in this introductory text. However, with a few well-chosen examples we can build up an appreciation for the inexactness of floating point arithmetic.

For example, it is possible for the assignment `x := x+y` not to change the value of `x` even if the value in y is nonzero. In the model floating point system that we have been using, if $x = 1.00$ and $y = .001$, then the computed sum of x and y is 1 because 1.001 rounds to 1.00. `Example2_6` illustrates this point.

```
program Example2_6;
{Explores floating point precision.}
var
   k:integer;
   small,OnePlusSmall:real;
begin
   ShowText;
   k := 0;
   Small := 1.0;
   OnePlusSmall := 2.0;
   writeln(' k 1 + 1/2^k');
   writeln('------------------');
   while OnePlusSmall <> 1 do
      {Small = 1/2^k}
      begin
         k := k + 1;
         Small := Small / 2.0;
         OnePlusSmall := 1.0 + Small;
         writeln(k : 2, OnePlusSmall : 16 : 7);
      end;
end.
```

Output:

```
 k      1 + 1/2^k
------------------
 1      1.5000000
 2      1.2500000
 3      1.1250000
 4      1.0625000
            .
            .
            .
20      1.0000010
21      1.0000005
22      1.0000002
23      1.0000001
24      1.0000000
```

The loop terminates as soon as the floating point addition of one and $1/2^k$ is one.

The length of the mantissa defines the precision of the floating point arithmetic. Roughly, the relative error in a floating point operation is about b^{-t+1} where b is the base and t is the length of the mantissa. We refer to b^{-t+1} as the *machine precision.* A typical mantissa might be comprised of 24 base-2 digits. This means that machine precision is about $2^{-23} \approx 10^{-6}$.

ThinkPascal also supports a *double precision* real data type called `double`. Variables of this type may be used in exactly the same fashion as `real` variables. However, the mantissa length is longer thereby permitting more accurate calculations.

PROBLEM 2.12. In `Example2_6`, change the declaration of `small` and `OnePlusSmall` to

```
small,OneplusSmall:double
```

Rerun the program and observe the change in output. Change the `20:7` format to `20:16`.

PROBLEM 2.13. Assume that `x` and `h` are real variables. Why might the fragments

```
h := 1.0/10.0;
x := 0.0;
k := 0;
while x <= 1.0 do
    begin
        writeln(k:3, x:13:7);
        x:= x + h;
        k:= k + 1;
    end;
```

and

```
h := 1.0/10.0;
x := 0.0;
for k:=0 to 10 do
    begin
        writeln(k:3, x:13:7);
        x:= x + h;
    end;
```

produce different output?

PROBLEM 2.14. Compare the output of the following program with what should be produced in exact arithmetic:

```
program Problem2_14;
var
    e,f,g,h,x,y,q:real;
begin
    h := 1.0/2.0;
    x := (2.0/3.0) - h;
    y := (3.0/5.0) - h;
    e := (x + x + x) - h;
    f := (y + y + y + y + y) - h;
    q := f/e;
    writeln(q:10:6)
end.
```

PROBLEM 2.15. Modify `Example2_6` so that just after the table it prints the value of the largest k so that the floating point addition of 1 and $1/2^k$ is bigger than 1.

So far we have spent most of the time discussing the mantissa and what its finiteness implies. But there are also a couple of things to discuss about the exponent part of the floating point

```
program Example2_7;
{Explores floating point underflow.}
var
   k:integer;
   small:real;
begin
   ShowText;
   small:=1.0;
   k:=0;
   writeln(' k 1/2^k');
   writeln('-------------------');
   while small <> 0 do
      {small = 1/2^k}
      begin
         small:= small/2.0;
         k:= k+1;
         writeln(k:3, ' ', small);
      end
end.
```

Output:

```
  k          1/2^k
-------------------
  1         5.0e-1
  2         2.5e-1
  3         1.2e-1
  4         6.2e-2
          .
          .
          .
147        5.6e-45
148        2.8e-45
149        1.4e-45
150         0.0e+0
```

number. When a floating point operation renders a nonzero answer that is too small to represent, then an *underflow* occurs and the result is set to zero. `Example2_7` computes $1/2^k$ for increasingly big k. Eventually, an underflow is produced.

At the other end of the scale, a floating point operation can result in an answer that is too big to represent. This is called *floating point overflow.* Overflows produce a special value called `inf`. Analogous to `Example2_7` that examines the effect of repeated halving, `Example2_8` performs repeated doubling until floating point overflow occurs:

PROBLEM 2.16. Modify `Example2_7` so that only the lines associated with $k = 10, 20, 30$,etc. are printed.

PROBLEM 2.17. Modify `Example2_8` so that instead of printing the table, it just prints the second to last line in the table. Hint: remove the `writeln` inside the loop, maintain a variable that houses the the "previous" value of `Big`, and print the value of that variable after the loop terminates.

```
program Example2_8;
{Explores floating point overflow.}
var
   k:integer;
   Big:real;
begin
   ShowText;
   Big:=1.0;
   k:=0;
   writeln(' k 2^k');
   writeln('-------------------');
   while big <> INF do
      {Big =2^k}
      begin
         k:= k+1;
         Big:= 2.0*Big;
         writeln(k:2,big);
      end
end.
```

Output:

```
  k          2^k
------------------
  1        2.0e+0
  2        4.0e+0
  3        8.0e+0
  4        1.6e+1
        :
125       4.3e+37
126       8.5e+37
127       1.7e+38
128           INF
```

PROBLEM 2.18. Write a program that prints the largest integer n so that `x:= exp(n)` does *not* assign the value `inf` to the real variable `x`.

2.4 Interactive Frameworks

In the preceding examples and problems, the loops execute without human intervention. Indeed, that is the power of the loop concept, for it makes it possible to specify a very extensive calculation with just a few lines of code. However, loops have an important role to play in the design of *interactive frameworks* that can be used to test computational ideas before they are encapsulated in "serious code."

To illustrate this, let us pretend that `sqrt` is not available and that we want to develop a method for computing $\sqrt{a}$ where a is a positive real number[4]. Here is our idea. Think of the $\sqrt{a}$ problem as the problem of producing a sequence of increasingly square rectangles each of which has area a. If x is an estimate of $\sqrt{a}$, then we associate with x a rectangle with base x and height a/x. Our geometric intuition tells us that $\sqrt{a}$ is in between x and a/x and so we conjecture that

$$x_{new} = \frac{1}{2}\left(x + \frac{a}{x}\right)$$

is a better approximation to $\sqrt{a}$. Here is a fragment that applies this refinement idea two times after prompting for a and an initial guess for its square root:

```
writeln('Enter positive real number a:  ');
readln(a);
writeln('Enter an initial guess for sqrt(a):  ');
readln(x);
x := (x + (a/x))/2;
x := (x + (a/x))/2;
AbsoluteError := abs(x - sqrt(a));
RelativeError := AbsoluteError/sqrt(a);
writeln('Computed root = ',root:10:7);
writeln('Absolute Error = ', AbsoluteError);
writeln('Relative Error = ', RelativeError)
```

It prints the approximate square root x together with its absolute error $|x - \sqrt{a}|$ and its relative error $|x - \sqrt{a}|/\sqrt{a}$.

A program that permits the trial of a single example isn't very handy. Most likely we would want to test our square root idea on a number of different a-values and to explore how the quality of the initial guess effects the accuracy of the computed square root. To that end, we can embed the above fragment in a while loop and obtain an "interactive framework" that supports repeated testing. See `Example2_11`.

The program makes use of the character type `char`. A `char` variable is able to store a single character like "y" or "n", or "3" or "+". In short, the value of any keyboard button may be stored in a `char` variable. In the above program, `AnotherEg` is a character variable whose value is used to determine if another example is to be run. Notice from the `while` condition that character values are enclosed in single quotes. The `while`-loop continues until the typist responds with 'n', the "magic character."

[4]What we are about to develop is in fact the root extraction method used by `sqrt`.

```
program Example2_9;
{Tests a method for computing square roots.}
var
   AnotherEg:char;
   a,x:real;
   AbsoluteError,RelativeError:real;
begin
   ShowText;
   AnotherEg = 'y';
   while AnotherEg <> 'n' do
      begin
         writeln;
         {*************************************}
         writeln('Enter a positive real number a:');
         readln(a);
         writeln('Enter an initial guess for the square root:');
         readln(x);
         x := (x + (a/x))/2;
         x := (x + (a/x))/2;
         AbsoluteError := abs(x - sqrt(a));
         RelativeError := AbsoluteError/sqrt(a);
         writeln('Computed root = ',root:10:6);
         writeln('Absolute Error = ',AbsoluteError);
         writeln('Relative Error = ',RelativeError);
         {*************************************}
         writeln;
         writeln('Another example?  Enter y (yes) or n (no)');
         readln(AnotherEg);
      end
end.
```

The while loop could also be controlled by comparing **AnotherEg** with 'y':

```
while AnotherEg = 'y' do
```

With this method of checking, the program terminates *unless* the "magic character" (y) is struck. However, this is not such a good thing since keystroke error brings about program termination.

Here is the output from a sample session with Example2_9:

```
Enter a positive real number a:
100
Enter an initial guess for the square root:
5
Computed root = 10.2500000
Absolute Error = 2.5e-1
Relative Error = 2.5e-2

Another example?  Enter y (yes) or n (no)
y

Enter a positive real number a:
100
Enter an initial guess for the square root:
9
Computed root = 10.0001535
Absolute Error = 1.5e-4
Relative Error = 1.5e-5

Another example?  Enter y (yes) or n (no)
n
```

Even with these few examples we see that the quality of our square root "idea" depends strongly on the quality of the initial guess. The natural thing to do after this brief computational experience is to go "back to the drawing boards" and figure a way to improve upon the method. A better initial guess or an increase in the number of refinements might be the recommended course of action.

The interactive framework used above is quite general. To explore some computational idea, you need only "fill in" the following template:

```
program InteractiveF;
{⟨Description of the computational idea.⟩}
var
   AnotherEg:char;

   ⟨Declarations associated with the fragment to be tested.⟩

begin
   AnotherEg = 'y';
   while AnotherEg <> 'n' do
      begin
         writeln;

         ⟨A fragment to be tested including writeln's⟩

         writeln;
         writeln('Another example?  Enter y (yes) or n (no)');
         readln(AnotherEg);
      end
end.
```

PROBLEM 2.16. A sphere with radius 1 has volume equal to $4\pi/3$. How long must the edge of a cube be so that it has the same volume? Use the interactive framework.

PROBLEM 2.17. Let a, b, and c be real numbers not all zero. How small can you make the quotient

$$Q = \frac{\sqrt{a^2 + b^2 + c^2}}{|a| + |b| + |c|}$$

Use the interactive framework.

PROBLEM 2.18. Use the interactive framework to estimate the area of the largest rectangle whose 4 vertices are on the curve defined by $x^4/19 + y^4/17 = 1$? Note: `f:= sqrt(sqrt(z))` assigns the value of $z^{1/4}$ to `f`.

Chapter 3

Elementary Graphics

It is hard to overstate the importance of graphics to computational science. Three reasons immediately come to mind:

- In most large applications, the amount of numerical data that makes up "the answer" is just too much for the human mind to assimilate in tabular form.
- The visual display of data often permits the computational scientist to spot patterns that would otherwise be hidden.
- Many computations have geometric answers and it is more effective to *show* the answer than to describe it in numerical terms.

To build an appreciation for computer graphics we need to do computer graphics. In this chapter we get started with a handful of ThinkPascal graphics procedures that are utilized throughout the text. Elementary graphical problems are solved that involve grids, rectangles, and ovals. A fringe benefit of the chosen applications is that they give us an opportunity to build up our iteration expertise. Visual patterns involve repetition and repetition requires the writing of loops. Screen granularity provides another setting for exploring the interplay between continuous and the discrete mathematics.

3.1 Grids

Let us consider the problem of drawing the pattern in FIGURE 3.1. We refer to an array of tiles like this as a *grid*. This particular grid is 10-by-12 meaning that it has 10 rows and 12 columns.

Although we may think of a grid as an array of 2-dimensional tiles, it is handier from the drawing point of view to regard it as a set of equally spaced horizontal lines and a set of equally

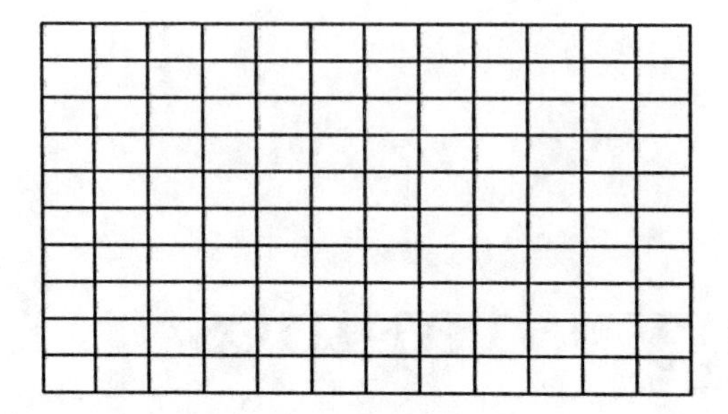

FIGURE 3.1 *A 10-by-12 Grid*

spaced vertical lines. This makes it possible to draw the grid using the built-in ThinkPascal procedures `MoveTo` and `LineTo` described below. These procedures can be used to draw straight lines in a special graphics window referred to as the *DrawWindow.* This window is a rectangular array of points with an integer coordinate system. The points are called *pixels*, a widely used term that stands for *picture element.* The origin (0,0) is situated in the upper left corner of the window. Any pixel in the DrawWindow is specified by an integer pair (h, v) where h is the number of pixels from the window's left edge and v is the number of pixels from the window's top edge as shown in FIGURE 3.2. Notice that in this coordinate system, we move *down* as the second coordinate increases. This is in contrast to the usual (x, y) coordinate system where increasing the y-value means moving up.

Let's see how to draw a line from (100,140) to (230,97), which we depict in Figure 3.3.[1] In the graphics setting, the "pen" that does the drawing is called a *mouse.* The act of drawing a line from screen point (h_1, v_1) to screen point (h_2, v_2) is a two-step process:

- "Lift" the mouse to (h_1, v_1).
- "Drag" the mouse in a straight line to (h_2, v_2) thereby producing a line.

`MoveTo` and `LineTo` handle these operations respectively and the fragment

```
MoveTo(100,140);
LineTo(230,97);
```

draws the line segment depicted above. In general, a reference to `MoveTo` has the form

`MoveTo(`⟨ *Integer-valued expression* ⟩`,`⟨ *Integer-valued expression* ⟩`)`

and positions the mouse at the pixel specified by the two expressions. Likewise,

`LineTo(`⟨ *Integer-valued expression* ⟩`,`⟨ *Integer-valued expression* ⟩`)`

draws a line from the current mouse position to the pixel specified by the two expressions.

Knowing now how to construct straight lines, we can address the grid drawing problem posed at the beginning of this section. To complete the specification of the problem, we need to stipulate where the 10-by-12 grid is to be located on the screen and the size of the individual tiles. Let

[1]Screen size varies from computer to computer, so do not be alarmed if our pictures differ slightly from what you observe on your own machine.

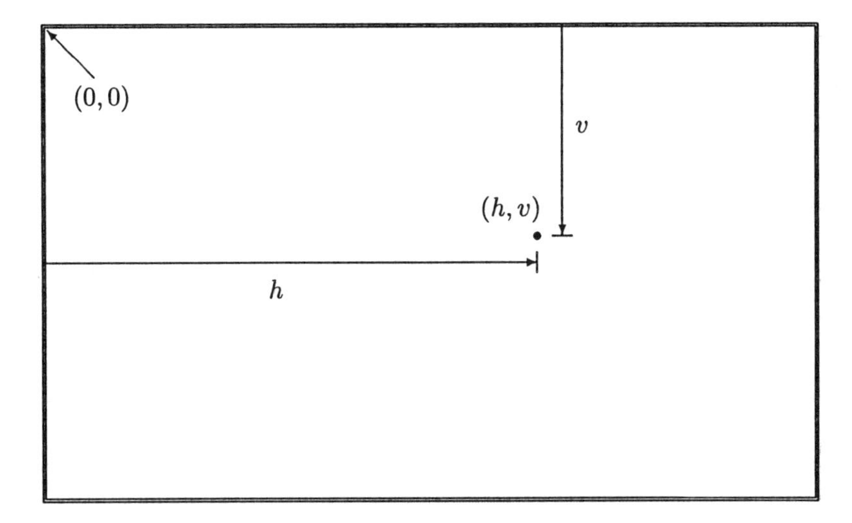

FIGURE 3.2 *The Screen Coordinate System*

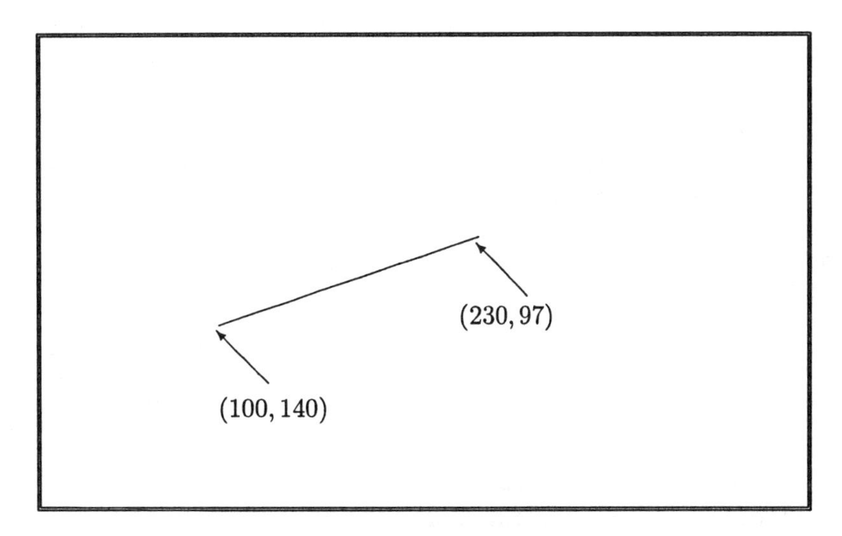

FIGURE 3.3 *A Line Segment*

us locate the upper left corner of the grid at (50,50) and assume that the tiles are 15 pixels high and 30 pixels wide.

Observing that the grid consists of 11 equally spaced horizontal lines and 13 equally spaced vertical lines, we aim for a solution fragment that has the following form:

⟨*Draw the horizontal lines*⟩
⟨*Draw the vertical lines.*⟩

Consider the horizontal lines first. The top edge extends from pixel (50,50) to pixel $(50+12\cdot30,50)$ = (410,50). Thus,

```
MoveTo(50,50);
LineTo(410,50);
```

draws the top edge. The remaining horizontals are produced in a similar fashion. Because the rectangle height is 15 pixels, each line is 15 pixels lower than the one above it. Thus, a fragment of the form

```
MoveTo(50,50);
LineTo(410,50);
MoveTo(50,65);
LineTo(410,65);
MoveTo(50,80);
LineTo(410,80);
    ⋮
MoveTo(50,200);
LineTo(410,200);
```

produces the required horizontal line segments in the grid. This calls for a `for`-loop. There are eleven `MoveTo`-`LineTo` pairs to execute. All lines start at $h = 50$ and end at $h = 410$. The vertical coordinate of the top line is given by $v = 50$. Each pass through the loop we add 15 to v to ensure that the next line is drawn at the proper "level" on the screen. Altogether we have

```
v := 50;
for k:=0 to 10 do
   {Draw the k-th horizontal line.}
   {v = vertical coordinate of the k-th line.}
   begin
      MoveTo(50,v);
      LineTo(410,v);
      v := v+15;
   end
```

Here is an equivalent fragment that uses a different approach to determine the v-values:

```
for k:=0 to 10 do
   {Draw the k-th horizontal line.}
   {v = vertical coordinate of the k-th line.}
   begin
      v := 50 + 15*k;
      MoveTo(50,v);
      LineTo(410,v);
   end
```

FIGURE 3.4 *The Horizontal Lines*

This version requires some extra "integer thinking." In particular, we have to deduce from the geometry of the grid that the k-th horizontal line (indexing from zero) has vertical coordinate $50 + 15 \cdot k$. A good way to come up with this recipe is to make a small table that relates k to v:

k	0	1	2	3	4	5	6	7	8	9	10
v	50	65	80	95	110	125	140	155	170	185	200

From this it is quite easy to deduce the two parameters that define v as a linear function of k.

Regardless of the method chosen, we have drawn the horizontal portion of the grid as shown in FIGURE 3.4. The vertical lines are generated similarly and we obtain

```
ShowDrawing;
v := 50;
for k:=0 to 10 do
   {Draw the k-th horizontal line.}
   {v = vertical coordinate of the k-th line.}
   begin
      MoveTo(50,v);
      LineTo(410,v);
      v := v+15;
   end
h := 50;
for k:=0 to 12 do
   {Draw the k-th vertical line.}
   {h = vertical coordinate of the k-th line.}
   begin
      MoveTo(h,50);
      LineTo(h,200);
      h := h+30;
   end
```

In order for the program to use the DrawWindow, it is necessary to include a `ShowDrawing` command at the beginning. This is analogous to the `ShowText` command that is required for programs to write in the TextWindow.

Notice that the fragment draws a specific grid. If we changed the dimension of the grid or the tile size or the position of the upper left corner, then the various integers that show up in the program would have to be revised.

To facilitate this process we identify the the following parameters which "define" a grid:

$hL =$ the horizontal coordinate of the left edge of the grid.

$vT =$ the vertical coordinate of the top edge of the grid.

$m =$ the number of rows of tiles in the grid.

$n =$ the number of columns of tiles in the grid.

$p =$ the spacing (in pixels) between the horizontal lines.

$q =$ the spacing (in pixels) between the vertical lines.

In our example, $hL = 50$, $vT = 50$, $m = 10$, $n = 12$, $p = 15$ and $q = 30$. If the integer variables `hL`, `vT`, `m`, `n`, `p` and `q` house these values, then the two loops that generate the grid may be rewritten as follows:

```
  {Draw the horizontal lines.}
  v := vT;
  for k:=0 to m do
     {Draw the k-th horizontal line.}
     {v = vertical coordinate of the k-th line.}
     begin
        MoveTo(hL,v);
        LineTo(hL+n*q,v);
        v := v+p;
     end;
  {Draw the vertical lines.}
  h := hL;
  for k:=0 to n do
     {Draw the k-th vertical line.}
     {h = vertical coordinate of the k-th line.}
     begin
        MoveTo(h,vT);
        LineTo(h,vT+m*p);
        h := h+q;
     end
end.
```

Since these values are not changed during program execution, it is appropriate to define the six parameters as *constants* rather than as variables. The program `Example3_1` shows how to do this. The `const` declaration should be placed before the `var` declaration. Notice that the equals symbol (not assignment symbol) is used in the definition of the constant. The type is inferred from the designated value. Thus,

```
program Example3_1;
{Draws a grid.}
const
   hL = 50; {the horizontal coordinate of grid's left edge.}
   vT = 50; {the vertical coordinate of grid's top edge.}
   m = 10; {the number of rows of tiles in the grid.}
   n = 12; {the number of columns of tiles in the grid.}
   p = 15; {the spacing (in pixels) between the horizontal lines.}
   q = 30; {the spacing (in pixels) between the vertical lines.}
var
   h,v:integer; {horizontal and vertical coordinates.}
   k:integer; {index used for counting lines.}
begin
   ShowDrawing
   {Draw the horizontal lines.}
   v := vT;
   for k:=0 to m do
      {Draw the k-th horizontal line.}
      {v = vertical coordinate of the k-th line.}
      begin
         MoveTo(hL,v);
         LineTo(hL+n*q,v);
         v := v+p;
      end;
   {Draw the vertical lines.}
   h := hL;
   for k:=0 to n do
      {Draw the k-th vertical line.}
      {h = vertical coordinate of the k-th line.}
      begin
         MoveTo(h,vT);
         LineTo(h,vT+m*p);
         h := h+q;
      end
end.
```

Output. See FIGURE 3.1.

```
const
    e = 2.718;
    DeckSize = 52;
```

establishes `e` as a real (because of the decimal point) and `Decksize` as an integer. Recall that in ThinkPascal, `pi` is a built-in constant.

The declaration of constants makes it easy to change a program that merely requires the alteration a constant value. In `Example3_1`, we can change the position and dimensions of the grid merely by changing the values of `hL`, `vT`, `m`, `n` , `p`, and `q` An ability to identify the parameters that 'define" an object is very important to the computer-aided design process and we shall return to the issue several times.

PROBLEM 3.1. Modify `Example3_1` so that the values for m, n, s, hL and vT, are acquired as input instead of encapsulated as constants. In particular, the body of the program should begin as follows:

```
ShowText;
ShowDrawing;
writeln('Enter rows,cols,tilesize,hL,vT');
readln(m,n,s,hL,vT);
```

Make the TextWindow a narrow band across the bottom of the screen and the DrawWindow everything above. Try the data $(m, n, s, hL, vT) = (200, 300, 2, 0, 0)$ and $(m, n, s, hL, vT) = 8, 8, 20, 150, 50)$.

PROBLEM 3.2. Write a program that draws

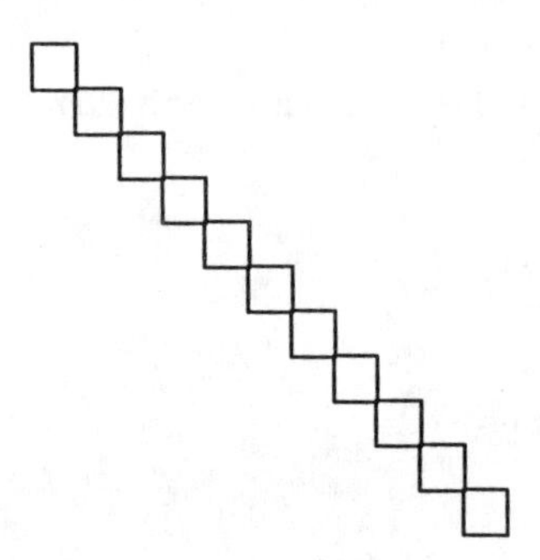

The tiles should be 15-by-15. The first tile in the upper left corner should have its upper left corner positioned at (150,50). For $k = 2..11$, the upper left corner of tile k should coincide with the lower right corner of tile k-1. Encapsulate the parameters of the pattern as constants.

PROBLEM 3.3. Write a program that draws

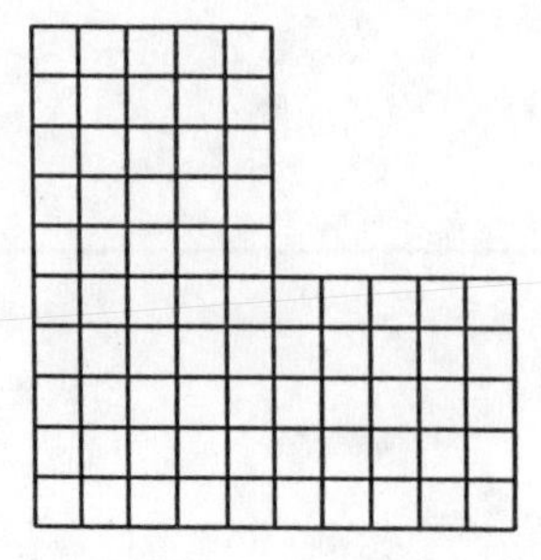

The tiles should be 15-by-15 with the upper left corner at (150,50). Encapsulate the parameters of the pattern as constants.

PROBLEM 3.4. Write a program that draws

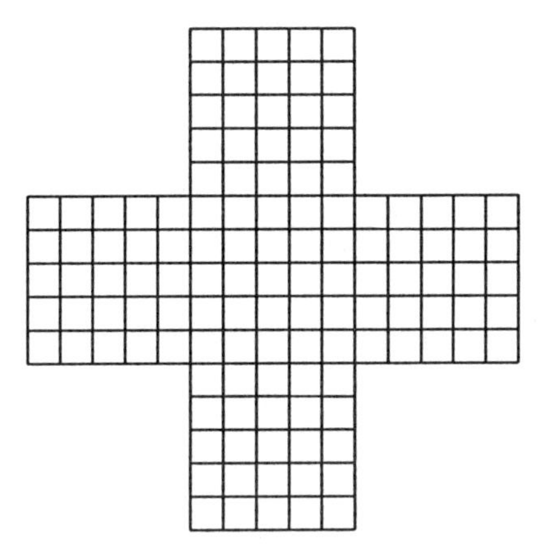

The tiles should be 10-by-10. The top edge of the pattern should have vertical coordinate $v = 50$ and the left edge should have horizontal coordinate $h = 50$. Encapsulate the parameters of the pattern as constants.

PROBLEM 3.5. Write a program that draws the following:

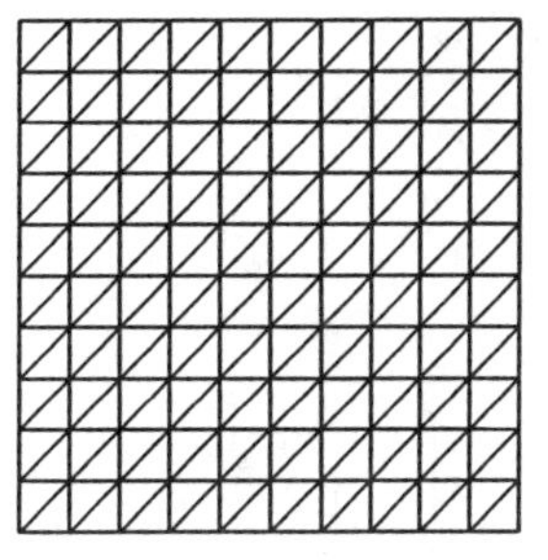

This is a 10-by-10 grid with upward sloping diagonals in each tile The tiles should be 15-by-15. Position the upper left corner at (150,50). Structure your initial solution so that it draws a conventional 10-by-10 grid first and then draws the diagonals. Encapsulate the parameters of the pattern as constants.

PROBLEM 3.6. Write a program that draws the following:

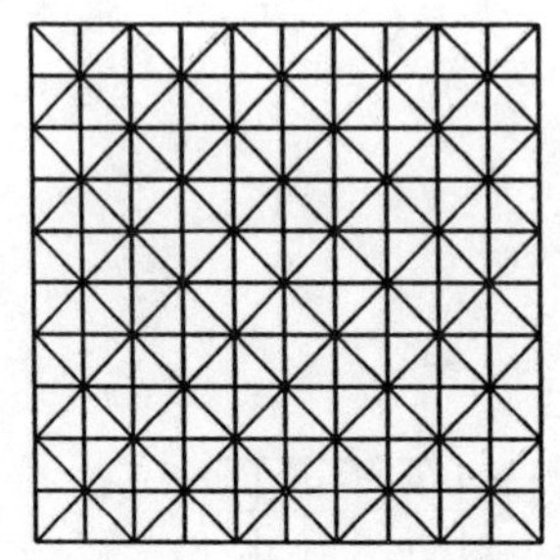

Make the tiles 15-by-15 and position the upper left corner at (150,50). Notice that each tile in the 10-by-10 grid has either an upwards-sloping or downwards-sloping diagonal. It is up to you to determine "the pattern." Try to encapsulate the parameters of the pattern as constants.

PROBLEM 3.7. Write a program that draws the following picture:

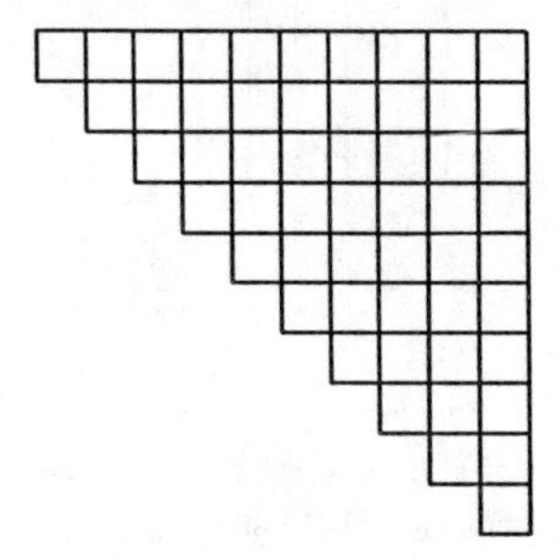

The tiles should be 15-by-15 and the upper left corner should be positioned at (100,50). Try to encapsulate the parameters of the pattern as constants.

PROBLEM 3.8. Write a program that draws the following picture without the vertex labels:

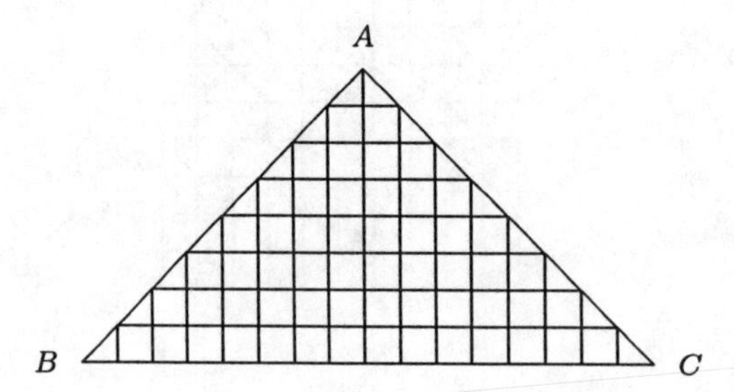

Vertex A should be at (200,20) and be a right angle. Vertices B and C should have the same vertical coordinates. The sides AB and AC should be inclined 45° to the horizontal. The vertical spacing between the horizontal lines should be 15 pixels. Make effective use of a `for`-loop. Encapsulate the parameters of the pattern as constants.

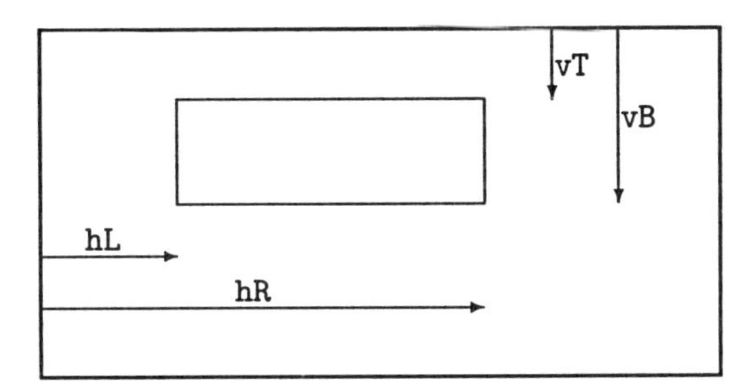

FIGURE 3.5 FrameRect(vT,hL,vB,hR)

3.2 Rectangles and Ovals

Rectangles can be drawn using the built-in procedures FrameRect and FrameOval. Their location and size is defined by a 4-tuple (T,L,B,R) of integers that define the top, left, bottom, and right edges. For example,

```
FrameRect(vT,hL,vB,hR)
```

produces a rectangle with vertices (hL, vT), (hL, vB), (hR, vT) and (hR, vB) as shown in FIGURE 3.5. PaintRect is identical only it shades the rectangle according to the "active" color. Three colors are available: white, gray, and black. Initially, the pen color is black. The color can be changed via the PenPat procedure. To illustrate, the fragment

```
PenPat(gray);
PaintRect(20,30,100,110);
PenPat(black);
PaintRect(40,50,80,90);
PaintRect(20,180,100,260);
PenPat(white);
PaintRect(40,200,80,240);
```

produces the picture displayed in FIGURE 3.6. The big gray square on the left is drawn first. The small black square is then drawn after a switch to black ink. The large black square on the right is then drawn followed by a switch to white ink. The drawing is completed with the drawing of the small white square on the right.

Sometimes, we may want to draw a rectangle that is not specified in the top-left-bottom-right "style." For example, we may need to draw a square that has center (h_c, v_c) and sides of length s that are parallel to the coordinate axes. Assume that all three parameters are integer and that (for clarity) s is even. FIGURE 3.7 helps us determine the values that must be passed to FrameRect. The conversion to top-left-bottom-right format is now obvious and we obtain the following solution fragment:

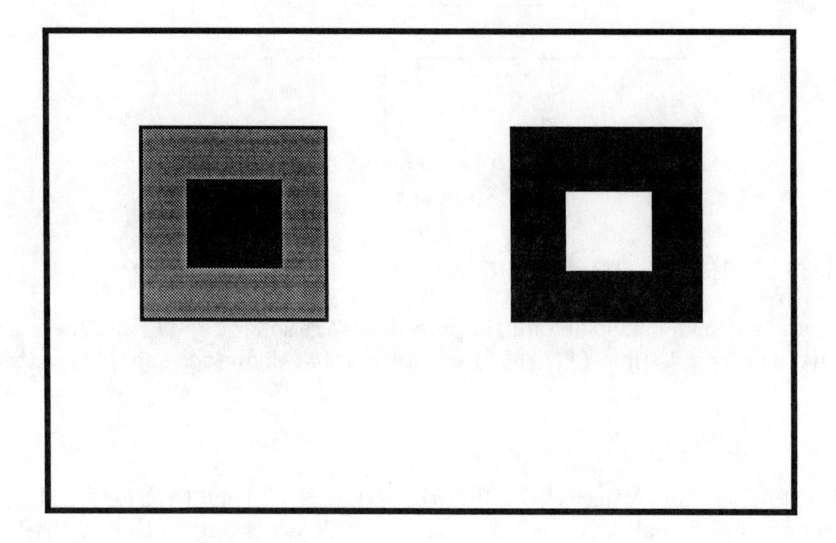

FIGURE 3.6 PaintRect *and* PenPat

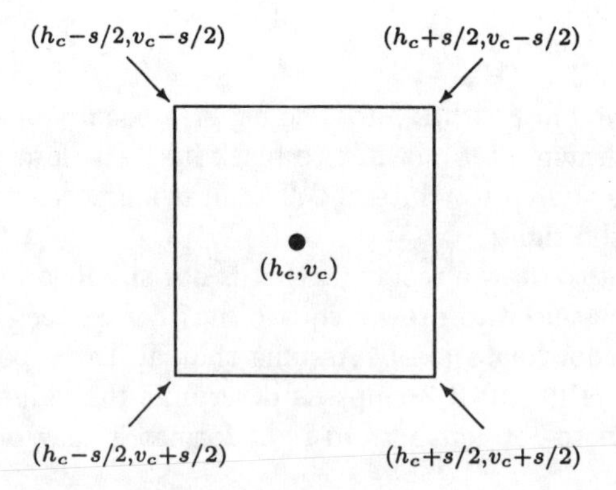

FIGURE 3.7 *A Square Centered at* h_c, v_c *with side* s.

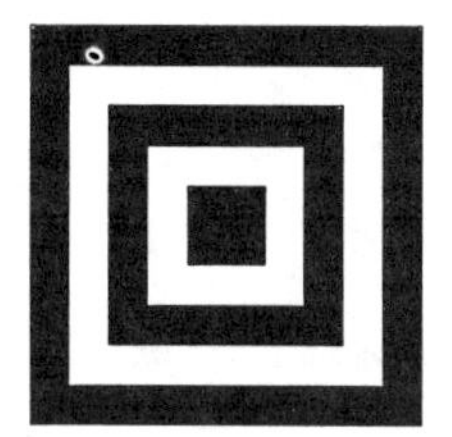

FIGURE 3.8 *A Square Bull's Eye*

```
r := s div 2;
vT := vc - r;
hL := hc - r;
vB := vc + r;
hR := hc + r;
FrameRect(vT,hL,vB,hR);
```

This is equivalent to

```
r := s div 2;
FrameRect(vc - r, hc - r, vc + r, hc + r);
```

In all this we assume that each variable has type `integer`.

We now turn to another problem that involves "positioning arithmetic" and which illustrates various aspects of shading. Let us draw the square "bull's eye" depicted in FIGURE 3.8. Assume that the center of each square is at $(h_c, v_c) = (200,120)$ and that the squares have sides 100, 80, 60, 40, and 20 respectively. Indexing the squares from one and starting on the outside we find:

Index	Side	Color
1	100	black
2	80	white
3	60	black
4	40	white
5	20	black

This gives us the following framework for solving the problem:

```
step := 20;
side := 100;
for i:=1 to 5 do
   {Draw square i.}
   begin
      ⟨⁝⟩
      side := side - step;
   end
```

```
program Example3_2;
{Draw a Square Bull's Eye}
const
   n = 5;
   step = 20;
   hc = 200;
   vc = 120;
   s:integer;
var
      i:integer; {Index of square being drawn.}
      side,r:integer; {The side and half side of the current square.}
begin
   ShowDrawing;
   side := 100;
   for i:=1 to n do
      {Draw square i.}
      begin
         r := s div 2;
         if odd(i) then
            PenPat(white)
         else
            PenPat(black)
         PaintRect(vc-r,hc-r,vc+r,hc+r);
         side := side - step;
      end
end.
```

Output: See FIGURE 3.8.

Whether or not we draw a white or a black square depends on the index. If the index is odd, then we draw a black square. Otherwise, a white square is drawn. See `Example3_2`.

The width of lines can be controlled by `Pensize`. The statement `Pensize(u,v)` gives the mouse a u-by-v "tip". The program `Example3_3` draws a nested sequence of squares using ever larger pen sizes.

PROBLEM 3.9. Write a program that draws 60 rectangles all with height 120 and width 100. The upper left corner of the first rectangle should be at (10,10). Each successive rectangle should be shifted to the right 2 pixels and down 2 pixels from its predecessor.

PROBLEM 3.10. Write a program that draws 80 squares each with center (250,125) and side 200. The k-th square should be drawn with `PenSize(k,k)`.

```
program Example3_3;
{Illustrates PenSize}
const
   hc = 240;
   vc = 130;
   s:integer;
var
      i:integer; {Index of square being drawn.}
      r:integer; {The half side of the current square.}
begin
   ShowDrawing;
   r:=5;
   for i:=1 to 10 do
      {Draw square i.}
      begin
         PenSize(i,2*i);
         FrameRect(vc-r,hc-r,vc+r,hc+r);
         r:=r+15;
      end
end.
```

Output: See FIGURE 3.9

FIGURE 3.9 *Output from* Example3_3

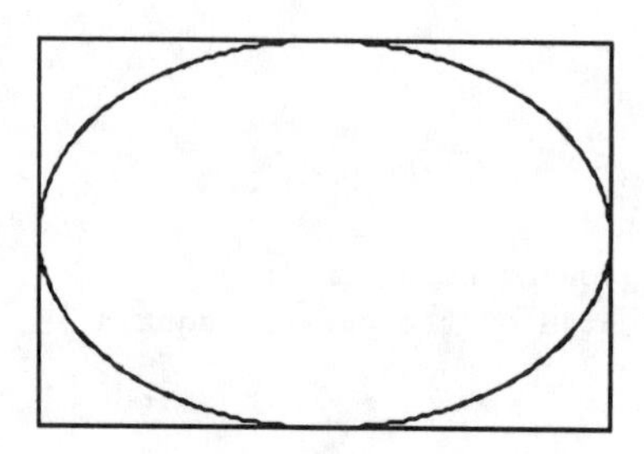

FIGURE 3.10 FrameOval(vT,hL,vB,hR)

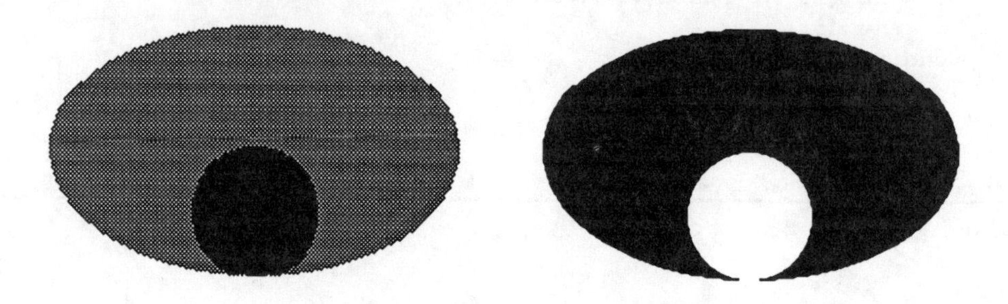

FIGURE 3.11 *Nested Ovals*

The drawing of ovals is analogous to the drawing of rectangles. The fragment

```
FrameRect(vT,hL,vB,hR);
FrameOval(vT,hL,vB,hR);
```

produces produces an oval that "just fits" inside the corresponding rectangle. See FIGURE 3.10. Analogous to the rectangle example above, the fragment

```
PenPat(gray);
PaintOval(20,30,140,230);
PenPat(black);
PaintOval(80,100,140,160);
PaintOval(20,270,140,470);
PenPat(white);
PaintOval(80,340,140,400);
```

draws the nested ovals in FIGURE 3.11

When drawing circles and disks, it is often more convenient to specify location and size in terms of the center and the radius. Thus, the statement `FrameOval(vc-r,hc-r,vc+r,hc+r)` draws a circle with center at (h_c, v_c) and radius r.

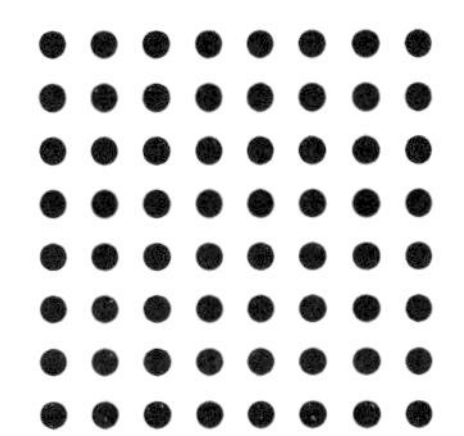

FIGURE 3.12 *Dot Array*

As an application of `PaintOval`, consider the problem of drawing the "dot array" in FIGURE 3.12. Let's assume that the dots have radius 5 and are 20 pixels apart, center-to-center. Assume also that the center of the upper left dot is at (100,50). We break the problem down at the "row-level" and outline a `for`-loop whose mission is to draw the dot-array row-by-row:

```
v:=50;
for row:=1 to 8 do
   {Draw a row of dots with vertical coordinate v.}
   begin
      ⟨:⟩
      v:=v+20;
   end
```

Code needs to be developed for drawing a particular row. The first dot in a row has horizontal coordinate $h = 100$. Each successive dot is shifted 20 pixels to the right. Thus, the $\langle \cdots \rangle$ is refined to

```
h := 100;
for col:=1 to 10 do
   begin
      PaintOval(v-5,h-5,v+5,h+5);
      h:=h+20;
   end
```

With suitable use of constants we obtain the program `Example3_4`. Notice how one loop is *nested* inside another. However, by virtue of the top-down approach, we never had to think of more than one loop at a time. The mission of the outer loop was established without regards to the details of its loop body. Then, as we focussed on those details, the inner loop was developed by itself.

```
program Example3_4;
{Array of dots.}
const
   r=5; {Radius of dots.}
   n=8; {Size of dot array.}
   hc=100; {Horizontal coordinate of upper left disk center.}
   vc=50; {Vertical coordinate of upper left disk center.}
var
   space:integer; { Spacing between rows and columns.}
   v,h,row,col:integer;
begin
   ShowDrawing;
   space := 4*r;
   v:=vc;
   for row:=1 to n do
      {Draw a row of dots with vertical coordinate v.}
      begin
         h := hc;
         for col:=1 to n do
            begin
               PaintOval(v-r,h-r,v+r,h+r);
               h := h+r;
            end
         v := v+15;
      end
end.
```

Output: See FIGURE 3.12.

PROBLEM 3.11. Write a program that draws the following picture:

Each disk should have radius 5. The center of the upper left disk should be at (100,50). Across each row and column, the disks should be spaced 15 pixels apart, center-to-center.

PROBLEM 3.12. Write a program that draws the following picture.

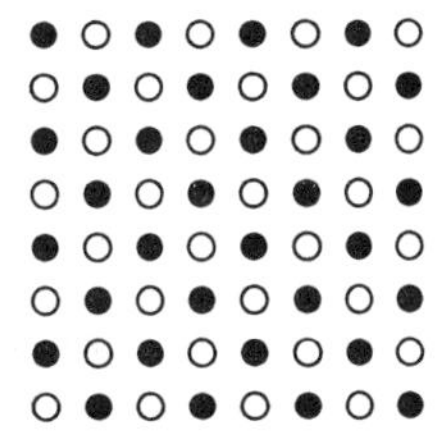

Each disk should have radius 5. The center of the upper left disk should be at (100,50). Across each row and column, the disks should be spaced 15 pixels apart, center-to-center.

PROBLEM 3.13. Write a program that draws the following picture:

Each disk should have radius 5. The center of the upper left disk should be at (100,50). Across each row and column, the disks should be spaced 15 pixels apart, center-to-center. Hint: if we index rows top to bottom, then there are i dots in the i-row.

3.3 Granularity

Unlike continuous "x-y graph paper," the screen is *discrete*. So far, this has posed no difficulties with our our carefully chosen set of initial examples and problems. But now we examine some

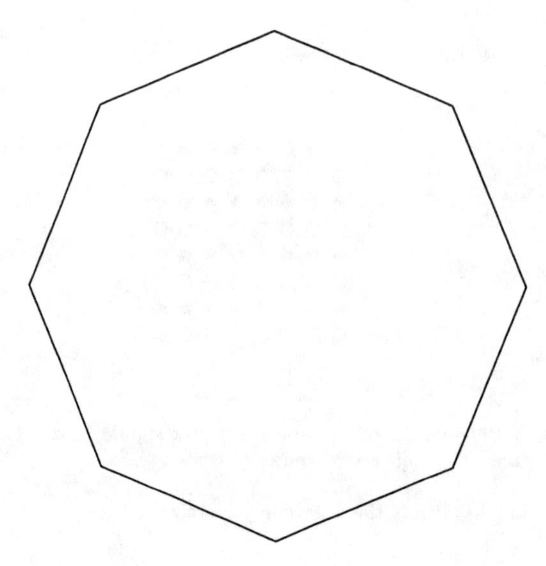

FIGURE 3.13 *A Near-Perfect Regular Octagon*

implications of *screen granularity.*

We start with a fragment that plots a regular octagon with vertices

$$v_k = (x_0 + r\cos(\pi k/4), y_0 + r\sin(\pi k/4)) \qquad k = 1, \ldots, 8.$$

Assume $(x_0, y_0) = (230, 120)$, and $r = 100$. The center (x_0, y_0) is a valid screen coordinate because it is integral. Likewise $v_0 = (330, 120)$, $v_2 = (230, 220)$, $v_4 = (130, 120)$ and $v_6 = (230, 20)$ are valid screen coordinates. But the remaining vertices have the form $(230 \pm \sqrt{100}, 120 \pm \sqrt{100})$ are non-integral and do not correspond to points on the screen. The fragment

```
r := 100; hc := 150; vc := 150;
MoveTo(hc+r,vc);
for k:=1 to 8 do
   LineTo(hc + r*cos(pi*k/4),vc - r*sin(pi*k/4));
```

does not work because `LineTo` expects integer arguments. One way to rectify this is merely to round the vertex coordinates:

```
r := 100; hc := 150; vc := 150;
MoveTo(hc+r,vc);
for k:=1 to 8 do
   LineTo(round(hc + r*cos(pi*k/4)),round(vc - r*sin(pi*k/4)));
```

See FIGURE 3.13. The bad news is that the octagon is no longer regular because four of the vertices are slightly shifted. The good news is that it probably doesn't really matter because to the human eye, the displayed octagon is "regular enough." However, in serious applications screen granularity can pose a major obstacle and complicated steps must be taken to minimize its effect. `Example3_5` can be used to explore further other ramifications of the screen's granularity.

It has two new features to discuss. The first is the `WriteDraw` statement. `WriteDraw` is identical to `Writeln` except that the result is printed in the DrawWindow at the current mouse

```
program Example3_5;
{Screen Granularity through Polygons}
const
   hc = 250;
   vc = 150; {(hc,vc) = polygon center}
   r = 100; {Polygon radius}
var
   AnotherEg:char; {Continuation indicator}
   n:integer; {Number of sides of the polygon}
   step:real; {2pi/n, the angular spacing between vertices}
   k:integer;
begin
   ShowText;
   ShowDrawing;
   AnotherEg := 'y';
   while AnotherEg <> 'n' do
      begin
         EraseRect(vc-r,hc-r,vc+r,hc+r);
         n := n+1;
         step := 2*pi/ n;
         MoveTo(round(hc+r),vc);
         for k := 1 to n do
            LineTo(round(hc+r*cos(k*step)),round(vc+r*sin(k*step)));
         MoveTo(hc-10,vc);
         WriteDraw('n = ',n:3);
         Writeln('Try next n?  Enter y (yes) or n (no)');
         readln(AnotherEg);
      end;
end.
```

Output: See FIGURE 3.13.

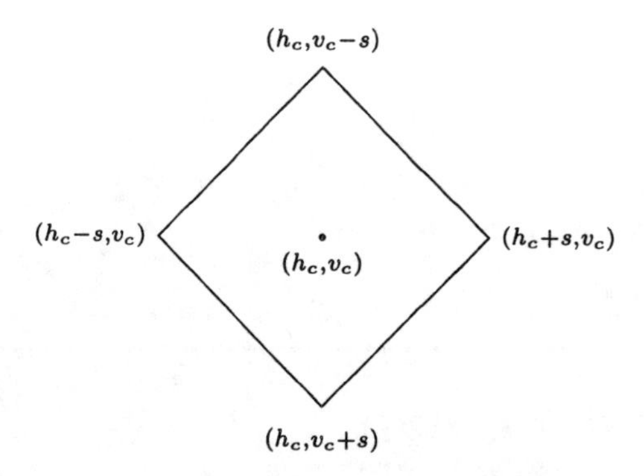

FIGURE 3.14 *A Diamond*

position instead of in the TextWindow. The preceding `MoveTo` statement ensures that the mouse is a little to the left of the polygon's center.

Another aspect of `Example3_5` deserving of comment is the use of both the TextWindow and the DrawWindow. The inclusion of both the `ShowText` and the `ShowDrawing` statements makes this possible. In programs like this it is important to position and size the two windows carefully. In `Example3_5`, the yes/no dialog associated with the interactive framework takes place in the TextWindow. By making this window small and in the lower right corner of the screen, it does not overlap with the display of the polygon in the DrawWindow. As in just about all of our graphics applications, the DrawWindow should be made as large as possible.

Notice the use of `EraseRect`. It effectively erases the zone where the polygons are displayed, thereby preventing their superpositioning as the loop progresses. It is interesting to run `Example_5` and discover the smallest n so that the displayed polygon may be regarded as a circle. The answer depends upon the chosen value of r, the screen's granularity, and the human eye.

An overtone in the above discussion is that the screen is an inferior version of the continuous xy plane. However, its discreteness permits calculations that would otherwise be impossible. Consider the shading of the diamond displayed in FIGURE 3.14 by drawing every possible horizontal line segment within its boundary. This can be accomplished by a loop of the following form:

```
for dv := 0 to r do
   {Draw shading lines at v = vc-dv and v = vc+dv.}
   begin
      dh := r - dv; {Halflength of the shading line segment.}
      MoveTo(hc - dh, vc - dv);
      LineTo(hc + dh, vc - dv);
      MoveTo(hc - dh, vc + dv);
      LineTo(hc + dh, vc + dv);
   end
```

The idea of drawing "every possible" line segment inside the diamond is only feasible because they are finite in number.

More general shading problems involve "point-slope" thinking and rounding. `Example3_6` shades the right triangle depicted in FIGURE 3.15. It works by drawing every possible horizontal line segment within its boundary. The loop counter `v` steps through the required range of vertical

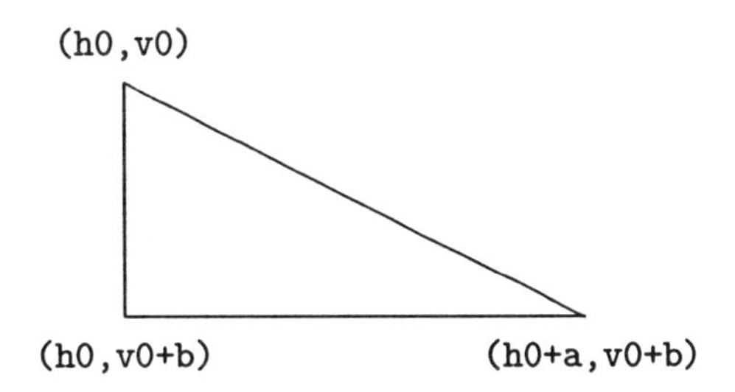

FIGURE 3.15 *Shading Boundary for* Example3_6

```
program Example3_6;
{Screen Granularity and Shading}
const
   {The triangle has vertices (h0,v0), (h0,v0+b), and (h0+a,v0+b).}
   h0 = 50;
   v0 = 150;
   a = 300;
   b = 100;
var
   v:integer; {Vertical coordinate of current shading line.}
   L:integer; {Length of the current shading line.}
begin
   ShowDrawing;
   for v := v0 to v0 + b do
      {Draw a horizontal shading line with vertical coordinate v.}
      begin
         MoveTo(h0,v);
         L := round(((v-v0)/b)*a);
         LineTo(h0+L,v);
      end;
end.
```

coordinates. The line segments are drawn left to right. The `MoveTo(h0,v)` positions the mouse on the triangle's "vertical leg." The `LineTo(h0+L,v)` draws the line over to the hypotenuse. Of critical importance is the quantity L, the length of the segment. Here is how to derive the assignment

```
L := round(((v-v0)/b)*a)
```

The fraction `(v-v0)/b` represents the fraction of the way that has been traversed from the top vertex to the base. That fraction, times the length of the base, i.e., `((v-v0)/b)*a`, is the length of the segment to be drawn. This follows from linearity. Finally, the round of this quantity is assigned to `L` because the expression `ho+L` must have type `integer` for `LineTo(h0+L)` to work.

PROBLEM 3.14. Write a program that draws a shaded hexagon whose screen vertices are rounded versions of

$$(h_c + r \cdot \cos(k\pi/3), v_c - r \cdot \sin(k\pi/3)) \qquad k = 0, 1, 2, 3, 4, 5$$

where $(h_c, v_c) = (200, 120)$ and $r = 100$. Encapsulate the center and radius as constants.

PROBLEM 3.15. Write a program that draws a line segment from (h_0, v_0) to

$$(h_0 + r, v_0 - r/2^k) \qquad k = 1, 2, \ldots, 9.$$

Set $r = 512$ and $(h_0, v_0) = (10, 250)$. Encapsulate these parameters as constants.

PROBLEM 3.16. Draw an eight inch ruler with hash marks corresponding to every sixteenth of an inch. Here are the details. The edge of the ruler should be displayed as a line from (h_0, v_0) to $(h_0 + L, v_0)$. The hash marks should be drawn as vertical line segments that extend above the edge of the ruler with the following lengths:

one-inch:	25 pixels
half-inch:	20 pixels
quarter-inch:	15 pixels
eighth-inch:	10 pixels
sixteenth-inch:	5 pixels

Define as constants $h_0 = 50$, $v_0 = 150$, and $L = 512$. This particular value of L produces an aesthetically pleasing ruler. But if L is not divisible by 128, then uneven spacings between the hash marks may result result. Experiment!

Think Pascal Graphics Procedures

- `MoveTo(h,v)` Moves the mouse to screen coordinate (h, v). Expects integer valued arguments.
- `LineTo(h,v)` Draws a line from the current mouse position to screen coordinate (h, v). Expects integer-valued arguments.
- `FrameRect(T,L,B,R)` Draws a rectangle whose top and bottom edges have vertical coordinates T and B and whose left and right edges have horizontal coordinates specified by L and R. Expects integer-valued arguments.
- `PaintRect(T,L,B,R)` Draws a solid rectangle whose top and bottom edges have vertical coordinates T and B and whose left and right edges have horizontal coordinates specified by L and R. Expects integer-valued arguments.
- `EraseRect(T,L,B,R)` Erases everything inside the rectangle whose top and bottom edges have vertical coordinates T and B and whose left and right edges have horizontal coordinates specified by L and R. Expects integer-valued arguments.
- `FrameOval(T,L,B,R)` Draws an ellipse. The ellipse is positioned in a rectangle whose top and bottom edges have vertical coordinates T and B and whose left and right edges have horizontal coordinates specified by L and R. Expects integer-valued arguments.
- `PaintOval(T,L,B,R)` Draws a solid ellipse. The ellipse is positioned in a rectangle whose top and bottom edges have vertical coordinates T and B and whose left and right edges have horizontal coordinates specified by L and R. Expects integer-valued arguments.
- `PenSize(u,v)` With this command, all subsequent lines are drawn as if with a pen having a u-by-v tip. Initial pen size is 1-by-1. Expects integer-valued arguments.
- `PenPat( · )` The effective "ink" in the pen can be changed with this command. `PenPat(white)`, `PenPat(grey)`, and `PenPat(black)` produces lines and shading as if from a white, gray, or black pen. Initially, black is the prevailing line and shading color.
- `WriteDraw(`⟨*List of things to print*⟩`)` works just like `writeln` only the list of things to print comes out in the DrawWindow beginning at the current mouse position.

Chapter 4

Sequences

§4.1 Summation

`While`-loops with compound termination criteria, computing n-th terms

§4.2 Recursions

One-term recursion, searching for a first occurrence, two-term recursion.

In Chapter 2 we played with the sequence of regular n-gon areas $\{a_n\}$ where

$$A_n = \frac{n}{2}\sin\left(\frac{2\pi}{n}\right).$$

We numerically "discovered" that

$$\lim_{n\to\infty} A_n = \pi\,,$$

a fact that is consistent with our geometric intuition.

In this chapter we build our "n-th term expertise" by exploring sequences that are specified in various ways. At the top of our agenda are sequences of sums like

$$S_n = 1 + \frac{1}{4} + \frac{1}{9} + \cdots + \frac{1}{n^2}.$$

Many important functions can be approximated by very simple summations, e.g.,

$$\exp(x) \approx 1 + x + \frac{x^2}{2!} + \cdots + \frac{x^n}{n!}.$$

The quality of the approximation depends upon the value of x and the integer n.

Sometimes a sequence is defined *recursively*. The n-term may be specified as a function of previous terms, e.g.,

$$f_n = \begin{cases} 1 & \text{if } n = 1 \text{ or } 2 \\ f_{n-1} + f_{n-2} & \text{if } n \geq 3 \end{cases}$$

Sequence problems of this variety give us the opportunity to practice some difficult formula-to-program transitions.

4.1 Summation

The summation of a sequence of "regularly scheduled" numbers is such a common enterprise that a special notation is used. This is the "sigma" notation and here is an example:

$$\sum_{k=0}^{n} \frac{x^k}{k!} = 1 + x + \frac{x^2}{2!} + \cdots + \frac{x^n}{n!}.$$

The value of many regular summations is known. For example, the sum of the first n positive integers is given by

$$\sum_{k=1}^{n} k = \frac{n(n+1)}{2}.$$

It is interesting to write programs that check rules like this. The program `Example4_1` confirms that the sum of the first n integers is given by the above rule for $n = 1$ to 20.

The program uses a `for`-loop to actually compute the summation. The result is then printed side-by-side with the value of the summation formula. The program is *not* a proof that (4.1.1). It merely checks its correctness for a small set of possible n.

PROBLEM 4.1. Modify `Example4_1` so that it confirms the following for $n = 1..20$:

$$\sum_{k=1}^{n} k^2 = \frac{n(n+1)(2n+1)}{6}.$$

PROBLEM 4.2. It is possible to find real numbers a, b, c, d, and e so that

$$\sum_{k=1}^{n} k^3 = an^4 + bn^3 + cn^2 + dn + e$$

for all n. Note that if n is large enough, then

$$\sum_{k=1}^{n} k^3 \approx an^4.$$

By dividing both sides by n^4 and assuming that n is large, we see that

$$a \approx \left(\sum_{k=1}^{n} k^3\right) / n^4$$

Write a program that estimates a using this approximation for $n = 1, \ldots, 50$.

PROBLEM 4.3. If $r \neq 1$, then it can be shown that

$$\sum_{k=0}^{n} r^k = \frac{1 - r^{n+1}}{1 - r}$$

Modify `Example4_1` so that it confirms this for $n = 1, \ldots, 20$. Design the modification so that the value of r is obtained as input.

PROBLEM 4.4. Modify `Example4_1` so that it confirms the following summation for $n = 1..12$:

$$(1 + 2 + \cdots + n)^2 = 1^3 + 2^3 + \cdots + n^3$$

for $n = 1, \ldots, 20$.

```
program Example4_1;
{Checks the rule for the summation of the first n integers.}
const
   nmax = 20; {The number of n-values used in the rule checking.}
var
   n:integer; {The number of terms.}
   s:integer; {The value of 1+...+n.}
   rhs:integer; {n(n+1)/2}
begin
   ShowText;
   writeln(' n Sum n(n+1)/2 ');
   writeln('------------------------');
   s := 0;
   for n:=1 to nmax do
      {s=1+...+n}
      begin
         s := s + n;
         rhs := (n*(n+1)) div 2;
         writeln(n:2,s:6,rhs:6);
      end
end.
```

Output:

```
 n   Sum   n(n+1)/2
-----------------
 1     1          1
 2     3          3
 3     6          6
         ⋮
19   190        190
20   210        210
```

Given a real number x, define S_n to be the summation

$$S_n = \sum_{k=0}^{n} \frac{(-1)^k x^{2k+1}}{(2k+1)!}$$

i.e.,

$$S_n = x - \frac{x^3}{3!} + \frac{x^5}{5!} - \frac{x^7}{7!} + \frac{x^9}{9!} - \ldots + (-1)^n \frac{x^{2n+1}}{(2n+1)!}$$

The general term

$$a_k = (-1)^k \frac{x^{2k+1}}{(2k+1)!}$$

for this summation looks formidable. It appears that each term requires (a) an exponentiation of -1, (b) an exponentiation of x, and (c) a factorial. But with a little thought, it is clear that we do not have to start these computations "from scratch" with the generation of each new term. For example, if

$$a_6 = \frac{x^{13}}{13!}$$

is available, then we can compute a_7 from the formula

$$a_7 = -\frac{x^{15}}{15!} = -\frac{x^2 x^{13}}{15 \cdot 14 \cdot 13!} = -a_6 \frac{x^2}{15 \cdot 14}$$

In general, since $x^{2k+1} = x^2 \cdot x^{2k-1}$ and

$$(2k+1)! = (2k+1)(2k)(2k-1)!,$$

we have

$$\frac{a_k}{a_{k-1}} = (-1)^k \frac{x^{2k+1}}{(2k+1)!} \Big/ (-1)^{k-1} \frac{x^{2k-1}}{(2k-1)!} = \frac{-x^2}{(2k+1)(2k)}$$

and so

$$a_k = \frac{-x^2}{(2k+1)(2k)} a_{k-1}.$$

Given this recursion, here is a fragment that assigns to s the value of S_n assuming that n and x are initialized:

```
s := x;
f := -sqr(x);
a := x;
for k:=1 to n do
   begin
      a := -a *(f/(2*k*(2*k+1)));
      s:= s + a;
   end
```

The key idea is to develop a *recursion* that relates the current term a_k to its predecessor a_{k-1}. A mathematically equivalent fragment that computes $(-1)^k$, x^{2k+1}, and $(2k+1)!$ from scratch for each k would look something like this:

```
s := 0;
for k:=0 to n do
   begin
      if k mod 2 = 0 then
         MinusOnePower := 1
      else
         MinusOnePower := -1;
      xPower:=1;
      for j:=1 to 2*k+1 do
         xPower:= x*xPower;
      factorial:=1;
      for j:=1 to k do
         factorial:= j*factorial;
      s := s + MinusOnePower*xPower/factorial;
   end
```

Although we shall study the issue of efficiency more formally later on, it is obvious that the "from scratch" approach is inferior.

It is known that for large n, $\sin(x) \approx S_n$. Thus, instead of building the summation for a fixed n, we may wish to continue the process of adding in terms until $|S_n - \sin(x)|$ is small. e.g.,

```
s:= x;
f:= -sqr(x);
a:= x;
k:= 0 ;
sx = sin(x);
while abs(s - sx) > 0.0001 do
   begin
      k := k+1;
      a:= -a *(f/(2*k*(2*k+1)));
      s:= s + a;
   end
```

A danger with this kind of exploration is that it may take an inordinate number of iterations before the termination criteria is satisfied. To guard against this possibility it is advisable to "put a lid" on the maximum number of steps. All we need to do is change the `while` statement to

```
while (abs(s - sx) > 0.0001) and (k < 20) do
```

Unlike the above example, it is usually the case that the limiting value of a summation is unknown and the termination criteria cannot be based upon its closeness to the limit. In this case, a reasonable course of action is to terminate when the value of the term about to be added is small, e.g.,

```
while (abs(a) > 0.0001) and (k < 20) do
```

It is sometimes preferable to terminate when the current term is small *relative* to the current sum, e.g.,

```
while (abs(a) > 0.0001*abs(s)) and (k < 20)  do.
```

Taking this last approach to termination, program **Example4_2** provides an interactive framework that enables us to examine S_n as an approximation to $\sin(x)$.

PROBLEM 4.5. For large n,

$$R_n = 1 - \tfrac{1}{3} + \cdots - \frac{(-1)^{n+1}}{2n-1} = \sum_{k=1}^{n} \frac{(-1)^{k+1}}{2k-1} \approx \frac{\pi}{4}$$

$$T_n = 1 + \frac{1}{2^2} + \cdots + \frac{1}{n^2} = \sum_{k=1}^{n} \frac{1}{k^2} \approx \frac{\pi^2}{6}$$

$$U_n = 1 + \frac{1}{2^4} + \cdots + \frac{1}{n^4} = \sum_{k=1}^{n} \frac{1}{n^4} \approx \frac{\pi^4}{90}$$

Write a single program (with three loops) that computes the smallest n so that

$$\left| R_n - \frac{\pi}{4} \right| \leq 0.001,$$

```
program Example4_2;
{Interactive exploration of the series for sin(x)}
var
   AnotherEg:char; {Continuation indicator}
   s:real; {The current sum.}
   f:real; {-x*x}
   k:integer; {Index of the current term.}
   a:real; {ratio between ak+1, ak}
   x:real; {Function argument}
   sx:real; {sin(x)}
begin
   AnotherEg := 'y';
   while AnotherEg <> 'n' do
      begin
         writeln; writeln('Enter x'); readln(x);
         s:= x;cf:= -sqr(x); a:= x; k:= 0 ;csx := sin(x);
         writeln;
         writeln(' n Approximation Error');
         writeln('----------------------------------');
         while (abs(a) > 0.0001*abs(s)) and (k < 20)  do
            begin
               k := k+1;
               a:= a *(f/(2*k*(2*k+1)));
               s:= s + a;
               writeln(k:2,s:10:6,abs(s-sx):10:6);
            end;
         writeln;
         writeln('Another example?  Enter y (yes) or n (no).');
         readln(AnotherEg);
      end
end.
```

Sample output:

```
Enter x
1

 n  Approximation       Error
----------------------------------
 1        0.833333        0.008138
 2        0.841667        0.000196
 3        0.841468        0.000003
 4        0.841471        0.000000

Another example?  Enter y (yes) or n (no).
n
```

the smallest n so that

$$\left|T_n - \frac{\pi^2}{6}\right| \leq 0.001,$$

and the smallest n so that

$$\left|U_n - \frac{\pi^4}{90}\right| \leq 0.001.$$

PROBLEM 4.6. Each of the following sequences converge to π:

$$\begin{aligned} a_n &= \frac{6}{\sqrt{3}}\sum_{k=0}^{n}\frac{(-1)^k}{3^k(2k+1)} \\ b_n &= 16\sum_{k=0}^{n}\frac{(-1)^k}{5^{2k+1}(2k+1)} - 4\sum_{k=0}^{n}\frac{(-1)^k}{239^{2k+1}(2k+1)} \end{aligned}$$

Write a single program that prints $a_0, \ldots, a_n$ where n is the smallest integer so $|a_n - \pi| \leq .000001$ and prints $b_0, \ldots, b_n$ where n is the smallest integer so $|b_n - \pi| \leq .000001$.

PROBLEM 4.7. For all positive n, define

$$a_n = \sum_{j=1}^{n^2}\frac{n}{n^2+j^2}.$$

Write a program that prints $a_2, \ldots, a_n$ where n is the smallest integer such that $|a_{n-1} - a_n| \leq .01$. Hint. Structure your solution as follows:

```
⟨Compute a1 and a2.⟩
n:=2;
while |a(n-1) - a(n)| > 0.01 do
    begin
        n:=n+1;
        ⟨Compute a(n).⟩
    end
```

The computation of a_n requires a loop itself and so this is a nested-loop problem.

PROBLEM 4.8. Explore the following approximations by modifying **Example4_2**. Warning: some of the approximations deteriorate very rapidly as $|x|$ gets large.

$$\begin{aligned} \text{(a) } \cos(x) &\approx \sum_{j=0}^{n}\frac{(-x^2)^j}{(2j)!} \\ \text{(b) } \csc(x) = 1/\sin(x) &\approx \frac{1}{x} + 2x\sum_{j=1}^{n}\frac{(-1)^j}{x^2 - k^2\pi^2} \\ \text{(c) } \sinh(x) = \frac{e^x - e^{-x}}{2} &\approx \sum_{j=0}^{n}\frac{x^{2j+1}}{(2j+1)!} \\ \text{(d) } \cosh(x) = \frac{e^x + e^{-x}}{2} &\approx \sum_{j=0}^{n}\frac{x^{2j}}{(2j)!} \\ \text{(e) } \exp(x) &\approx \sum_{j=0}^{n}\frac{x^k}{k!} \end{aligned}$$

PROBLEM 4.9. Write a program that verifies the inequalities

$$\frac{2}{3}n\sqrt{n} \leq \sum_{k=1}^{n}\sqrt{k} \leq \frac{4n+3}{6}\sqrt{n}$$

for $n = 1, \ldots, 100$.

PROBLEM 4.10. Define

$$E_n = \left(\sum_{k=1}^{n} \frac{1}{k}\right) - \ln(n)$$

It is known that E_n converges to the *Euler constant* for large n. Write a program that prints E_{100k} for $k = 1, \ldots, 100$.

Analogous to the summation problem is the "product problem." Consider the following sequence:

$$P_0 = 2, \quad P_1 = 2\left(\frac{2}{1}\frac{2}{3}\right), \quad P_2 = 2\left(\frac{2}{1}\frac{2}{3}\right)\left(\frac{4}{3}\frac{4}{5}\right), \quad P_3 = 2\left(\frac{2}{1}\frac{2}{3}\right)\left(\frac{4}{3}\frac{4}{5}\right)\left(\frac{6}{5}\frac{6}{7}\right), \text{etc.}$$

In general,

$$P_k = P_{k-1}\frac{2k}{2k-1}\frac{2k}{2k+1} = P_k\frac{4k^2}{4k^2-1}$$

and we have

```
prod := 2;
for k:=1 to n do
   begin
      factor:=4*k*k/(4*k*k -1);
      prod:=prod*factor;
   end
```

In order to specify products succinctly, there is a notation analogous to the Σ-notation. If $a_0, a_1, \ldots$ then

$$P_k = \prod_{j=0}^{k} a_j = a_0 a_1 a_2 \cdots a_k.$$

Thus, in the above example, $a_0 = 2$ and $a_k = 4k^2/(4k^2 - 1)$ for $k \geq 1$.

PROBLEM 4.11. Using the interactive framework, explore the quality of the approximation

$$\sin(x) \approx x \prod_{j=1}^{n} \left(1 - \frac{x^2}{j^2\pi^2}\right)$$

Use $n = 50$ and print all the partial products and their errors.

PROBLEM 4.12. Numerically determine the value of

$$P_n = \prod_{k=2}^{n} \frac{k^3 - 1}{k^3 + 1}$$

as n gets large. To avoid integer overflow, make sure the variable `k` has type `longint`.

4.2 Recursions

The simplest way that a sequence $\{a_n\}$ can be specified is with an explicit recipe for each term, e.g., $a_n = 2^{-n}$. Sometimes a sequence is defined by giving the first term and then a rule for all the successors:

$$a_n = \begin{cases} 1 & \text{if } n = 0 \\ n \cdot a_{n-1} & \text{if } n \geq 1 \end{cases} .$$

This is an example of a *one-term* recurrence and we see that

$$\begin{aligned} a_1 &= 1 \cdot a_0 &= 1 \\ a_2 &= 2 \cdot a_1 &= 2 \\ a_3 &= 3 \cdot a_2 &= 6 \\ a_4 &= 4 \cdot a_3 &= 24 \end{aligned}$$

The fragment

```
a:=1;
for n:=1 to 4 do
   begin
      a:=n*a;
      writeln(n,a);
   end
```

produces a short table with the same values. It is not hard to see that values of the factorial function are being reported:

$$a_n = n! = 1 \cdot 2 \cdot 3 \cdots n$$

The n-th term for this particular one-term recursion can be specified explicitly, but this typically not the case. An interesting example that does not permit the "closed formula" expression for the general term is the "up and down" sequence:

$$a_n = \begin{cases} \text{any positive integer} & \text{if } n = 1 \\ a_{n-1}/2 & \text{if } n > 1 \text{ and } a_{n-1} \text{ is even} \\ 3a_{n-1} + 1 & \text{if } n > 1 \text{ and } a_{n-1} \text{ is odd} \end{cases}$$

Thus, if $a_1 = 17$, then the sequence

$$17,\ 52,\ 26,\ 13,\ 40,\ 20,\ 10,\ 5\ 16,\ 8,\ 4,\ 2,\ 1,\ 4,\ 2,\ 1,\ 4,\ 2,\ 1, \ldots$$

is produced. Notice that once the number one "is reached", the cycle 1,4,2,1,4,2,.. begins. It is known that the up and down sequence always reaches one no matter what the choice of a_1. There is no simple, explicit recipe for a_n as in the case for the factorial sequence.

Let $f(m)$ designate the smallest integer so $a_{f(m)} = 1$ given that $a_1 = m$. From the above example we see that $f(17) = 13$. We may also conclude from that same example that $f(1) = 1$, $f(4) = 3$, and $f(52) = 12$. If `m` houses the starting integer m, then the fragment

```
a := m;
k := 1;
while a <> 1 do
   {a = a(k)}
   begin
      if a mod 2 = 0 then
         a := a div 2
      else
         a := 3*a + 1;
      k := k + 1;
   end;
writeln('f(',m,') = '), k:3);
```

prints the value of $f(m)$.

Now let's augment this fragment so that it also prints the largest value encountered along the up-and-down route from m to 1. This is the first of many look-for-the-max problems that we shall encounter. It requires the maintenance of a variable whose mission is to keep track of the largest integer encountered "so far." The program **Example4_3** presents the details. Notice that **amax** is initially set to m, the starting value. Each time a new a-value is generated, it is compared to the value of **amax**. The variable **amax** always houses the largest a-value that has arisen since the loop started. More precisely,

$$\texttt{amax} = \max\{a_1, \ldots, a_k\}$$

If `a > amax` is true, then a new largest value has been encountered and **amax** is revised accordingly.

PROBLEM 4.13. Modify **Example4_3** so that for starting values $m = 1, 2, \ldots, 100$, it prints m, $f(m)$, and the associated max value.

The *Fibonacci sequence* gives us an opportunity to see what an iteration looks like that is based upon a *two-term recurrence*. Here is a table of the first 8 Fibonacci numbers $f_1, \ldots, f_8$:

k	1	2	3	4	5	6	7	8
f_k	1	1	2	3	5	8	13	21

The pattern should be clear. Once we "get going", each Fibonacci number is the sum of its two predecessors:

$$f_k = \begin{cases} 1 & \text{if } k = 1 \\ 1 & \text{if } k = 2 \\ f_{k-1} + f_{k-2} & \text{if } k > 2 \end{cases}$$

This is an example of a *two-term recurrence* and a loop that generates such a sequence requires the maintenance of two variables. One variable (call it `a`) is needed for the current Fibonacci number and another (call it `b`) is needed for its predecessor. The triplet

```
c:=a+b; b:=a; a:=c;
```

updates these two variables so that they respectively house the next Fibonacci number and *its* predecessor. FIGURE 4.1 depicts the changes these variables undergo assuming that initially `a = 5`, `b = 3`, and `c = 5`. **Example4_4** puts these updates under the control of a `while`-loop and

```
program Example4_3;
{The Up-and-Down Sequence}
var
   m:longint; {Starting value}
   k:longint;
   a:longint; {The k-th integer in sequence, i.e., a(k)}
   amax:  longint; {max of a(1),...,a(k)}
begin
   ShowText;
   writeln(' m f(m) max');
   writeln('-------------------------------');
   for m:=50 to 60 do
      begin
         a := m;
         k := 1;
         amax := m;
         while a <> 1 do
            {a = a(k)}
            begin
               k := k + 1;
               if a mod 2 = 0 then
                  a := a div 2
               else
                  a := 3*a + 1;
               if a>amax then
                  amax := a;
            end;
         writeln(m:10, k:10, amax:10);
      end
end.
```

Output:

m	f(m)	max
50	25	88
51	25	232
52	12	52
53	12	160
54	113	9232
55	113	9232
56	20	56
57	33	196
58	20	88
59	33	304
60	20	160

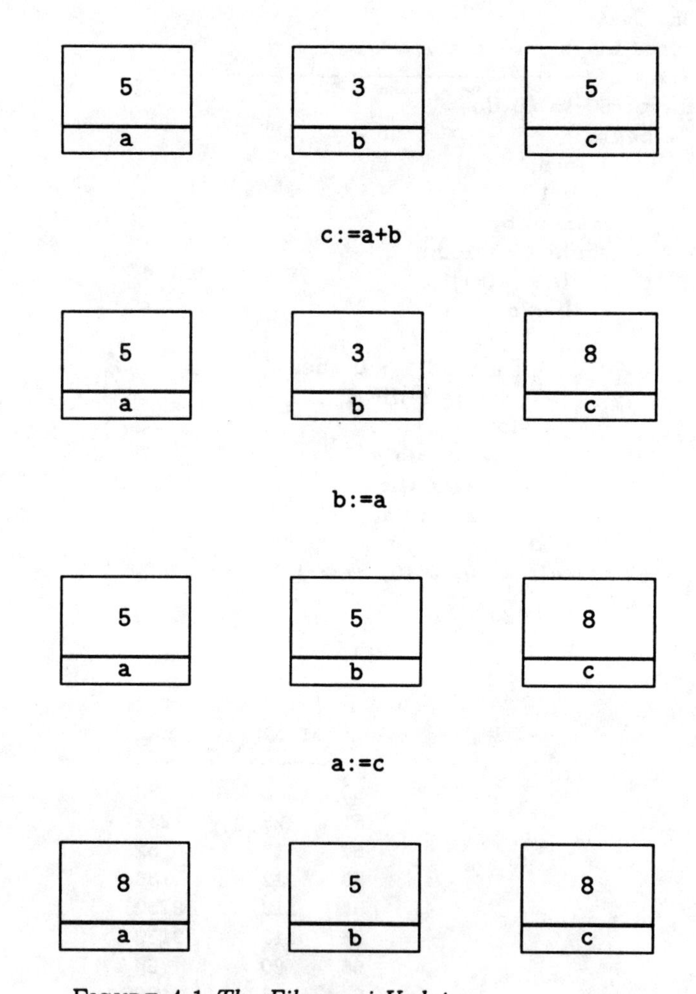

FIGURE 4.1 *The Fibonacci Update*

```
program Example4_4;
{Print all Fibonacci numbers less than a given upper bound.}
const
   m = 1000000; {The upper bound.}
var
   k:longint; {The number of Fibonacci numbers so far printed}
   a:longint; {f(k) = latest Fibonacci number}
   b:longint; {f(k-1) = next to latest Fibonacci number}
   c:longint; {f(k+1) = next Fibonacci number}
begin
   ShowText;
   writeln(' k f(k)');
   writeln('--------------');
   a := 1;
   b := 1;
   writeln(1:3,a:8);
   writeln(2:3,b:8);
   k := 2;
   while a+b < m do
      {b=f(k-1), a=f(k), c=f(k+1)}
      begin
         c:=a+b;
         b := a;
         a := c;
         k := k + 1;
         writeln(k:3, a:8);
      end;
end.
```

Output:

```
 k      f(k)
--------------
 1         1
 2         1
 3         2
 4         3
 5         5
 6         8
       ⋮
28    317811
29    514229
30    832040
```

prints a list of all the Fibonacci numbers that are less than one million. Notice that the first two Fibonacci numbers are set up before the loop begins.

PROBLEM 4.14. Modify **Example4_4** so that it reads in an integer $x > 1$ and prints Fibonacci numbers f_k and f_{k+1} where $f_k \le x < f_{k+1}$.

PROBLEM 4.15. Modify **Example4_4** so that it reads in a positive integer and prints a message that indicates whether or not it is a Fibonacci number.

PROBLEM 4.16. Define $r_k = f_k/f_{k-1}$, the ratio of f_k to its predecessor. Modify **Example4_4** so that it prints $r_3, \ldots, r_n$ where n is the smallest integer with the property that $|r_n - r_{n-1}| \le 0.000001$.

PROBLEM 4.17. Define

$$\begin{aligned}
t_0 &= \sqrt{1+0} \\
t_1 &= \sqrt{1+1} \\
t_2 &= \sqrt{1+2} \\
t_3 &= \sqrt{1+2\sqrt{1+3}} \\
t_4 &= \sqrt{1+2\sqrt{1+3\sqrt{1+4}}} \\
t_5 &= \sqrt{1+2\sqrt{1+3\sqrt{1+4\sqrt{1+5}}}}
\end{aligned}$$

Pick up the pattern and develop a program that prints $t_1, \ldots, t_{26}$. A loop is required for each t_k.

PROBLEM 4.18. Let m be a positive integer and consider the sequence

$$\begin{aligned}
t_1 &= \sqrt{m} \\
t_2 &= \sqrt{m-\sqrt{m}} \\
t_3 &= \sqrt{m-\sqrt{m+\sqrt{m}}} \\
t_4 &= \sqrt{m-\sqrt{m+\sqrt{m-\sqrt{m}}}} \\
t_5 &= \sqrt{m-\sqrt{m+\sqrt{m-\sqrt{m+\sqrt{m}}}}}
\end{aligned}$$

Pick up the pattern and write a program that helps you determine the limit of t_n as n gets large. Use the interactive framework, soliciting m and iterating until $|t_n - t_{n-1}| \le .0001$. For your information, the limit is an integer if $m = 7, 13, 21, 31$, or 43. A loop is required for each t_k.

Chapter 5

Random Simulations

Many phenomena have a random, probabilistic aspect: the role of the dice, the diffusion of a gas, the number of customers the enter a bank between noon and 12:05. Some special tools are needed to simulate events like these with the computer. This section is about random number generation and how to write programs that answer questions about random phenomena.

5.1 Generating Random Reals and Integers

Suppose we have a function `rand(x)` with the property that if the value of `x` is in the interval (0,1) and `y := rand(x)`, then the value of `y` is also in (0,1). We say that `rand` is a *pseudo-random number generator* if the fragment

```
r := r0;
for k := 1 to N do
   begin
      r := rand(r);
      writeln(r:10:6);
   end
```

produces a sequence of *uniformly distributed* random numbers between 0 and 1.[1] We assume that the value of `r0` is in (0,1) and that the value of the integer `N` is large. A rigorous definition of what "random" means is beyond the scope of this book. However, we offer two intuitive definitions based upon the output of the above fragment:

- there is no discernible pattern in the list that is printed.

[1]This just means that the numbers chosen are equally likely. Other distributions are possible including the *normal* distribution that is associated with "bell-shaped" curves.

- if p is in between 0 and 1, then approximately pN of the printed numbers are less than p.

The value of r0 that starts the generation of the random sequence is called the *seed*. Different seeds produce different random sequences. Notice that the value returned by rand becomes the input value for the next call.

The program Example5_1 illustrates how to access and use the rand function. It solicits the seed and an integer n and displays a random sequence of n numbers. We repeat, the same seed produces the same sequence.

The program includes a new type of declaration:

```
uses
   DDcodes;
```

This has the effect of making rand a built-in function. DDCodes is a file that contains a small set of "tools" that are used throughout the text. *Whenever you have a program that uses one of these tools like the* rand *function, you must follow two rules*:

1. Include the declaration uses DDcodes right after the program heading.
2. Add the file DDcodes to the project before the file that contains your program.

It is not our concern how rand works. The design of a good pseudo-random number generator is an advanced topic.

Through scaling and shifting, it is possible to generate random reals that range over any prescribed interval. For example, if the value of r0 is in (0,1), then

```
r := r0;
for k := 1 to n do
   begin
      r := rand(r); {random real between 0 and 1}
      z1 := (b-a)*r; {random real between 0 and b-a}
      z := a + (b-a)*r; {random real between a and (b-a)+a = b
      writeln(z);
   end
```

produces a sequence of random real numbers from the interval (a, b).

Similarly, if Lo and Hi are initialized integers and the value of r0 is in (0,1), then the following fragment prints a list of integers that are selected randomly from the set $\{Lo, Lo + 1, \ldots, Hi\}$:

```
r := r0;
for k := 1 to n do
   begin
      r := rand(r); {random real between 0 and 1}
      x := (Hi-Lo)*r; {random real between 0 and Hi-Lo}
      s := trunc(x); {random integer from [0,Hi-Lo-1]}
      i := Lo + s; {random integer from [Lo,Hi]}
      writeln(i);
   end
```

The program Example5_2 shows how to generate random integer sequences. It prints the outcome of 600 simulated dice rolls.

Two features of Example5_2 require elaboration. If you think of the simulation as an experiment, then for the sake of experiment repeatability it is "good science" to package the seed as a constant. To conduct a new experiment, you merely change its value.

```
program Example5_1;
{Random Sequences}
uses
   DDcodes;
var
   AnotherEg:char; {Continuation indicator}
   r0:real; {The seed.}
   n:integer; {Length of displayed sequence.}
   k:integer;
   r:real; {The k-th number in the sequence.}
begin
   ShowText;
   AnotherEg := 'y';
   while AnotherEg <> 'n' do
      begin
         writeln;
         writeln('Enter seed r0 between 0 and 1:');
         readln(r0);
         writeln('Enter length of sequence:');
         readln(n);
         r:=r0;
         for k := 1 to n do
            begin
               r := rand(r);
               writeln(r:10:6)
            end;
         writeln;
         writeln('Another example?  Enter y (yes) or n (no)');
         readln(AnotherEg);
      end
end.
```

Output:

```
Enter seed r0 between 0 and 1:
.123456
Enter length of sequence:
8
0.668800
0.056538
0.921371
0.856747
0.866997
0.846836
0.072571
0.126904

Another example?  Enter y (yes) or n (no)
n
```

```
program Example5_2;
{Simulates dice rolls.  }
uses
   DDcodes;
const
   seed = 0.123456;
   n = 600;
var
   r:real;
   throw:integer;
   k:integer;
begin
   ShowText;
   r := seed;
   for k := 1 to 1000 do
      begin
         r := rand(r);
         throw := trunc(6*r) + 1;
         if k mod 25 = 0 then
            writeln(throw:2)
         else
            write(throw:2);
      end
end.
```

Output:

```
 5 1 6 6 6 6 1 1 6 6 3 1 2 1 3 1 6 6 3 2 5 1 3 1 6 6 6 2 6 5
 4 1 1 3 6 3 3 5 1 2 5 2 2 2 6 1 4 3 5 3 5 2 2 2 3 5 2 3 3 1
 5 2 6 2 5 3 2 2 2 4 4 6 1 5 6 2 4 4 5 5 3 4 3 4 1 2 5 6 2 3
 4 5 3 2 2 3 2 1 4 3 3 3 3 2 6 4 1 1 5 3 5 2 3 3 1 2 1 3 4 5
 4 3 3 3 6 4 3 5 6 4 1 1 6 2 5 1 3 4 1 6 6 6 5 6 2 6 1 2 3 3
 6 2 4 6 6 5 1 3 4 2 4 5 6 1 5 1 3 4 3 5 4 3 6 5 4 5 2 2 1 5
 1 4 4 3 1 3 2 1 2 6 5 5 3 5 5 3 5 6 4 5 6 2 6 6 4 1 3 4 2 3
 2 3 1 4 4 1 2 2 1 2 1 5 5 3 1 3 6 1 1 5 1 3 1 6 5 2 1 6 1 3
 2 2 5 4 3 4 2 3 6 2 4 2 5 1 5 1 6 6 1 1 5 4 1 6 5 5 3 3 2 2
 5 4 3 5 2 6 6 5 1 2 1 4 5 1 1 6 4 6 5 1 4 3 5 2 3 5 6 6 5 4
 3 1 2 5 4 1 1 5 5 5 3 5 5 1 4 6 3 2 2 6 5 6 1 2 1 1 2 3 5 2
 4 3 2 3 5 6 5 1 1 1 6 5 2 6 1 6 1 3 3 1 1 1 5 4 6 3 4 5 4 4
 6 6 2 2 1 2 2 3 6 5 1 4 3 4 3 4 5 6 3 6 2 2 4 2 5 6 6 6 3 4
 2 2 5 5 4 6 3 4 4 6 2 2 3 3 1 4 1 6 2 1 2 2 6 1 2 4 5 1 6 5
 4 2 2 5 2 6 4 2 5 2 5 3 4 3 6 5 1 4 1 6 1 4 5 3 2 3 4 6 6 6
 2 6 1 6 5 5 1 2 2 5 6 4 6 2 6 3 3 3 3 3 4 3 4 2 1 3 4 1 1 3
 1 3 5 3 5 4 5 3 4 5 4 6 5 2 1 1 3 1 2 2 1 3 3 4 4 1 6 2 4 2
 3 1 2 5 3 2 5 2 1 4 4 4 3 6 3 3 1 6 6 4 2 5 1 2 5 3 1 2 3 1
 3 1 4 6 3 5 1 4 1 6 6 3 2 4 1 6 6 2 1 1 3 5 3 3 2 2 4 5 2 5
 5 4 1 2 5 5 5 4 1 2 5 4 1 6 2 6 3 4 1 3 6 2 6 4 5 3 5 1 5 5
```

Example5_2 makes us of the **write** statement. The **write** statement is just like **writeln** except that their is no "carriage return" after the printing. (The carriage return forces the next output to begin on a new line.) By using **write**, the next simulated dice roll is reported on the same line. This is handy when a a lot of output is anticipated which is the case in **Example5_2**.

There are roughly equal numbers of 1's, 2's, 3's, 4's, 5's, and 6's. With the above choice of seed we find

Outcome	Occurrences
1	107
2	105
3	106
4	84
5	102
6	96

To reinforce again the notion of randomness, suppose you were presented with two lists of dice roll outcomes, one produced by running **Example5_2** and one produced by 600 physical dice rolls. If **rand** is a good pseudo-random number generator, then it would be impossible for you to identify the computer-produced list.

PROBLEM 5.1. Write a program that generates a list of 100 integers selected randomly from the set

$$\{-20,\ -10,\ 0,\ 10,\ 20,\ 30\}.$$

PROBLEM 5.2. Write a program that prints a list of 100 real numbers selected randomly from the set $\{x : 0 < x < 2 \text{ or } 7 < x < 10\}$.

Our intuition about random sequences can be further enriched through graphics. The program **Example5_3** randomly places a large number of dots inside a given square. As can be seen from FIGURE 5.1, a "scoreboard" is centered beneath the square and reports on the number of dots that have been drawn. Notice that the scoreboard area is erased before it is updated and that the updates are after every 100th dot.

The uniformity of the dots over the square is a direct consequence of the uniform distribution of the generated random numbers. The uniformity becomes more striking as the number of displayed dots increases.

Notice that the program "lives" off a single stream of random numbers even though two numbers are required for the positioning of each dot. This ensures that there is not a correlation between the chosen h and v. *Make it a habit in all* **rand** *applications to use just a single sequence of random numbers.*

The topic of randomness, creativity, and visual aesthetics is fascinating and deep. We'll not delve into the issues and instead just offer **Example5_4** as to suggest that random patterns can be visually interesting. The program superimposes 500 "random" squares on the screen. See FIGURE 5.2. Notice how each square is defined by three random integers and that a single stream of random numbers is used to generate them.

PROBLEM 5.3. Modify **Example5_3** so that it places dots randomly inside the disk defined by

```
FrameOval(vT,hL,vT+s,hL+s);
```

To determine the location of a dot, randomly select a real number ρ from the interval $(0, s/2)$ and a real number θ from $(0, 2\pi)$. With these two numbers available, display a dot at the rounded version of $(h_c+\rho\cos(\theta), v_c+\rho\sin(\theta))$ where (h_c, v_c) is the circle's center. Explain why the displayed dots are not uniform across the disk.

```
program Example5_3;
{Random Dots in a Square}
uses
   DDcodes;
const
   MaxDots = 10000; {Number of Dots}
   s = 200; {Length of square's side}
   hL = 150; {Left edge of square}
   vT = 30; {Top edge of square}
   seed = 0.123456;
var
   n:longint;
   h,v:integer; {(h,v) = coordinate of nth dot}
   r:real; {Random number}
begin
   ShowDrawing;
   r := seed;
   {Draw the square}
   Pensize(2,2); FrameRect(vT,hL,vT+s,hL+s); PenSize(1,1);
   for n := 1 to MaxDots do
      begin
      {Display a random dot.}
      r := rand(r); h := trunc(hL+s*r);
      r := rand(r); v := trunc(vT+s*r);
      MoveTo(h,v); LineTo(h,v);
      if n mod 100 = 0 then
         {Update the "scoreboard".}
         begin
            EraseRect(vT+s+20,0,vT+s+50,1000);
            MoveTo(hL+(s div 2)-50,vT+s+40);
            WriteDraw('Dots drawn = ', n:5);
         end;
   end;
end.
```

Output: See FIGURE 5.1.

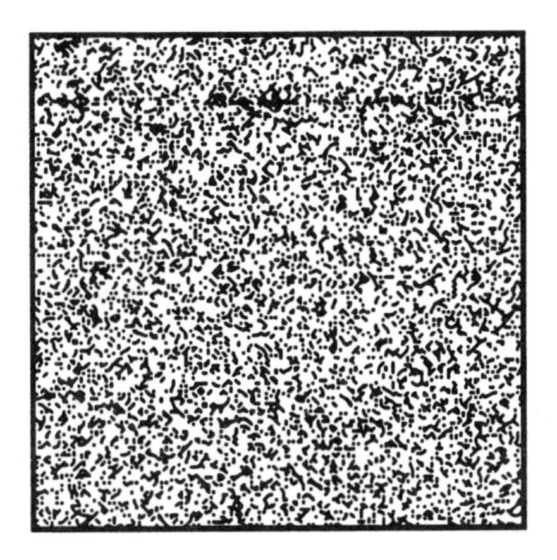

Dots drawn = 10000

FIGURE 5.1 *Random Dots in a Square*

```
program Example5_4;
{Draws random squares.}
uses
   DDcodes;
const
   seed = 0.123456;
   n = 500; {The number of rectangles.}
var
   r:real;
   hc,vc,half:integer;
   k:integer;
begin
   ShowDrawing;
   r := seed;
   for  k:=1 to n do
      {Draw a random square.  Center is randomly positioned in}
      {the rectangle defined by 100<=h<=500, 50<=v<=250.}
      {The side of the square is randomly chosen from [1,20].}
      begin
         r := rand(r); hc := trunc(400*r) + 100;
         r := rand(r); vc := trunc(200*r) + 50;
         r := rand(r); half := 1 + trunc(20 * r);
         FrameRect(vc - half, hc - half, vc + half, hc + half)
      end
end.
```

Output: See FIGURE 5.2.

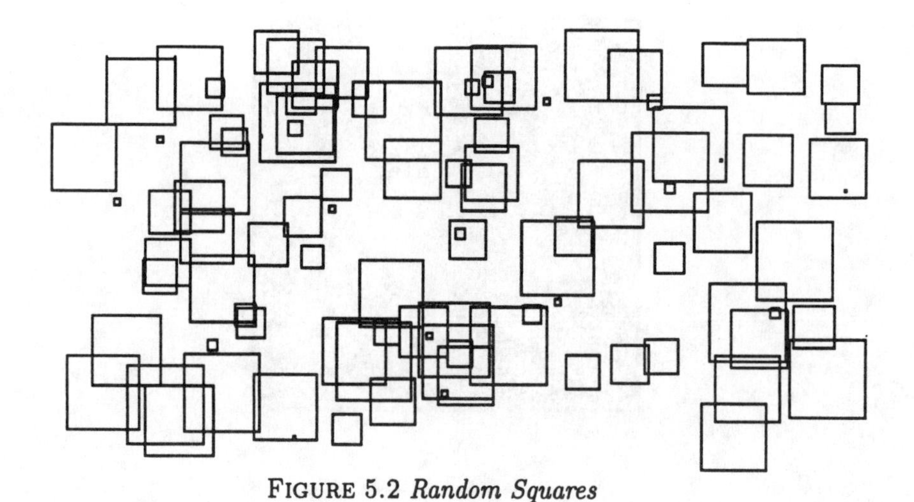

FIGURE 5.2 *Random Squares*

PROBLEM 5.4. Modify Example5_4 so that it draws black squares, gray squares, and white squares with equal probability.

5.2 Estimating Probabilities and Averages

Our next examples show how to simulate events that involve more than one "random happening." The first example is discrete and involves dice rolls. The other example is continuous and involves roots of a quadratic equation.

Suppose that we would like to know the probability that when three dice are rolled, all three outcomes are different. In the program Example5_5 we take an interactive approach to this question. Notice that all three random events are determined from a single stream of numbers that are randomly produced by rand. Different output results if either the seed or the value of n is changed. With respect to the latter, the larger the number of trials, the greater will be our confidence in the computed probability. For example, here is the output for three different simulations, all with seed .123456:

n	Computed Probability
100	.520
1000	.548
10000	.557

It can be shown that the correct probability is $5/9 = .555$.

Likewise, changing the seed changes the computed probability even with fixed n. Foi example, with $n = 1000$ we find

seed	Computed Probability
.123456	.548
.933402	.565

```
program Example5_5;
{Three-Dice Rolls}
uses
   DDcodes;
var
   AnotherEg:char; {indicates if another experiment is to be performed.}
   r:real; {a number between 0 and 1.}
   dice1,dice2,dice3:integer; {the outcomes of the three rolls.}
   k:integer; {counts the number of 3-dice rolls.}
   n:integer; {total number of rolls}
   count:integer; {counts .}
begin
   ShowText;
   AnotherEg := 'y';
   while AnotherEg <> 'n' do
      begin
         writeln;
         writeln('Enter the seed.  (Must be in (0,1).');
         readln(r);
         writeln;
         writeln('Enter the number of 3-dice rolls to perform.');
         readln(n);
         count:=0;
         for k:=1 to n do
            begin
               r:= rand(r); dice1:= 1 + trunc(6*r);
               r:= rand(r); dice2:= 1 + trunc(6*r);
               r:= rand(r); dice3:= 1 + trunc(6*r);
               if (dice1 <> dice2) and (dice1 <> dice3) and (dice2 <> dice3) then
                  count := count +1;
            end;
         writeln('Probability of 3 different values =',(count/n):10:6);
         writeln;
         writeln('Another Example?  Enter y (yes) or n (no).');
         readln(Anothereg);
      end
end.
```

Output:

```
Enter the seed. (Must be in (0,1).
.123456

Enter the number of 3-dice rolls to perform.
1000
Probability of 3 different values = 0.548000

Another Example? Enter y (yes) or n (no).
n
```

Example5_6 is concerned with the probability that the quadratic equation $ax^2+bx+c=0$ has complex roots given that the three coefficients are randomly selected from (0,1). The program Example5_6 can be used to estimate this probability. The idea behind the program is simply to generate three random coefficients a, b, and c and then examine the sign of $b^2 - 4ac$. If this quantity is negative, then the quadratic has complex roots since the roots are specified by $(-b \pm \sqrt{b^2 - 4ac})/(2a)$.

PROBLEM 5.5. Run a modified version of Example5_5 that can be used to answer the following questions. (a) What is the probability that at least two of the dice have the same value? (b) What is the probability that the value of the third dice roll is strictly in between the values of the first two rolls?

PROBLEM 5.6. Run modified versions of Example5_6 that can be used to answer the following questions. (a) What is the probability of complex roots if $a = 1$ and b and c are randomly selected from (-1,1)? (b) What is the probability of complex roots if all three coefficients are randomly selected from (-1,1)? (c) What is the probability that the two roots of the quadratic are real and within .5 of each other given that $a = 1$ and b and c are randomly selected from (-1,1)?

PROBLEM 5.7. A dart, thrown at random, hits a square target. Assuming that any two parts of the target of equal area are equally likely to be hit, find the probability that the point hit is nearer to the center than to any edge. Note that the answer does not depend upon the size of the square.

PROBLEM 5.8. A coin of diameter 1 inch is thrown onto a surface that is an n-by-n array of square tiles that are each two inches on a side. What is the probability that the coin is entirely within a tile? Assume that the center of the coin lands randomly on the surface.

PROBLEM 5.9. Two points on the unit circle are randomly selected. What is the probability that the length of the connecting chord is greater than 1?

PROBLEM 5.10. A point (x, y) is randomly selected on the semicircle $S = \{(x, y) : x^2 + y^2 = 1, y \geq 0\}$. What is the expected value of the area of the right triangle formed by (x, y), (1,0), and (-1,0)? Note that the random selection of (x, y) is tantamount to selecting a random angle θ from $(0, \pi)$ and setting $x = \cos(\theta)$ and $y = \sin(\theta)$.

PROBLEM 5.11. A stick of unit length is broken into two pieces. Assume that the breakpoint is randomly situated. On average, how long is the shorter piece? Write a program to answer the question.

Many estimate-the-average problems involve nested loops. As an example, let's estimate the expected value of $|H - T|$ where H and T are the number of heads and tails that result when a fair coin is tossed 100 times. Taking the top-down approach, assume that our program will perform n experiments where an experiment consists of 100 simulated coin tosses. We then aim for a program body of the following form:

```
s := 0;
for k:=1 to n do
   {Compute diff = |H-T| for k-th experiment.}
   ⟨⋮⟩
   s = s + diff;
end
Ave = s/n;
writeln('expected value of |H-T| = ',Ave:5:2);
```

The calculation of diff itself requires a loop as seen from Example5_7.

PROBLEM 5.12. On average, how many times must a dice be thrown until one gets a 6? Write a program to answer the question.

```
program Example5_6;
{Roots of random quadratics.}
uses
   DDcodes;
var
   a,b,c:real; {Coefficients of the quadratic.}
   r:real; {A number between 0 and 1.}
   k:integer; {index}
begin
   ShowText;
   writeln;
   writeln('Enter the seed.  (Must be in (0,1).');
   readln(r);
   count = 0;
   writeln(' Trials Probability');
   writeln('----------------------');
   for k:=1 to 1000 do
      begin
         r := rand(r); a := r;
         r := rand(r); b := r;
         r := rand(r); c := r;
         if sqr(b) < 4*a*c then
            count := count + 1;
         if k mod 100 = 0 then
            writeln(k:4,(count/k):13:4);
      end
end.
```

Output:

```
Enter seed.  (Must be in (0,1).)
.123456

 Trials    Probability
 ----------------------
    100         0.7500
    200         0.7450
    300         0.7433
    400         0.7550
    500         0.7420
    600         0.7433
    700         0.7443
    800         0.7487
    900         0.7478
   1000         0.7390
```

```
program Example5_7;
{Expected value of |H-T| in 100 coin tosses.}
uses
   DDcodes;
const
   n = 50; {Number of experiments.}
   seed = 0.123456;
var
   r:  real;
   s, diff, Ave:  real;
   k, i:  integer;
   H, T: integer;
begin
   s := 0;
   for k:=1 to n do
      {Compute diff = |H-T| for k-th experiment.}
      {s houses the sum of the diff's for the first k experiments.}
      begin
         H := 0; T := 0;
         for i:=1 to 100 do
            {H and T house the number of heads and tails observed after i tosses.}
            begin
               r := rand(r);
               if r > 0.5 then
                  H := H+1
               else
                  T := T+1;
            end;
         diff := abs(H-T);
         s = s + diff;
      end
   Ave = s/n;
   writeln('expected value of |H-T| = ',Ave:5:2);
end.
```

Output:

```
Expected value of |H-T| = 7.16
```

5.3 Monte Carlo

The solution to certain mathematical questions can be answered by setting up a "game" that has the same solution. To illustrate, let's estimate π by simulating a carefully constructed dart throw game. In particular, we shall throw darts randomly at a square target with vertices (1,1), (-1,1), (-1,-1), and (1,-1). If the darts land anywhere on the square with equal probability, then for a large number of throws the following approximation is reasonable:

$$\frac{\text{Number of throws inside unit circle}}{\text{Number of throws}} \approx \frac{\text{Area of circle}}{\text{Area of square}}.$$

Since the area of the circle is π (unknown) and the area of the square is 4 (known), we obtain the approximation

$$\pi \approx 4\frac{\text{Number of throws inside circle}}{\text{Number of throws}}.$$

The program **Example5_8** uses **rand** to simulate this game. This kind of simulation is called *Monte Carlo* and it is a widely used technique in a number of areas.

PROBLEM 5.12. In three dimensions, the unit sphere is defined by

$$S_3 = \{(x_1, x_2, x_3) : x_1^2 + x_2^2 + x_3^2 \leq 1\}$$

This is just a ball of radius 1 and its volume is given by a formula of the form has the form $V = a\pi r^3$ where a is a constant and r is the radius. Our goal is to estimate a. Start by generalizing **Example5_9** so that it prints estimates of the volume of a sphere with radius 1. (Hint: set up a 3-dimensional dart game in which the dart lands randomly in a cube of side 2.) For each printed volume estimate $\tilde{V}$, print the a-estimate $\tilde{a} = \tilde{V}/pi$.

PROBLEM 5.13. A length L needle where $L < 1$ is tossed n times onto the xy-plane. Each toss has the following properties:

1. The needle center lands randomly on the x-axis between 0 and m where m is a positive integer.
2. The needle makes an angle of θ radians with the x-axis where θ is a random number between 0 and $\pi/2$.

Let c be the number of tosses where the needle touches one of the parallel lines $x = 0$, $x = 1, \ldots, x = m$. Write a program that confirms

$$\pi \approx \frac{2Ln}{c}$$

Organize your program so that L, m, and n are constants. Set $n = 10000$ and print the π estimates for every 100 tosses. For $L = 0.5$, try $m = 10$, 100, and 1000. Do the estimates improve with increasing m? Also explore the quality of the estimate by varying L.

This method for computing π was proposed by the French mathematician Buffon over 200 years ago.

```
program Example5_8;
{Monte Carlo estimation of pi.}
uses
   DDcodes;
const
   n = 10000; {Number of dart throws.}
   seed = 0.319291;
var
   k:integer;
   r:real; {random number}
   hits:integer; {Number of throws inside unit circle}
   x,y:real; {(x,y) = coordinate of k-th throw.}
   OurPi:real; {Estimate of pi.}
begin
   r := seed;
   hits := 0
   writeln(' Throws Pi Estimate');
   writeln('-----------------------');
   for k := 1 to 1000 do
      {Hits is the number of darts inside unit circle after k throws.}
      begin
         r := rand(r); x := -1 + 2*r;
         r := rand(r); y := -1 + 2*r;
         if sqr(x) + sqr(y) <= 1 then
            hits := hits + 1;
         if (k mod 100) = 0 then
            begin
               OurPi := (hits/k)*4;
               writeln(k:5, OurPi:13:4);
            end
      end
end.
```

Output:

```
Throws   Pi Estimate
--------------------
  100         3.1600
  200         3.0600
            ⋮
10000         3.1432
```

Chapter 6

Fast, Faster, Fastest

How fast a program runs is usually of interest and so the intelligent acquisition of timing data, called *benchmarking*, is important. Benchmarking serves many purposes:

- It can be used to identify program bottlenecks.
- It can be used to the quantify how hard it is to solve a problem.
- It can be used to determine whether one solution process is more favorable than another.
- It can be used to calibrate the performance of a particular machine architecture.

However, in this chapter we merely illustrate the mechanics of benchmarking and show how it can be used to assess efficiency improvements as a program undergoes development.

Two examples are used to illustrate the design of efficient code. The plotting of an ellipse is used to show how to remove redundant arithmetic and function evaluation in a loop context. The computation of a rational approximation to π is used to show how the reduction of a doubly-nested fragment to a single loop can result in an order-of-magnitude speed-up.

Behind all the discussion is a quiet, but very important ambition: to build an aesthetic appreciation for the fast program. Programs that run fast are creations of beauty. This is widely accepted in practical settings where time is money. But in addition, program efficiency is something to revel in for its own sake. It should be among the aspirations for every computational scientist who writes programs.

6.1 Benchmarking

The time required to execute a program fragment can be measured by referencing a built-in function called `TickCount`. This function returns a "snapshot" of the computer's internal clock. Before and after snapshots are needed to time how long it takes the computer to execute a fragment:

```
program Example6_1;
{Integer Arithmetic Benchmark}
const
   MyComputer = 'Macintosh Powerbook 170, System 7.1'; {Modify accordingly}
   n=500000; {Repetition factor.  May need adjustment.}
var
   k:longint;
   x,y,z:integer; {For the sample additions.}
   StartTime, StopTime:real; {Clock Snapshots.}
   TimeInSec:real; {Total elapsed time in seconds for n additions.}
   m:longint; {The number of integer additions performed in 1 second.}
begin
   ShowText;
   writeln('Computer used = ', MyComputer);
   writeln;
   y:=2; z:=3;
   StartTime := TickCount;
   for k:=1 to n do
      x := y+z;
   StopTime := TickCount;
   TimeInSec := (StopTime-StartTime)/60;
   m := round(n/TimeInSec);
   writeln('Number of integer additions in 1 second = ',m:6);
end.
```

Output:

Computer used = Macintosh Powerbook 170, System 7.1.

Number of integer additions in 1 second = 89820

```
StartTime := TickCount;
⟨Fragment to be Timed⟩
StopTime := TickCount;
TimeInSec := (StopTime - StartTime)/60;
```

Elapsed time in seconds is obtained by subtracting the clock snapshots and dividing by sixty. The scaling by sixty is necessary because the clock is digital and "ticks" every 60th of a second.

`Example6_1` uses this technique to time how long it takes to execute 500,000 integer additions. This program has a number of important features to discuss. To begin with, the clock snapshots are stored in a pair of `real` variables `StartTime` and `StopTime`. The value assigned to `TimeInSec` may vary from run to run simply because the clock is discrete. However, if the fragment being clocked involves several second's worth of computation, then the computed elapsed time will be correct to within a percent and that is almost always sufficient in a benchmarking context. Bear in mind that what a computer can do in several seconds varies tremendously. A large supercomputer performs arithmetic hundreds of thousands times faster than the typical personal computer. If the task to benchmark is sufficiently small, then it has to be repeated in order to get a meaningful measure of execution time. For the computer being used to run `Example6_1`, $n = 500,000$ integer additions are timed. On a much faster machine, a larger number of additions would be required to get a sufficiently reliable timing. On a much slower machine, a smaller value of n could be used without jeopardizing the quality of the reported benchmark. Notice the encapsulation of n as a constant to facilitate easy change.

The program also makes use of a *string-valued* constant called `MyComputer`:

```
MyComputer = 'Macintosh Powerbook 170, System 7.1';
```

We cover strings in detail in Chapter 12. For now, just think of `MyComputer` as a message that is printed by the statement

```
writeln('Computer used = ', MyComputer);
```

Because the process of benchmarking is machine and system dependent, it is "good science" for the benchmarking program to report on the hardware and software used during the timing experiments[1]. The reason for using a constant string, like a constant of any type, is that it makes changes easy. The information in this string should report on the computer and system being used during the benchmarking.

Finally we mention that the time required for an integer addition does not depend upon the operands and so there is no loss in generality in timing how long it takes to add the integers 2 and 3. In more complicated situations, care must be exercised to ensure that one isn't benchmarking a "special case" that does not reflect general performance.

The benchmarking approach implemented in `Example6_1` is flawed because it fails to take into account the integer arithmetic that is associated with the actual running of the loop. In particular, the count variable must be incremented and compared to `n` each pass through the loop. (Comparisons usually involve a subtraction and a check of the sign.) Thus, the "loop overhead" contaminates the reported benchmark. To remove its effect, we can time the "empty loop" and subtract it from the original benchmark as illustrated in `Example6_2`. Although it depends upon the computer used, this typically doubles the speed that is reported for integer addition.

Usually, the operation to be timed is more involved than just a single integer addition and the underlying loop overhead is irrelevant. For example, a `PaintOval` reference is several hundred times slower than an integer addition and so the benchmarking of this graphics procedure need not take loop overhead into consideration. See the program `Example6_3` where `PaintRect` and `PaintOval` are compared. Note that this program reports a *relative benchmark* and in effect tells us how much slower `PaintOval` is compared to `PaintRect`. Relative timings are often easier to assimilate than absolute timings. However, in either case the machine used should be reported *since all benchmarks are machine dependent.*

PROBLEM 6.1. Expand `Example6_2` so that it also reports the number of integer multiplications, `mod` operations, and `div` operations that can be performed in one second. You may have to adjust the repetition factor for your computer.

PROBLEM 6.2. Write a program that benchmarks floating point addition, multiplication, and division as well as the functions `sqrt`, `sin`, `cos`, `arctan`, `exp`, and `ln`. For each of these functions, express the time required for a single call in "A-units," where an A-unit is the time required to perform a single integer addition. Loop overheads should be accounted for and it will be up to you to determine an appropriate repetition factor for your computer.

PROBLEM 6.3. Write a program that prints a table that indicates how long it takes to execute `PaintOval(0,0,B,R)`. The table should have 25 entries and report results for $B = 40, 80, 120, 160, 200$ and $R = 40, 80, 120, 160, 200$. Express time as a multiple of the time it takes to execute the "base case" `PaintOval(0,0,40,40)`.

PROBLEM 6.4. Let α_n be the time required to execute the empty `for`-loop

[1] In complicated computing environments, the timing of a computation may also be effected by the number of other people using the computer and the actual fraction of the computer that has been allocated to the computation.

```
program Example6_2;
{Improved Integer Arithmetic Benchmark}
const
   MyComputer = 'Macintosh Powerbook 170, System 7.1'; {Modify accordingly}
   n=500000; {Repetition factor.  May need adjustment.}
var
   k:longint;
   x,y,z:integer; {For the sample additions.}
   StartTime, StopTime:real; {For clock snapshots.}
   TimeInSec:real; {Total elapsed time in seconds for n additions.}
   OverheadTime:real; {Time to execute a length n empty for-loop.}
   m:longint; {The number of integer additions performed in 1 second.}
begin
   ShowText;
   writeln('Computer used = ', MyComputer);
   writeln;
   StartTime := TickCount;
   for k:=1 to n do
      ;
   StopTime := TickCount;
   OverheadTime := StopCount - StartCount;
   y:=2; z:=3;
   StartTime := TickCount;
   for k:=1 to n do
      x := y+z;
   StopTime := TickCount;
   TimeInSec := (StopTime-StartTime-OverheadTime)/60;
   m := round(n/TimeInSec);
   writeln('Number of integer additions in 1 second = ',m:6);
end.
```

Output:

```
Computer used = Macintosh Powerbook 170, System 7.1.

Number of integer additions in 1 second = 177515
```

```
program Example6_3;
{PaintOval vs PaintRect}
const
   MyComputer = 'Macintosh Powerbook 170, System 7.1'; {Modify accordingly}
   n=1000; {Repetition factor.  May need adjustment.}
var
   k:longint;
   StartTime, StopTime:real; {For clock snapshots.}
   OvalTime,RectTime:real; {PaintOval and PaintRect times.}
begin
   ShowText;
   ShowDrawing;
   Writeln('Computer used = ', MyComputer);
   StartTime := TickCount;
   for k:=1 to n do
      PaintOval(50,50,200,150);
   StopTime := TickCount;
   OvalTime := StopTime-StartTime;
   StartTime := TickCount;
   for k:=1 to n do
      PaintRect(50,250,200,350);
   StopTime := TickCount;
   RectTime := StopTime-StartTime;
   Writeln('PaintOval Time / PaintRect Time = ', (OvalTime/RectTime):5:2);
end.
```

Computer used = Macintosh Powerbook 170, System 7.1.
PaintOval Time / PaintRect Time = 52.76

```
for k:=1 to n do
    ;
```

and let β_n be the time reouired to execute the empty while-loop

```
k:=1;
while k<=n do
    k:=k+1;
```

Write a program that reports on the ratio α_n/β_n for $n = 1000, 2000, \ldots, 30000$.

PROBLEM 6.5. Assume that x and y are initialized real variables. Write a program that determines which of the following assignment statements is most efficient:

```
Method1 := x*x - y*y;
Method2 := sqr(x) - sqr(y);
Method3 := (x+y)*(x-y)
```

6.2 Efficiency

Benchmarking is the best way to quantify the efficiency of different programs that are designed to compute the same thing. To illustrate this point, let us examine the drawing of the tilted ellipse depicted in FIGURE 6.1. The parametric equations that define this ellipse are given by

$$\begin{aligned} x(t) &= x_c + \cos(\phi)r_x\cos(t) + \sin(\phi)r_y\sin(t) \\ y(t) &= y_c - \sin(\phi)r_x\cos(t) + \cos(\phi)r_y\sin(t) \end{aligned}$$

where $0 \leq t \leq 2\pi$. Assume that the constants

```
n = 100; {Number of ''sides'' in the drawn ellipse.}
rx = 100; ry = 50; {rx and ry are the semiaxes}
xc = 200; yc = 125; {(hc,vc) is the center}
phi = 1; {tilt factor (radians)}
```

are declared. Here is a fragment that draws the ellipse defined by these parameters:

```
{Fragment A}
for k := 0 to n do
   begin
      h := xc + round(cos(phi)*rx*cos(2*k*pi/n)+sin(phi)*ry*sin(2*k*pi/n));
      v := yc + round(-sin(phi)*rx*cos(2*k*pi/n)+cos(phi)*ry*sin(2*k*pi/n));
      if k = 0 then
         MoveTo(h,v)
      else
         LineTo(h,v);
   end;
```

Note that it uses a very literal transcription of the parametric equations given above. Although it is mathematically correct, Fragment A does something quite foolish from the human point of view. It computes $\cos(\phi)$ and $\sin(\phi)$ n times. A more sensible strategy is to compute and store these two quantities *before* the loop:

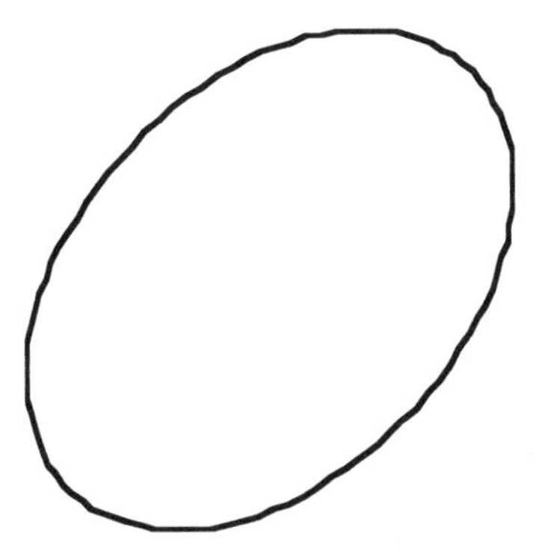

FIGURE 6.1 *Tilted Ellipse*

```
{Fragment B}
c:= cos(phi); s:= sin(phi);
for k := 0 to n do
   begin
      h := xc + round(c*rx*cos(2*k*pi/n)+s*ry*sin(2*k*pi/n));
      v := yc + round(-s*rx*cos(2*k*pi/n)+c*ry*sin(2*k*pi/n));
      if k = 0 then
         MoveTo(h,v)
      else
         LineTo(h,v);
   end;
```

Along the same lines, $\cos(2k\pi/n)$ and $\sin(2k\pi/n)$ appear to be computed twice each time though the loop. A better strategy is to compute and store these quantities at the beginning of the loop body:

```
{Fragment C}
c:= cos(phi); s:= sin(phi);
for k := 0 to n do
   begin
      ck := cos(2*k*pi/n); sk:= sin(2*k*pi/n));
      h := xc + round(c*rx*ck + s*ry*sk); v := yc + round(-s*rx*ck + c*ry*sk);
      if k = 0 then
         MoveTo(h,v)
      else
         LineTo(h,v);
   end;
```

Fragments B and C remove the superfluous sine and cosine evaluations. Turning our attention to superfluous arithmetic, we notice that the products `c*rx`, `s*ry`, `s*rx` and `c*ry` are computed each time through the loop. This suggests the following modification:

```
{Fragment D}
c:= cos(phi);
s:= sin(phi);
crx := c*rx;
srx := s*rx;
cry := c*ry;
sry :=s*ry;
Theta := 2*pi/n;
for k := 0 to n do
   begin
      ck := cos(k*Theta);
      sk := sin(k*Theta);
      h := xc + round(crx*ck + sry*sk);
      v := yc + round(-srx*ck + cry*sk);
      if k = 0 then
         MoveTo(h,v)
      else
         LineTo(h,v);
   end;
```

The program Example6_4 benchmarks this succession of improvements. Relative times are reported. FIGURE 6.2 indicates the number of flops and trig evaluations needed by each fragment.

Fragment	F	E
A	24	8
B	24	4
C	18	2
D	9	2

FIGURE 6.2 *Work Associated With Fragments A, B, C, and D. F = flops/iteration, E = trigonometric evaluations/iteration.*

PROBLEM 6.6. Fragment D can be improved by exploiting the additive formulae

$$\begin{aligned}\cos((k+1)\Delta) &= \cos(k\Delta)\cos(\Delta) - \sin(k\Delta)\sin(\Delta)\\ \sin((k+1)\Delta) &= \sin(k\Delta)\cos(\Delta) + \cos(k\Delta)\sin(\Delta).\end{aligned}$$

where $\Delta = 2\pi/n$. If $\cos(\Delta)$ and $\sin(\Delta)$ are available, then $\cos(2\Delta)$ and $\sin(2\Delta)$ follow from

$$\begin{aligned}\cos(2\Delta) &= \cos(\Delta)\cos(\Delta) - \sin(\Delta)\sin(\Delta)\\ \sin(2\Delta) &= \sin(\Delta)\cos(\Delta) + \cos(\Delta)\sin(\Delta).\end{aligned}$$

With $\cos(2\Delta)$ and $\sin(2\Delta)$ available, we get $\cos(3\Delta)$ and $\sin(3\Delta)$ via

$$\begin{aligned}\cos(3\Delta) &= \cos(2\Delta)\cos(\Delta) - \sin(2\Delta)\sin(\Delta)\\ \sin(3\Delta) &= \sin(2\Delta)\cos(\Delta) + \cos(2\Delta)\sin(\Delta).\end{aligned}$$

Add a "Fragment E"to Example6_5 that computes the necessary sines and cosines through this process. Quantify its efficiency in the same way as Fragments B, C, and D.

```
program Example6_4;
{Efficient Ellipse Drawing}
const
   trials = 20; {Repetition factor}
   MyComputer = 'Macintosh Powerbook 170, System 7.1.';
   n = 100; {Number of ''sides'' in the drawn ellipse.}
   rx = 100; ry = 50; {rx and ry are the semi-axes}
   xc = 200; yc = 125; {(xc,yc) is the center}
   phi = 1; {tilt factor in radians}
var
   ⟨⋮⟩
begin
   ShowText;
   ShowDrawing;
   writeln('Computer used = ', MyComputer);
   writeln;
   StartTime := TickCount;
   ⟨ Fragment A⟩
   StopTime := TickCount;
   T0 := StopTime - StartTime;
   writeln('Time for Fragment A = ', 1.000:5:3);
   StartTime := TickCount;
   ⟨ Fragment B⟩
   StopTime := TickCount;
   writeln('Time for Fragment B = ', ((StopTime - StartTime)/T0):5:3);
   StartTime := TickCount;
   ⟨ Fragment C⟩
   StopTime := TickCount;
   writeln('Time for Fragment C = ', ((StopTime - StartTime)/T0):5:3);
   StartTime := TickCount;
   ⟨ Fragment D⟩
   StopTime := TickCount;
   writeln('Time for Fragment D = ', ((StopTime - StartTime)/T0):5:3);
end.
```

Output:

```
Computer used = Macintosh Powerbook 170, System 7.1.

Time for Fragment A = 1.000
Time for Fragment B = 0.736
Time for Fragment C = 0.577
Time for Fragment D = 0.538
```

We close with an example that shows how *order of magnitude* changes in efficiency can result through algorithmic insight. The problem we consider is the approximation of π by a *rational number.* A rational number is just the quotient of two integers. A widely used rational approximation to π is 22/7. Better approximations exist and to organize our the search for them, we define R_n to be the closest rational number to π whose numerator and denominator are less than or equal to n. The program Example6_5 computes R_{512}. Its a classic look-for-the-min computation. All possible quotients of the form p/q are formed where $p \leq n$ and $q \leq n$. Notice that a double loop is required to cycle through all the possibilities. Bestp and Bestq house the numerator and denominator of the "currently best" rational approximation. BestErr is its error. Every time a better approximation is found, the triplet (Bestp,Bestq,BestErr) is updated. The rational number 3/1 is the basis for the initialization.

A writeln records each "improved" approximation. It should be stressed that the error is based upon comparison with the built-in constant pi and not the true mathematical π. The error in pi is approximately 10^{-7}.

Example6_5 has the property that if n is doubled, then the amount of work is approximately quadrupled. To see this we merely observe that the fragment

```
OurPi := p/q; Err := abs(OurPi - pi);
if Err < BestErr then
   begin
      Bestp := p; Bestq := q; BestErr := Err;
      writeln(Bestp:4, Bestq:4, OurPi:12:6, BestErr:12:7);
   end;
```

is executed n^2 times. For even modest n, the computer spends the vast majority of its time executing these statements. Programs whose running time depends upon the square of a parameter n are said to be $O(n^2)$, or *quadratic.*

PROBLEM 6.7. Write a program that benchmarks the method of search implemented in Example6_5 for $n = 25$, 50, 100, 200, and 400. Your program should print a table that specifies how long it takes to carry out the search for each of these n-values. Remove the writeln's from Example6_5 before benchmarking. The table should reveal that if n is doubled, then the required time goes up by a factor of 4 (or thereabouts).

The search for R_n can be made more efficient with the observation that for a given denominator q, the "best" numerator is given by round(q*pi). The program Example6_6 incorporates this observation.

The inner loop (the p-loop) is replaced by a single assignment. The resulting program is no longer $O(n^2)$. If we double n, the running time is *not* quadrupled simply because the key fragment

```
q:=q+1; p:= round(q*pi); OurPi:=p/q; Err:=abs(OurPi-pi);
if Err < BestErr then
   begin
      Bestp:=p; Bestq:=q; BestErr:=Err;
      writeln(Bestp:4, Bestq:4, OurPi:12:6, BestErr:12:7);
   end;
```

is executed n times. We therefore anticipate that if n is doubled, then the running time of Example6_6 should be approximately doubled. Programs whose running time depends linearly upon a parameter n are said to be $O(n)$.

```
program Example6_5;
{Rational approximations to pi of the form p/q where both p and q are <= n.}
const
   n = 512; {Bound}
var
   p,q:integer; {Numerator and denominator of approximation.}
   Bestp,Bestq:integer; {Best numerator and denominator so far.  }
   OurPi:real; {Bestp/Bestq}
   Err,BestErr:real; {Error in p/q and Bestp/Bestq}
begin
   ShowText;
   writeln(' p q p/q Error';
   writeln('---------------------------------');
   OurPi := 3; Bestq := 1; Bestp := 3; BestErr := abs(OurPi - pi);
   for q := 1 to n do
      for p := 1 to  n do
         begin
            OurPi := p/q;
            Err := abs(OurPi - pi);
            if Err < BestErr then
               begin
                  Bestp := p; Bestq := q; BestErr := Err;
                  writeln(Bestp:4, Bestq:4, OurPi:12:6, BestErr:12:7);
               end;
         end;
end.
```

Output:

p	q	p/q	Error
13	4	3.250000	0.1084073
16	5	3.200000	0.0584074
19	6	3.166667	0.0250741
22	7	3.142857	0.0012644
179	57	3.140351	0.0012418
201	64	3.140625	0.0009677
223	71	3.140845	0.0007476
245	78	3.141026	0.0005671
267	85	3.141176	0.0004162
289	92	3.141304	0.0002884
311	99	3.141414	0.0001785
333	106	3.141510	0.0000831
355	113	3.141593	0.0000003

```
program Example6_6;
{Rational approximations to pi of the form p/q where both p and q are <= n.}
const
   n = 512; {Bound}
var
   p,q:integer; {Numerator and denominator of approximation.}
   Bestp,Bestq:  integer; {Best numerator and denominator so far.}
   OurPi:real; {Bestp/Bestq}
   Err,BestErr:  real; {Error in p/q and Bestp/Bestq}
begin
   ShowText;
   writeln(' p q p/q Error');
   writeln('---------------------------------');
   q:=1; p:=3; OurPi:=p/q; Bestq:=1; Bestp:=3; BestErr:=abs(OurPi-pi);
   while q<n/pi do
      begin
         q:=q+1; p:=round(q*pi); OurPi:=p/q; Err:=abs(OurPi-pi);
         if Err < BestErr then
            begin
               Bestp:=p; Bestq:=q; BestErr:= Err;
               writeln(Bestp:4, Bestq:4, OurPi:12:6, BestErr:12:7);
            end;
      end
end.
```

Output: Same as for Example6_5.

PROBLEM 6.8. Write a program that benchmarks the method of search implemented in Example6_6 for $n = 25$, 50, 100, 200, and 400. Your program should print a table that specifies how long it takes to carry out the search for each of these n-values. Remove the writeln's from Example6_6 before benchmarking. The table should reveal that if n is doubled, then the required time goes up by a factor of 2 (or thereabouts). In other words, Example6_6 is a linear time program.

Chapter 7

Exponential Growth

§7.1 Powers

function declarations, **real**-valued functions, preconditions and post conditions, parameter lists, formal and actual parameters, functions that call other functions, scope rules, development through generalization, **integer**-valued functions.

§7.2 Binomial Coefficients

longint-valued functions, weakening the precondition, the **uses** declaration, setting up a **unit**, the **interface** and **implementation** declarations.

There are a number of reasons why the built-in **sin** function is so handy. To begin with, it enables us to compute sines *without having a clue* about the method used. It so happens that the design of an accurate and efficient sine function is somewhat involved. But by taking the "black box" approach, we are able to be effective **sin**-users while being blissfully unaware of how the built-in function works. All we need to know is that **sin** expects a real input value and that it returns the sine of that value interpreted in radians.

Another advantage of **sin** can be measured in keystrokes and program readability. Instead of disrupting the "real business" of a program with lengthy compute-the-sine fragments, we merely invoke **sin** as required. The resulting program is shorter and reads more like traditional mathematics.

A programming language like ThinkPascal always comes equipped with a *library* of built-in functions. The designers of the language determine the library's content by anticipating who will be using the language. If that group includes scientists and engineers, then invariably there will be built-in functions for the sine, cosine, log, and exponential functions because they are of central importance to work in these areas.

It turns out that if you need a function that is not part of the built-in function library, then *you can write your own*. The art of being able to write efficient, carefully organized functions is an absolutely essential skill for the computational scientist because it suppresses detail and permits a higher level of algorithmic thought.

To illustrate the mechanics of function writing we have chosen a set of examples that highlight a number of important issues. On the continuous side we look at powers, exponentials, and logs. These functions are monotone increasing and can be used to capture different rates of growth. Factorials and binomial coefficients are important for counting combinations. We bridge the continuous/discrete dichotomy through a selection of problems that involve approximation.

7.1 Powers

If x is an initialized real variable and n is an integer variable with nonnegative value, then the following fragment assigns x^n to xpower:

```
xpower:=1;
for k:=1 to n do
   xpower:=x*xpower;
   {xpower = x^k}
```

Each pass through the loop raises the "current" power of x by one[1]. In a program that requires the computation of a power in just a few places, it is not unreasonable to insert this single-loop calculation as required. However, it is not hard to imagine a situation where exponentiations are required many times throughout a program. It is then a major inconvenience to be personally involved with each and every powering. The program Example7_1 reinforces the point. The

```
program Example7_1;
{Examines |x^n + y^n - z^n| for real x,y,z and whole number n.}
var
   x,y,z:real; {The values to be powered.}
   n:integer; {The required exponent.}
   xpower,ypower,zpower:real;
   d:real; {|x^n + y^n - z^n|}
   k:integer;
begin
   ShowText;
   writeln('Enter x, y, and z:');
   readln(x,y,z);
   writeln('Enter nonnegative integer n:');
   readln(n);
   xpower:=1;
   ypower:=1;
   zpower:=1;
   for k:=1 to n do
      begin
         xpower:=x*xpower;
         ypower:=y*ypower;
         zpower:=z*zpower;
         {xpower = x^k, ypower = y^k, zpower = z^k}
      end;
   d:= abs(xpower+ypower-zpower);
   writeln(d:20:6, ' = |x^n + y ^n - z^n| ');
end.
```

```
Enter x, y, and z:
3 2 5
Enter nonnegative integer n:
3
        90.000000 = |x^n + y^n - z^n|
```

[1]Recall that x^k means x^k and is *not* a Pascal operator.

program illustrates the kind of tedium that is involved when the same computation is repeated over and over again. There is a threefold application of the exponentiation "idea." Ideally, we would like to specify once and for all how powers are computed and then just use that specification in a handy way to get x^n, y^n, and z^n.

Fortunately, there is a way to do this and it involves the creation and use of a programmer-defined function. The concept is illustrated in Example7_2. In this program, the function is called power and it is *declared* right after the var section of the program. Declarations always come before the body of the main program because of their set-the-stage role:

```
program:⟨ name⟩;
⟨ Lead Comment⟩
⟨ Constant Declarations (if any)⟩
⟨ Variable Declarations⟩
⟨ Function Declarations (if any)⟩
begin
   ⟨ Main Program Body⟩
end.
```

The order of variable, constant, and function declarations is somewhat flexible, but for now we recommend putting the function declarations after the constant and variable declarations to avoid trouble.

Let us look at how a function declaration is structured. A casual glance at

```
function power(a:real; n:integer):real;
{Pre:n >= 0.}
{Post:a^n.}
var
   k:integer;
   apower:real;
begin
   apower:=1;
   for k:=1 to n do
      apower:=a*apower;
      {apower = a^k}
   power:=apower;
end
```

shows that a function declaration resembles a main program. It has a name, it has its own declaration section, and it has its own body. But in addition, it has a *parameter list* and a *type*. These are designated in the *function heading*:

$$\textbf{function}\ \underbrace{\texttt{power}}_{name}(\underbrace{\texttt{a:real;n:integer}}_{parameter\ list}):\underbrace{\texttt{real}}_{type}$$

The parameter list is made up of the function's *formal parameters*. The function power has two: a and n. Each formal parameter must be typed. The parameter a is real and so we write a:real. As in the var context, a colon separates the formal parameter's name and its type. The second parameter n and its type is similarly established: n:integer. Notice how a semicolon is used to separate the formal parameters. Formal parameters are sometimes called *arguments*. Thus, power is a 2-argument function. The type of value returned by the functions is designated at the end of the function heading.

```
program Example7_2;
{Examines |x^n + y^n - z^n| for real x,y,z and whole number n.}
var
   x,y,z:real; {Values to exponentiate.}
   n:integer; {The required exponent.}
   d:real; {|x^n + y^n - z^n|}

function power(a:real; n:integer):real;
{Pre:n >= 0.}
{Post:a^n.}
var
   k:integer;
   apower:real;
begin
   apower:=1;
   for k:=1 to n do
      apower:=a*apower;
      {apower = a^k}
   power:=apower;
end;

begin
   ShowText;
   writeln('Enter x, y, and z:');
   readln(x,y,z);
   writeln('Enter nonnegative n:');
   readln(n);
   d := abs(power(x,n)+power(y,n)-power(z,n));
   writeln(d:20:6, ' = |x^n + y^n - z^n| ');
end.
```

Output:

```
Enter x, y, and z:
3 2 5
Enter nonnegative integer n:
3
     90.000000 = |x^n + y^n - z^n|
```

After the function heading comes the *specification*. This is a comment that communicates all one needs to know about using the function. It has a *pre-condition* part and a *post-condition* part identified with the abbreviations "Pre" and "Post":

```
function power(a:real; n:integer):real;
{Pre:n >= 0.}
{Post:a^n.}
```

The pre-condition indicates properties that must be satisfied for the function to work correctly. Apparently, power does not work with negative n. Since there is no restriction on the value of a, there is no mention of this parameter in the precondition.

The post-condition describes the value that is produced by the function. To say that the post-condition is "a^n" is to say that the function returns the value a^n.

The specification should *not* tell us what we already know from the function heading. For example, the specification

```
{Pre:n is a nonnegative integer, a is any real.}
{Post:The real value a^n is returned.}
```

is redundant because we already know from the function heading statement that n is an integer, a is real, and that a real value is returned. The specification should not detail the method used to compute the returned value. The goal is simply to provide enough information so that the function can be used.

Now let's go "inside" the function and see how the required computation is carried out. To produce a^n, power needs its own variables to carry out the looping and repeated multiplication:

```
var
   k:integer;
   apower:real;
```

In this regard, power is just like the main program which has *its* **var** section. To stress the distinction between the main program's variables and those used "inside" a function, we refer to the latter as *local variables*. Thus, k and apower are *local* to power.

Finally, we have reached the *body* of power where the recipe for exponentiation is set forth:

```
begin
   apower:=1;
   for k:=1 to n do
      apower:=a*apower;
      {apower = a^k.}
   power:=apower;
end;
```

The function body is placed between a **begin** and an **end**. Notice how the powers are built up in apower. Once the required a^n is available, it is "assigned" to what looks like a variable having the name of the function:

```
power:=apower;
```

We say "assigned" because the action prescribed is not one of true assignment–there is no variable called power. This is merely a way of returning the value produced by the function. The function body

```
begin
   power:=1;
   for k:=1 to n do
      power:=a*power;
end;
```

is illegal because power is not a variable and may not participate in an expression like a*power.

The last item on our agenda concerns the use of power by the main program. Recall that a built-in function such as sin returns a value that can be used in an arithmetic expression, e.g., `v := sin(3*x) + 4*sin(2*x)`. The same is possible with power:

$$\text{d:} = \underbrace{\texttt{power(x,n)}}_{x^n} + \underbrace{\texttt{power(y,n)}}_{y^n} + r\,\underbrace{\texttt{power(z,n)}}_{z^n}$$

In any arithmetic expression that calls for a power, we merely insert an appropriate reference to power. These references are called *function calls*. The assignment to d includes three function calls to power. Suppose x, y, z, and n have value 2, 5, 4 and 3 respectively. It follows that

$$\left.\begin{array}{l}\texttt{power(x,n)}\\ \texttt{power(y,n)}\\ \texttt{power(z,n)}\end{array}\right\} \text{has the value} \left\{\begin{array}{l}2^3 = 8\\ 5^3 = 125\\ 4^3 = 64\end{array}\right.$$

and so d is assigned the value 8 + 125 + 64 = 197.

Every time a function is referenced, make sure that the number of arguments and their type agrees with what is specified in the function declaration. Thus, power(n,z) results in a type error if n is integer or if z is real.

A functions like power is conveniently thought of as a factory. The "raw materials" are a and n and the "finished product" is a^n. Thus, an "order" to produce $(2.5)^3$ involves (a) the receipt of the 2.5 and 3, (b) the production of the "consumer product" 2.5^3, and (c) the "shipment" of the computed result 15.625. See FIGURE 7.1.

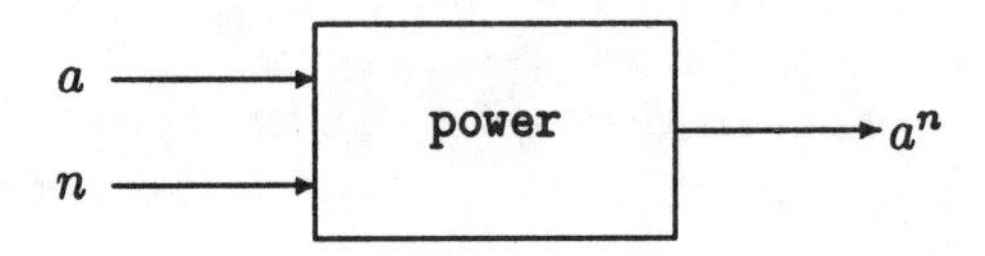

FIGURE 7.1 *Visualizing a Function*

An important substitution mechanism attends each function call and it is essential that you master the underlying dynamics. We motivate the discussion by considering how we use the Centigrade-to-Fahrenheit formula

$$F = \frac{9}{5}C + 32.$$

If we substitute "20" for "C" and evaluate the result, then we conclude that $F = 68$. The formula is merely a template with F and C having placeholder missions.

The situation is similar in a ThinkPascal function. The function is merely a formula *in algorithmic form* into which values are substituted. Let's step through a call to power and trace what happens. Consider the following main program fragment:

```
x:=3;
m:=4;
p:=power(x,m);
```

After the first two assignment statements we have the following situation:

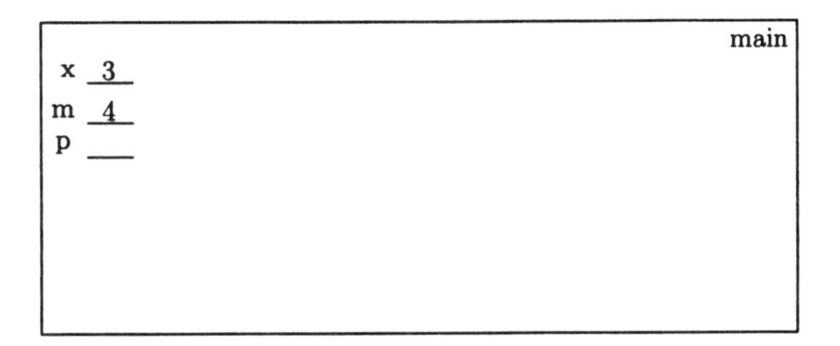

We have enclosed the main program in a box to stress the concept of *current context*. Next comes the reference to **power**. To trace what happens, we draw a "function box" inside the main program box:

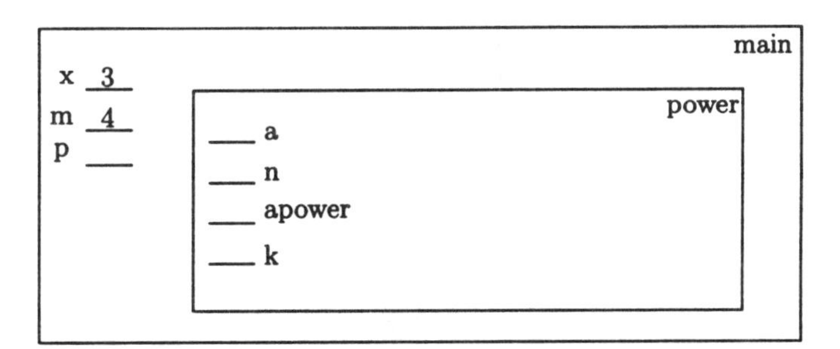

The function box for **power** indicates its formal parameters and its local variables. At the time of the call, values are placed in the parameter boxes:

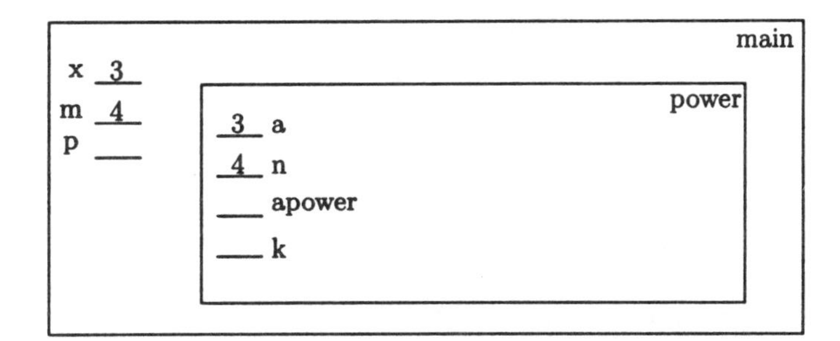

Execution proceeds *inside* the function box because **power** is now the current context. At the end of the first pass through the loop in **power** we reach the following state:

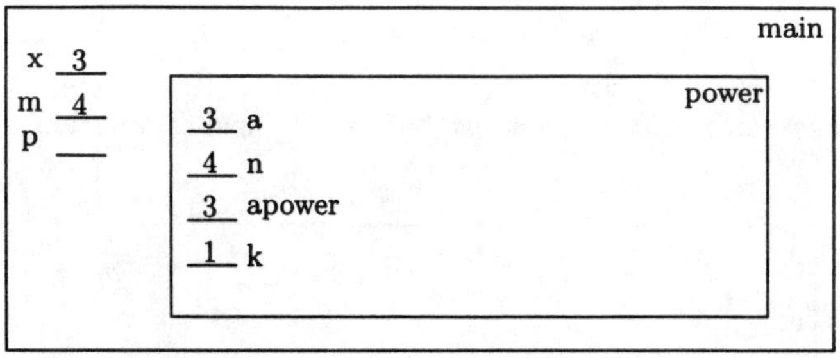

```
main
x  3
m  4
p  __
        power
        3   a
        4   n
        3   apower
        1   k
```

Execution inside **power** continues until at the end of the fourth and final pass we obtain:

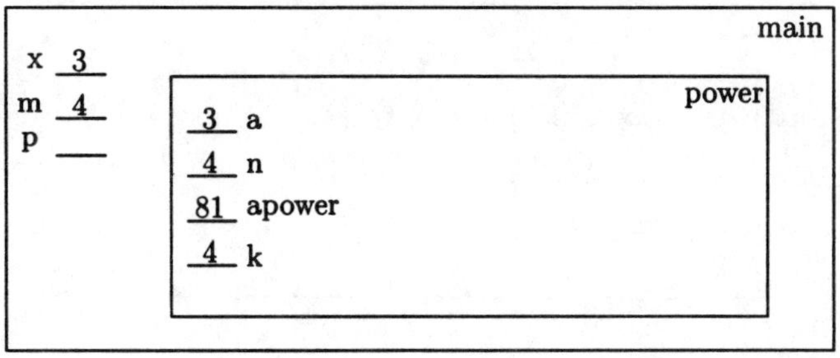

```
main
x  3
m  4
p  __
        power
        3   a
        4   n
        81  apower
        4   k
```

At this time, the function box "closes" and control is passed back to the main program with the value 81 placed in p:

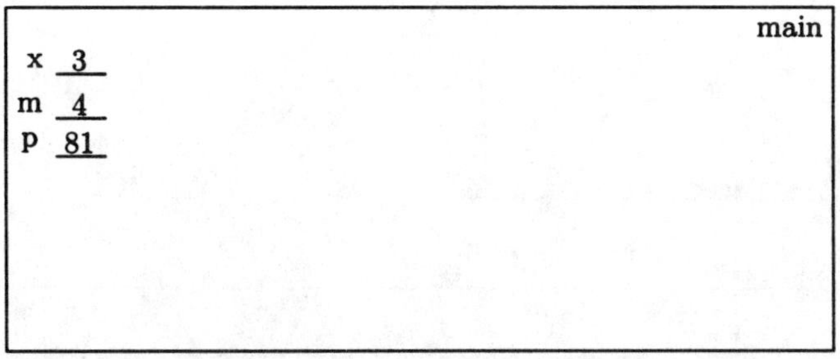

```
main
x  3
m  4
p  81
```

The type returned by a function must be taken very seriously. If `m` has type integer, then the assignment `m := power(3,4)` is illegal because **power** returns a real type. The fact that the computed power is the integer 81 is irrelevant. It is represented in **real** form and cannot be assigned to an integer variable. In an application dealing with powers of integers, it may be more appropriate to develop an integer-valued power function:

```
function powerI(a:integer; n:integer):integer;
{Pre:n>=0.}
{Post:a^n.}
var
   k:integer;
   apower:integer;
begin
   apower:=1;
   for k:=1 to n do
      apower:=a*apower;
      {apower = a^k}
   powerI:=apower;
end;
```

The essential difference between this and `power` is the switching of all `real` types to `integer` types.

PROBLEM 7.1. Note that x^{64} can be obtained through repeated squaring:

$$x \to x^2 \to x^4 \to x^8 \to x^{16} \to x^{32} \to x^{64}.$$

Thus, x^{64} can be "reached" with only six multiplications in contrast to the sixty-three products that are required if we go down the repeated multiplication path:

$$x \to x^2 \to x^3 \cdots \to x^{63} \to x^{64}.$$

Using the repeated squaring idea, write a function `TwoPower(k:integer; x:real):real` that computes x^n where $n = 2^k$. Write a good specification.

PROBLEM 7.2. The Fibonacci numbers $f_1, f_2, \ldots$ are defined as follows:

$$f_n = \begin{cases} 1 & \text{if } n = 1 \\ 1 & \text{if } n = 2 \\ f_{n-1} + f_{n-2} & \text{if } n \geq 3 \end{cases}. \qquad (1)$$

See §4.2. It can be shown that

$$f_n = \frac{1}{\sqrt{5}}\left(\left(\frac{1+\sqrt{5}}{2}\right)^n - \left(\frac{1-\sqrt{5}}{2}\right)^n\right). \qquad (2)$$

Write a program that prints a table showing the first 32 Fibonacci numbers computed in two ways. In the first column of the table should be the values produced by the recursion (1) and in the second column the values obtained by using (2). In the latter case, make effective use of `power` and print the real values to six decimal places so that the effects of floating point arithmetic can be observed.

A more general exponentiation function can be obtained by exploiting the formula

$$x^y = (e^{\ln(x)})^y = e^{\ln(x)*y}.$$

Since a programmer-defined function can reference a built-in functions we may write

```
function powerGR(x,y:real):real;
{Pre:x>0.}
{Post:x^y}
begin
   powerGR := exp(ln(x)*y);
end
```

The example shows that a function need not have any local variables. We also see that parameters with the same type may be "typed together" in the parameter list. In particular, `powerGR(x,y:real):real` is equivalent to `powerGR(x:real;y:real):real`.

Functions can also have their own local constants. So instead of

```
function Bell(x,y:real):real;
{Post:exp(-(x*x+y*y))}
begin
   Bell := exp(-sqr(x)-sqr(y))/sqrt(pi);
end
```

we may write

```
function Bell(x,y:real):real;
{Post:exp(-(x*x+y*y))}
const
   c = 0.5641896; {1/sqrt(pi)}
begin
   Bell := c*exp(-sqr(x)-sqr(y));
end
```

By "precomputing" the constant $1/\sqrt{\pi}$, a square root is saved each call to `Bell`. This example shows that a function need not have a precondition. For `Bell`, this just means that `x` and `y` can house any real value.

PROBLEM 7.3. Using `powerGR`, write a program that discovers the smallest positive integer n so that $(\sqrt{n})^{\sqrt{n+1}}$ is larger than $(\sqrt{n+1})^{\sqrt{n}}$.

PROBLEM 7.4. How big must n be before the following fragment prints `inf`?

```
x:=1;
for k := 1 to n do
   x := powerRG(2,x);
writeln(x)
```

Try to guess the answer before writing a program.

A program can have more than one programmer-defined function. Moreover, it is possible for one programmer-defined function to call another programmer-defined function. To illustrate, `Example7_3` prints a table of values for the function

$$f(x) = x^n(1-x)^m$$

for the case $n = 3$ and $m = 4$. In programs like this, there is an important rule to follow: *if function "A" calls function "B", then function "B" must be declared first.* In `Example7_3`, an error would result if the declaration of `prod` came before the declaration of `power`.

A problem-solving strategy known as *function generalization* often leads to a situation where one function calls another. Consider the a^n problem again where now n can be *any* integer. Mathematically, there is no problem if $a \neq 0$ whenever $n < 0$. Suppose for some reason we find the nonnegative n case "easy" and the negative n case "hard." We proceed to develop `power` as above with the precondition $n \geq 0$. We the realize through the formula

$$a^{-n} = \frac{1}{a^n}$$

```
program Example7_3;
{The function x^n (1-x)^m}
var
   k:integer;
   z,fval:real; {fval = f(z) where f(x) = x^3 (1-x)^4}

function power(a:real; n:integer):real;
{Pre:n>=0.}
{Post:a^n.}
var
   k:integer;
   apower:real;
begin
   apower:=1;
   for k:= 1 to n do
      apower:= a*apower; {apower := a^k}
   power:=apower;
end;

function Prod(z:real; n,m:integer):real;
{Pre:  0<=z<=1, m>=0, n>=0.}
{Post:  (z^n)*(1-z)^m}
begin
   if n <= m then
      Prod := power(z*(1-z),n)*power(1-z,m-n)
   else
      Prod := power(z*(1-z),m)*power(z,n-m);
end;

begin
   ShowText;
   writeln(' Let f(z) = z^3 (1-z)^4'); writeln;
   writeln(' z f(z) '); writeln('----------------------');
   for k:= 0 to 10 do
      begin
         z := k/10; fval := Prod(z,3,4);
         writeln(z:5:3, ' ', fval:10:6);
      end
end.
```

Output:

```
Let f(z) = z^3 (1-z)^4

    z          f(z)
--------------------
0.000     0.00000000
0.100     0.00065610
0.200     0.00327680
0.300     0.00648270
             ⋮
```

that a-to-a-negative-power is merely the reciprocal of a raised to the corresponding positive power:

$$3^{-4} = \frac{1}{81}.$$

This suggests that we build our more general power function upon the restricted version already developed:

```
function powerG(a:real;n:integer):real;
{Pre:a is nonzero if n is negative.}
{Post:a^n}
begin
   if n>=0 then
      powerG := power(a,n);
   else
      powerG := 1/power(a,-n)
end
```

A main program that uses powerG would have to be structured as follows:

```
program main;
   ⟨:⟩
function power(a:real;n:integer):real;
   ⟨:⟩
function powerG(a:real;n:integer):real;
   ⟨:⟩
begin
   ⟨:⟩
end.
```

Later, after our understanding of the the negative n case is well enough understood, we may dispense with the reference to power and handle the repeated multiplication explicitly. For example, we can compute $a^{|n|}$ and then reciprocate the result based upon a check of n's sign:

```
function powerG(a:real;n:integer):real;
{Pre:a is nonzero if n is negative.}
{Post:a^n}
var
   apower:real;
   k:integer;
begin
   apower:=1;
   for k:=1 to abs(n) do
      apower:=a*apower;
      {apower = |a|^k.}
   if n>=0 then
      powerG:=apower
   else
      powerG := 1/apower
end
```

With this implementation, a main program that references powerG could be organized without power:

```
program main;
    ⟨:⟩
function powerG(a:real;n:integer):real;
    ⟨:⟩
begin
    ⟨:⟩
end.
```

PROBLEM 7.5. Make the following three modifications to Example7_3. (a) Rewrite prod so that it does not call power.(b) Add a function

```
function DerProd(z:real; n,m:integer):real;
{Pre:m,n>0}
{Post:Derivative of x^n (1-x)^m at x = z.}
```

(c) Modify the program's body so that it prints a table reporting the values $f(z)$ and $f'(z)$ for $z = 0.0, 0.1, \ldots, 0.9, 1.0$ where $f(x) = x^2(1-x)^3$.

7.2 Binomial Coefficients

A number of integer-valued functions have an eminent role to play in the area of *combinatorics*. This branch of discrete mathematics is concerned with the counting of combinations. The simplest combinatoric function is the factorial function $n! = 1 \cdot 2 \cdots (n-1) \cdot n$. The number of ways that n people can stand in a line is given by $n!$. If $n = 4$ and a, b, c, and d designate the four individuals, then here are the $24 = 1 \cdot 2 \cdot 3 \cdot 4$ possibilities:

$$\begin{array}{cccccc} abcd & abdc & acbd & acdb & adbc & adcb \\ bacd & badc & bcad & bcda & bdac & bdca \\ cabd & cadb & cbad & cbda & cdab & cdba \\ dabc & dacb & dbac & dbca & dcab & dcba \end{array}$$

The factorial function grows very rapidly:

n	$n!$
8	40320
9	362880
10	3628800
11	39916800
12	479001600
13	6227020800

If we write a function for computing $n!$, then it is appropriate to have it return a longint value in order to maximize the range of acceptable input values. Since maxlongint = 2,147,483,647, it follows from the table that $n = 12$ is the largest acceptable input value and we obtain

```
function Factorial(n:integer):longint;
{Pre:0<=n<=12}
{Post:n!}
var
   k:integer;
   f:longint;
begin
   f := 1;
   for k := 1 to n do
      f := f*k;
      {f = k!}
   Factorial := f;
end;
```

Reasonable approximations of "large-n" factorials can be obtained via the *Stirling formula:*

$$n! \approx S_n \equiv \sqrt{2\pi n}\left(\frac{n}{e}\right)^n .$$

If we encapsulate this estimate in the form of a real-valued function, then we obtain

```
function Stirling(n:integer):real;
{Pre:  n>=0}
{Post:Stirling approximation to n!}
const
   e = 2.718282; {exp(1)}
var
   k:integer; {Loop counter}
   s:real; {n/e}
   f:real; {For building p powers of s}
begin
   if n = 0 then
      Stirling := 1
   else
      begin
         f := 1;
         s := n/e;
         for k := 1 to  n do
            f := f*s; {f = (n/e)^k}
         Stirling := sqrt(2*pi*n)*f;
      end
end
```

The case $n = 0$ is handled separately and is included to simplify the use of `Stirling`.

We could make `Stirling` a `longint`-valued function by changing the heading to

```
function Stirling(n:integer):longint;
```

and the last statement to

```
Stirling := round( sqrt(2*pi*n)*f);
```

```
program Example7_4;
  {Stirling Approximation}
  function Factorial(n:integer):longint;
    ⟨:⟩
  function Stirling(n:integer):real;
    ⟨:⟩
  var
    n:integer;
  begin
    ShowText;
    writeln(' n n!  Stirling Appx.');
    writeln('-----------------------------------');
    for n:=0 to 12 do
      writeln(n:3, Factorial(n):13, round(Stirling(n)):13);
  end.
```

Output:

```
 n           n!       Stirling Appx.
-----------------------------------
 0            1                1
 1            1                1
 2            2                2
 3            6                6
 4           24               24
 5          120              118
 6          720              710
 7         5040             4980
 8        40320            39902
 9       362880           359537
10      3628800          3598693
11     39916800         39615616
12    479001600        475687328
```

```
program Example7_5;
{Stirling approximation}
uses
   Chap7Codes;
var
   n:integer;
begin
   ShowText;
   writeln(' n n!  Stirling Appx.');
   writeln('----------------------------------');
   for n:=0 to 12 do
      writeln(n:3, Factorial(n):13, round(Stirling(n)):13);
end.
```

The functions Factorial and Stirling are declared in the unit Chap7Codes. Output: Same as output of Example7_4.

However, we would then have to impose the same $n \leq 12$ precondition.

The program Example7_4 compares the Stirling approximation with the exact value of the factorial function for $n = 0, 1, \ldots, 12$. For this range of n, the relative error in the Stirling approximation S_n is about one percent.

Example7_4 points to a software problem that will become increasingly acute as we write programs that involve more and more functions:

- The main program body, presumably of great interest to us, is buried under a mountain of function declarations.

- If we write several main programs that involve the same set of functions, then those functions must be declared in each main program.

To solve this problem, ThinkPascal supports the notion of a `unit`. By placing functions in a unit, and then "linking" the main program to it, we can reference the functions in the unit as if they were actually declared in the main program. Two things are required for this to work. (1) The main program must include a `uses` declaration that names the unit in question. (2) The file that contains the unit must be added to the project before the file that includes the main program.

As an example, if `Factorial` and `Stirling` are properly declared in the unit Chap7Codes, then the program Example7_5 is equivalent to Example7_4.

The declaration

```
uses
   Chap7Codes
```

enables us to reference both `Factorial` and `Stirling`. as if they were declared in the usual way. The unit Chap7Codes is shown in FIGURE 7.2. Notice that a unit has an "interface" part that includes just the headings and specifications and an implementation part, where the full declarations appear.

```
unit Chap7Codes;

interface

function Factorial(n:integer):longint;
{Pre:0<=n<=12}
{Post:n!}
   •
function Stirling(n:integer):real;
{Pre:  n>=0}
{Post:Stirling approximation to n!}

implementation

function Factorial(n:integer):longint;
{Pre:0<=n<=12}
{Post:n!}
var
   k:integer;
   f:longint;
begin
   f := 1;
   for k := 1 to n do
      f := f*k;
      {f = k!}
   Factorial := f;
end;

function Stirling(n:integer):real;
{Pre:  n>=0}
{Post:Stirling approximation to n!}
const
   e = 2.718282; {exp(1)}
var
   k:integer; {Loop counter}
   s:real; {n/e}
   f:real; {For building p powers of s}
begin
   if n = 0 then
      Stirling := 1
   else
      begin
         f := 1;
         s := n/e;
         for k := 1 to  n do
            f := f*s; {f = (n/e)^k}
         Stirling := sqrt(2*pi*n)*f;
      end
end;

end.
```

FIGURE 7.2 *The Unit* Chap7Codes

The general form of a unit is as follows:

unit ⟨ *Name* ⟩
  interface
    ⟨ *First function heading with pre and post conditions.* ⟩
    ⟨ *Second function heading with pre and post conditions.* ⟩
      ⟨:⟩
    ⟨ *Last function heading with pre and post conditions.* ⟩

  implementation
    ⟨ *Complete declaration of first function.* ⟩
    ⟨ *Complete declaration of second function.* ⟩
      ⟨:⟩
    ⟨ *Complete declaration of last function.* ⟩

end.

Note that by merely reading the `interface` portion you can learn all that is required just to *use* the functions in the unit. The `implementation` portion has the complete function declarations and is only of interest if you want to change how the function works.

PROBLEM 7.6. If a positive integer x is written in base-10 notation, then the number of digits required is given by 1 plus the trunc of $\log_{10} x$. Using the identity

$$\log_{10}(n!) = \sum_{k=1}^{n} \log_{10}(k),$$

complete the following function:

```
function FactorialDigits(n:integer):integer;
{Pre:n>=1}
{Post:The number of digits in n!}
```

Write a program that uses `FactorialDigits` and prints a 50-line table. On line n, the table should contain the following values: n, the number of digits in $n!$ as determined by `FactorialDigits`, the number of digits in the Stirling approximation S_n, and S_n. Make use of the unit `Chap7Codes`.

PROBLEM 7.7. Complete the following function

```
function f(n:longint;d:integer):longint;
{Pre:n>=1, 0<=d<=9}
{Post:The number of nonnegative integers <= n that end with the digit d.}
```

Using `f`, write a program that prints a 20-line table. On line k should appear k and the smallest n so that $n!$ is divisible by 10^k. $k = 1, 2, \ldots, 20$. Hint: Think about $f(n,0) + f(n,5)$.

The number of ways that k objects can be selected from a a set of n objects is given by the binomial coefficient

$$\binom{n}{k} \equiv \frac{n!}{k!(n-k)!}.$$

Thus, there are

$$\binom{5}{2} = \frac{5!}{2!3!} = \frac{1\cdot 2\cdot 3\cdot 4\cdot 5}{(1\cdot 2)(1\cdot 2\cdot 3)} = \frac{5\cdot 4}{1\cdot 2} = 10$$

possible chess matches within a pool of 5 players. Similar cancellations permit us to compute

$$\binom{26}{4} = \frac{26!}{4!22!} = \frac{26 \cdot 25 \cdot 24 \cdot 23}{1 \cdot 2 \cdot 3 \cdot 4} = 14950$$

which is the number of four-letter words with distinct letters and

$$\binom{52}{5} = \frac{52!}{5!47!} = \frac{52 \cdot 51 \cdot 50 \cdot 49 \cdot 48}{1 \cdot 2 \cdot 3 \cdot 4 \cdot 5} = 2,303,000$$

which is the number of 5-card poker hands. From the definition we may write

```
function BinCoeff0(n,k:integer):longint;
{Pre:0<=k<=n<=12}
{Post:  Number of ways to select k objects from a set of n objects.}
begin
   BinCoeff0 := (factorial(n) div factorial(k)) div factorial(n - k)
end
```

The trouble with this implementation is that n cannot exceed 12 because the function `Factorial` breaks down for larger values. However, from the above examples we see that there is considerable cancellation between the factorials in the numerator and denominator. Indeed, it can be shown that

$$\binom{n}{k} \equiv \frac{n \cdot (n-1) \cdots (n-k+1)}{1 \cdot 2 \cdots k}.$$

Using this formula for the binomial coefficient we obtain

```
function BinCoeff(n,k:integer):longint;
{Pre:0<=k<=29}
{Post:Number of ways to select k objects from a set of n objects}
var
   j:integer;
   b:longint;
begin
   b := 1;
   for j:=1 to  k  do
      b := (b*(n-j+1)) div j;
      {b = n choose j}
   BinCoeff := b
end
```

Because n-choose-j is integer-valued, there is no remainder when the integer division is carried out during each pass through the loop. Thus, the exact binomial coefficients results. The restriction that n be less than or equal to 29 has to do with `maxlongint`. The program `Example7_6` prints out an array of binomial coefficients.

Binomial coefficients arise in many situations. The term itself comes from the fact that if

$$(x+y)^n = a_0 x^n + a_1 x_{n-1} + a_2 x_{n-2} y^2 + \cdots + a_{n-1} x y^{n-1} + a_n y^n,$$

then

$$a_k = \binom{n}{k} \qquad k = 0, \ldots, n.$$

```
program Example7_6;
{The Pascal Triangle of Binomial Coefficients}
uses
   Chap7Codes; {Contains BinCoeff.}
const
   nmax = 15;
var
   n,k:integer;
begin
   ShowText;
   writeln;
   for n := 0 to nmax  do
      begin
         writeln;
         for k := 0 to n  do
            write(BinCoeff(n,k):5);
      end;
end.
```

The function BinCoeff is declared in Chap7Codes. Output:

```
    1
    1    1
    1    2    1
    1    3    3    1
    1    4    6    4    1
    1    5   10   10    5    1
    1    6   15   20   15    6    1
    1    7   21   35   35   21    7    1
    1    8   28   56   70   56   28    8    1
    1    9   36   84  126  126   84   36    9    1
    1   10   45  120  210  252  210  120   45   10    1
```

For example,

$$\begin{aligned}(x+y)^4 &= x^4+4x^3y+6x^2y^2+4xy^3+y^4\\ &= \binom{4}{0}x^4+\binom{4}{1}x^3y+\binom{4}{2}x^2y^2+\binom{4}{3}xy^3+\binom{4}{4}y^4\end{aligned}$$

PROBLEM 7.8. For a given n, the binomial coefficient $\binom{n}{k}$ attains its largest value when $k = n div 2$. Write a program that prints the value of BinCoeff(n,n div 2) for n = 1 to 30. Explain why an incorrect value is returned when $n = 30$. Complete the following function

```
function FloatBC(n,k:integer):real;
{Pre:0<=k<=n}
{Post:  Number of ways to select k objects from a set of n objects}
```

and find out what happens if floating point arithmetic is used to compute the binomial coefficients. Make use of Chap7Codes.

PROBLEM 7.9. Let S_n be the Stirling approximation to $n!$ and define the

$$\beta(n,k) = \frac{S_n}{S_k S_{n-k}}$$

(Assume that $S_0 = 1.$). Write a real-valued function StirlingBC(n) that returns $\beta(n,k)$. Make effective use of Stirling(n). Write a program that of StirlingBC to compute

$$e_n = \max_{0\le k\le n} \frac{\left|\beta(n,k)-\binom{n}{k}\right|}{\binom{n}{k}}$$

for $n = 1$ to 29. Make use of Chap7Codes.

PROBLEM 7.10. Note that

$$\binom{n}{k} = \binom{n}{n-k}.$$

Modify BinCoeff so that it uses the the expression on the right hand side if $2k > n$. Rerun Example7_6 with the modified function. Explain why the modified program is more efficient. Make use of Chap7Codes.

PROBLEM 7.11. Imagine writing n letters and addressing (separately) the n envelopes. The number of ways that all n letters can be placed in incorrect envelopes is given by the Bernoulli-Euler number

$$B_n = \sum_{k=0}^{n}(-1)^k\binom{n}{k}(n-k)!$$

Thus,

$$B_4 = \binom{4}{0}4!-\binom{4}{1}3!+\binom{4}{2}2!-\binom{4}{3}1!\binom{4}{4}0! = 24-24+12-4+1 = 9$$

Write a function

```
function BernEuler(n:integer):longint;
{Pre:1<=n<=12}
{Post:Number of ways to put n letters all in the wrong n envelopes.}
```

and use it to print a table that shows $B_1,\ldots,B_{12}$. Make use of Chap7Codes.

PROBLEM 7.12. The number ways a set of n objects can be partitioned into m nonempty subsets is given by

$$\sigma_n^{(m)} = \sum_{j=1}^{m}\frac{(-1)^{m-j}j^n}{(m-j)!j!}.$$

It is not obvious but the summation always renders an integer as long as $1 \le m \le n$. For example,

$$\sigma_4^{(2)} = \frac{(-1)^{2-1}1^4}{(2-1)!1!}+\frac{(-1)^{2-2}2^4}{(2-2)!2!} = -1+8 = 7.$$

Thus, there are 7 ways to partition a 4-element set like $\{a,b,c,d\}$ into two non-empty subsets:

1	:	{a},{b,c,d}
2	:	{b},{a,c,d}
3	:	{c}, {a,b,d}
4	:	{d},{a,b,c}
5	:	{a,b},{c,d}
6	:	{a,c},{b,d}
7	:	{a,d},{b,c}

Note that $\sigma_n^{(1)} = \sigma_n^{(n)} = 1$. Print a table with 10 lines. On line n should be printed the numbers $\sigma_n^{(1)}, \sigma_n^{(2)}, \ldots, \sigma_n^{(n)}$. Make use of `Chap7Codes`.

PROBLEM 7.13. If j and k are nonnegative integers that satisfy $j + k \leq n$, then the coefficient of $x^k y^j z^{n-j-k}$ in $(x + y + z)^n$ is given by the *trinomial coefficient*

$$T(n, j, k) = \binom{n}{j}\binom{n-j}{k}$$

Write a function `TriCoeff(n,j,k)` that computes $T(n, j, k)$ and use it to print a list of all trinomial coefficients of the form $T(10, j, k)$ where $0 \leq j \leq k$ and $j + k \leq 10$. Make use of `BinCoeff`. Make use of `Chap7Codes`.

PROBLEM 7.14. Modify the fragment

```
for a := 1 to 4 do
   for b := 1 to 4 do
      for c := 1 to 4 do
         for d := 1 to 4 do
            writeln(a:1, b:1, c:1, d:1);
```

so that it prints a list of all possible permutations of the digits 1,2,3, and 4, i.e.,

1234	1243	1324	1342	1423	1432
2134	2143	2314	2341	2413	2431
3124	3142	3214	3241	3412	3421
4123	4132	4213	4231	4312	4321

(The order of the 24 numbers in the list is not important.)

Chapter 8

Patterns

§8.1 Encapsulation

The procedure, formal and actual parameters, pre and post conditions, local variables and constants.

§8.2 Hierarchy

Boolean variables, Procedures that call other procedures.

Procedures hide computational detail and in that regard they are similar to functions. The procedures discussed in this chapter draw objects.[1] Once such a procedure is written, it can be used as a "black box."

Writing and using procedures that draw geometric patterns is symbolic of what engineers and scientists do. Geometric patterns are defined by parameters and deciding what the "right" parameters are requires a geometric intuition. Similar is the design of an alloy that requires a metallurgist's intuition or the building of a model to predict crop yield that requires a biologist's intuition. What are to be the constituent metals? What are the factors effecting the growth? Once the parameters are identified, construction is possible by setting their value. A pattern is drawn. An alloy is mixed. A model is formulated. Optimality can then pursued: What choice of parameter values renders the most pleasing pattern, the strongest alloy, the most accurate model of crop yield?

Our use of graphics procedures to shed light on the processes of engineering design and scientific discovery begins in this chapter. We start by showing how to "package" the computations that produce the pattern. It's an occasion to practice the writing of clear specifications that define what a piece of software can do. Patterns can be built upon other, more elemental patterns, a fact that we use to motivate the design of procedure hierarchies. Optimization issues are discussed further in Chapters 13, 23, and 24.

8.1 Encapsulation

Consider the display of two grids in the DrawWindow as shown in FIGURE 8.1. Assume that upper left corner of the lefthand grid is situated at (50,60) and that its tiles are 25-by-25. Assume that the upper left corner of the righthand grid is located at (200,85) and that its tiles are 50-by-50. Using the grid-drawing expertise developed in §3.1, we obtain the program `Example8_1`. Notice the similarity between the left-grid fragment and the right-grid fragment. Indeed, to

[1] Procedures that return values are covered in Chapter 10.

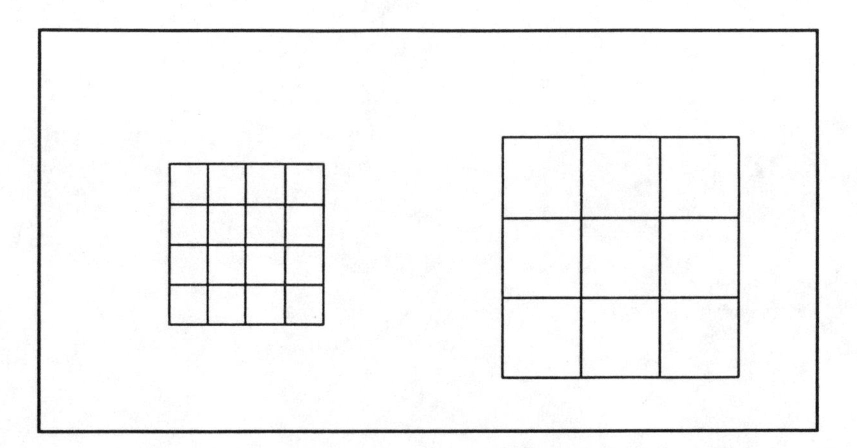

FIGURE 8.1 *Two Grids*

obtain the latter from the former you need only substitute **nR** for **nL**, **sR** for **sL**, **aR** for **aL**, and **bR** for **bL**. We begin to wish that we could merely substitute the four values that define the grid into some established grid drawing "formula." Having to write a pair of loops for every grid is a major distraction.

Fortunately, grid-drawing can be "packaged" in the same way that rectangle-drawing is "packaged" by the built-in procedure `FrameRect`. Programmer-defined procedures make this possible and `Example8_2` shows how they simplify the two-grid computation. The syntactic rules for procedures are similar to those that apply to functions. First, a procedure must be *declared* in the main program. All programmer-defined functions and procedures are declared just before the **begin** that indicates the start of the main program body:

program:⟨ *name* ⟩;
{⟨ *Program description.* ⟩}
⟨ *Constant declarations (if any).* ⟩
⟨ *Variable declarations (if any).* ⟩
⟨ *Function and procedure declarations (if any).* ⟩
begin
  ⟨ *Main program body.* ⟩
end.

The procedure declaration begins with a heading:

procedure `DrawGrid`(`n,s,hL,vT:integer`)

name — `DrawGrid`; *parameter list* — `n,s,hL,vT:integer`

The parameter list indicates each formal parameter and its type. `DrawGrid` has four integer parameters that define the size and location of the grid. Together with the procedure heading, the *specification* supplies all the information needed by someone who is going to use the procedure:

```
{Pre:n>0,s>0.}
{Post:Draws an n-by-n grid with s-by-s tiles.  The upper left corner}
{of the upper left tile has screen coordinate (hL,vT).}
```

```
program Example8_1;
{Draws a pair of grids.}
const
   nL = 4; {Grid size of left grid.}
   sL = 25; {Tile size of left grid.}
   aL = 50;
   bL = 60; {(aL,bL) = upper left corner of left grid.}
   nR = 3; {Grid size of right grid.}
   sR = 50; {Tile size of right grid.}
   aR = 200;
   bR = 85; {(aR,bR) = upper left corner of right grid.}
var
   v,h:integer; {Vertical and horizontal coordinates.}
   k:integer;
begin
   ShowDrawing;
   {Draw the lefthand grid.}
   v:=bL;
   for k := 0 to nL do
      {The kth horizontal line has vertical coordinate v.}
      begin
         MoveTo(aL,v); LineTo(aL+nL*sL,v);
         v:= v+sL;
      end;
   h:=aL;
   for k:= 0 to  nL do
      {The kth vertical line has horizontal coordinate h.}
      begin
         MoveTo(h,bL); LineTo(h,bL+nL*sL);
         h := h+sL;
      end;
   {Draw the righthand grid.}
   v:=bR;
   for k := 0 to nR do
      {The kth horizontal line has vertical coordinate v.}
      begin
         MoveTo(aR,v); LineTo(aR+nR*sR,v);
         v:=v+sR;
      end;
   h:=aR;
   for k := 0 to nR do
      {The kth vertical line has horizontal coordinate h.}
      begin
         MoveTo(h,bR); LineTo(h,bR+nR*sR);
         h:=h+sR;
      end;
end.
```

Output: See FIGURE 8.1.

```
program Example8_2;
{Draws a pair of grids.}
const
   nL = 4; {Grid size of left grid.}
   sL = 25; {Tile size of left grid.}
   aL = 50;
   bL = 60; {(aL,bL) = upper left corner of left grid.}
   nR = 3; {Grid size of right grid.}
   sR = 50; {Tile size of right grid.}
   aR = 200;
   bR = 85; {(aR,bR) = upper left corner of right grid.}

procedure DrawGrid(n,s,hL,vT:integer);
{Pre:n>0,s>0.}
{Post:Draws an n-by-n grid with s-by-s tiles.  The upper left corner}
{ of the upper left tile has screen coordinate (hL,vT).}
var
   v,h,k:integer;
begin
   v := vT;
   for k := 0 to n do
      {The kth horizontal line has vertical coordinate v.}
      begin
         MoveTo(hL,v); LineTo(hL+n*s,v);
         v:= v+s;
      end;
   h := hL;
   for k:=0 to n do
      {The kth vertical line has horizontal coordinate h.}
      begin
         MoveTo(h,vT); LineTo(h,vT+n*s);
         h := h + s;
      end
end;

begin
   ShowDrawing;
   {Draw the left hand grid.}
   DrawGrid(nL,sL,aL,bL);
   {Draw right hand grid.}
   DrawGrid(nR,sR,aR,bR);
end.
```

Output: See FIGURE 8.1.

The precondition stipulates requirements that must be satisfied at the time of the procedure call. The postcondition describes what the procedure *does*. The rest of the procedure looks like a main program in itself. Local variables are declared:

```
var
   v,h,k:integer;
```

The *procedure body*, is set off with a **begin**-**end** pair:

```
begin
   v:=vT;
   for k:=0 to n do
      begin
         MoveTo(hL,v); LineTo(hL+n*s,v); v:=v+s;
      end;
   h:=hL;
   for k:=0 to n do
      begin
         MoveTo(h,vT); LineTo(h,vT+n*s); h:=h+s;
      end
end;
```

This completes the declaration of DrawGrid. In general, the structure of a procedure declaration is as follows:

procedure ⟨ *Name* ⟩ (⟨ *parameter list* ⟩);
⟨ *Specification* ⟩
⟨ *Constant Declarations (if any)* ⟩
⟨ *Variable Declarations (if any)* ⟩
begin
   ⟨ *Body* ⟩
end

We now turn our attention to the use of DrawGrid in the body of the main program. In Example8_2, the main program body consists of two *procedure calls*. Recalling the values of the constants nL, sL, aL and bL we see that the first call, DrawGrid(nL,sL,aL,bL), effectively says "draw a 4-by-4 grid with 25-by-25 tiles with the upper left corner of the upper left tile situated at (50,60)." It is appropriate to think of DrawGrid as a formula into which are substituted the values 4, 25, 50, and 60. These four values are substituted for the formal parameters n, s, hL and vT respectively. To clarify the substitution mechanism, we use the "box notation" developed in §7.1 to facilitate the tracing of function execution. The program starts and the current context is the main program:

	main
nL 4	
sL 25	
aL 50	
bL 60	
nR 3	
sR 30	
aR 200	
bR 85	

With the call to **DrawGrid**, we open up a box that indicates the value parameters and the local variables:

```
nL  4                                          main
sL  25
aL  50   DrawGrid
bL  60     ___ n       ___ v
nR  3      ___ s       ___ h
sR  30     ___ hL      ___ k
aR  200    ___ vT
bR  85
```

The values indicated in the call `DrawGrid(nL,sL,aL,bL)` are assigned:

```
nL  4                                          main
sL  25
aL  50   DrawGrid
bL  60     4   n       ___ v
nR  3      25  s       ___ h
sR  30     50  hL      ___ k
aR  200    60  vT
bR  85
```

Execution now takes place within the `DrawGrid` context. In particular, the body of `DrawGrid` is executed, with `n`, `s`, `hL`, and `vT` behaving as if they were "regular" integer variables. Thus, the assignment `v := vT` assigns the value of `vT` to `v`:

```
nL  4                                          main
sL  25
aL  50   DrawGrid
bL  60     4   n       60  v
nR  3      25  s       ___ h
sR  30     50  hL      ___ k
aR  200    60  vT
bR  85
```

And so it goes through the execution of the rest of `DrawGrid`. After the grid is drawn, `DrawGrid` has completed its mission and control passes back to the main program. We signify this in our trace by closing the `DrawGrid` box, indicating that the current context is once again the main program. However, the next line in the main program is `DrawGrid(nR,sR,aR,bR)`. The whole process repeats itself, with of course different values being passed to `DrawGrid`.

PROBLEM 8.1. The Swedish flag has the following proportions:

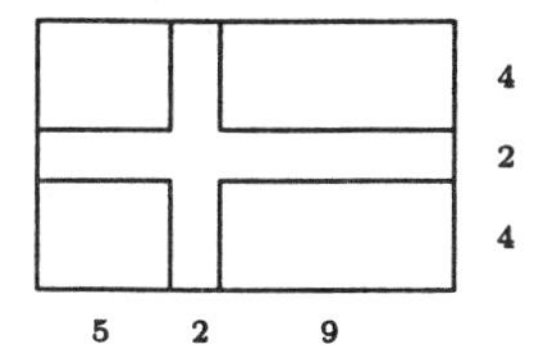

Write a procedure

```
procedure DrawSwedenFlag(hL,vT,p:integer);
{Pre:p>0}
{Post:Draws a Swedish flag with horizontal width 16p pixels and}
{vertical height 10p pixels.  The upper left corner = (hL,vT).}
```

PROBLEM 8.2. A *diamond* with center (h_c, v_c), horizontal radius a, and vertical radius b is the quadrilateral obtained by connecting (in order) the vertices $(h_c + a, v_c)$, $(h_c, v_c + b)$, $(h_c - a, v_c)$ and $(h_c, v_c - b)$. Complete the procedure

```
procedure DrawDiamond(hc,vc,hr,vr:integer);
{Post:Draws diamond with vertices }
{(hc-hr,vc),(hc,vc+hr),(hc+hr,vc), (hc,vc-hr).}
```

and use it to draw the following picture:

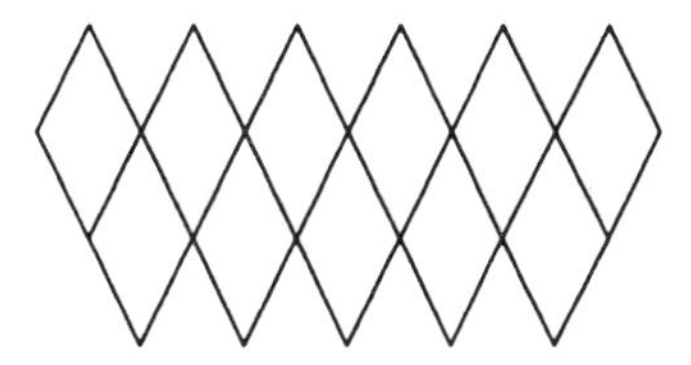

Assume that each diamond has horizontal radius 20 and vertical radius 40 and that the center of the leftmost diamond in the top row is (70,90).

As we move into more involved graphical computations, it is handy to have a set of procedures and functions that facilitate the acquisition of visual data and the display of results. These will be presented over the next few chapters and they all may be accessed through the unit DDCodes. Two of these "tools" are required now:

```
procedure Wait;
{Post:Returns as soon as the mouse is clicked.}

procedure ClearScreen;
{Post:Clears the entire DrawWindow}

Declared in DDCodes
```

They happen to illustrate that it is legal to have procedure with no parameters. But it will be our policy not to discuss *how* the procedures and functions in DDCodes work. Instead, we merely state their specifications as above.

The reason for presenting `Wait` and `ClearScreen` now is that they can be used to set up frameworks that cycle through a sequence of graphical operations. In `Example8_3` we examine the behavior of `DrawGrid` for various choices of tile size. `DrawGrid` itself is made available through

```
program Example8_3;
{Tests DrawGrid.  }
uses
   DDCodes, Chap8Codes;
var
   s:integer; {Tile size.}
   n:integer; {Grid size.}
begin
   ShowDrawing;
   s := 64;
   while (s >= 1) do
      {Draws (192/s)-by-(192/s) grid with grid with s-by-s tiles}
      begin
         ClearScreen;
         MoveTo(20,20);
         WriteDraw('Click mouse to continue');
         n := 192 div s;
         DrawGrid(n,s,100,30);
         MoveTo(170,250);
         writeDraw('n = ', n:2, ' , s = ', s:3);
         s := s div 2;
         Wait;
      end;
   EraseRect(0,0,20,200);
end.
```

The procedures `ClearScreen` and `Wait` are declared in `DDCodes`. The procedure `DrawGrid` is declared in `Chap8Codes`.

`Chap8Codes`. Note that the **uses** declaration names the two **units** that must be linked to the main program. `DDCodes` contains `Wait` and `ClearScreen` and `Chapt8Codes` contains `DrawGrid`.

The framework illustrated by `Example8_3` permits the exploration of *graphical ideas*. It is analogous to the numerical exploration framework presented in §2.4.

PROBLEM 8.3. A football field looks like this:

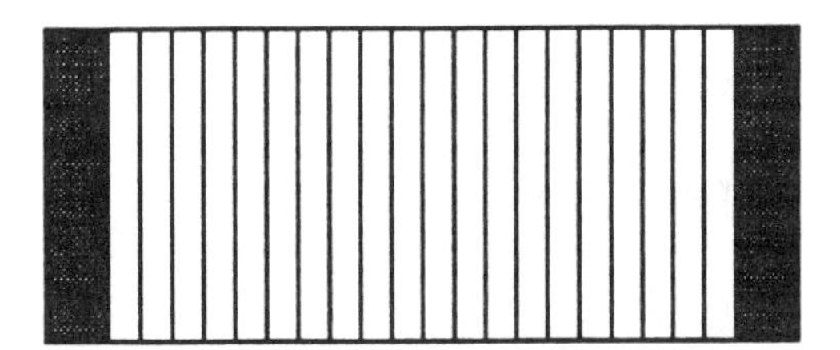

It is 120 yards long and 50 yards wide. The shaded grey areas are called *end zones* and they are 10 yards wide. In between the end zones and spaced every 5 yards are the *yard lines*. Complete the following procedure:

```
procedure DrawFootball(hL,vT,p:integer);
{Pre:p>0}
{Post:Draws a football field at a scale of p pixels per yard.}
{The upper left corner of the field located at (hL,vT).}
```

The order in which you draw the outer boundary, the yard lines, and the end zones will effect appearance of the displayed field. Make it look nice! To that end, you may have to "fine tune" some of the values that are passed to the built-in graphics procedures.

PROBLEM 8.4. The specification of a tennis court (in feet) is as follows:

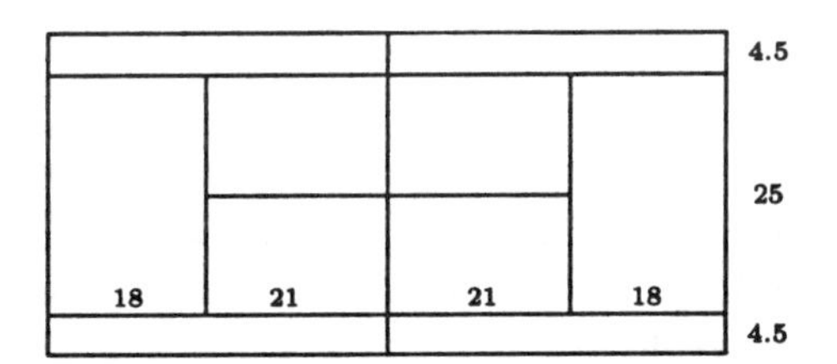

Write a procedure

```
procedure DrawTennisCourt(L,T,s:integer);
{Post:Draws a tennis court to scale that is s pixels long.}
{The upper left corner is at (L,T). }
```

The displayed court will not be perfect. For example, if the left and right baselines have horizontal coordinates 10 and 101, then ideally the centerline should have horizontal coordinate $h = 55.5$. However, because of the screen's discreteness, we have to settle with a centerline displayed at $h = 55$ or 56. This kind of rounding will have to be part of the scaling process, but it should not effect adversely the appearance of the displayed court.

8.2 Hierarchy

Let's consider the problem of drawing a checkerboard shown in FIGURE 8.2. A procedure that encapsulates this computation requires four parameters:

FIGURE 8.2 *A Checkerboard*

```
procedure DrawCheckerboard(n,s,hL,vT:integer);
{Pre:n>0,s>0}
{Post:Draws an n row, n column checkerboard with sxs tiles }
{and upper left corner (hL,vT).}
```

Checkerboards are made up of rows and this suggests the following approach to the computation:

```
v := vT;
for k := 1 to m do
   begin
      ⟨Draw the k-th row whose top edge has vertical coordinate v.⟩
      v := v + s;
   end;
```

A row begins with either a dark tile or a light tile and there are several ways that the alternation can be handled. One way is to use a *boolean variable*. A boolean variable is able to store one of the values **true** or **false**. The declaration

```
var
   Dark:boolean
```

establishes **Dark** as a boolean variable. To illustrate how such a variable might be used, consider the fragment:

```
Dark:=true;
h:=10;
for k:=1 to 4 do
   begin
      if Dark then
         PaintRect(100,h,140,h+40)
      else
         FrameRect(100,h,140,h+40);
      Dark := not Dark;
      h:=h+40;
   end
```

FIGURE 8.3 *A Checkered Row*

It produces a checkered row as shown in FIGURE 8.3. The assignment `Dark := not Dark` negates the value of `Dark`. `True` becomes `false` and `false` becomes `true`. The `if` is used to generate the right kind of tile. If the value of `Dark` is `true`, then a call to `PaintRect` is made. If the value of `Dark` is `false`, then the `else` option is taken and a call to `FrameRect` is made.

We mention that the above fragment is equivalent to

```
h:=10;
for k:=1 to 4 do
   begin
      if k mod 2 = 1 then
         PaintRect(100,h,140,h+40)
      else
         FrameRect(100,h,140,h+40);
      h:=h+40;
   end
```

However, the boolean solution is preferable in terms of style and gets you thinking about the alternation the "right way."

Returning to the checkerboard problem, if we had a procedure of the form

```
procedure DrawCheckerRow(n,s,hL,vT:integer; DarkFirst:boolean);
{Pre:n>0,s>0.}
{Post:Draws an n-by-1 row of s-by-s tiles.  The upper left corner}
{ of the upper left tile is at (hL,vT). Every other tile is dark.}
{The leftmost tile is dark if DarkFirst is true.  Otherwise,}
{the leftmost tile is not dark}
```

then we can implement `DrawCheckerBoard` as follows:

```
procedure DrawCheckerBoard(n,s,hL,vT:integer);
{Pre:n>0,s>0.}
{Post:Draws an n-by-n checkerboard with s-by-s tiles.  The upper left corner}
{ of the upper left tile has screen coordinate (hL,vT).}
var
   v:integer; {Top vertical coordinate of current row being drawn.}
   Dark:boolean; {Indicates whether first tile in current row is dark or not dark.}
   k:integer;
```

```
begin
   v := vT;
   Dark := false;
   for k := 1 to n do
      begin
         DrawCheckerRow(n,s,hL,v,Dark);
         v := v + s;
         Dark := not Dark;
      end;
end;
```

Just as one programmer-defined function can call another programmer-defined function, so it is with procedures. Notice how we can develop DrawCheckerBoard before DrawCheckerRow as long as the latter is completely specified. This is yet another example of the "top-down" approach to problem-solving. It is an extremely important example because it is the standard way of doing business in settings where programmer teams are used. You can develop DrawCheckerBoard and your colleague can develop DrawCheckerRow *at the same time* as long as you both agree on the specification for the latter procedure.

As to the implementation of DrawCheckerRow, we may proceed as follows:

```
procedure DrawCheckerRow(n,s,hL,vT:integer; DarkFirst:boolean);
{Pre:n>0,s>0.}
{Post:Draws an n-by-1 row of s-by-s tiles.  The upper left corner}
{ of the upper left tile has screen coordinate (hL,vT). Every other tile}
{is dark.  The leftmost tile is dark if DarkFirst is true.  }
{Otherwise it is not dark.}
var
   h:integer; {Left horizontal coordinate of current tile being drawn.}
   Dark:boolean; {Indicates whether current tile is dark or not dark.}
   k:integer;
begin
   h := hL;
   Dark := DarkFirst;
   for k := 1 to n do
      begin
         if Dark then
            {Draw a black tile.}
            PaintRect(vT,h,vT+s, h+s)
         else
            {Draw a white tile.}
            FrameRect(vT,h,vT+s,h+s);
         h := h + s;
         Dark := not Dark;
      end
end
```

The program Example8_4 puts it all together. It is essential that DrawCheckerRow be declared *before* DrawCheckerBoard.

```
program Example8_4;
{Draws Checkerboards.}
uses
   DDCodes;
var
   n:integer; {Number of columns in checkerboard.}
   s:integer; {Tile size.}
procedure DrawCheckerRow(n,s,hL,vT:integer;DarkFirst:boolean);
   ⟨:⟩
procedure DrawCheckerBoard(n,s,hL,vT:integer);
   ⟨:⟩
begin
   ShowDrawing;
   s := 128;
   while s >= 4 do
      begin
         ClearScreen;
         MoveTo(20,20);
         WriteDraw('Click mouse to continue');
         n := 256 div s;
         DrawCheckerBoard(n,s,20,30);
         s := s div 2;
         Wait;
      end;
   EraseRect(0,0,20,200);
end.
```

The procedures ClearScreen and Wait are declared in DDCodes.

PROBLEM 8.5. Generalize `DrawCheckerBoard` so that it can accommodate rectangular grids with an option to draw the light tiles gray or white.

PROBLEM 8.6. Complete the following procedure making effective use of `DrawGrid`:

```
procedure DrawGridWithEdge(n,s,hL,vT:integer);
{Pre:n>0,s>0.}
{Post:Draws an n-by-n grid with s-by-s tiles.  The upper left corner}
{ of the upper left tile has screen coordinate (hL,vT). The edge tiles}
{ are further partitioned into 2-by-2 grids.}
```

Here is the $n = 8$ case:

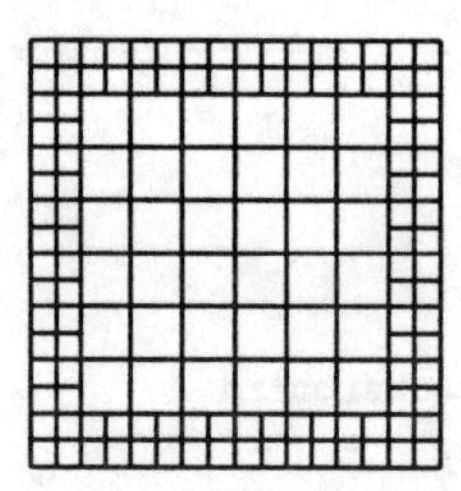

PROBLEM 8.7. Suppose `hc`, `vc` and `s` are positive integers. The fragment

```
MoveTo(hc-s,vc);
LineTo(hc,vc+s); LineTo(hc+s,vc);
LineTo(hc,vc-s); LineTo(hc-s,vc);
```

draws a diamond with center (hc, vc) and "radius" s. (a) Write a procedure of the following form that can draw a shaded version of this:

```
procedure PaintDiamond(hc,vc,s:integer; shaded boolean);
{Pre:s >=0.}
{Post:Draws a shaded square whose four vertices on the screen are}
{given by (hc-s,vc), (hc+s,vc), (hc,vc-s), and (hc,vc+s).}
```

Shade by drawing a sufficient number of horizontal line segments in the diamond's interior. (b) using `PaintDiamond`, complete the following procedures:

```
procedure TiltedCheckerRow(n,s,hL,vB:integer;DarkFirst:boolean);
{Pre:n>0,s>0}
{Post:  Draws a diamonds with radii s and centers at (hL+ks,vB-ks),}
{k=0,..,n-1.  Every other diamond is shaded.  The k=0 diamond is }
{shaded if DarkFirst is true.}

procedure TiltedCheckerBoard (n,s,hL,vc:integer);
{Pre:n>0,s>0.}
{Post:Draws a 45 degree tilted n-by-n grid with s-by-s tiles.  }
{The leftmost tile has center (hL,vc).}
```

This hierarchy corresponds to the one developed for `CheckerBoard`.

PROBLEM 8.8. Complete the procedures

```
procedure FrameDiamond(h0,v0:integer; r:real; Side:char);
{Pre:Draws a diamond with vertices at the rounded versions of}
{(h0,v0), (h0-c*r,v0-s*r), (h0-c*r,v0+s*r), (h0,v0+r) where}
{s = 1/2 and c = sqrt(3)/2 if Side = 'L' and -sqrt(3)/2 otherwise.}
```

```
procedure PaintDiamond (h0,v0:integer; r:real);
{Pre:r>0}
{Post:Draws a shaded diamond with vertices at the rounded versions of}
{(h0,v0), (h0+c*r,v0-s*r), (h0,v0-r), (h0-c*r,v0-s*r)}
```

and make effective use of them to write

```
procedure DrawCube(hc,vc,r:integer);
{Post:Shades a hexagon with vertices }
{(hc+r*cos(pi/6 + k*pi/3),vc - sin(pi/6 + k*pi/3)) where k=0,1,2,3,4,5.  }
{Line segments are drawn from the center to vertices 0,2, and 4 thereby}
{producing 3 diamond-shaped sectors.  The top one of these is shaded.}
```

PROBLEM 8.9. Complete the following procedure

```
procedure PaintTriangle(h1,v1,h2,v2,h3,v3:integer);
{Post:Draws a shaded triangle with vertices}
{(h1,v1), (h2,v2), and (h3,v3).}
```

Proceed by drawing a large number of lines from (h_1, v_1) to the opposite side. Then write

```
procedure PaintPolygon(n,r,hc,vc:integer);
{Post:Draws a shaded regular n-gon with center at (hc,vc) and radius r.}
```

PROBLEM 8.10. A phased moon with radius r and center (x_c, y_c) is a partially shaded disk

$$D = \{(x,y) : (x - x_c)^2 + (y - y_c)^2 \leq r^2\}$$

The shading depends upon the time T and the period P. If $0 \leq T \leq P/2$, then a point $(x, y) \in D$ is in the dark region if

$$x \leq x_c + \cos(2\pi T/P)\sqrt{r^2 - (y - y_c)^2}$$

On the other hand, if $P/2 \leq T \leq P$, then it is in the shaded region if

$$x \geq x_c - \cos(2\pi T/P)\sqrt{r^2 - (y - y_c)^2}.$$

Complete the following procedure using $P = 28$:

```
procedure PaintMoon(hc,vc,r:integer; T:real);
{Post:Draws the moon as it appears T days after the new moon.}
{T = 0 or 28 corresponds to the new moon.  T= 14 corresponds to the full moon.}
{The moon is centered at (hc, vc) and has radius r .  }
```

Accomplish this by drawing a sufficiently large number of horizontal line segments. Using `PaintMoon`, write

```
procedure WeekOfMoons(hc,vc,r:integer;T:real);
{Post:Draws 7 moons with radius r.}
{The k-th moon has phase T+k and center (hc,vc+3kr), k=0,1..,6}
```

and use it to draw a 4-row display that shows all the phases during a 28-day cycle.

We close with an example that gives us an opportunity to exercise two very important skills associated with the use of procedures:

- The ability to use procedures that have already been written.
- The ability to look at a problem and know what procedures to write.

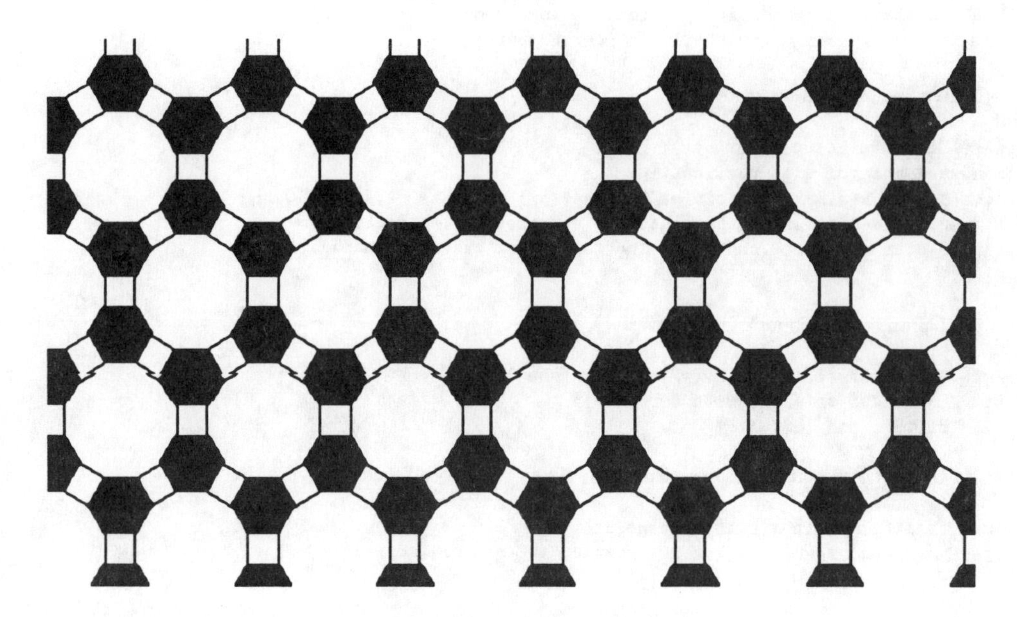

FIGURE 8.4 *A Tessellation*

Our goal is to draw the *tessellation* that is shown in FIGURE 8.4. The first step is to identify the basic pattern being repeated. This consists of a regular dodecagon with filled hexagons placed along every other side. It is convenient to think of this as a "tile." See FIGURE 8.5.

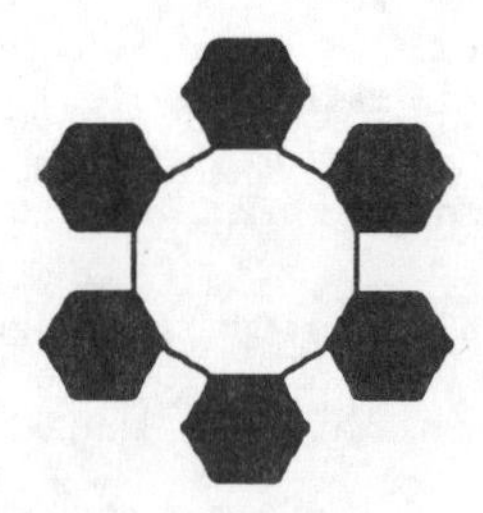

FIGURE 8.5 *The Unit Tile*

Let r be the radius of the dodecagon, i.e., the distance from its center to any of its 12 vertices. This is the key parameter and we see that the sides of the dodecagon have radius $r_{hex} = 2r\sin(\pi/12)$.

The centers of the six hexagons are on a circle of radius $r\cos(\pi/12) + r_{hex}\cos(\pi/6)$. With this trigonometry we are set to write a procedure for displaying a single unit tile:

```
procedure HexWheel(h0,v0:integer; r:real);
{Pre:r>=0}
{Post:Draws a radius r dodecagon centered at (h0,v0) with }
{six black hexagonal knobs.}
var
   k:  integer;
   theta:real; {Angle associated with k-th knob.}
   h,v:integer; {(h,v) = center of k-th knob}
   rHex:real; {radius of hexagonal knob.}
   rTot:real; {radius of circle through the hexagon centers.}
begin
   rHex := 2*r*sin(pi/12); rTot := r*cos(pi/12)+rhex*cos(pi/6);
   theta := pi/2;
   for k := 1 to 6 do
      {Draw the k-th knob.}
      begin
         h := round(h0+rTot*cos(theta));
         v := round(v0-rTot*sin(theta));
         PaintHex(h,v,rhex);
         theta := theta + pi/3;
      end;
   FramePoly(12,h0,v0,r,pi/12);
end;
```

This procedure makes use of

```
procedure FramePoly(n,h0,v0:integer; r,theta:real);
{Pre:n>=3}
{Post:Draws a polygon whose vertices are rounded versions of}
{(h0+rcos(theta +2pi*k/n),v0-r*sin(theta+2pi*k/n)), k=0,..,n-1.}

procedure PaintHex(h0,v0:integer; r:real);
{Pre:r>=0}
{Post:Draws a shaded hexagon with vertices that are rounded versions}
{of (h0+r*cos(k*pi/3),v0-r*sin(k*pi/3)) for k=0,1,..,5}
```

both of which are declared in `Chap8Codes`.

The tessellation is made up of rows and the center-to-center spacing of the tiles along a row is given by $2r(\cos(\pi/12) + \sin(\pi/12))$. See FIGURE 8.6. Building upon the `HexWheel` we obtain

```
procedure HexWheelRow(n,h0,v0:integer; r:real);
{Pre:r>=0}
{Post:Draws a row of n hexagon wheels.  Leftmost wheel centered at (h0,v0).}
var
   k:integer;
   h:real; {Horizontal coordinate of k-th wheel.}
   hspace:real; {Horizontal spacing}
```

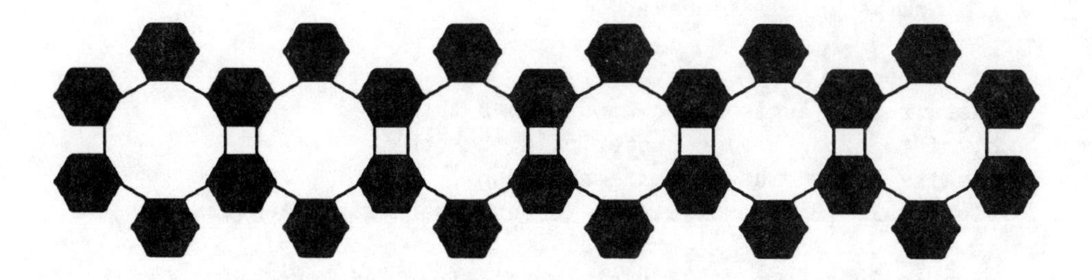

FIGURE 8.6 *Row of Tiles*

```
begin
   h := h0; hspace := 2*r*(cos(pi/12) + sin(pi/12));
   for k := 1 to  n do
      begin
         HexWheel(round(h),v0,r);
         h := h + hspace;
      end;
end
```

Putting together the rows to produce the array requires two more trigonometric observations. The spacing between the tessellation rows is given by

$$v_{space} = r\left(\cos\left(\frac{\pi}{12}\right) + \sin\left(\frac{\pi}{12}\right)\right)$$

and the indentation associated with the odd-indexed rows is given by

$$h_{indent} = r\left(\cos\left(\frac{\pi}{12}\right) + \sin\left(\frac{\pi}{12}\right)\left(4\cos\left(\frac{\pi}{6}\right) + 1\right)\right)$$

At last we obtain:

```
procedure HexWheelArray(m,n,h0,v0:integer;r:real);
{Pre:r>=0}
{Post:Draws m, length-n hexagon rows.  }
{Upper left hexagon wheel is centered at (h0,v0).}
{Inner dodecagon has radius r.}
var
   i:integer;
   v:real; {Vertical coordinate of i-th row.}
   hindent:  integer; {Horizontal starting value for indented rows.}
   vspace:  real; {Vertical spacing.}
```

```
begin
   v := v0;
   hindent := round(h0+ r*(cos(pi/12)+sin(pi/12)));
   vspace := r*(cos(pi/12)+sin(pi/12)*(4*cos(pi/6)+1));
   for i := 1 to  m  do
      begin
         if i mod 2 = 1  then
            HexwheelRow(n,h0,round(v),r)
         else
            HexWheelRow(n,hindent,round(v),r);
         v := v + vspace;
      end;
end
```

The program Example8_5 illustrates the use of the procedures HexWheel, HexWheelRow, and HexWheelArray

```
program Example8_5;
{Tessellation with hexagons.}
uses
   DDCodes,Chap8Codes;
const
   r = 30;
procedure HexWheel(h0,v0:integer; r:real);
      ⟨:⟩
procedure HexWheelRow(n,h0,v0:integer; r:real);
      ⟨:⟩
procedure HexWheelArray(m,n,h0,v0:integer;r:real);
      ⟨:⟩
begin
   ShowDrawing;
   HexWheel(100,150,r);
   MoveTo(20,20); WriteDraw('Click Mouse');
   Wait;
   ClearScreen;
   HexWheelRow(6,100,150,r);
   MoveTo(20,20); WriteDraw('Click Mouse');
   Wait;
   ClearScreen;
   HexWheelArray(6,10,0,0,r);
end.
```

The procedures ClearScreen and Wait are declared in DDCodes. Output: See FIGURE 8.4.

PROBLEM 8.11. Write procedures to generate the following tessellations. Model your solution after Example8_5. To force the development of hierarchies, no procedure should contain a nested loop. Use the procedures in Chap8Codes whenever applicable including PaintHex, FramePoly, and

```
procedure PaintETriangle(h0,v0:integer;r:real; Orient,shade:char);
{Pre:r>=0, Orient has the value 'T','B','L',or'R'. Shade has the value}
{ 'w', 'b', or 'g.'}
{Post:Draws an equilateral triangle with one vertex at (h0,v0) and sides of length r.}
{If Orient = 'T' then the side opposite (h0,v0) is horizontal and below.}
{If Orient = 'B' then the side opposite (h0,v0) is horizontal and above.}
{If Orient = 'L' then the side opposite (h0,v0) is vertical and to the right.}
{If Orient = 'R' then the side opposite (h0,v0) is vertical and to the left.}
{If shade = 'w' then the triangle is white.}
{If shade = 'b' then the triangle is black.}
{If shade = 'g' then the triangle is gray.}

procedure PaintDiamond(h0,v0:integer; r:real; Orient,Shade:char);
{Pre:r>=0, Orient = 'T', 'L', 'B', 'R'}
{Post:  Draws a diamond with one vertex at (h0,v0).  This vertex is }
{is either at the top (Orient = 'T'), the left (Orient = 'L'), the bottom}
{(Orient = 'B', or the right (Orient = 'R'). }
{Diamond is gray if shade = 'g' and black otherwise.}
```

(a)

(b)

(c)

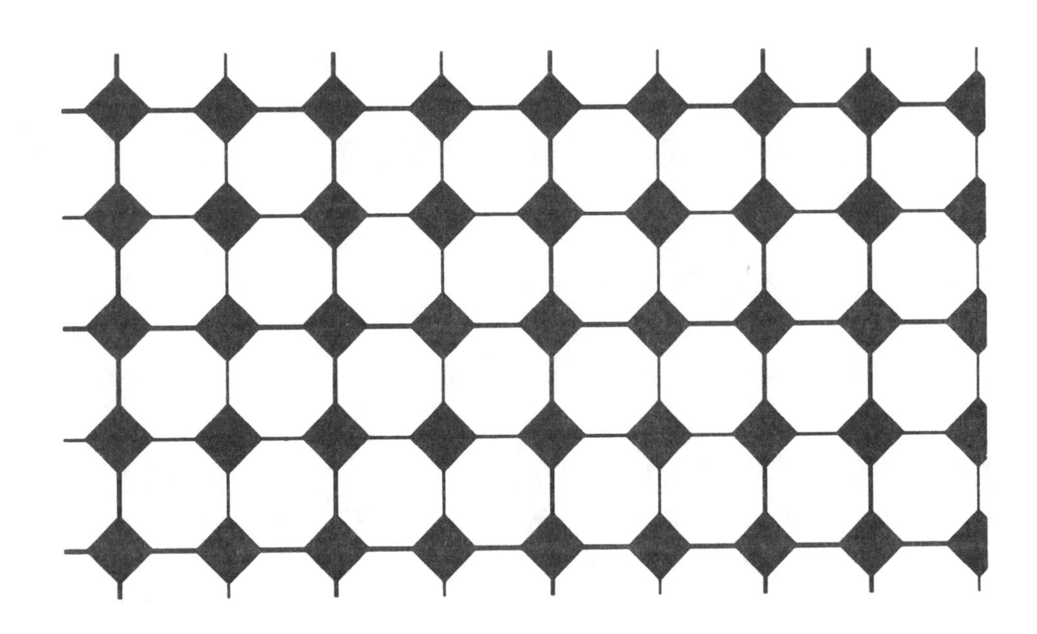

(d)

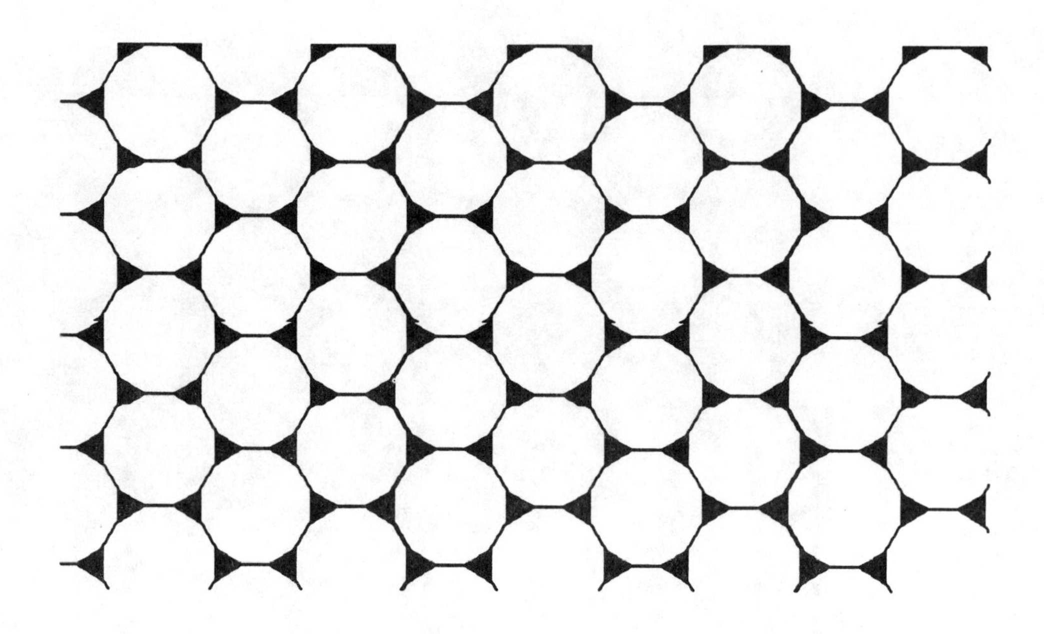

(e)

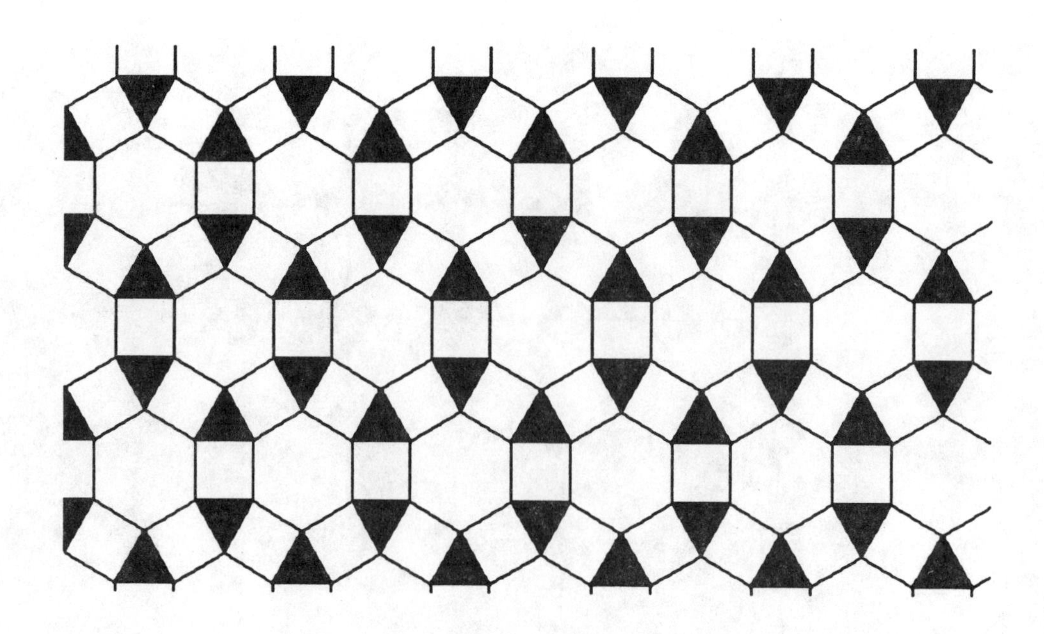

(f)

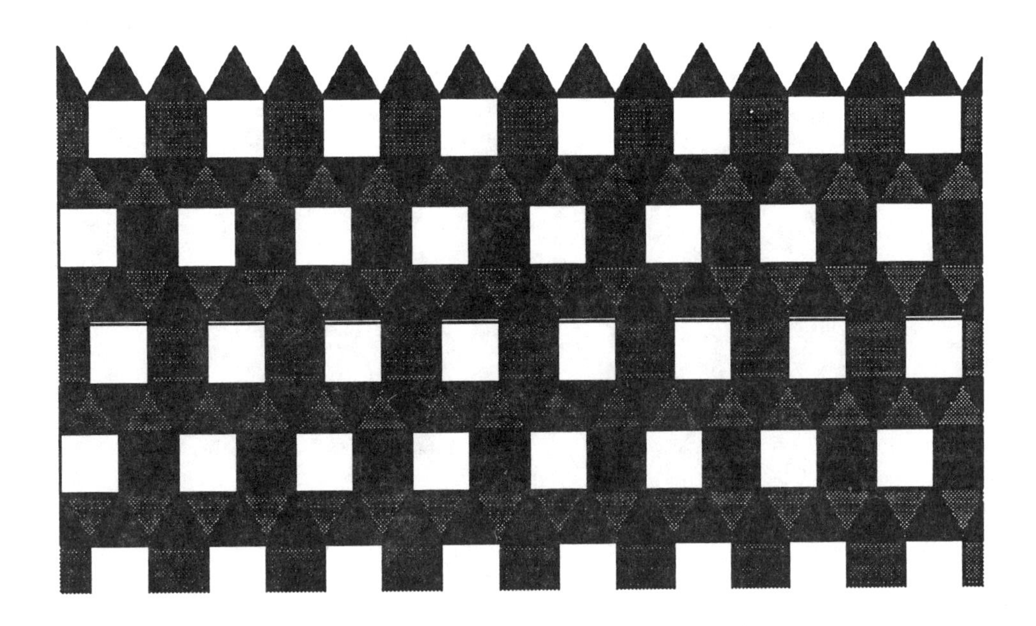

(g)

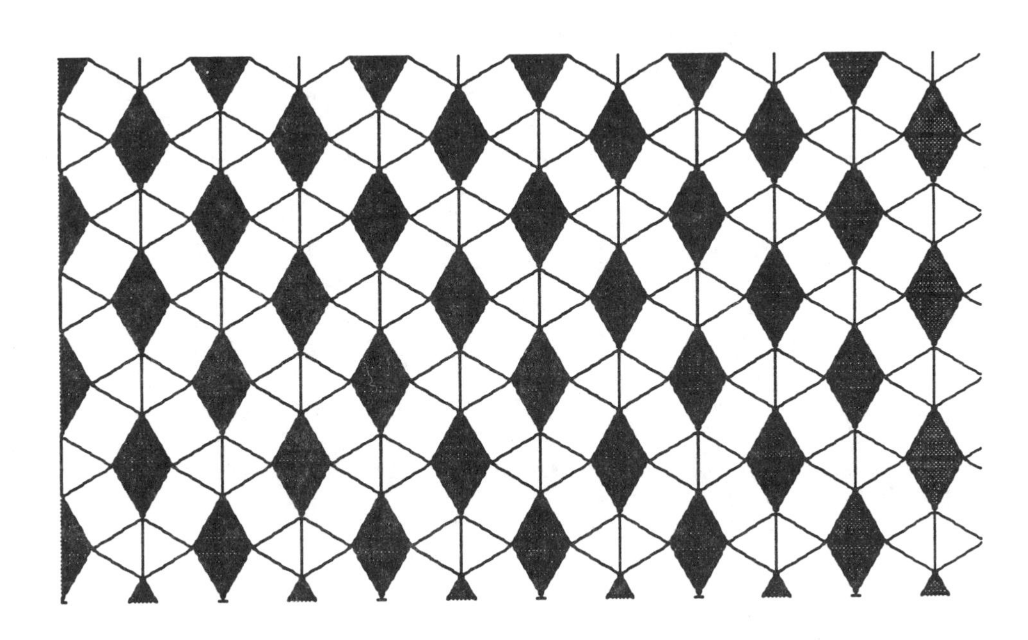

(h)

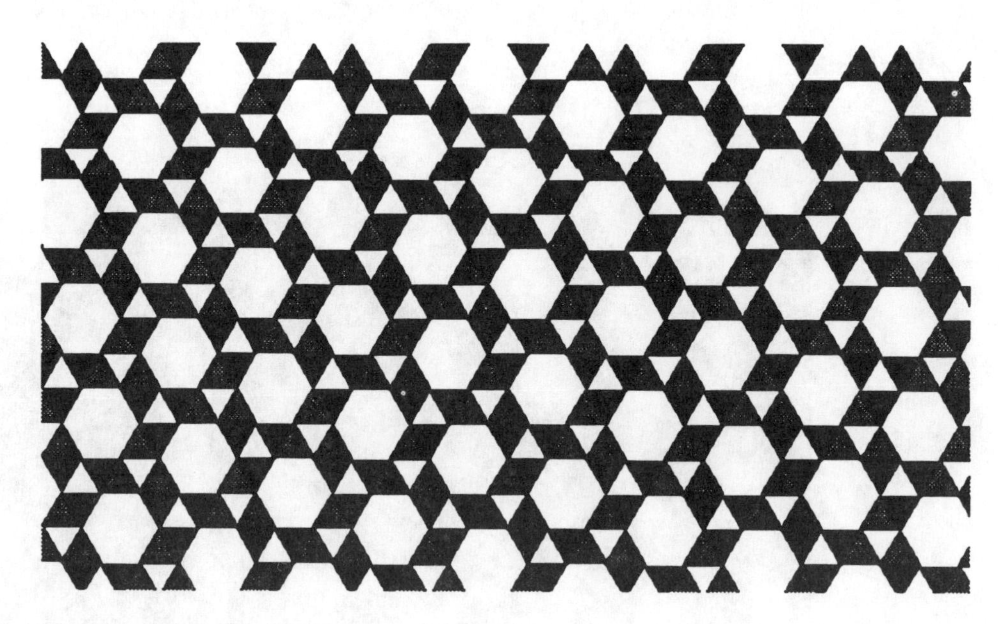

PROBLEM 8.12. Assume the availability of

```
procedure FrameQuad(h0,v0,rT,rL,rB,rR:integer);
{Pre:rT,rL,rB,rR>0}
{Post:Draws the quadrilateral with vertices }
{(h0+rR,v0), (h0,v0-rT),(h0-rL,v0),(h0,v0+rB).}
```

Use this procedure to produce quadrilateral-based tessellation of the form:

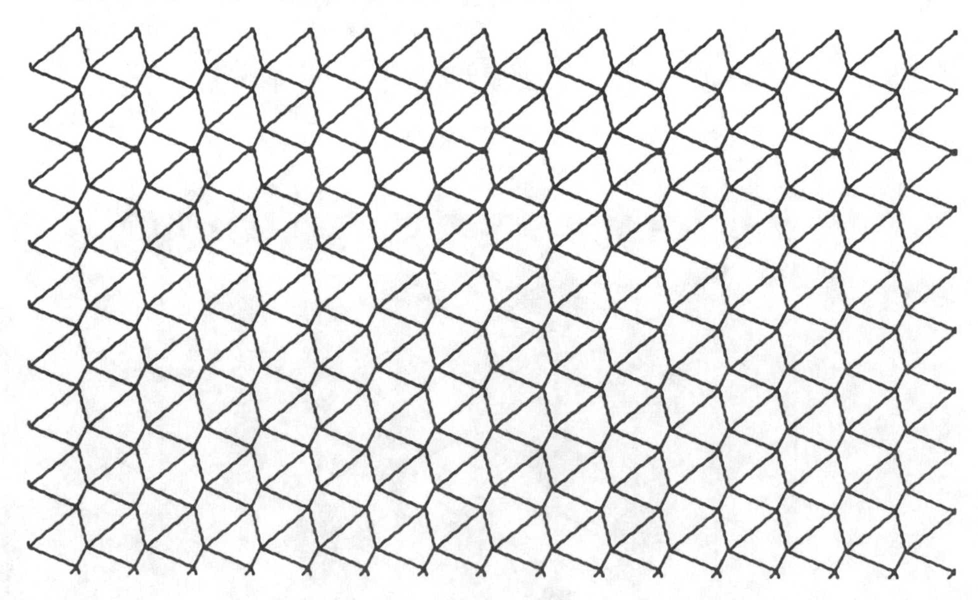

The idea is to augment the "base quadrilateral" with its "reflection", and then consider that as the "tile" upon which the tessellation is based.

Chapter 9

Proximity

Questions of proximity are of central importance in computational science. How *near* is a given mechanical system to wild oscillation? How *near* is a given fluid flow to turbulence? How *near* is a given configuration of molecules to a state of minimal energy? How near is one digitized picture to another? The key word is "near" and the recognition that a "distance function" is required to measure "nearness." The notion of distance is familiar to us in geometric settings:

- What is the distance between two points in the xy plane?
- What is the distance from a point to a line segment?
- What is the distance from a point to a polygon?

Our plan is to cut our "nearness" teeth on planar distance problems of this variety, illustrating the distinction between constrained and unconstrained optimization and the complicated boundary between exact mathematics and practical computation.

In the geometric setting, extreme nearness "turns into" inclusion. Instead of asking how near one rectangle is to another, we may ask whether one rectangle is inside another. Questions like this have yes/no answers. Distance questions, on the other hand, have a continuity about them and culminate in the production of a single, nonnegative real number.

The problem of when three points are collinear gives us a snapshot of just how tricky it can be to handle a yes/no geometric question. In theory, three points either line up or they do not. In practice, fuzzy data and inexact arithmetic muddy the waters. For example, we may be using a telescope and a computer to determine the precise moment when both members of a binary star system line up. But both tools have limited precision. Stars and numbers that are too close together are impossible to resolve, and so the computational scientist formulates a distance function that can be used to investigate how near the astronomical situation is to exact collinearity.

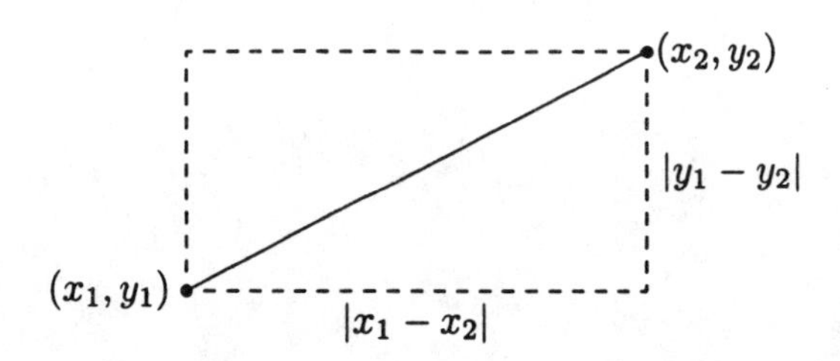

FIGURE 9.1 *Distance Between Two Points*

9.1 Distance

The distance between the points (x_1, y_1) and (x_2, y_2) is given by

$$d = \sqrt{(x_1 - x_2)^2 + (y_1 - y_2)^2}\,.$$

This is the recipe for *Euclidean* distance. See FIGURE 9.1. Other point-to-point distance measures are possible. For example, the "taxicab" distance is given by

$$d_{taxicab} = |x_1 - x_2| + |y_1 - y_2|.$$

The term is appropriate because it is the relevant measure of distance in cities that have a rectilinear grid of streets.

In applications, the choice of a distance function may be obvious. However, there are occasions when several reasonable possibilities exist and it is then up to the computational scientist to decide which is the most appropriate. For simplicity, we stick to Euclidean distance in the text and build upon the following function:

```
function Distance(x1,y1,x2,y2:real):real;
{Post:The distance from (x1,y1) to (x2,y2).}
begin
   d := sqrt(sqr(x1-x2) + sqr(y1-y2));
end
```

The program **Example9_1** computes the average distance of a point to the vertices of the unit regular n-gon:

PROBLEM 9.1. What is the expected distance between two points that are randomly placed inside the square with vertices (0,0), (1,0), (1,1), and (0,1)?

PROBLEM 9.2. Complete the following function

```
function DistanceToCorner(x0,y0,a:real):real;
{Pre:  a>=0, 0<=x<=a, 0<=y<=a}
{Post:  The distance from (x,y) to the closest vertex of the square}
{whose vertices are (0,0), (a,0), (a,a), and (0,a).}
```

Make effective use ofDistance(x1,y1,x2,y2).

```
program Example9_1;
{Average distance of a point to vertices of unit regular n-gon.}
uses
   Chap9Codes;
var
   AnotherEg:char; {Continuation indicator.}
   n:integer; {Number of sides in the polygon.}
   k:integer;
   x,y:real; {(x,y) = point}
   s:real; {Running sum of distances.}
   Ave:real; {The average distance.}
begin
   ShowText;
   AnotherEg := 'y';
   while AnotherEg <> 'n' do
      begin
         writeln('Enter x and y:');
         readln(x,y);
         writeln('Enter n:');
         readln(n);
         s := 0;
         for k := 0 to  n - 1  do
            s := s + Distance(x,y, cos(2*k*pi/n), sin(2*k*pi/n));
         Ave := s/n;
         writeln('Average = ', Ave:10:6);
         writeln;
         writeln('Another Example?  Enter y (yes) or n (no).');
         readln(AnotherEg)
      end
end.
```

The function `Distance` is declared in the unit `Chap9Codes`. Sample output:

```
Enter x and y:
2 3
Enter n:
6
Average = 3.675168

Another Example?  Enter y (yes) or n (no).
n
```

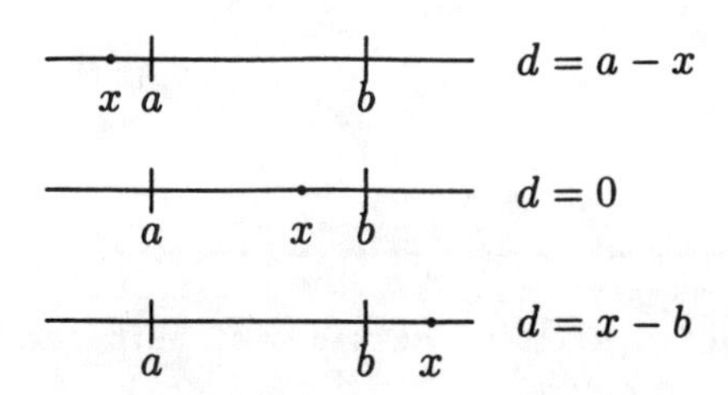

FIGURE 9.2 *The Distance from a Point to an Interval*

The distance from a real number x to an interval $[a, b]$ is prescribed by

$$d = \begin{cases} a - x & x < a \\ 0 & a \le x \le b \\ x - b & b < x \end{cases} .$$

See FIGURE 9.2. A literal encoding of this definition leads to

```
function PointToInterval1(x,a,b:real):real;
{Pre:a<=b}
{Post:The distance from x to [a,b].}
begin
   if x<a then
      PointToInterval1:=a-x;
   if (a<=x) and (x<=b) then
      PointToInterval1:=0;
   if b<x then
      PointToInterval1:=x-b;
end
```

Note that each call to `PointToInterval1` involves three comparisons and that some of these may be superfluous. For example, if `x<a` is true, then there is no need to execute the second and third `if` statements. The superfluous checking can be avoided by nesting:

```
function PointToInterval(x,a,b:real):real;
{Pre:a<=b}
{Post:The distance from x to [a,b].}
begin
   if x < a then
      PointToInterval:= a - x
   else if x > b then
      PointToInterval:=x - b
   else
      {a<=x<=b}
      PointToInterval:=0;
end
```

This reduces the number of required comparisons. Only one is required if $x < a$. Otherwise, two comparisons are required. The benchmarking of PointToInterval1 and PointToInterval would not reveal a dramatic improvement because the amount of arithmetic involved in comparison is so small. However, many problems in computational geometry involve a sequence of tests each of which may be very expensive. In this kind of situation, it is important to organize the computation in a way that avoids redundant calculation.

PROBLEM 9.3. Define the distance from a point (x, y) to the rectangle

$$\mathcal{R} = \{(x, y) : L \le x \le R,\ B \le y \le T\}$$

is given by $\sqrt{(u-x)^2 + (v-y)^2}$ where (u, v) is the closest point in R to (x, y). The appropriate u and v values to substitute into the distance function depends upon which of nine regions houses (x, y):

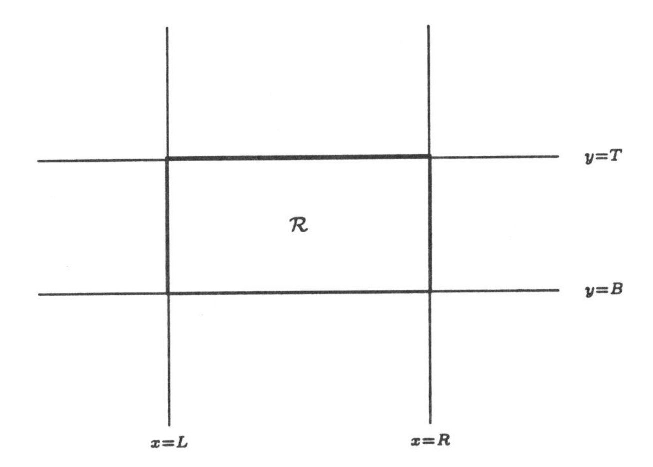

In particular,

$$d = \begin{cases} y - T & L \le x \le R,\ T < y \\ B - y & L \le x \le R,\ y < B \\ x - R & B \le y \le T,\ R < x \\ L - x & B \le y \le T,\ x < L \\ \sqrt{(x-R)^2 + (y-T)^2} & R < x,\ T < y \\ \sqrt{(x-R)^2 + (y-B)^2} & R < x,\ y < B \\ \sqrt{(x-L)^2 + (y-T)^2} & x < L,\ T < y \\ \sqrt{(x-L)^2 + (y-B)^2} & x < L,\ y < B \\ 0 & L \le x \le R,\ B \le y \le T \end{cases}$$

In view of these possibilities, complete the following function:

```
function PointToRectangle(x0,y0,L,R,B,T:real):real;
{Pre:L<=R, B<=T.}
{Post:The distance from (x0,y0) to the rectangle whose vertices are}
{(L,B), (R,B), (L,T), and (R,T).}
```

Use nested if's for efficiency.

Let L be the line that passes through the points $P_1 = (x_1, y_1)$ and $P_2 = (x_2, y_2)$ which we assume to be distinct. The points on L are characterized by the following pair of *parametric equations*:

$$\begin{aligned} x(t) &= x_1 + t(x_2 - x_1) \\ y(t) &= y_1 + t(y_2 - y_1) \end{aligned} \qquad -\infty < t < \infty.$$

It is useful to think of $(x(t), y(t))$ as the location at time t of a "particle" that is moving along the line. At $t = 0$ the particle is at P_1 and at time $t = 1$ it is at P_2. The parametric representation facilitates the solution to various geometric problems that involve L. For example, to compute the minimum distance between a point $P_0 = (x_0, y_0)$ and a point on L, we set the derivative of

$$d(t) = \sqrt{(x(t) - x_0)^2 + (y(t) - y_0)^2}$$

to zero. A calculation shows that if

$$t_* = \frac{(x_0 - x_1)(x_2 - x_1) + (y_0 - y_1)(y_2 - y_1)}{(x_2 - x_1)^2 + (y_2 - y_1)^2}.$$

then $d'(t_*) = 0$ and so $d(t_*)$ is the minimizing distance. This gives

```
function PointToLine(x0,y0,x1,y1,x2,y2:real):real;
{Pre:(x1,y1) and (x2,y2) are distinct points.}
{Post:Distance from (x0,y0) to the line through (x1,y1) to (x2,y2).}
var
   u,v:real; {(u,v) = point on the line closest to (x0,y0).}
   tStar:real; {The ''time'' associated with the closest point.}
begin
   tStar := ((x0-x1)*(x2-x1) + (y0-y1)*(y2-y1))/(sqr(x2-x1)+sqr(y2-y1));
   u := x1 + tStar*(x2-x1);
   v := y1 + tStar*(y2-y1);
   PointToLine := Distance(u,v,x0,y0);
end
```

Now consider the problem of finding the distance from P_0 to the *line segment* that connects P_1 and P_2. The value of t_* given above can be used to solve this problem. If $t_* \leq 0$, then P_1 is the closest point. If $t_* \geq 1$ then P_2 is the closest point. Otherwise, the closest point to the line through P_1 and P_2 is on the connecting line segment. The three cases are depicted in FIGURE 9.3. These observations lead to the following distance function:

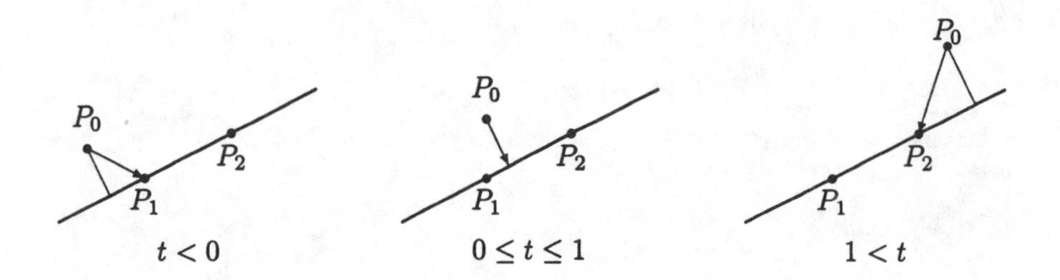

FIGURE 9.3 *Distance from a Point to a Line Segment*

```
function PointToLineSeg(x0,y0,x1,y1,x2,y2:real):real;
{Pre:(x1,y1) and (x2,y2) are distinct points.}
{Post:Distance from (x0,y0) to line segment from (x1,y1) to (x2,y2).}
var
   tStar:real; {The ''time'' associated with the closest point.}
   u,v:real; {(u,v) = point on line segment closest to (x0,y0).}
begin
   tStar := ((x0-x1)*(x2-x1) + (y0-y1)*(y2-y1))/(sqr(x2-x1)+sqr(y2-y1));
   if tstar<=0 then
      begin
         u := x1;
         v := y1;
      end
   else if tstar >=1 then
      begin
         u := x2;
         v := y2;
      end
      else
         begin
            u := x1 + tStar*(x2-x1);
            v := y1 + tStar*(y2-y1);
         end;
   PointToLineSeg := Distance(u,v,x0,y0);
end
```

It is possible to do away with the variables u and v:

```
   tStar := ((x0-x1)*(x2-x1) + (y0-y1)*(y2-y1))/(sqr(x2-x1)+sqr(y2-y1));
   if tstar<=0 then
      PointToLineSeg := Distance(x1,y1,x0,y0);
   else if tstar >=1 then
      PointToLineSeg := Distance(x2,y2,x0,y0);
   else
      PointToLineSeg := Distance(x1 + tStar*(x2-x1), y1 + tStar*(y2-y1),x0,y0);
```

However, in the problem solving process, the point (u, v) has a central role to play. This is obscured in the cryptic implementation. The program **Example8_3** illustrates both of these distance functions.

The point-to-line and point-to-line-segment problems illustrate the difference between unconstrained and constrained *optimization.* In an optimization problem one sets out to minimize (or maximize) the value of some *objective function.* In practical problems this might be a function that defines the potential energy of a physical system or a function that specifies the cost of some manufactured product. For us, it is the function $d(t)$. In the point-to-line problem, t is unconstrained and we seek

$$d_{min} = \min_{-\infty<t<\infty} \sqrt{(tx_1 + (1-t)x_2 - x_0)^2 + (ty_1 + (1-t)y_2 - y_0)^2}.$$

```
program Example9_2;
{Distance from the vertices of the unit Hexagon}
{to a given line and line segment.}
uses
   Chap9Codes;
const
   n = 6; {Number of sides in the polygon.}
var
   AnotherEg:char; {Continuation indicator.}
   k:integer;
   x0,y0:real; {(x0,y0) = k-th vertex.}
   x1,y1,x2,y2:real; {(x1,y1) and (x2,y2) define the line}
   d1,d2:real; {Distance to line and line segment.}
begin
   ShowText;
   AnotherEg := 'y';
   while AnotherEg <> 'n' do
      begin
         writeln('Enter x1 and y1:'); readln(x1,y1);
         writeln('Enter x2 and y2:'); readln(x2, y2);
         writeln; writeln(' x0 y0 Line Line Segment');
         writeln('--------------------------------------');
         for k := 0 to n - 1 do
            {Compute distances to k-th vertex.}
            begin
               x0 := cos(2*pi*k/n); y0 := sin(2*pi*k/n);
               d1 := PointToLine(x0,y0,x1,y1,x2,y2);
               d2 := PointToLineSegment(x0,y0,x1,y1,x2,y2);
               writeln(x0:8:3, y0:8:3, d1:8:3, d2:12:3);
            end;
         writeln; writeln('Another Example?  Enter y (yes) or n (no).');
         readln(AnotherEg)
      end
end.
```

The functions `PointToLine` and `PointToLineSegment` are declared in the unit `Chap9Codes`. Sample output:

```
Enter x1 and y1:
2   0
Enter x2 and y2:
2   1

      x0      y0    Line  Line Segment
  -----------------------------
   1.000   0.000   1.000         1.000
   0.500   0.866   1.500         1.500
  -0.500   0.866   2.500         2.500
  -1.000  -0.000   3.000         3.000
  -0.500  -0.866   2.500         2.646
   0.500  -0.866   1.500         1.732
```

In the point-to-line-segment-line problem, t is constrained to lie in the interval [0,1] and we compute

$$d_{min} = \min_{0<t<1} \sqrt{(tx_1 + (1-t)x_2 - x_0)^2 + (ty_1 + (1-t)y_2 - y_0)^2}.$$

The solution of constrained and unconstrained optimization problems is central to computational science. See Chapter 23.

PROBLEM 9.4. Write a function `PointToTriangle(x0,y0,x1,y1,x2,y2,x3,y3:real):real` that computes the distance from (x_0, y_0) to the nearest point on the boundary of the triangle defined by (x_1, y_1), (x_2, y_2), and (x_3, y_3). Make use of `PointToLineSegment`.

9.2 Inclusion

Many mathematical questions have "yes-no" answers. Consider whether or not a number z is in an interval $[a, b]$. If the boolean expression

```
(a <= z) and (z <= b)
```

is true, then $z \in [a, b]$. Boolean variables are sometimes handy for recording the results of true-false tests. If `InAB` is boolean then

```
InAB := (a <= z) and (z <= b)
```

records "for future use", the result of our interval inclusion test.

Just as arithmetic expressions can be rearranged into equivalent forms, so can boolean expressions. In our example, we know that z is *not* in the interval if z is to the left of a *or* if z is to the right of b. It follows that we may also write

```
InAB := not ((z < a) or (b < z))
```

The two equivalent expressions illustrate *de Morgan's law.* This famous rule states that if $b1$ and $b2$ are boolean expressions, then `not (b1 or b2)` is equivalent to `(not b1) and (not b2)`. In our example, $b1 = (z < a)$ and $b2 = (b < z)$.

Boolean-valued functions are essential for the packaging of more complicated logical tests. Here is a function that can be used to indicate whether or not a point is inside a rectangle:

```
function InRectangle(a,b,c,d,u,v:real):boolean;
{Pre:a<=b and c<=d.}
{Post:(u,v) is in the rectangle with vertices}
{(a,c),(b,c),(a,d),(b,d).}
var
   uInAB,vInCD:boolean;
begin
   uInAB := (a <= u) and (u <= b);
   vInCD := (c <= v) and (v <= d);
   InRectangle := uInAB and vInCD;
end
```

Note the use of the boolean variables **uInAB** and **vInCD** to "hold" our observations about u's location relative to $[a,b]$ and v's location relative to $[c,d]$. Of course, we could dispense with these two local variables altogether and write

```
InRectangle := (a <= u) and (u <= b) and (c <= v) and (v <= d)
```

However, the two-dimensional nature of the checking process is lost in this long expression and so we prefer the original formulation.

From the standpoint of minimizing the required number of comparisons, it is better to organize the body of InRectangle as follows:

```
InRectangle:=false;
if a<=u then
   if u<=b then
      {a<=u<=b}
      if c<=v then
         if v<=d then
            {a<=u<=b and c<=v<=d}
            InRectangle:=true;
```

Notice the use of assertions to facilitate the reading of the nested if's. This implementation is an example of an "innocent until proven guilty" programming style. We "tentatively" assume that (u,v) is not in the rectangle and "change our mind" if the four conditions are satisfied.

A function like InRectangle can be used in any situation where a boolean value is expected. Thus,

```
if InRectangle(x1,x2,y1,y2,3.3,1.2) then
   writeln('Inside')
else
   writeln('Outside');
```

prints "Inside" if (3.3,1.2) is inside the rectangle

$$R = \{(x,y) : x_1 \leq x \leq x_2,\ y_1 \leq y \leq y_2\}$$

and "Outside" if not.

It is possible for one boolean-valued function to call another. Thus, if

```
function InInterval(x,a,b:real):boolean;
{Pre:a<=b}
{Post:x is in [a,b].}
begin
   InInterval := (a<=x) and (x<=b)
end
```

has been declared, then InRectangle can be implemented as follows:

```
function InRectangle(a,b,c,d,u,v:real):boolean;
{Pre:a<=b and c<=d.}
{Post:(u,v) is in the rectangle with vertices}
{(a,c),(b,c),(a,d),(b,d).}
begin
   InRectangle := InInterval(u,a,b) and InInterval(v,c,d);
end
```

An ability to encapsulate boolean computations in a boolean function is as important as being able to package numerical computations in a real or integer valued function.

As an illustration, let's write a boolean-valued function that determines whether or not two chords on a semicircle intersect. See FIGURE 9.4. The points A, B, C, and D may be specified

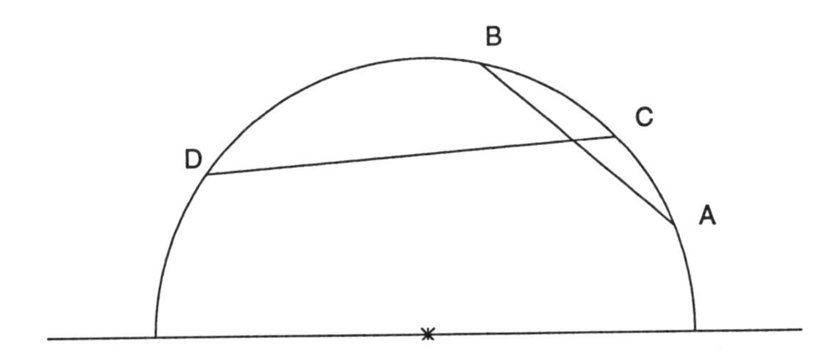

FIGURE 9.4 *Intersecting Chords*

by angles a, b, c, and d:

$$\begin{aligned} A &= (\cos(a), \sin(a)) \\ B &= (\cos(b), \sin(b)) \\ C &= (\cos(c), \sin(c)) \\ D &= (\cos(d), \sin(d)) \end{aligned}$$

Assuming that $0 \leq a \leq b \leq \pi$ and $0 \leq c \leq d \leq \pi$, we see that intersection occurs if A is in between C and D and B is not:

```
((c <= a) and (a <= d)) and (d < b)
```

or if B is in between C and D and A is not,

```
((c <= b) and (b <= d)) and (a < c)
```

Thus, we obtain the following function:

```
function ChordIntersect(a,b,c,d:real):boolean;
{Pre:0<=a<=b<=pi and 0<=c<=d<=pi}
{Post:The line segment from (cos(a),sin(a)) to (cos(b),sin(b)) intersects}
{the line segment from (cos(c),sin(c)) to (cos(d),sin(d)).}
var
   BToRight,AToLeft:boolean; {The two situations for intersection.}
begin
   BToLeft := ((c <= a) and (a <= d)) and (d < b)
   AToRight := ((c <= b) and (b <= d)) and (a < c)
   ChordIntersect := BToLeft or AToRight;
end;
```

Notice the use of the local boolean variables BToLeft and AToRight. They have a clarifying influence because they reflect the fact that there are two ways for intersection to occur.

With an advanced expertise in mathematics and logic, it is possible to *prove* the correctness of ChordIntersect. However, even without that expertise it is possible to build confidence in the function by applying it to a wide range of examples. Example9_3 presents a testing framework for this purpose. The four angles are chosen randomly. The parameters of the randomization were determined (through experimentation) to ensure that all possible cases are "hit" during the simulation.

PROBLEM 9.5. Complete the following functions:

```
function Contain(a,b,c,d:real):boolean;
{Pre:a<=b, c<=d.}
{Post:[a,b] is contained in [c,d].}

function Overlap(a,b,c,d:real):boolean;
{Pre:a<=b, c<=d.}
{Post:c is in [a,b] and d is not.}

function Empty(a,b,c,d:real):boolean;
{Pre:a<=b, c<=d.}
{Post:No point in [a,b] is in [c,d].}

function Include(a,b,c,d:real):boolean;
{Pre:a<=b, c<=d.}
{Post:Either [a,b] is in [c,d] or [c,d] is in [a,b]}
```

PROBLEM 9.6. Complete the following function:

```
function Intersect(a1,b1,s1,a2,b2,s2:real):boolean;
{Pre:s1>=0,s2>=0.}
{Post:The square with vertices (a1,b1), (a1+s1,b1),}
{(a1+s1,b1+s1), and (a1,b1+s1) intersects the square with}
{vertices (a2,b2), (a2+s2,b2), (a2+s2,b2+s2), and (a2,b2+s2).}
```

PROBLEM 9.7. Complete the following function:

```
function CircleIntersect(a1,b1,r1,a2,b2,r2:real):boolean;
{Pre:r1>=0, r2>=0.}
{ Post:The circle with center (a1,b1) and radius r1}
{intersects the circle with center (a2,b2) and radius r2.}
```

PROBLEM 9.8. Complete the following function:

```
function Corner(a,b,c,d,x,y:real):boolean;
{Pre:a<=b, c<=d.  }
{Post:Let R be the rectangle with vertices (a,c), (b,c), (a,d), (b,d).}
{The closest point in R to (x,y) is one of R's vertices.}
```

PROBLEM 9.9. Complete the following function:

```
function IsTriangle(a,b,c:real):boolean;
{Post:It is possible to form a triangle with sides}
{that have length a, b, and c.}
```

```
program Example9_3;
{Chord Intersection}
uses
   DDcodes, Chap9Codes;
const
   seed = 0.123456;
   n = 20; {Number of trials.}
   rad = 150;
   xc = 200;
   yc = 200; {Use upper semicircle with radius rad and center (xc,yc).}
var
   r:real;
{Next random number.}
   a,b,c,d:real; {Angles that define the four points.}
   k:integer;
begin
   r := seed;
   for k := 1 to n  do
      begin
         ShowDrawing;
         {Generate a,b,c, and d.}
         r := rand(r); a := r*pi/2;
         r := rand(r); b := a + 0.2*pi + r*(0.8*pi-a);
         r := rand(r); c := r*pi/2;
         r := rand(r); d := c + 0.2*pi + r*(0.8*pi-c);
         {Depict the semicircle and the two chords.}
         FrameOval(yc - rad, xc - rad, yc + rad, xc + rad);
         EraseRect(yc, xc - rad, yc + rad, xc + rad);
         MoveTo(round(xc + rad * cos(a)), round(yc - rad * sin(a)));
         LineTo(round(xc + rad * cos(b)), round(yc - rad * sin(b)));
         MoveTo(round(xc + rad * cos(c)), round(yc - rad * sin(c)));
         LineTo(round(xc + rad * cos(d)), round(yc - rad * sin(d)));
         MoveTo(160,10);
         {Perform the test.}
         if ChordIntersect(a,b,c,d)  then
            WriteDraw('Intersect')
         else
            WriteDraw('Do Not Intersect');
         Wait;
         ClearScreen;
   end;
end.
```

PROBLEM 9.10. A triangle is Pythagorean if the square of one side is the sum of the squares of the two other sides. Complete the following function:

```
function IsPythagorean(a,b,c:integer):boolean;
{Post:It is possible to form a Pythagorean triangle with sides}
{that have length a, b, and c.}
```

9.3 Collinearity

Boolean-valued functions are usually "not the way to go" in settings where fuzzy data and inexact arithmetic prevail. Consider the following definition:

Three distinct points A, B, and C are collinear if they lie on the same line.

One way to test for this situation is to use the function `PointToLine`. If the distance is zero from one point to the line defined by the other two points, then the three points are collinear.

```
function IsCollinear1(x1,y1,x2,y2,x3,y3:real):boolean;
{Pre:(x1,y1), (x2,y2), and (x3,y3) are distinct.}
{Post:(x1,y1),(x2,y2),and (x3,y3) are collinear.}
begin
   IsCollinear1 := (PointToLine((x1,y1,x2,y2,x3,y3) = 0);
end
```

This looks reasonable enough, but there is a definite problem having to do with the stringency of the test. A simple example makes the point. The value of `IsCollinear1(0,0,1/3,1,1,3)` is false because the computed version of $(1/3, 1)$, is not on the line through (0,0) and (1,3). (The floating point divide 1/3 will produce something like .333333..)

Clearly, we need a "fuzzy" definition of collinearity that is consistent with the inexactitudes of floating point arithmetic. One possibility is to regard the three points as collinear if the value returned by `PointToLine(x1,y1,x2,y2,x3,y3)` is small as depicted in FIGURE 9.5. This

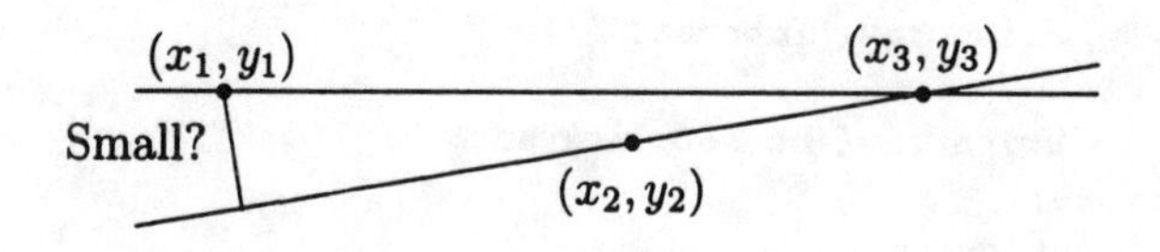

FIGURE 9.5 *Testing for Near-Collinearity*

suggests that we test for collinearity with a function such as `IsCollinear2`:

```
function IsCollinear2(x1,y1,x2,y2,x3,y3:real):boolean;
{Pre:(x1,y1), (x2,y2), and (x3,y3) are distinct.}
{Post:(x1,y1),(x2,y2), and (x3,y3) are collinear.}
const
   tol := 0.005;
begin
   IsCollinear2 := (PointToLine((x1,y1,x2,y2,x3,y3) <= tol);
end
```

In effect, this implementation says that the points are "nearly collinear" if (x_1, y_1) is "nearly" on the line defined by (x_2, y_2) and (x_3, y_3). The value of the constant `tol` is used to quantify the definition of "nearly."

Unfortunately, `IsColinear2` is flawed because it does not treat the three points symmetrically. For example, FIGURE 9.5 depicts a situation where (let's pretend) (x_1, y_1) is not close (in the tol sense) to the line defined by (x_2, y_2) and (x_3, y_3). On the other hand, it may well be that (x_2, y_2) is close to the line through (x_1, y_1) and (x_3, y_3). In that case, `IsCollinear2(x2,y2,x1,y1,x3,y3)` would be `true`. To avoid his problem and to ensure that all three points "participate equally" in the decision process, we can insist that each point be close to the line defined by the other two points.

```
function IsCollinear3(x1,y1,x2,y2,x3,y3:real):boolean;
{Pre:(x1,y1), (x2,y2), and (x3,y3) are distinct.}
{Post:(x1,y1),(x2,y2), and (x3,y3) are colinear.}
const
   tol := 0.005;
var
   Check1,Check2,Check3:boolean;
begin
   Check1 := IsCollinear2(x1,y1,x2,y2,x3,y3);
   Check2 := IsCollinear2(x2,y2,x1,y1,x3,y3);
   Check3 := IsCollinear2(x3,y3,x1,y1,x2,y2);
   IsColinear3 := Check1 and Check2 and Check3;
end
```

However, we *still* have a problem. `IsCollinear3` identifies as colinear any three points that are sufficiently close together. In particular, if the three points fit inside a circle with diameter `tol`, then `IsCollinear3` is `true` *even if the three points define an equilateral triangle!.*

The moral of this discussion is that we must abandon our distance-based approach to collinearity and work instead with the angles defined by the three points. Note that if the sine of any of three angles is small, then the points are nearly collinear. If a triangle has sides a, b, and c, then the three sines are given by

$$\begin{aligned}\sin(\theta_{ab}) &= 2\sqrt{h(h-a)(h-b)(b-c)}/ab\\ \sin(\theta_{ac}) &= 2\sqrt{h(h-a)(h-b)(b-c)}/ac\\ \sin(\theta_{bc}) &= 2\sqrt{h(h-a)(h-b)(b-c)}/bc\end{aligned}$$

See FIGURE 9.6. We therefore obtain

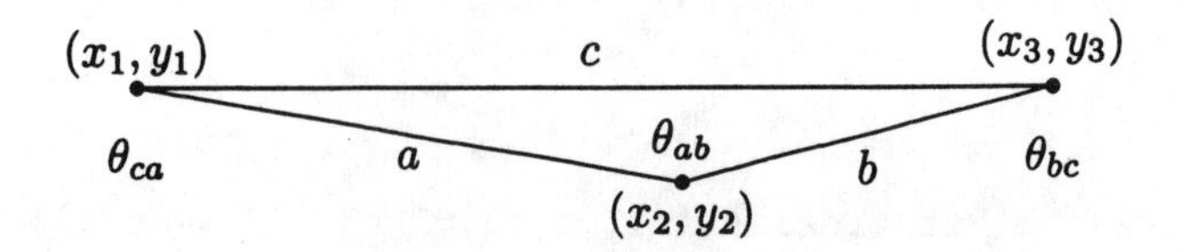

FIGURE 9.6 *Angle-Based Colinearity Testing*

```
function IsCollinear(x1,y1,x2,y2,x3,y3:real):boolean;
{Pre:(x1,y1),(x2,y2), and (x3,y3) are distinct.}
{Post:(x1,y1),(x2,y2), and (x3,y3) are collinear.}
const
   tol = 0.005;
var
   a,b,c:real;{The sides of the triangle.}
   h:real; {(a+b+c)/2}
   sab,sbc,sca:real;{The sines of the three angles,}
   TwoArea:real; {Twice the area of the triangle.}
begin
   a := Distance(x1,y1,x2,y2);
   b := Distance(x2,y2,x3,y3);
   c := Distance(x3,y3,x1,y1);
   h := (a+b+c)/2;
   TwoArea := 2*sqrt(h*(h-a)*(h-b)*(h-c));
   sab := TwoArea/(a*b);
   sbc := TwoArea/(b*c);
   sca := TwoArea/(c*a);
   IsCollinear := (sab < tol) or (sbc < tol) or (sca < tol);
end
```

The definition of near-collinearity embodied by this function is distance-independent and is immune to the difficulties that plague the other implementations. The program Example9_4 illustrates its use.

```
program Example9_4;
{Alignment}
uses
   Chap9Codes;
var
   k:integer; {Index of the day.}
   xA,yA:real; {(xA,yA) = location of Planet A on day k.}
   xB,yB:real; {(xB,yB) = location of Planet B on day k.}
begin
   ShowText;
   writeln('Collinear Days');
   writeln('-----------------');
   for k := 1 to 2000 do
      {Is there approximate alignment of sun (0,0), (xA,yA), and (xB,yB)?}
      begin
         xA := 95*cos(2*pi*k/365);
         yA := 91*sin(2*pi*k/365);
         xB := 130*cos(2*pi*k/687);
         yB := 124*sin(2*pi*k/687);
         if IsCollinear(0,0,xA,yA,xB,yB)  then
            writeln(k:5);
      end;
end.
```

The function IsCollinear is declared in the unit Chap9Codes. Output:

```
Collinear Days
-----------------
              388
              389
              390
              778
              779
             1167
             1168
             1169
             1557
             1558
             1946
             1947
             1948
```

There still is the sticky problem of setting the tolerance. The appropriate value may be application dependent. For example, if the points in question are experimentally determined and accurate to only two significant digits, then `tol = 0.01` may be appropriate. In a setting like this, it may be preferable to let the calling program set the tolerance value:

```
function HowCollinear(x1,y1,x2,y2,x3,y3:real):real;
{Pre:(x1,y1),(x2,y2), and (x3,y3) are distinct.}
{Post:Sine of the smallest angle in the triangle defined by.}
{(x1,y1),(x2,y2),(x3,y3)}
var
   a,b,c:real;{The sides of the triangle.}
   h:real; {(a+b+c)/2}
   sab,sbc,sca:real;{The sines of the three angles,}
   TwoArea:real; {Twice the area of the triangle.}
   MinSine:real; {The minimum of sab, sbc, and sca.}
begin
   a := Distance(x1,y1,x2,y2);
   b := Distance(x2,y2,x3,y3);
   c := Distance(x3,y3,x1,y1);
   h := (a+b+c)/2;
   TwoArea := 2*sqrt(h*(h-a)*(h-b)*(h-c));
   sab := TwoArea/(a*b);
   sbc := TwoArea/(b*c);
   sca := TwoArea/(c*a);
   MinSine:= sab;
   if sbc<MinSine then
      MinSine := sbc;
   if sca<MinSine then
      MinSine:= sca
   HowCollinear := MinSine;
end
```

This function returns the numerical value that determines the boolean value returned by `IsCollinear`.

PROBLEM 9.11. A triangle with sides a, b, and c is *nearly equilateral* if $|a-\mu| \leq tol$, $|b-\mu| \leq tol$, and $|c-\mu| \leq tol$ where $\mu = (a+b+c)/3$. Write a boolean-valued function that tests for this property.

PROBLEM 9.12. Based upon the kind of reasoning behind `IsCollinear`, develop the following "nearness" functions:

```
function NearEqual(x1,y1,x2,y2,x3,y3,x4,y4:real):boolean;
{Post:The line segment from (x1,y1) to (x2,y2) is nearly equal to}
{the length of the line segment from (x3,y3) to (x4,y4).}

function NearParallel(x1,y1,x2,y2,x3,y3,x4,y4:real):boolean;
{Pre:(x1,y1) and (x2,y2) are distinct and (x3,y3) and (x4,y4) are distinct.}
{Post:The line through (x1,y1) and (x2,y2) is nearly parallel to the line}
{through (x3,y3) and (x4,y4).}
```

Using `NearEqual` and `NearParallel` as required, complete the following functions:

```
function NearTrapezoid(x1,y1,x2,y2,x3,y3,x4,y4:real):boolean;
{Pre:P = (x1,y1), Q = (x2,y2), R = (x3,y3), and S = (x4,y4)}
{are distinct and line segment PR intersects line segment QS.}
{Post:quadrilateral PQRS has at least one pair of nearly}
{parallel sides.}
```

```
function NearParallelogram(x1,y1,x2,y2,x3,y3,x4,y4:real):boolean;
{Pre:P = (x1,y1), Q = (x2,y2), R = (x3,y3), and S = (x4,y4)}
{are distinct and line segment PR intersects line segment QS.}
{Post:The quadrilateral PQRS has two pairs of nearly}
{parallel sides.}

function NearRhombus(x1,y1,x2,y2,x3,y3,x4,y4:real):boolean;
{Pre:P=(x1,y1), Q=(x2,y2), R=(x3,y3), S=(x4,y4) are distinct.}
{and the line segment PR does not intersect line segment QS.}
{Post:The quadrilateral PQRS is a parallelogram with 4 equal sides.}

function NearSquare(x1,y1,x2,y2,x3,y3,x4,y4:real):boolean;
{Pre:P= (x1,y1), Q = (x2,y2), R = (x3,y3), S = (x4,y4) are distinct }
{and the line segment PR does not intersect line segment QS.}
{Post:The quadrilateral PQRS is a rhombus with diagonals of equal length}
```

Chapter 10

Roots

Our first experiences with root-finding are typically with "easy" functions that permit exact, closed-form solutions like the quadratic equation:

$$ax^2 + bx + c = 0 \quad \Rightarrow \quad x = \frac{-b \pm \sqrt{b^2 - 4ac}}{2a},\ a \neq 0$$

However, even the implementation of such a math book formula involves interesting computational issues.

In practical problems we are rarely able to express roots in closed form and this pushes us once again into the realm of the approximate. Just as we had to develop the notion of approximate collinearity to make "computational progress" in §9.3, so must we develop the notion of an approximate root. Two definitions are presented and exploited in the methods of bisection and Newton that we develop. The discussion of these implementations lead to some larger software issues. For example, these root finders expect the underlying function to be specified in a certain way. This may require the "repackaging" of an existing implementation that does not meet the required specification. The exercise of modifying "your" software so that it can interact with "someone else's" software is typical in computational science, where so many techniques are embodied in existing software libraries.

We use the development of a modest Newton method root-finder to dramatize the difference between a math book implementation of a formula and a finished, usable piece of software. It is absolutely essential for the computational scientist to appreciate the difficulties associated with software development.

10.1 Quadratic Equations

An equation of the form

$$q(x) = ax^2 + bx + c = 0$$

with $a \neq 0$ is called a *quadratic equation.* Quadratic equations have two roots:

$$r_1 = \frac{-b + \sqrt{b^2 - 4ac}}{2a} \qquad r_2 = \frac{-b - \sqrt{b^2 - 4ac}}{2a}.$$

These roots are complex if the *discriminant* $d = b^2 - 4ac$ is negative. We postpone the discussion of this case until later.

Restricting our attention to quadratics that have real roots, here is a function that can be used to compute the larger of the two real roots:

```
function LargerRoot(a,b,c:real):real;
{Pre:a<>0, b*b - 4ac >= 0.}
{Post:The larger root (in absolute value) of ax*x + bx + c = 0.}
var
   r1,r2:real;{The two roots.}
   d:real; {The discriminant d = b*b-4ac.}
begin
   d := sqrt(sqr(b) - 4*a*c);
   r1 := (-b+d)/(2*a);
   r2 := (-b-d)/(2*a);
   if abs(r1) > abs(r2) then
      LargerRoot := r1
   else
      LargerRoot := r2;
end
```

The function mechanism works because only one value needs to be returned. What if we wanted to encapsulate the entire root-finding process by returning *both* roots? To do this we must use a procedure with a new type of parameter called a *variable parameter.*

```
procedure QuadRoots(a,b,c:real; var r1,r2:real);
{Pre:a<>0, b*b - 4ac >= 0.}
{Post:r1 and r2 are the roots of ax*x + bx + c = 0.}
var
   d:real; {The discriminant d = b*b-4ac.}
begin
   d := sqrt(sqr(b) - 4*a*c);
   r1 := (-b+d)/(2*a);
   r2 := (-b-d)/(2*a);
end
```

The "var" that precedes `r1` and `r2` identifies those two parameters as variable parameters. Variable parameters are used to communicate values back to the calling program. `QuadRoots` uses `r1` and `r2` to transmit the value of the two roots. A call to a procedure that has variable parameters initiates a new kind of substitution process that we now discuss using `Example10_1`. Let us assume that the values 2, -3, and 1 are read into the coefficients `a0`, `a1` and `a2`. With the context box notation developed in §7.1, we have

```
program Example10_1;
{Illustrate quadratic equation solving.}
uses
   Chap10Codes;
var
   a0,a1,a2:real; {Coefficients of the quadratic.}
   root1,root2:real; {Roots of a0 + a1*x + a2*x*x.}
begin
   ShowText;
   writeln('Enter coefficients a0,a1,a2 of q(x) = a0 + a1*x + a2*x^2:');
   readln(a0,a1,a2);
   QuadRoots(a2,a1,a0,root1,root2);
   writeln;
   writeln(' root q(root)');
   writeln('-----------------------------');
   writeln(root1:10:6, (a2*sqr(root1) + a1*root1 + a0):15:6);
   writeln(root2:10:6, (a2*sqr(root2) + a1*root2 + a0):15:6);
end.
```

The procedure `QuadRoots` is declared in the unit `Chap10Codes`. Sample output:

```
Enter coefficients a0,a1,a2 of q(x) = a0 + a1*x + a2*x^2 :
2     -3     1

   root      q(root)
 ---------------------
 2.000000    0.000000
 1.000000    0.000000
```

main
a0 2
a1 -3
a2 1
root1 ___
root2 ___

At the time of the call `QuadRoots(a2,a1,a0,root1,root2)`, the three value parameters `a`, `b`, and `c` are assigned the values of `a2`, `a1`, and `a0` respectively. No values are assigned to the variable parameters `r1` and `r2`. Instead, the actual parameters `root1` and `root2` are *identified* with `r1` and `r2`. We use arrows to depict the connection:

main
QuadRoots
a0 2 — 2 a
a1 -3 — -3 b
a2 1 — 1 c
root1 ___ ← r1
root2 ___ ← r2
___ d

Control passes to the procedure `QuadRoots` which we now regard as the current context. The local variable `d` is the object of the first assignment

```
d := sqrt(sqr(b) - 4*a*c);
```

Since $\sqrt{3^2 - 4 \cdot 1 \cdot 2} = 1$ we obtain

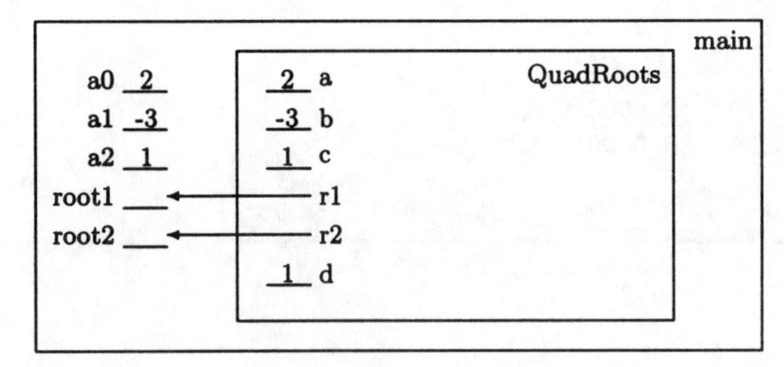

The next statement looks like an assignment to a variable called `r1`:

```
r1 := (-b+d)/(2*a);
```

However, `r1` is just a formal variable parameter. In this call to `QuadRoots`, it is identified with the main program variable `root1`. (Observe the arrow.) This means that all references to `r1` in the current context are, in effect, references to `root1`. Thus, the assignment `r1 := (-b+d)/(2*a)` changes the value of the main program variable `root1`:

main
QuadRoots
a0 2 2 a
a1 -3 -3 b
a2 1 1 c
root1 2 r1
root2 r2
1 d

Likewise, the statement `r2 := (-b-d)/(2*a)` results in an assignment to the main program variable `root2`:

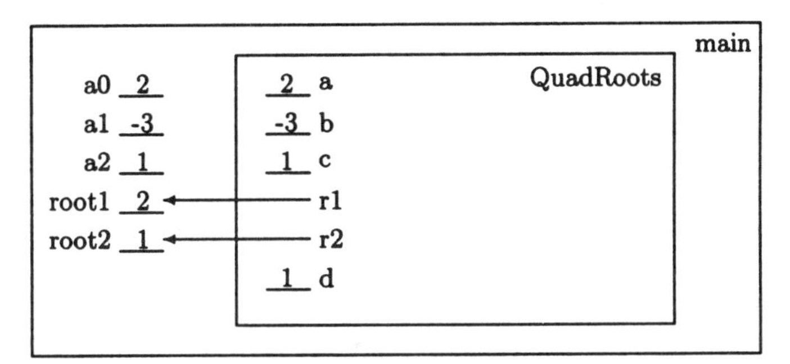

At this point the execution of the procedure ends and control passes back to the main program with `root1` and `root2` housing the two root values:

main
a0 2
a1 -3
a2 1
root1 2
root2 1

The box associated with the procedure call is closed and the main program is once again the current context. The roots, now resident in the variables `root1` and `root2`, are printed.

Here some issues to bear in mind when writing and using procedures that have variable parameters. (1) Make sure that the name of a calling program variable is used for the actual parameters that correspond to `var` parameters and check for type agreement. (2) It is legal to write a procedure that has a single variable parameter. However, in this case it is usually preferable to write the procedure as a function. (3) Variable parameters of the same type may by grouped in the procedure heading, e.g.,

```
...;var r1,r2:real;...
```

However, it is also legal to write

```
...;var r1:real; var r2:real;...
```

(4) Forgetting to put the "var" in front of an intended variable parameter is not a syntactic error. However, with such a mistake the parameter in question becomes a value parameter and the procedure call will behave accordingly. In particular, the procedure will not alter the value of the variable passed in that position. (5) Most of the time it is useful to think of value parameters and variable parameters as input and output parameters. Usually, the value parameters provide the procedure with a set of input values. These are manipulated by the procedure to produce the output values which are communicated to the calling program through the **var** parameters. However, we shall encounter examples later on where a **var** parameter is used for both input and output. In such cases, it is not right to equate value parameters with input and variable parameters with output.

Now let us drop the assumption that the quadratic equation $ax^2 + bx + c = 0$ has real roots. If $b^2 - 4ac < 0$, then the roots are given by $r_1 = s + it$ and $r_2 = s - it$ where $s = -b/(2a)$ and $t = \sqrt{-d}/(2a)$, with $i^2 = -1$. By adding a boolean **var** parameter we can indicate whether or not the quadratic has real roots. If it does, then r1 and r2 can be used to send back the roots. Otherwise, we can use these two parameters to send back the real and imaginary parts of the roots:

```
procedure QuadRoots2(a,b,c:real;var r1,r2:real; var complex:boolean);
{Pre:a<>0.}
{Post:  If both roots of ax^2 + bx + c = 0 are real, then }
{ complex = false, and r1 and r2 are the roots.}
{ If the roots are complex, then complex = true, and r1 and r2 }
{ return the real and imaginary parts of the roots, i.e., the two}
{ roots are r1 + i*r2 and r1 - i*r2 where i = sqrt(-1).}
var
   d:real; {The discriminant.}
begin
   d:= sqr(b)-4*a*c;
   if d >= 0 then
      {Real root case.}
      begin
         complex:=false;
         d:= sqrt(d);
         r1 := (-b-d)/(2*a);
         r2 := (-b+d)/(2*a)
      end
   else
      {Complex root case}
      begin
         complex:=true;
         d := sqrt(-d);
         r1 := -b/(2*a);
         r2 := d/(2*a)
      end
end
```

```
program Example10_2;
{Quadratic equation solving.}
uses
   Chap10Codes;
var
   a0,a1,a2:real;{The coefficients.}
   root1,root2:real; {Root information.}
   flag:boolean; {Indicates complex or real roots.}
begin
   ShowText;
   writeln('Enter coefficients a0,a1,a2 of q(x) = a0 + a1*x + a2*x^2:');
   readln(a0,a1,a2);
   GenQuadRoots(a2,a1,a0,root1,root2,flag);
   if not flag then
      writeln('Real Roots:  ', root1:10:6, ',',root2:10:6)
   else
      writeln('Complex Roots:  ', root1:10:6, ' +or- ', root2:10:6,'i');
end.
```

The procedure **GenQuadRoots** is declared in the unit **Chap10Codes**. Sample output:

```
Enter coefficients a0,a1,a2 for q(x) = a0 + a1*x + a2*x^2 :
25   -6   1
Complex Roots: 3.000000 +or- 4.000000i
```

A boolean parameter like `complex` is sometimes called a *boolean flag*. `Example10_2` shows how a calling program can use such a variable.

PROBLEM 10.1. Modify `Example10_1` so that it keeps soliciting the three coefficients the conditions `a2 <>0` and `sqr(a1) >= a2*a0` are satisfied.

PROBLEM 10.2. Modify `Example10_2` so that it also prints the value of the quadratic at each computed root. In the complex case, this means printing the real and imaginary parts of the quadratic evaluated at the two roots.

PROBLEM 10.3. By equating coefficients in $ax^2 + bx + c = a(x - r_1)(x - r_2)$ and manipulating the results, we obtain the following recipes:

$$r_1 = \frac{2c}{-b - \sqrt{b^2 - 4ac}} \qquad r_2 = \frac{2c}{-b + \sqrt{b^2 - 4ac}}$$

Write a function `SmallRoot(p,q:real):real` that returns the smaller root (in absolute value) of the quadratic $x^2 + 2px - q = 0$ where p and q are positive real numbers. Use the appropriate root formula above.

PROBLEM 10.4. Consider problem of computing the minimum value of a cubic polynomial

$$q(x) = ax^3 + bx^2 + cx + d \qquad a \neq 0$$

in the interval $[L, R]$. The zeros r_1 and r_2 of its derivative

$$q'(x) = 3ax^2 + 2bx + c$$

are involved in the computation. Recall that if $q'(r) = 0$ and $q''(r) >= 0$, then $q(x)$ has a local minimum at $x = r$. Thus, to find the smallest value of $q(x)$ on $[L, R]$, we must compare $q(L)$, $q(R)$, and the value of q at any local minimum that is inside $[L, R]$. The following sequence of checks does the job:

If the quadratic $q'(x)$ has complex roots, then
$q_{min} = \min\{q(L), q(R)\}$.
Else
{The quadratic $q'(x)$ has two real roots r_1 and r_2.}
If $r_1 \in [L, R]$ and $p''(r_1) \geq 0$ then
{r_1 is a local minimum in the interval}
$q_{min} = \min\{q(L), q(r_1), q(R)\}$.
Else
If $r_2 \in [L, R]$ and $q''(r_2) \geq 0$ then
{r_2 is a local minimum in the interval}
$q_{min} = \min\{q(L), q(r_2), q(R)\}$.
Else
{$q(x)$ has no critical points in the interval.}
$q_{min} = \min\{q(L), q(R)\}$.

Using these facts, complete the following procedure:

```
procedure CubicMin(a,b,c,d,L,R:real; var qmin,xmin:real);
{Pre:L <= R.}
{Post:qmin is the minimum value of the cubic q(x) = ax^3 + bx^2 = cx + d}
{on the interval [L,R] and xmin satisfies q(xmin) = qmin.}
```

10.2 The Method of Bisection

We now discuss the first of two methods that can be used to compute a zero of a general real-valued function $f(x)$. However, the boundaries of the "playing field" need to be redefined before we start because it is unreasonable to expect the production of a floating point number x_* so $f(x_*) = 0$. For example, $\sqrt{2}$ is a root for $f(x) = x^2 - 2$, but any floating point representation of this irrational number is approximate. This suggests the following definition:

Definition 1. Given a small positive number ϵ, we say that z is an approximate root of $f(x)$ if there is an exact root r so that $|z - r| \leq \epsilon$.

See FIGURE 10.1. Thus, if $f(x) = \sin(x)$ and $\epsilon = .001$, then $z = 3.1416$ is an approximate root.

A simple method that can be used to compute an approximate root is the method of *bisection*. Bisection is typical of a wide range of *divide and conquer* techniques that are characterized by a repeated halving of the "search space." Here is how it works. Suppose $f(x)$ is continuous on $[L, R]$ and that $f(L)f(R) < 0$. In this case we say that $[L, R]$ is a *bracketing interval* because f must have at least one root somewhere in between L and R as illustrated in FIGURE 10.2. Informally, it is impossible to draw the graph from $(L, f(L))$ to $(R, f(R))$ without crossing the x-axis.

Our goal is to find a zero of f on $[L, R]$. [1] Let *mid* be the midpoint of $[L, R]$. If $f(L)f(mid) \leq 0$, then we know that $[L, mid]$ has a root and we can pursue our search in that half-interval. On the other hand, if $f(L)f(mid) > 0$, then we know that $f(mid)f(R) <= 0$. From this we conclude that $[mid, R]$ contains a zero and we continue the search for a root on that half-interval. Repetition of this process is the idea behind the method of bisection. It produces a sequence of ever-shorter bracketing intervals. When their length is sufficiently small, then an endpoint of the final interval can be accepted as an "approximate zero."

Here is a fragment that does just this. It assumes that eps is positive and that the values in L and R define a bracketing interval.

[1] Of course, there may be more than one root in $[L, R]$. So, we cannot maintain that we have found *the the* root in $[L, R]$.

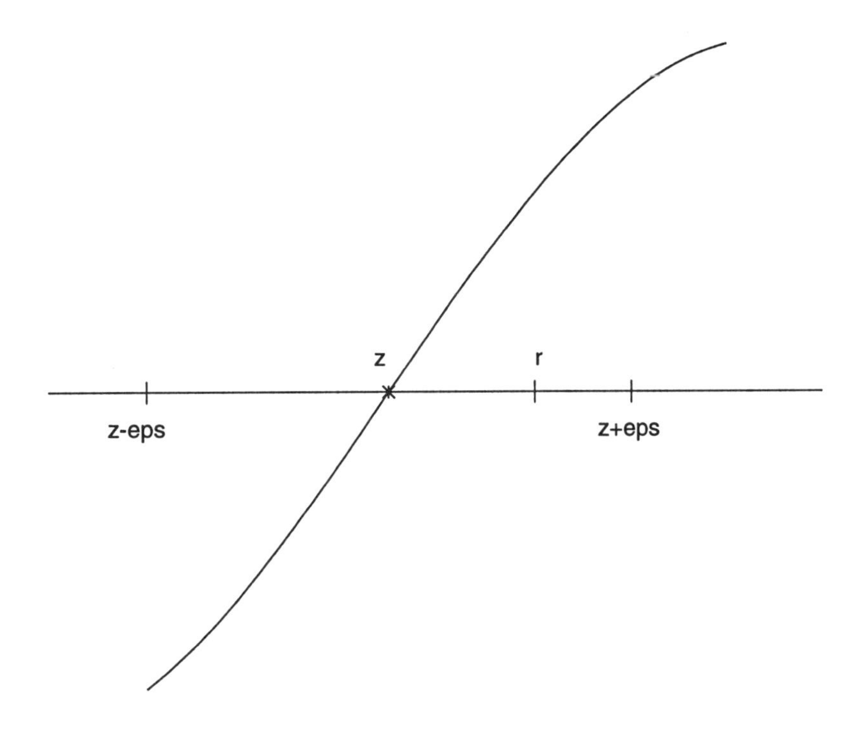

FIGURE 10.1 *An Approximate Root (Definition 1)*

```
while (R-L) > eps do
   {[L,R] is a bracketing interval.}
   begin
      mid :=(L+R)/2;
      if f(L)*f(mid) <= 0 then
         {Continue with interval [L,mid].}
         R:=mid
      else
         {Continue with interval [mid,R].}
         L:=mid;
   end;
root := L;
froot := f(root);
```

Upon termination we have an approximate root as depicted in FIG.10.1. Note that `root` is assigned a value that is within `eps` of a true zero of $f(x)$. Thus, the while-loop termination criteria is based upon *Definition 1* above.

The value of f at the approximate root is assigned to `froot`. The number of required iterations N is a function of `eps` and the length of the original bracketing interval. In particular, N is the smallest integer such that $d/2^N$ is less than or equal to the value of `eps` where d is the length of the original bracketing interval.

By saving the function evaluations in variables, we are able to reduce the number of function evaluations per iteration from two to one:

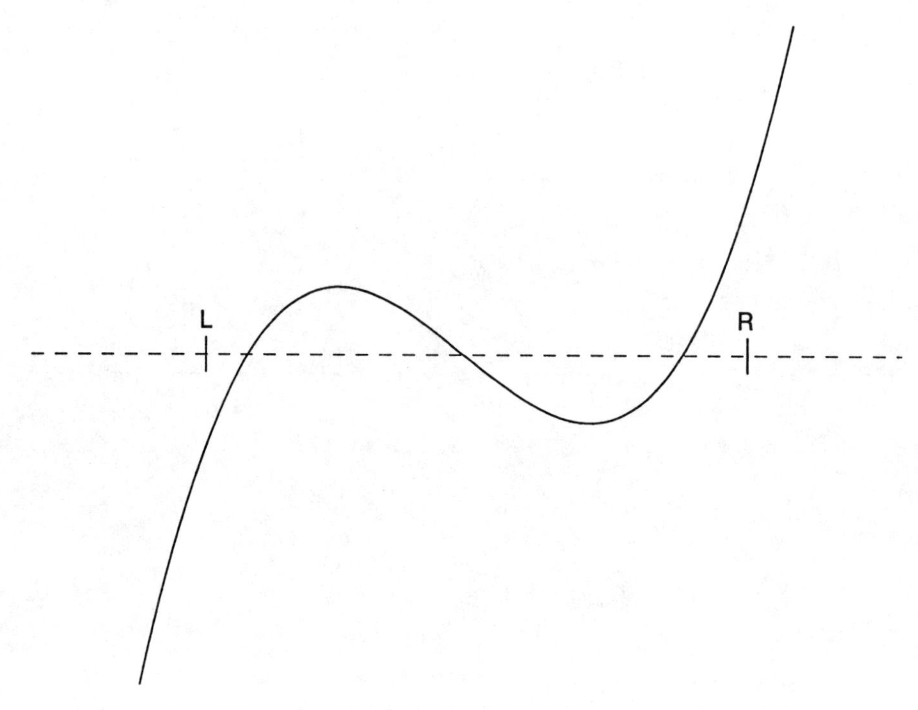

FIGURE 10.2 *Bracketing Interval*

```
fL:=f(L);
fR:=f(R);
while (R-L) > eps do
   {[L,R] is a bracketing interval, fL = f(L), fR = f(R).}
   begin
      mid:=(L+R)/2; fmid:=f(mid);
      if fL*fmid <= 0 then
         {Continue with interval [L,mid].}
         begin
            R:=mid; fR:=fmid
         end
      else
         {Continue with interval [mid,R].}
         begin
            L:=mid; fL:=fmid
         end
   end;
root := L; froot := fL;
```

This is an important saving because the time it takes to carry out the bisection process is typically dominated by the time it takes process the `f`-evaluations. (Imagine a function so complicated that takes the computer one hour to evaluate.)

Another point to bring up concerns the boolean expression `fL*fmid <= 0`. In many `real`-arithmetic comparisons, the choice between `<=` and `<` is usually unimportant. This is *not* the case here. If `fL` is zero, then

```
if fL*fmid < 0 then
   begin
      R:=mid;
      fR:=fMid
   end
else
   begin
      L:=mid;
      fL:=fMid
   end
```

establishes $[mid, R]$ as the "next" bracketing interval. Thus, the zero at L is abandoned and an infinite loop may result because $[mid, R]$ need not contain a root.

We now write a procedure `Bisection` that encapsulates the whole process. There are four inputs and two outputs:

`f`	:	The function.
`L`	:	The left endpoint of the initial bracketing interval.
`R`	:	The right endpoint of the initial bracketing interval.
`eps`	:	The tolerance that defines the quality of the approximate root.
`root`	:	The approximate root.
`froot`	:	The value of the function at the approximate root.

See FIGURE 10.3 This is a classic situation for a procedure, but it presents a new situation. Can

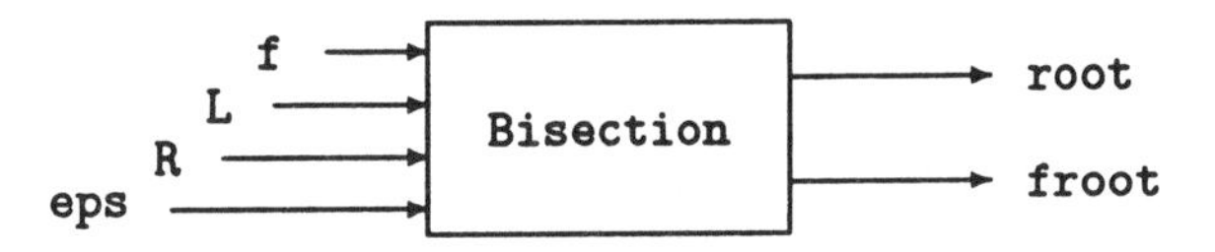

FIGURE 10.3 *Visualizing* `Bisection`

a procedure have a function as a parameter? The answer is "yes" and here is how it is handled:

```
procedure Bisection (function f(x:real):real;L,R:real;
                eps:real; var root,froot:real);
{Pre:f(x) is continuous on [L,R], f(L)f(R)<0, and eps > 0}
{Post:root is is within eps of a true root and froot = f(root).}
var
   mid:real; {[L,R] = current bracketing interval, mid = its midpoint.  }
   fL,fR,fmid:real; {value of f at L,R, and mid }
begin
   fL:=f(L);
   fR:=f(R);
```

```
    while (R - L) > eps do
      {[L,R] is a bracketing interval, fL = f(L), fR = f(R).}
      begin
        mid:=(L+R)/2; fmid:=f(mid);
        if fL*fmid < 0 then
          {Continue with interval [L,mid].}
          begin
            R:=mid;
            fR:=fmid
          end
        else
          {Continue with interval [mid,R].}
          begin
            L:=mid;
            fL:=fmid
          end
      end;
    root:=L;
    froot:=fL;
  end;
```

Note the complete typing of the functional parameter. Simply writing `function f` is not enough. `Bisection` "needs to know" that the function has a single real parameter and returns a real value. In general, if a formal parameter is a function, then at that position in the parameter list must appear a complete typing of the form

function ⟨ *Name* ⟩(⟨ *parameter list* ⟩):⟨*type*⟩

The program `Example10_3` shows how to use `Bisection`. It computes an approximate zero for two functions: $g(x) = \sin(x/4) - \cos(x/4)$ and $h(x) = e^x - 10$. A couple of syntactic details before we outline how the program works. First, notice that *just the name* of the actual function is given in a call. A reference of the form

```
Bisection(g(x),0,4,0.000001,MyPi,gval)
```

is incorrect. Recognize also that we could rewrite g as

```
function g(z:real):real;
begin
   g := sin(z/4) - cos(z/4)
end;
```

and everything would work the same. The names used for the formal parameters do not matter. Also, in the main program it is not necessary to declare `Bisection` before g and h. The definition of `Bisection` does not depend upon these other functions and so the order of their placement in the main program is irrelevant.

To trace a call to `Bisection` we need to expand our context box notation. In designating the "players" in the main program, we also include the names of all functions and procedures that are declared there. When a call to a procedure like

```
Bisection(g,0,4,0.000001,MyPi,gval)
```

```
program Example10_3;
{The bisection process.}
uses
   Chap10Codes;
var
   MyPi,gval:real;
   Log10,hval:real;
function g(x:real):real;
begin
   g := sin(x/4) - cos(x/4)
end;

function h(x:real):real;
begin
   h := exp(x) - 10.0;
end;

begin
   ShowText;
   Bisection(g,0,4,0.000001,MyPi,gval);
   writeln('pi = ',MyPi:10:7,' g(MyPi) = ', gval:10:7);
   Bisection(h,0,5,0.000001,Log10,hval);
   writeln('log(10) = ',Log10:10:7,' h(Log10) = ', hval:10:7);
end.
```

The procedure `Bisection` is declared in the unit `Chap10Codes`. Output:

```
     pi = 3.1415920    g(MyPi) = -0.0000002
log(10) = 2.3025846   h(Log10) = -0.0000044
```

is made, we open up a new context box making the usual substitutions for the value and variable parameters. But for the functional parameter, we draw an arrow back to the main program function that is participating in the call:

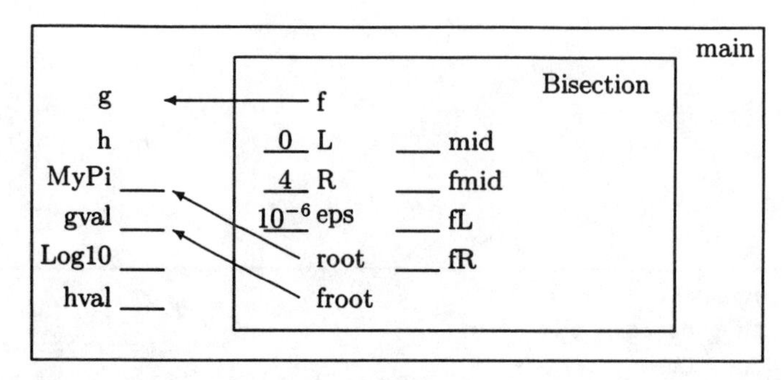

We are now set to execute. The arrow back to `g` reminds us that `f` is just a placeholder in the body of `Bisection`. During this particular call, all references to `f` are references to `g`. In effect, the body of `Bisection` becomes

```
fL := g(L); fR := g(R);
while (R - L) > eps do
begin
   mid := (L+R)/2;
   fmid := f(mid);
        ⟨:⟩
```

Note that without `Bisection`, `Example10_3` would have to be written to include two in-line, function specific bisection fragments, one for the `g` and one for `h`.

PROBLEM 10.5. Modify `Bisection` so that it prints the sequence of bracketing intervals. Try it on the problem $f(x) = x^2 - 2000000$ with initial interval $[0, 1000000]$. Set `eps = 0.001` , but experiment with smaller values to observe the danger of an infinite loop.

10.3 Newton's Method

Now let us look at another root-finding method that uses both the value of the function and its derivative at each step. Suppose f and its derivative are defined at x_c. (The "c" is for current.) The intersection of the tangent line to f at x_c has a zero at $x_c - f(x_c)/f'(x_c)$ assuming that $f'(x_c) \neq 0$ as shown in FIGURE 10.5. *Newton's method* for finding a zero of the function f is based upon the repeated use of the formula

$$x_{next} = x_c - \frac{f(x_c)}{f'(x_c)}$$

Assume that f and f' are available through the real-valued functions `f(x)` and `fp(x)` respectively. Using a `while`-loop to handle the repetition and agreeing to terminate as soon as $|f(x_c)|$ is smaller than the value of a small positive number `eps`, we obtain the following "math book" implementation:

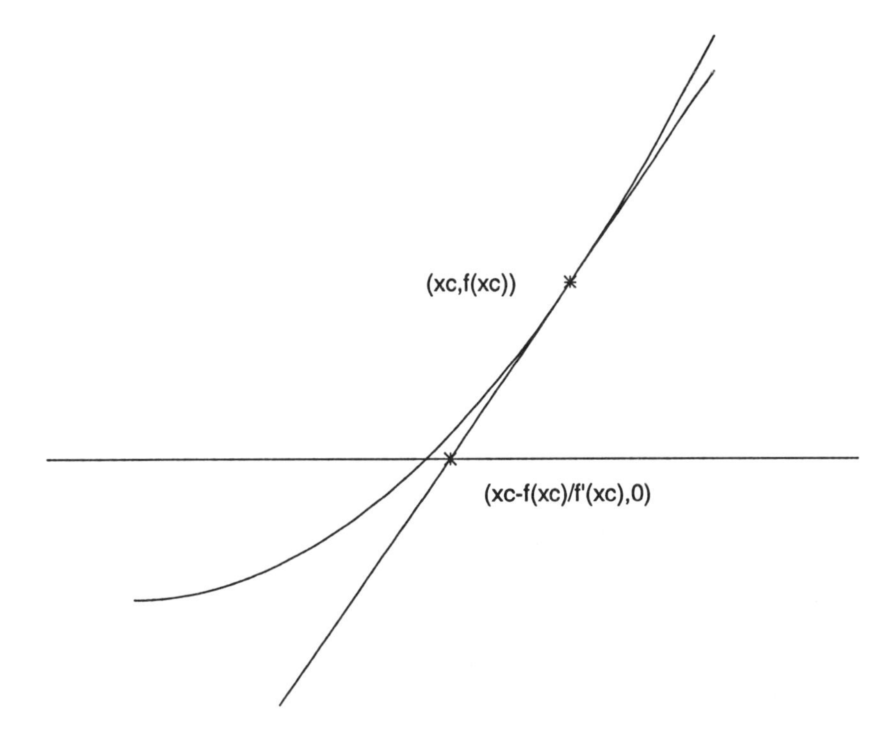

FIGURE 10.4 *Newton's Method*

```
xc:=InitialGuess;
fval:= f(xc);
fpval := fp(xc);
while abs(fval) > eps do
   begin
      xc := xc - (fval/fpval);
      fval := f(xc);
      fpval := fp(xc);
   end
```

This root-finding framework is based upon a different definition of an approximate root then what we used to terminate the bisection process:

Definition 2. Given a small positive number ϵ, we say that z is an approximate root of $f(x)$ if $|f(z)| \leq \epsilon$.

See FIGURE 10.5 Thus, if $f(x) = x^2$ and $\epsilon = .0001$, then $z = .01$ is an approximate root. according to Definition 2 but not according to Definition 1.

Unlike bisection, the Newton process is not guaranteed to terminate. If $f'(x_x)$ is zero, then the Newton step is not even defined. If $f'(x_c)$ is small, then a Newton step can take us very far away from a root. See FIGURE 10.6. To guard against these possibilities we establish the following boolean-valued function:

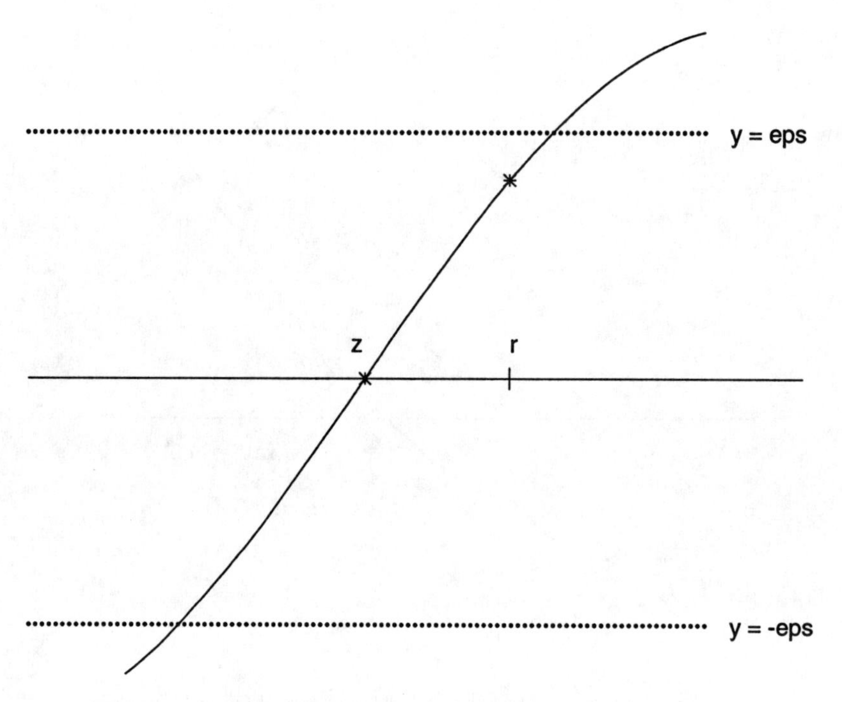

FIGURE 10.5 *An Approximate Root (Definition 2)*

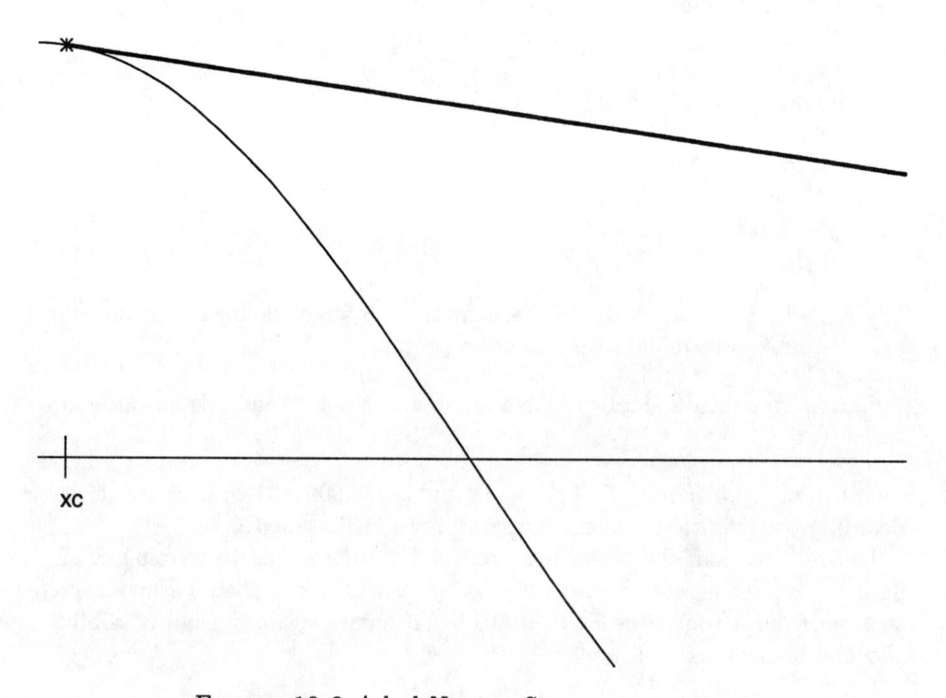

FIGURE 10.6 *A bad Newton Step*

```
function NextInInterval(xc,fval,fpval,L,R:real):boolean;
{Pre:L<=R.}
{Post:xc - fval/fpval is in the interval [L,R].}
var
   xnext:real;
begin
   if fpval = 0 then
      NextInInterval:=false
   else
      begin
         xnext := xc - fval/fpval;
         NextInInterval := (L <= xnext) and (xnext <= R);
      end
end
```

In this context, $[L, R]$ should be thought of as an "interval of interest." Presumably it contains the sought-after root. If the next Newton step is outside that interval, then it is time to shut down Newton process:

```
xc:=InitialGuess;
fval:= f(xc);
fpval := fp(xc);
while (abs(fval) > eps) and NextInInterval(xc,fval,fpval,L,R) do
   begin
      xc := xc - (fval/fpval);
      fval := f(xc);
      fpval := fp(xc);
   end
```

The guard effectively says, "take another step because x_c is not an approximate root and x_{next} is defined and falls inside $[L, R]$." It does *not* mean that the Newton brings us closer to an approximate root.

This brings us to another difficulty. Even if the iteration stays inside $[L, R]$, the sequence of approximate roots may not be converging fast enough. Perhaps the value of **eps** is too small or the function f is highly nonlinear. To guard against the method taking too many steps we incorporate an iteration bound **Itmax**:

```
xc:=InitialGuess;
fval:= f(xc);
fpval := fp(xc);
Its:=0;
while (abs(fval) > eps) and NextInInterval(xc,fval,fpval,L,R)
           and (Its <= ItMax) do
   begin
      xc := xc - (fval/fpval);
      fval := f(xc);
      fpval := fp(xc);
      Its:=Its+1;
   end
```

Now the loop guard says "take another step because x_c is not an approximate root, x_{next} is defined and falls inside $[L, R]$, *and* we can afford it."

What we have done with these two modifications is make the original "math book" implementation more robust. To appreciate this point further we encapsulate our safeguarded Newton iteration as a procedure:

```
procedure Newton(function f(x:real):real; function fp(x:real):real;
         InitialGuess:real; ItMax:integer; eps:real; L,R:real;
         var root,froot:real; var its:integer);
{Pre:ItMax>0, L<= R, the functions f and fp are defined on [L,R],}
{      eps>0, InitialGuess is in [L,R] and is an estimate of a root.}
{Post:froot=f(root).  Either |froot| <= eps or:}
{      Its = ItMax and the iteration max is reached, or}
{      Its < ItMax and the next iterate would be out of [L,R].}
var
   xc:real; {The current x.}
   fval,fpval:real; {fval = f(xc), fpval = fp(xc).}
function NextInInterval(xc,fval,fpval,L,R:real):boolean;
{Pre:L<=R}
{Post:  xc - fval/fpval is in the interval [L,R].}
      ⟨:⟩

begin
   Its := 0;
   xc := InitialGuess;
   fval := f(xc);
   fpval := fp(xc);
   while (abs(fval) > eps) and NextInInterval(xc,fval,fpval,L,R)
         and (Its < ItMax) do
      begin
         xc:= xc - (fval/fpval);
         fval:=f(xc);
         fpval:=fp(xc);
         Its:=Its+1;
      end;
   root := xc;
   froot := fval;
end;
```

The actual number of Newton steps that are taken is returned in the **var** parameter `Its`. This is useful information and provides a commentary on the quality of the initial guess. If `Its=ItMax` is true, then you know that the search for a root is called off because too many steps are required. If budget permits, `Newton` can be called again with the value of `root` serving as the initial guess. If `Its<ItMax`, then one of two situations holds:

- `abs(froots) <= eps` is true in which case an approximate root is found in the sense of Definition 2.
- `abs(froots) <= eps` is false implying that the iteration terminated because the next step takes the search out of $[L, R]$.

To illustrate the use of `Newton`, `Example10_4` sets up an interactive environment that can be used to find zeros of

$$g(x) = \sin(x/4) - \cos(x/4).$$

This function has a zeroes at $\pi, 5\pi, 9\pi$, etc.

It is important to remember that both approximate root definitions can be "fooled" as shown in FIGURE 10.6. A Definition 1 root need not make $f(x)$ particularly small and a Definition 2 root need be particularly close to an actual root. Sophisticated root finders that involve a blend of bisection and Newton ideas are gracefully able to close these loop holes.

PROBLEM 10.6. Modify `Example10_4` so that it prints $x_1, \ldots, x_n$ where x_i is the value of `xc` after i passes through the `while`-loop in `Newton` and n is the total number of steps taken. The value of $f(x_i)$ should also be printed. Hint:repeatedly call `Newton` `ItMax` equal to one. If the returned value of `Its` is one, then a step was taken and the value of `root` and `froot` should be printed.

```
program Example10_4;
{The Newton process.}
uses
   Chap10Codes;
const
   eps = 0.000001; {Termination criteria.}
   ItMax = 20; {Iteration maximum.}
var
   Its:integer; {Number of required iterations.}
   root,froot:real; {For the computed root and f at that root.}
   InitialGuess:real; {The initial guess.}
   AnotherEg:char; {Indicates another try.}
   L,R:real; {The left and right endpoints of the search interval.}
function g(x:real):real;
   begin
      g := sin(x/4) - cos(x/4)
   end;

function gp(x:real):real;
   begin
      gp := (sin(x/4) + cos(x/4))/4
   end;
begin
   ShowText;
   AnotherEg := 'y';
   while AnotherEg <> 'n' do
      begin
      writeln('Enter starting value, L, and R ');
      readln(InitialGuess,L,R);
      Newton(g,gp,InitialGuess,ItMax,eps,L,R,root,froot,its);
      writeln;
      writeln(' Initial guess =', InitialGuess :  10 :  3);
      writeln(' root =', root :  10 :  6);
      writeln(' f(root) =', froot :  12 :  7);
      writeln(' iterations = ', its :  3);
      writeln(' eps = ', eps);
      writeln;
      writeln('Another example?  (y/n)');
      readln(AnotherEg);
   end
end.
```

The procedure `Newton` is declared in the unit `Chap10Codes`. Sample results:

InitialValue	L	R	root	froot	its
1.0	0	10	3.141593	0.0000002	3
3.1	0	10	3.141594	0.0000005	1
8.0	0	20	8.000000	1.3254421	0
8.0	-100	100	40.840706	-.0000005	5
1000	-10000	10000	995.884888	-.0000058	20

Chapter 11

Area

The breaking down of large complex problems into smaller, solvable subproblems is at the heart of computational science. The calculation of area is symbolic of this enterprise and makes an interesting case study. When the region is simple, there may be a formula, e.g., $A = base \cdot height$. Otherwise, we may have to partition the region into simpler regions for which there are area formulas. For example, we can cover a polygon with triangles and then sum their areas. Other times we may have to resort to approximation. If we can pack (without overlap) N h-by-h tiles inside a shape that is bounded by curves, then Nh^2 approximates its area. A variation of this idea, with limits, leads to the concept of integration in the calculus. By exploring these limits we obtain yet another glimpse of the boundary between exact mathematics and approximate calculation.

11.1 Triangulation

The area of a triangle with sides a, b, and c is given by

$$T = \sqrt{h(h-a)(h-b)(h-c)} \qquad h = (a+b+c)/2$$

Assuming the availability of the distance function

```
function Distance(x1,y1,x2,y2:real):real;
{Post:The distance from (x1,y1) to (x2,y2).}
begin
   Distance := sqrt(sqr(x1-x2)+sqr(y1-y2))
end
```

we obtain

```
function T(x1,y1,x2,y2,x3,y3:real):real;
{Post:  The area of the triangle with vertices (x1,y1), (x2,y2), and (x3,y3).}
var
   a,b,c:real;
   h:real;
begin
   a:= Distance(x1,y1,x2,y2);
   b:= Distance(x2,y2,x3,y3);
   c:= Distance(x3,y3,x1,y1);
   h := (a+b+c)/2;
   T := sqrt(h*(h-a)*(h-b)*(h-c));
end
```

By building upon T, we can develop area functions for more complicated regions. Consider the problem of computing the area of the quadrilateral $ABCD$ displayed in FIGURE 11.1. Our

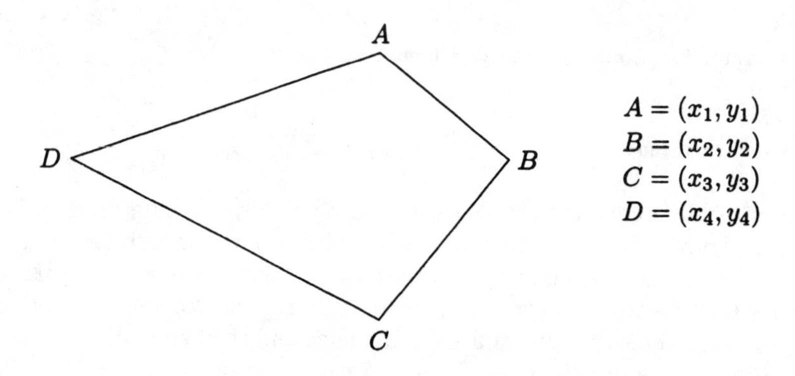

FIGURE 11.1 *A Quadrilateral*

approach is to "partition" $ABCD$ into a pair of triangles ACB and ACD as shown in FIGURE 11.2. If T_{ACB} and T_{ACD} are their respective areas, then $Q_{ABCD} = T_{ACB} + T_{ACD}$ and the area of

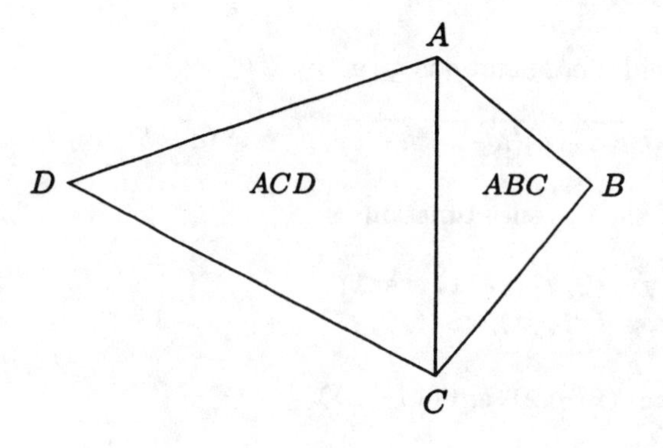

FIGURE 11.2 *Partitioning a Quadrilateral*

ABCD can be obtained by calling T twice with appropriate arguments and summing the values returned.

The trouble with this approach is that it doesn't work with certain vertex configurations. The quadrilateral *ABCD* may have sides that cross or a vertex may be "pushed in" too far. See FIGURE 11.3. These situations can be ruled out by insisting that the diagonals *AC* and *BD*

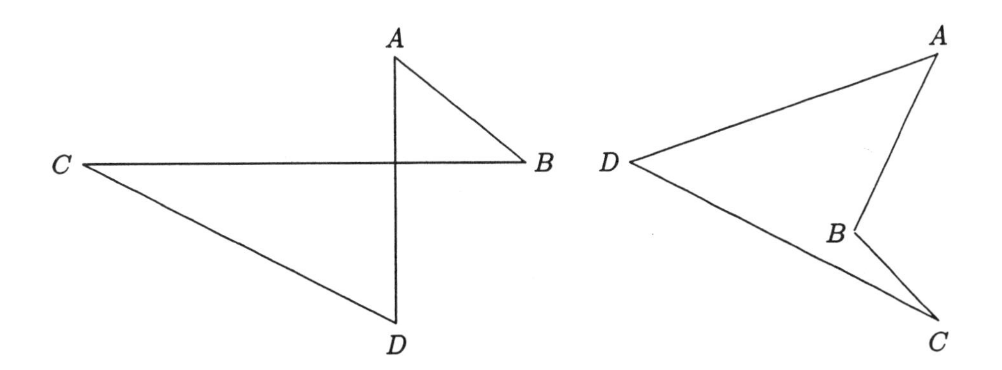

FIGURE 11.3 *Bad Quadrilaterals*

cross. See FIGURE 11.4. Incorporating this criteria into the precondition we obtain

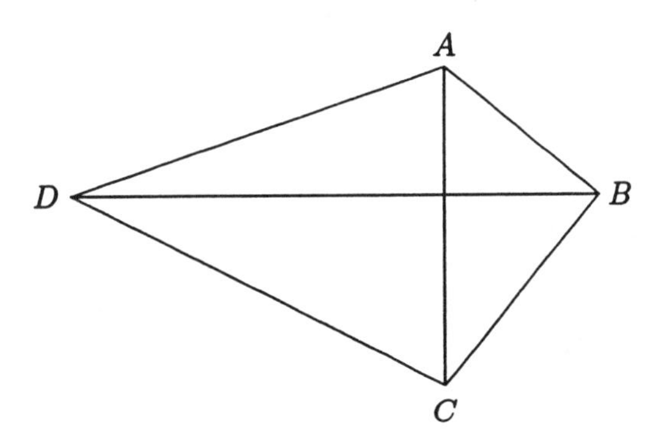

FIGURE 11.4 *Crossing Diagonals*

```
function Q(x1,y1,x2,y2,x3,y3,x4,y4:real):real;
{Pre:The line segment that connects (x1,y1) and (x3,y3) }
{intersects the line segment that connects (x2,y2) and (x4,y4).}
{Post:The area of the quadrilateral obtained ABCD by connecting the points}
{A=(x1,y1), B=(x2,y2), C=(x3,y3), and D=(x4,y4) in order.}
begin
   Q := T(x1,y1,x2,y2,x3,y3) + T(x1,y1,x3,y3,x4,y4);
end
```

Testing whether the diagonals of a general quadrilateral cross is discussed in Chapter 18, where more involved geometric calculations are covered. The program Example11_1 uses Q to estimate the expected area of a quadrilateral whose four vertices are randomly placed on successive sides of the unit square. A nice feature of this implementation is Q's brevity and how well the underlying partition idea ($Q_{ABCD} = T_{ABC} + T_{ACD}$) is displayed in the program. However, the clarity hides a modest inefficiency. Each call to T in turn involves three calls to Distance. Thus, six Distance calls are required altogether. However, only five calls are required in the following T-free alternative:

```
function Q1(x1,y1,x2,y2,x3,y3,x4,y4:real):real;
{Pre:The line segment that connects (x1,y1) and (x3,y3) }
{intersects the line segment that connects (x2,y2) and (x4,y4).}
{Post:The area of the quadrilateral obtained ABCD by connecting the points}
{A=(x1,y1), B=(x2,y2), C=(x3,y3), and D=(x4,y4) in order.}
var
   dAB,dBC,dCA,dCD,dDA:real; {Vertex-toVertex distances}
   hABC,hACD:real; {Area formula factors}
   TABC,TACD:real; {Triangular areas}
begin
   dAB := Distance(x1,y1,x2,y2);
   dBC := Distance(x2,y2,x3,y3);
   dCA := Distance(x3,y3,x1,y1);
   dCD := Distance(x3,y3,x4,y4);
   dDA := Distance(x4,y4,x1,y1);
   hABC := (dAB+dBC+dCA)/2;
   TABC := sqrt(hABC*(hABC-dAB)*(hABC-dBC)*(hABC-dCA));
   hACD := (dCA+dCD+dDA)/2;
   TACD := sqrt(hACD*(hACD-dCA)*(hACD-dCD)*(hACD-dDA));
   Q1 := TABC + TACD
end
```

The sacrifice in clarity is not great here and so the modest speed-up brought about by the reduction in Distance-calls can be defended. In other settings, the efficiency/clarity tradeoff is much more complicated and you need serious training in software design to deal successfully with the "function versus in-line fragment" alternative.

```
program Example11_1;
{Random quadrilaterals with one vertex on each side of the unit square.}
uses
   DDCodes,Chap11Codes;
const
   n = 1000; {The number of trials.}
   seed = 0.123456;
var
   r:real; {random number}
   k:integer;
   x1,y2,x3,y4:real; {kth quadrilateralhas vertices (x1,0), (1,y2),(x3,1), and (0,y4).}
   s:real;
begin
   ShowText;
   r := seed;
   s := 0;
   writeln('Trials Expected Area');
   writeln('------------------------');
   for k := 1  to n  do
      begin
         {s = sum of areas of first k quadrilaterals.}
         r := rand(r); x1 := r;
         r := rand(r); y2 := r;
         r := rand(r); x3 := r;
         r := rand(r); y4 := r;
         s := s + Q(x1,0,1,y2,x3,1,0,y4);
         if k mod 100 = 0 then
            writeln(k:5,(s/k):16:4, ' ');
      end;
end.
```

The function `rand` is declared in the unit `DDCodes`. The function `Q` is declared in the unit `Chap11Codes`. Output:

```
Trials  Expected Area
---------------------
  100       0.5124
  200       0.5052
  300       0.5023
  400       0.4997
  500       0.5003
  600       0.5042
  700       0.5028
  800       0.5032
  900       0.5019
 1000       0.5017
```

PROBLEM 11.1. Suppose the quadrilateral $ABCD$ has the property that the sides only intersect at the vertices. This rules out the left quadrilateral in FIGURE 11.3. In this case the area may be obtained from the formula

$$Q_{ABCD} = \min\{\, T_{ABC} + T_{DBC}\ ,\ T_{ABD} + T_{CBD}\}.$$

Generalize Q so that it can handle quadrilaterals of this form. What is the expected area of a quadrilateral whose four vertices are randomly selected from successive quadrants of the square $\{(x,y)\colon -1 \le x \le 1, -1 \le y \le 1\}$.

PROBLEM 11.2. Suppose $0 \le \theta_1 < \theta_2 < \theta_3 < \theta_4 < 2\pi$ and that we want to compute the area Q of the quadrilateral whose vertices (in order) are $A = (\cos(\theta_1), \sin(\theta_1)$, $B = (\cos(\theta_2), \sin(\theta_2)$, $C = (\cos(\theta_3), \sin(\theta_3)$, and $D = (\cos(\theta_4), \sin(\theta_4)$. If the origin is inside the quadrilateral, then we have a triangulation of the form

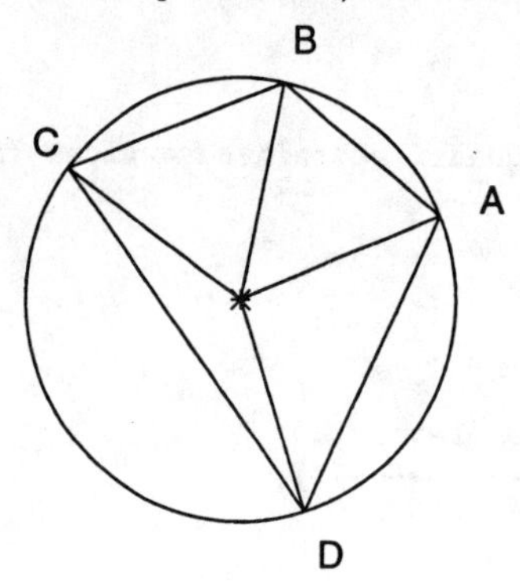

In this case,

$$Q = T_{ABO} + T_{BCO} + T_{CDO} + T_{DAO}.$$

Bearing this in mind, complete the following procedure:

```
function QCircle(theta1,theta2,theta3,theta4:real):real;
{Pre:0<=theta1<=pi/2<=theta2<=pi<=theta3<=(3pi/2)<=theta4<=2pi.}
{Post:The area of the quadrilateral with vertices}
{ (cos(theta1),sin(theta1)) }
{ (cos(theta2),sin(theta2)) }
{ (cos(theta3),sin(theta3)) }
{ (cos(theta4),sin(theta4)) }
```

11.2 Tiling

Consider the problem of counting the number of h-by-h "tiles" that can be placed (without overlap) inside the the unit disk $\{(x,y) : x^2 + y^2 \le 1\}$. Assume that a "base tile" is positioned in the lower left corner of the first quadrant. See FIGURE 11.5. By symmetry we know that the total number of tiles that fit in the disk is four times the number that fit in the first quadrant. Since the tiles are h-by-h, it follows that there are at most trunc$(1/h)$ first-quadrant rows of tiles to tabulate. Taking the top-down approach, we start with the following framework:

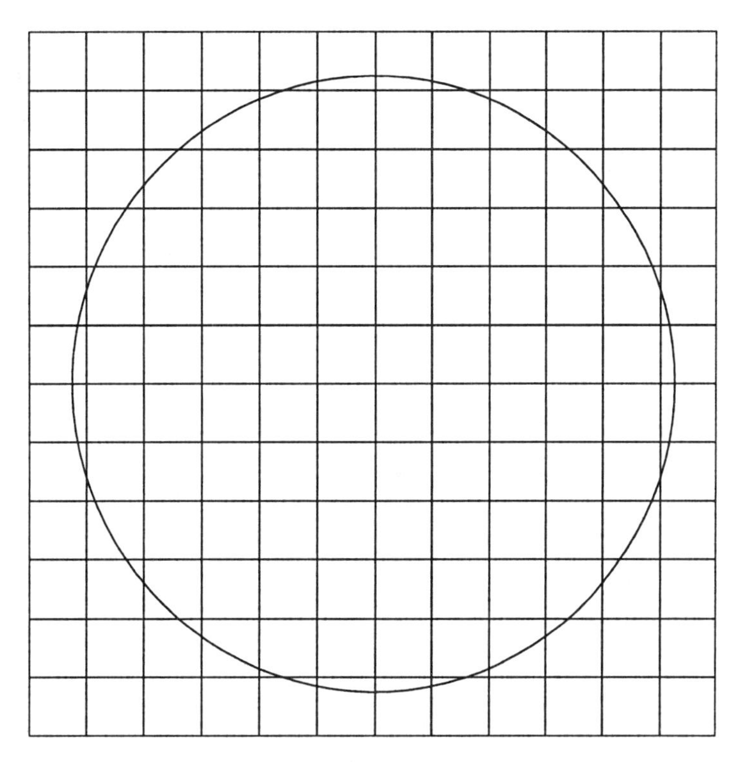

FIGURE 11.5 *Tiling a Circle*

```
N:=0;
for k:=1 to trunc(1/h) do
   begin
      TilesInRow := ⟨Number of first quadrant tiles in k-th row⟩
      N := N + TilesInRow;
   end;
TilesOnDisk := 4*N;
```

To compute the number of complete tiles that fit in row k, we observe that $y = kh$ along the top edge of the k-th tile row. From FIGURE 11.6 we conclude that the number of tiles in this row is the whole number portion of $\sqrt{1-(kh)^2}/h$. This leads to the following function

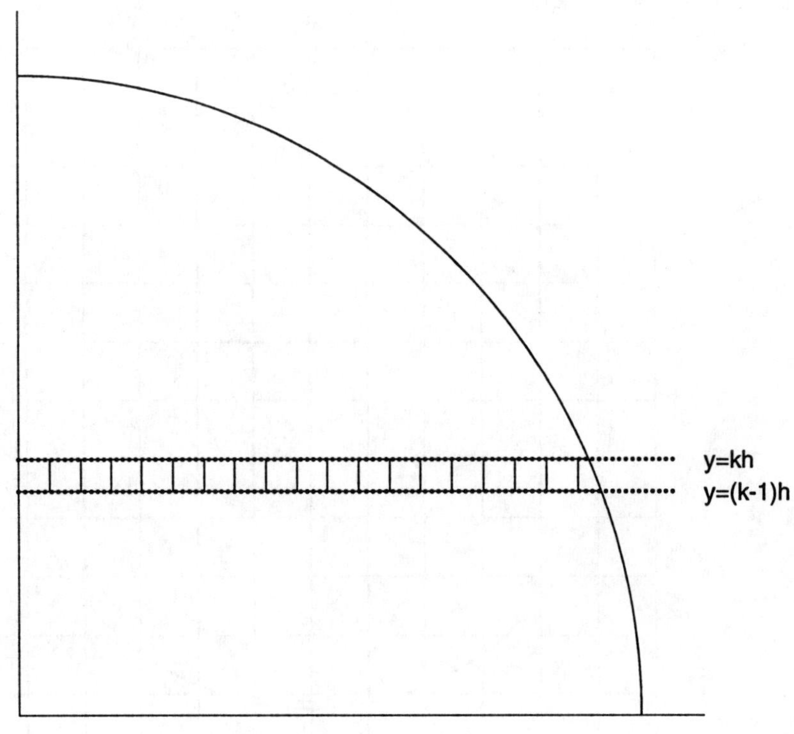

FIGURE 11.6 *Tiles in a Row*

```
function TilesInDisk(h:real):longint;
{Pre:h>0}
{Post:The number of h-by-h tiles that fit entirely within a circle}
{of radius 1 centered at the origin.  The base tile has vertices (0,h),}
{(h,h), (h,0), and (0,0).}
var
   N:longint; {Running sum for first quadrant tiles.}
   d:real; {The length of k-th row.}
   k:integer;
begin
   N := 0;
   for k:=1 to trunc(1/h) do
      begin
         d := sqrt(1 - sqr(h*k));
         N := N + trunc(d/h);
      end;
   TilesOnDisk := 4*N;
end
```

Since the area of the disk is π, it follows that the value of `TilesOnDisk(h)*sqr(h)` is an approximation to π. The program `Example11_2` explores the quality of this estimate as h gets small. The value of h passed to `TilesInDisk` cannot be too small, for otherwise the number of tiles exceeds `maxlongint`. Thus, this method for computing π has restricted accuracy.

```
program Example11_2;
{Pi estimates via unit disk tiling.}
var
   h:real; {The edge of a tile.}
   N:longint; {Number of tiles inside unit disk.}
   k:integer;
   PiEstimate:real; {Estimate of pi.}
function TilesInDisk(h:real):longint;
      ⟨:⟩
begin
   ShowText;
   h:=1;
   writeln(' h Estimate of Pi');
   writeln('----------------------------');
   for k:=1 to 12 do
      begin
         h:=h/2;
         N:=TilesInDisk(h);
         PiEstimate := N*sqr(h);
         writeln(h:8:6,PiEstimate:8:6);
      end
end.
```

The function `TilesInDisk` is declared in the unit `Chap11Codes`. Output:

```
     h        Estimate of Pi
----------------------------
0.500000        1.000000
0.250000        2.000000
0.125000        2.562500
0.062500        2.859375
0.031250        3.007812
0.015625        3.075195
0.007812        3.107910
0.003906        3.125549
0.001953        3.133484
0.000977        3.137589
0.000488        3.139624
0.000244        3.140601
```

PROBLEM 11.3. Write a function `CenteredTilesOnDisk(h:real):integer` that is identical to `TilesOnDisk` except that the base tile has vertices at $(h/2, h/2)$, $(-h/2, h/2)$, $(-h/2, -h/2)$, and $(h/2, -h/2)$.

PROBLEM 11.4. Complete the following function

```
function TilesInEllipse(hx,hy,rx,ry:real):longint;
{Pre:hx,hy,rx,ry>0}
{Post:The number of hx-by-hy tiles that fit entirely within ellipse}
{(sqr(x/rx) + sqr(y/ry) = 1.  The base tile has vertices (0,hx),}
{(hx,hy), (hx,0), and (0,0).}
```

PROBLEM 11.5. The unit sphere is defined by $\{(x, y, z) \mid x^2 + y^2 + z^2 \leq 1\}$. Complete the function

```
function CubesinSphere(h:real):longint;
{Pre:h>0}
{Post:  The number of h-by-h-by-h cubes that fit entirely}
{within the unit sphere.  Assumes that the}
{base cube has vertices at the points}
{(0,0,0),(h,0,0),(0,h,0), (h,h,0), (0,0,h), (h,0,h), (0,h,h), (h,h,h)}
```

exploiting the following facts:

- The number of $h \times h \times h$ cubes that fit inside a cylinder of radius r and height h is given by `TilesInDisk(h/r)`. (Assuming that the cubes are cornered at the center.)
- If the "northern hemisphere" is sliced horizontally at $z = h$, $z = 2h$, $z = 3h$, etc, then the number of cubes that fit inside the k-th slice can be computed using the previous fact.
- The number of cubes that fit inside the unit sphere is twice the number of cubes that fit inside its northern hemisphere.

PROBLEM 11.6. Complete the following function

```
function CubesInPyramid(alpha,beta,h:real):longint;
{Pre:alpha,beta>=0,h>0}
{Post:Number of h-by-h-by-h cubes that fit inside a pyramid}
{with base vertices are at (s,s,0), (-s,s,0), (-s,-s,0), (s,-s,0) }
{and whose apex is at (0,0,beta).  Here, s = sqrt(alpha)/2.  }
{Assumes that the base cube is cornered at the origin.  }
```

The volume of the pyramid described in the post-condition has the form constant·base·height. Write a program that uses `CubesInPyramid` to estimate the constant.

11.3 Integration

Our third look at area computation involves the approximation of the integral

$$I = \int_a^b f(x)dx$$

with a summation. The value of the integral I is the (signed) area underneath the curve $y = f(x)$. This area may be approximated by the sum of "skinny rectangle" areas as shown in FIGURE 11.7. If there are to be n of these rectangles of equal width, then the width is given by

$$h = (b - a)/n$$

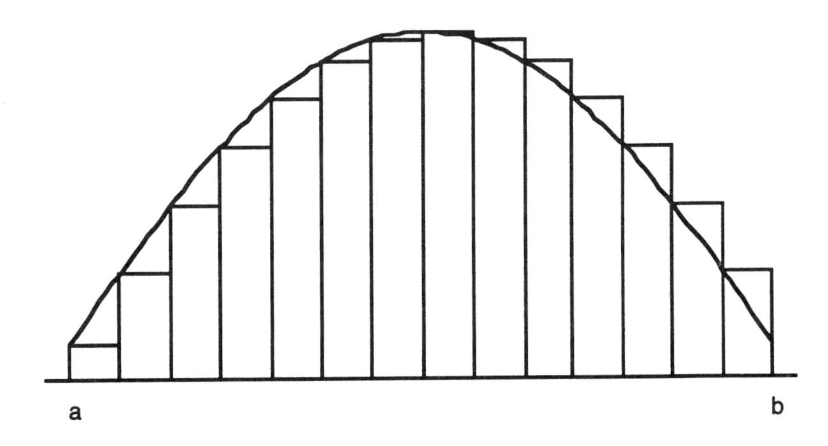

FIGURE 11.7 *Rectangle Rule*

and the i-th rectangle is situated between $x = a + ih$ and $x = a + (i + 1)h$. If we sample $f(x)$ at first of these two values, then the i-th rectangle has height $f(a + ih)$ and area $hf(a + ih)$. Summing from $i = 0$ to n we obtain the *rectangle rule*:

$$I \approx R = \sum_{i=0}^{n-1} \text{Area of rectangle } i = \sum_{i=0} n - 1h \cdot f(a + ih) = h \sum_{i=0} n - 1 f(a + ih).$$

If `f(x:real):real` is a declared function, then the fragment

```
s:=0;
h := (b-a)/n;
for i:=0 to n-1 do
   s := s + f(a+i*h);
Area := s*h
```

assigns the value of the rectangle rule to **Area**.

Let's consider the encapsulation of the rectangle rule as a function. It expects four things (the integrand, the lower limit a, the upper limit b, and the number of rectangles) and produces a single number (the signed area). See FIGURE 11.8. The implementation requires a procedure with a functional parameter:

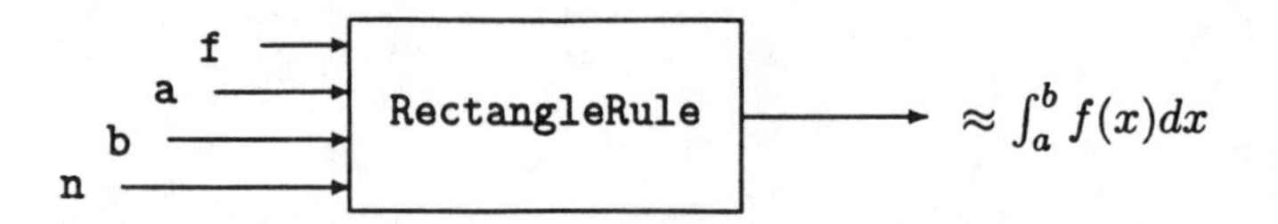

FIGURE 11.8 RectangleRule(f,a,b,n)

```
function RectangleRule(function f(x:real):real; a,b:real; n:longint):real;
{Pre:f(x) is defined on [a,b] and n > 0.}
{Post:The integral of f from a to b using the rectangle rule with n approximating rectangle
var
   s:real; {The sum of rectangular areas.}
   h:real; {rectangle width.}
   k:integer;
begin
   s:=0;
   h:=(b - a)/n;
   for k:=0 to n - 1 do
      s := s + f(a + k*h);
   RectangleRule := h*s;
end;
```

To illustrate how RectangleRule can be used, let us compute π by applying it to the integral in the equation

$$\pi = 4\int_0^1 \sqrt{1-x^2}dx.$$

See Example10_3.

PROBLEM 11.7. The midpoint rule and the trapezoidal rule approximations to

$$I = \int_a^b f(x)dx$$

are defined by

$$M = \sum_{k=0}^{n-1} hf(a+(k+1/2)h)$$

$$T = \sum_{k=0}^{n-1} \frac{h}{2}(f(a+kh)+f(a+(k+1)h))$$

where $h = (b-a)/n$. Write functions for both of these rules and augment Example10_3 so that it produces a table of π estimates based upon the rectangle, midpoint, and trapezoidal rules.

PROBLEM 11.8. Let h_m be the area enclosed by the *hyperellipse* $|x|^m + |y|^m = 1$. Note from symmetry that this area is four times the area under the curve $y = (1-x^m)^{1/m}$ in the first quadrant. Use RectangleRule to estimate h_m for $n = 2, 4, 8, 16, 32$.

```
program Example11_3;
{Estimate pi by via integral from 0 to 1 of sqrt(1-x^2).}
uses
   Chap11Codes;
var
   PiEstimate real; {Estimate of pi.}
   n:longint;
function g(x:real):real;
{Pre:  abs(x)<=1.  }
{Post:The square root of 1 - sqr(x).}
begin
   g := sqrt(1 - sqr(x))
end;

begin
   ShowText;
   n := 10;
   writeln; writeln(' n PiEstimate');
   writeln('--------------------------------');
   while n <= 10000 do
      begin
         PiEstimate := 4 * RectangleRule(g,0,1,n);
         writeln(n:8, PiEstimate:12:6);
         n := 10*n;
      end
end.
```

The function **RectangleRule** is declared in the unit **Chap11Codes** Output:

```
       n   PiEstimate
--------------------
      10      3.04518
     100     3.160418
    1000     3.143554
   10000     3.141791
```

Chapter 12

Encoding Information

This chapter is about the *representation* of information, a term already familiar to us. For example, a real number has a *floating point representation* when stored in a `real` variable. Numbers stored in `integer` variables have a different kind of representation. In this chapter we advance our understanding of the representation "idea" by looking at several "conversion problems" that involve numbers as strings of characters. A greater appreciation for the place-value notation is obtained.

12.1 Notation and Representation

A date is a triplet of integers and in the "m/d/y" notation we use slashes to separate the parts:

"10/4/93" $\leftrightarrow$ "October 4, 1993"

In an application that involves dates, we could use three integer variables `day`, `month` and `year` to represent the appropriate information. Alternatively, we could use a *string*. A string is just a sequence of characters. ThinkPascal comes equipped with a string type and `Example12_1` shows how to declare a string variable and use it in a simple read/write setting. It solicits a date in m/d/y form and echoes the input. If we type

```
10/4/93
```

then `date` is assigned the string value '10/4/93' *without the quotes.* As `Example12_1` shows, string variables are typed as follows:

⟨*Variable Name*⟩ : **string**

In keeping with all variable declarations, this sets up "a box." Boxes of type `string` can hold string data.

String variables can also be assigned values. For example,

```
date := '10/4/93';
```

```
program Example12_1;
{String reads and writes.}
var
   date:string;
begin
   writeln('Enter a date in m/d/y form:');
   readln(date);
   writeln('The date is ',date,'.')
end.
```

Sample output:

```
Enter a date in m/d/y form:
7/4/1994
The entered string is 7/4/1994.
```

assigns '10/4/93' to `date`. The single quotes serve as *delimiters* and are not part of the string[1]. Note that if `date` has type `string`, then `date:='7'` is legal but `date:=7` is not. This is because you cannot put an integer in a variable designated for strings.

When discussing particular strings, we enclose the string within quotes, with the understanding that the delimiting quotes are *not* part of the string itself.

In string problems it is important to bear in mind that a blank is a character. (It is obtained by striking the space bar.) To designate blanks in the text we use the symbol b̸. Thus, if we run `Example 12_1` and respond with 10b̸/4/22b̸, then the string 10b̸/4/22b̸ is assigned to `date`.

ThinkPascal includes a small library of functions that facilitate string manipulation. We begin our survey of these with a discussion of the `length` function. The length of a string is the number of characters (including blanks) that make up the string. The integer-valued `length` function may be used to compute string length. If `d1` and `d2` are string variables and `n1` and `n2` are integer variables, then the fragment

```
d1 := '10/4/93';
n1 := length(d1);
d2 := '10/20/93';
n2 := length(d2);
```

assigns the value 7 to `n1` and 8 to `n2`. A string with no characters is called the *empty string* and it has length zero. Thus, the fragment

```
d:='';
n:=length(d);
```

assigns the empty string to `d` and 0 to `n`.

It is legal to assign the contents of a character-valued expression to a string. For example, if `letter` has type `char` and `s` has type `string`, then `s:=letter` is valid. When reasoning about the `char` and `string` types, it is appropriate to think of a character as a string with unit length.

[1]To include a single quote in a string, double it: `s:='In harm''s way'`.

A "square bracket" notation is used to designate specific characters in a string. For example, if `s` is a string and `k` is an integer whose value is between 1 and the length of `s`, then `s[k]` references the k-th position in the string. Thus,

```
date := '10/4/93';
slash:= date[3];
decade := date[6];
```

assigns the value '/' to `slash` and '9' to `decade`. For this to make sense, `slash` and `decade` must have type `string` or `char`. In general, "square-bracket" references have the form

⟨*String Variable Name*⟩ [⟨*Integer-Valued Expression*⟩]

It is important that the expression inside the square brackets has a value that is associated with an existing string position.

In doing computations with m/d/y-specified dates, we need to be able to "get hold" of the month, day, and year portions of a string like '10/4/93'. This can be done using the string-valued `copy` function that extracts *substrings*. Substrings of '10/4/93' include '10', '/93', '4', '10/4/93', and the empty string. Two integers specify a substring: its starting position and its length. Thus, the starting position of the substring '/93' in '10/4/93' is 5 and its length is 3. References to `copy` have the form

`copy(`⟨*source string*⟩`,`⟨*starting position of substring*⟩`,`⟨*length of substring*⟩`)`

For example,

```
date := '10/4/93';
m := copy(date,1,2);
d := copy(date,4,1);
y := copy(date,6,2);
```

assigns the strings '10', '4', and '93' to the string variables `m`, `d`, and `y` respectively.

When using `copy`, make sure that the integer arguments name a valid substring. For example, in a reference of the form `copy(s,n,m)`, if `s` has length L, then n and m should satisfy $1 \leq n \leq L$ and $0 \leq m \leq L - n + 1$.

Let's see how `copy` can be used to extract the month, day, and year from a valid m/d/y string housed in `date`. Starting with the year substring, observe that the assignment

```
y := copy(date,6,2)
```

used above only works if the lengths of the month and day substrings sum to 3. (In this case, three digits and two slashes precede the start of the year substring.) To cope with the variable length of the month and day substrings, we note that the year substring is always comprised of the last two characters in the string. These may be "copied out" as follows:

```
n:=length(date);
y:=copy(date,n-1,2)
```

The extraction of the month substring begins by determining the location of the first '/'. The integer-valued `pos` function can be used for this purpose. If `s1` and `s2` are strings, then `pos(s1,s2)` checks to see if $s1$ is a substring of $s2$. If $s1$ is not a substring of $s2$, then `pos(s1,s2)` is zero. If $s1$ is a substring of $s2$, then `pos(s1,s2)` is the starting position of the *first occurrence* of $s1$ in $s2$. Thus, if the value of `date` is '10/4/93', then `pos('/',date)` is 3. Even though `date` has two distinct copies of the substring '/', the first occurrence begins in position 3 and that is the value returned by `pos`. Using `pos`, we can extract the month substring as follows:

```
p:=pos('/',date);
m:=copy(date,1,p-1);
```

The extraction of the day substring requires a little integer manipulation. We know that if the first '/' is in position p, then the day substring starts in position $p+1$. Similarly, if n is the length of `date`, then the day substring ends in position $n-3$. The length of a substring whose start and end positions are known is given by (End position) - (Start Position) + 1. Applying this rule to our problem gives

```
p:=pos('/',date);
n:=length(date);
d:=copy(date,p+1,n-p-3);
```

Combining this with the month and year substring computations gives the following procedure:

```
procedure TakeApart(s:string; var d,m,y:string);
{Pre:s is a string of the form 'mm/dd/yy' where:}
{    mm is one of the strings '1','2',...,'12' and indicates month.}
{     dd is one of the strings '1','2',...,'31' and indicates day.}
{     yy is one of the strings '00','01', ...,'99' and indicates year.}
{Post:  d=dd, m=mm, and y=yy.}
var
   p:  integer; {The location of the first /.}
   n:  integer; {The length of s.}
begin
   p := pos('/',s);
   n := length(s);
   d := copy(s,p+1,n-p-3);
   m := copy(s,1,p-1);
   y := copy(s,n-1,n);
end;
```

Note that `TakeApart` returns three separate strings that collectively represent the date.

Another way to *represent* a date is in the *mnemonic d/m/y notation:*

'4OCT93' $\leftrightarrow$ 'October 4, 1993'

Let us develop a fragment that converts an m/d/y specified date into its mnemonic d/m/y form, e.g.,

'10/4/93' $\rightarrow$ '4OCT93'

Three steps are involved. We start by using `TakeApart` to extract the month, day, and year substrings from the m/d/y date housed in `date`:

```
TakeApart(date,day,month,year)
```

Variables `day`, `month`, and `year` must have type `string`.

To obtain the mnemonic for a given month we define a string constant

```
ListOfMonths = 'JANFEBMARAPRMAYJUNJULAUGSEPOCTNOVDEC'
```

and then "copy out" the appropriate length-3 substring. For example, the value of

```
copy(ListOfMonths,3*4-2,3)
```

```
program Example12_2;
{Conversion from m/d/y representation to mnemonic d/m/y representation.}
uses
   Chap12Codes;
const
   ListOfMonths = 'JANFEBMARAPRMAYJUNJULAUGSEPOCTNOVDEC';
var
   date:string; {Test date.}
   s:string; {Mnemonic forms produced by Convert}
   AnotherEg:char;
   MonthIndex:integer; {Index of month.}
   month,day,year,MonthName:string;
begin
   ShowText;
   AnotherEg := 'y';
   while AnotherEg <> 'n' do
      begin
         writeln('Enter date in m/d/y form:');
         readln(date);
         writeln;
         TakeApart(date,day,month,year);
         if length(month) = 1  then
            MonthIndex := Char2Int(month)
         else
            MonthIndex := 10*Char2Int(month[1]) + Char2Int(month[2]);
         MonthName := copy(ListOfMonths, 3*p-2, 3);
         s := concat(day, MonthName, year);
         writeln('d/m/y mnemonic representation:',s);
         writeln; writeln('Another date?  Enter y (yes) or n (no).');
         readln(AnotherEg);
      end;
end.
```

The procedure `TakeApart` and the function `Char2Int` are declared in the unit `Chap12Codes`. Sample output:

```
Enter date in m/d/y form:
7/4/94

d/m/y mnemonic notation:  4JUL94

Another date?  Enter y (yes) or n (no).
n
```

is 'APR', the mnemonic for the 4th month. This approach requires the computation of the integer-valued index of the month under consideration. For this task we define the function

```
function Char2Int(c:char):integer;
{Pre:c is one of the ten digits.}
{Post:The value of c as an integer}
begin
   Char2Int := pos(c,'0123456789') - 1
end
```

This function works by searching the string '0123456789' for an occurrence of c. The position of the match is one greater than the value of c interpreted as an integer. This explains the "minus 1." To determine the value of the month string as an integer, we need to convert the one's place character and (if necessary) the ten's place character:

```
if length(month) = 1 then
   MonthIndex := Char2Int(month)
else
   MonthIndex := 10 * Char2Int(month[1]) + Char2Int(month[2]);
```

Noting that the mnemonic for the p-th month begins in position $3p - 2$ of `ListOfMonths` we see that

```
MonthName := copy(ListOfMonths, 3*p-2, 3);
```

gives us the required mnemonic for the month.

The conversion to the mnemonic form is completed by "gluing" day, `MonthName` and `year` together. The formal name for this is *concatenation* and the `concat` function can be used for this purpose:

```
s:= concat(day,MonthName,year)
```

If day, `MonthName`, and `year` contain the strings '4', 'OCT', and '93' respectively, then `s` is the single string '4OCT93'. This completes the conversion to the mnemonic date form and `Example12_2` provides a framework for exploring its correctness.

PROBLEM 12.1. Suppose `ListOfMonths` is a string with the value

```
'1JAN2FEB3MAR4APR5MAY6JUN7JUL8AUG9SEP10OCT11NOV12DEC';
```

Note that if `m` is one of the strings '1', '2',...,'12', then it is followed by the corresponding month mnemonic. Thus,

```
n := pos('7',ListOfMonths);
month := copy(ListOfMonths,n+1,3)
```

assigns 'JUL' to `month`. Rewrite `Example12_2` so that it uses this idea.

PROBLEM 12.2. (a) Complete the following function:

```
function DigitString(s:string):boolean;
{Post:  s is string made up entirely of digits.}
```

Hint: every character in `s` must be a substring of '0123456789'. (b) Complete the following function:

```
procedure TwoSlash(s:string; var p1,p2:integer; var Valid:boolean;
{Post:  If s has exactly two slashes (/), then p1 is the position of}
{the first and p2 is the position of the second, and Valid is true.}
{If s does not have exactly two slashes, then Valid is false.}
```

Hint: If p is the position of the first slash, then `pos('/',copy(s,p+1,length(s)-p))` tells you something about the second slash. **(c)** Using `DigitString` and `TwoSlash`, complete the following function:

```
function IsValid (s: string): boolean;}
{Post:Reading left to right, s is the concatenation of (1) a digit}
{string of length 1 or 2, (2) a slash, (3) a digit string of length 1}
{or 2, (4) a slash, and (5) a digit string of length 2.}
```

Hint: Look at the substrings that are separated by the slashes.

As a more involved problem of encoding information, let's consider how we might use a single string to represent the following six attributes of a given country;

Name
Population
Land Area
Name of Capital City
Latitude of Capital City
Longitude of Capital City

If we use the pound symbol # to separate these items of interest, then the string representation of

United States
258,300,000
3,536,341 square miles.
Washington D.C.
38 degrees, 53 minutes north.
77 degrees, 2 minutes west

has the form

```
'United States#258,300,000#3,536,341#Washington D.C.#38 53N#77 02W'
```

For the latitude and longitude, we adopt the convention that a single blank separates the degree and minute values. (There are sixty minutes in a degree.) In the above example, the 'N' and 'W' stand for "north" and "west." Southern latitudes and eastern longitudes are denoted by 'S' and 'E', e.g.,

```
'Australia#25,300,000#3,536,341#Canberra#38 53S#150 02E'
```

In anticipation of calculations that involve country data, it is handy to have a procedure that can take a string with the above format and split it into its six parts:

```
procedure CountryParts(s:string; var Name,Pop,Area,Capital,Lat,Long: string);
{Pre:s has the form s1#s2#s3#s4#s5#s6 where s1..s6 are substrings}
{ that do not contain a #.}
{Post:These substrings are assigned to Name, Pop, Area, }
{Capital, Lat, and Long respectively.}
```

The main challenge is to locate the delimiters and to that end we use the pos function. If s contains a country string, then

```
p1 := pos('#',s);
Name := copy(s,1,p1-1);
```

assigns the position of the first # to p1 (an integer variable) and the name substring to Name (a string variable). The second delimiter in s is the first delimiter in the the string copy(s,p1+1,length(s)-p1) and so the fragment

```
p2 := pos('#',copy(s,p1+1,length(s)-p1));
Pop := copy(s,p1+1,p2-p1-1);
```

assigns the population substring to Pop. Similar maneuvers can be used to isolate the area, capital, latitude, and longitude substrings.

However, a simpler approach that involves less "integer thinking" is to define the function

```
procedure Split(s:string; var s1,s2:string);
{Pre:If p = pos('#',s), then 1 < p < length(s)}
{Post:s1 = s[1..p-1] and s2 = s[p+1..length(s)]}
var
   p:integer;
begin
   p := pos('#',s);
   s1 := copy(s,1,p-1);
   s2 := copy(s,p+1,length(s)-p)
end
```

Here $s[i..j]$ designates the substring between and including the i-th and j-th characters. Split returns two strings. Everything *before* but not including the first # is assigned to s1 while everything *after* but not including the first # is assigned to s2. This permits the following implementation of CountryParts:

```
procedure CountryParts(s:string; var Name,Pop,Area,Capital,Lat,Long:  string);
{Pre:s has the form s1#s2#s3#s4#s5#s6 where s1..s6 are substrings}
{ that do not contain a #.}
{Post:These substrings are assigned to Name, Pop, Area, }
{Capital, Lat, and Long respectively.}

   procedure split(s:string; var s1,s2:string);
      ⟨:⟩
begin
   split(s,Name,s);
   split(s,Pop,s);
   split(s,Area,s);
   split(s,Capital,s);
   split(s,Lat,Long);
end
```

Notice that we have nested split inside CountryParts.

Example12_3 makes use of CountryParts and a new method for acquiring the input data that is contained in a file. Here are the key points behind this method of acquiring input data. First, the name of the file is established with the reset instruction:

```
reset(f,'World');
```

```
program Example12_3;
{Prints the data in the file World}
uses
   Chap12Codes;
var
   k:integer;
   f:text; {File name}
   s:string; {Country data}
   name, pop, Area, Capital, Lat, Long:  string; {Substrings of s}
begin
   ShowText;
   reset(f, 'World');
   writeln(' Country Population Area Capital Lat.  Long.');
   writeln('---------------------------------------------------------------');

   while not eof(f) do
      begin
         readln(f,s);
         CountryParts(s, Name, Pop, Area, Capital, Lat, Long);
         writeln(Country:20, Pop:12, Area:12, Capital:20, Lat:10, Long:10);
      end
end.
```

The procedure `CountryParts` is declared in the unit `Chap12Codes`. To run this program, the text file `World` must be in the same folder as the project. Output:

```
     Country  Population      Area    Capital     Lat.           Long.
----------------------------------------------------------------------
 Afghanistan  14,800,000   250,000      Kabul   34 30N          67 40E
     Albania   3,200,000    11,100     Tirana   41 20N          19 50E
     Algeria  24,900,000   919,595    Algiers   36 50N          03 00E
                                   .
                                   .
                                   .
       Zaire  41,200,000   905,365  Kinshasha    4 18S          15 17E
      Zambia   8,600,000   290,586     Lusaka   15 30S          28 10E
    Zimbabwe  10,700,000   150,698     Harare   33 33N          44 44W
```

This assigns the name of the file, 'World', to the variable f which is declared to have the type text. Once this is accomplished we use readln as always except that the first argument is now f. *For this to work, it is necessary to have the project and the data file in the same folder.*

Another new feature in Example12_3 is the use of the built-in boolean-valued function Eof (End-of-File). This function is false after the last line in the file has been read. Thus, the while iteration continues until the last line in the file is read. It is equivalent to equivalent to

```
for k:=1 to n do
   begin
      readln(f,s);
      CountryParts(s, Name, Pop, Area, Capital, Lat, Long);
      writeln(Country:20, Pop:12, Area:12, Capital:20, Lat:10, Long:10);
   end
```

assuming that n contains the number of lines in World. The while/EoF strategy is better because data files are sometimes very long and it is inconvenient to count the number of lines.

PROBLEM 12.3. Modify Example12_3 so that it prints a list of all countries whose capitals are in the southern hemisphere.

PROBLEM 12.4. Modify Example12_3 so that it prints a list of all countries that have at least 100 million people.

12.2 Place Value

Consider the modification of Example12_3 so that it prints the sum of the country land areas. Even if A has type longint, a fragment of the form

```
A := 0;
for k := 1  to n do
   begin
      readln(f,s);
      CountryParts(s, Name, Pop, Area, Capital, Lat, Long);
      A := A + Area;
   end
```

does *not* work because Area contains a string representation of the country's land area, e.g., '3,536,341'. A string value like this must be converted to its numerical equivalent before it can participate in an arithmetic summation.

The conversion of the string Area begins with the removal of the commas and so we first develop a function with that mission If s is a string with a comma in between its first and last position, then

```
n:=pos(',',s);
s := concat(copy(s,1,n-1), copy(s,n+1,length(s)-n));
```

removes the first occurrence of a comma. By repeating this process under the control of a while loop, all commas can be removed:

```
function RemoveCommas(s:string):string;
{Pre:s[1] and s[length(s)] are not commas.}
```

```
{Post:The string obtained by deleting all commas from s.}
var
   n:integer;
begin
   n := pos(',', s);
   while n > 0 do
      begin
         s := concat(copy(s,1,n-1), copy(s,n+1,length(s)-n));
         n := pos(',',s)
      end;
   RemoveCommas := s
end
```

It follows that

```
s := RemoveCommas(s)
```

overwrites s with a comma-less version of itself.

To complete the conversion of a digit string to its numerical counterpart, we need to determine the value of each digit in accordance with base-10, place-value notation. When we write a number like "2703829" the value of each digit depends upon its place, e.g., 9 ones plus 2 tens plus 8 hundreds etc. The value of a number represented in this style is a summation with terms that are a multiple of a power of ten:

$$2703829 = 9 \cdot 10^0 + 2 \cdot 10^1 + 8 \cdot 10^2 + 3 \cdot 10^3 + 0 \cdot 10^4 + 7 \cdot 10^5 + 2 \cdot 10^6.$$

Encapsulating the process we obtain

```
function Str2Int(s:string):longint;
{Pre:s is a string encoding of an integer x that satisfies}
{0<=x<=2^31 - 1.}
{Post:x}
var
   n:integer; {The length of the string s}
   k:integer;
   TenPower:longint;{10^k}
   x:longint; {The numerical value of the last k digits in s.}
   digit:char; {The (n-k)-th character in s.}
   value:integer; {The numerical value of digit.}
begin
   n:=length(s); k:=0; TenPower:=1; x:=0;
   while k < n do
      {TenPower = 10^k, x = value of s[n-k+1..n]}
      begin
         digit := s[n-k];
         value := Char2Int(digit);
         x:= x + value*TenPower;
         k:=k+1;
         TenPower:= TenPower*10;
      end;
   Str2Int := x;
end
```

The idea is to peel off the characters in `s` in right-to-left order. When `k=0`, the assignment to `digit` has the form `digit:=s[n]` and the ones-place is obtained. When `k=1`, `digit` gets the 10's place since in effect we have `digit:=s[n-1]`. As this process continues, `TenPower` proceeds from 1 to 10 to 100 etc. The summation is built up in `x`.

The program `Example12_4` uses `Str2Int` and `RemoveCommas` to obtain a sum of land areas.

```
program Example12_4;
{Total Land Area}
uses
   Chap12Codes;
var
   k:integer;
   f:text; {File name}
   s:string; {Country data}
   A:longint;
begin
   ShowText;
   reset(f,OldFileName('World'));
   A:=0;
   while not eof(f) do
      begin
         readln(f,s);
         CountryParts(s,Name,Pop,Area,Capital,Lat,Long);
         A := A + Str2Int(RemoveCommas(Area));
      end
   writeln(A:10, ' = Total Area');
end.
```

The procedure `CountryParts` and the functions `Str2Int` and `RemoveCommas` are declared in the unit `Chap12Codes`. To run this program, the text file `World` must be in the same folder as the project. Output:

```
51095218 = Total Area
```

PROBLEM 12.5. Modify `Example12_3` so that it prints the country information for countries whose capital cities are in the tropics, meaning that their latitude is in between 23°30′S and 23°30′N.

PROBLEM 12.6. Modify `Example12_3` so that it prints a table of those countries whose population is greater than one million and whose population density exceeds 500. Each line in the table should have the country's name, population, area, and population density.

PROBLEM 12.7. Complete the following functions:

```
function DaysInMonth (m:integer):  integer;
{Pre:1<=m<=12}
{Post:Number of days in month m.  (1=Jan, 2=Feb, etc).  Leap year not assumed.}
function DaysSinceBase(s:string):longint;
{Pre:s is a string of the form mm/dd/yy where:}
{ mm is one of the strings '1' , '2' , ...,'12' and indicates month.}
{ dd is one of the strings '1' , '2' , ...,'31' and indicates day.}
```

```
{ yy is one of the strings '00' , '2' , ...,'99' and indicates year.}
{Post:The number of days from January 1, 1900 to the date specified by s.}
{Leap years are taken into account.}
```

PROBLEM 12.8. Define a "Roman String" to be string that is comprised of the characters I, V, X, L, C, D, and M. Here are the rules that specify the value of these symbols in the Roman Numeral system:

$$
\begin{aligned}
M &= 1000 \\
D &= 500 \\
C &= \begin{cases} -100 & \text{if followed by a D or M} \\ +100 & \text{otherwise.} \end{cases} \\
L &= 50 \\
X &= \begin{cases} -10 & \text{if followed by a L or C} \\ +10 & \text{otherwise.} \end{cases} \\
V &= 5 \\
I &= \begin{cases} -1 & \text{if followed by a V or X} \\ +1 & \text{otherwise.} \end{cases}
\end{aligned}
$$

(Note: A Roman string need not represent a valid Roman Numeral, e.g., 'MCCCM'. However, even in such a case, the rules given produce an unambiguous value, e.g., 1000 + 100 + 100 -100 + 1000 = 2100. Complete the following function

```
function NumeralValue(s:string):integer;
{Pre:s is a Roman string.}
{Post:The numerical value of s.}
```

You mind find it handy to the base the implementation upon

```
function NextIsBigger(s:string; k:integer):boolean;
{Pre:k>=1}
{Post:k<length(s) and copy(s,k,2) = 'IV', 'IX','XL','XC','CD', or 'CM'.}
```

PROBLEM 12.9. *Bar codes* are a graphical means by which to represent numerical information, e.g.,

A 7-strip field is used to encode a single digit. Each field is either black or white. Each digit has a "left code" and a "right code" depending upon whether it is to the left or right of the center. Here is a table that shows the twenty possibilities

Digit	Left	Right
0	0001101	1110010
1	0011001	1100110
2	0010011	1101100
3	0111101	1000010
4	0100011	1011100
5	0110001	1001110
6	0101111	1010000
7	0111011	1000100
8	0110111	1001000
9	0001011	1110100

A "1" indicates black and a "0" indicates white. Complete the following function:

```
function DigitCode(n:integer; side:char):string;
{Pre:side = 'L' or 'R'}
{Post:The left (L) or right (R) binary UPC code for n}
```

PROBLEM 12.10. Complete the following function

```
function Str2Real(s:string):real;
{Pre:  s = concat(s1,s2,s3,s4) where:}
{ s1 = empty string or '-'}
{ s2 = digit string}
{ s3 = '.'}
{ s4 = digit string}
{Post:The value of s as a real number.}
```

Now let's turn the tables and consider the representation of a number as a string. Our goal will be to develop the string-valued function

```
function SciNot(x:real):string;
{Post:x in 6-digit scientific notation}
```

The definition 6-digit scientific notation is amply illustrated with a few examples:

23	⇔	.230000x10^2
.0012345678	⇔	.123457x10^-2
0	⇔	.000000x10^0
123456100000	⇔	.123456x10^12

Thus, if x is a real variable, then the fragment

```
x:=pi;
writeln('x = ', SciNot(x))
```

produces the output

```
.314159x10^1
```

The development of SciNot is a good occasion to practice top-down problem solving. The production of the required string involves (a) computing a real number m that satisfies

$$\frac{1}{10} \le m < 1$$

so that $x = m \cdot 10^e$ where e is an integer, (b) the conversion of m and e to their "string equivalents," and (c) the concatenation of the mantissa and exponent strings with other symbols that are part of the notation system. Our plan is to package the solution to (a) and (b) in

```
procedure MantExp(x:real;var m:real; var e:integer);
{Post:if x<>0, x = m*10^e where 0<m<1.}
{Otherwise m = 0 and e = 0}

function Int2Str(n:longint):string;
{Post:The string 'equivalent' of n.}
{Thus, 31 becomes '31', -578 becomes '-578', etc.}
```

Assuming that these are available, then `SciNot` need only take care of the concatenations that make up part (c) of the problem solving process:

```
function SciNot(x:real):string;
{Post:x in 6-digit scientific notation}
var
   m:real; {The mantissa of x.}
   e:integer; {The exponent of x.}
   PosX:string; {abs(x) in scientific notation.}
begin
   if x = 0 then
      SciNot := '.000000x10^0'
   else
      begin
         MantExp(x,m,e);
         PosX:=concat('.', Int2Str(round(1000000 * abs(m))),'x10^', Int2Str(e));
         if x < 0 then
            SciNot := concat('-', PosX)
         else
            SciNot := PosX
      end;
end;
```

Thus, if m = .1234567 and e = 3, then the string representation of $round(1000000 * m)$ gives '123456', the string representation of e gives '3'), and `SciNot` produces '.123456x10^3' via `concat('.','123456','x10^','3')`.

The development of `MantExp` and `IntToString` are exercises in base-10, positional notation. For the former there are three cases to consider: $x < 0$, $x = 0$, and $x > 0$.

```
procedure MantExp(x:real;var m:real; var e:integer);
{Post:x = m*10^e where 0<m<1.}
begin
   m := abs(x);
   e := 0;
   if m <> 0 then
      begin
         while m >= 1 do
            {Divide m by 10 and increase e by 1.}
            {Note:abs(x) = m*10^e}
            begin
               e:= e+1;
               m:= m/10;
            end;
         while m < 0.1 do
            {Multiply m by 10 and decrease e by 1.}
            {Note:abs(x) = m*10^e}
            begin
               e := e-1;
               m := 10*m
            end;
      end;
```

```
    if x < 0 then
        m := -m;
end;
```

The idea is to initially set $m = |x|$ and then to repeatedly scale its value by 10 or 1/10 until it is in the proper range. In conjunction with this, the value of e (initially equal to zero) is modified to maintain the relationship $|x| = m \cdot 10^e$. The final value of m is negated if the input value is negative. Zero is represented as $0 \cdot 10^0$.

The task of the integer-to-string conversion function Int2Sttr is the inverse of Str2Int developed earlier. The idea is to peel off the digits that make up n one by one, converting them to their character equivalents. The required string can be obtained by the concatenation of these characters. Digit-to-character conversion can be accomplished by using the digit to extract the right character from the string '0123456789':

```
function Digit2Char(n:integer):char;
{Pre:0<=n<=9}
{Post:The character equivalent of n.}
const
    Digits = '0123456789';
begin
    Digit2Char:=copy(n,1,Digits)
end
```

This allows us to write the following integer-to-string conversion function:

```
function Int2Str(n:longint):string;
{Pre:n>=0.}
{Post:The string equivalent of n.  Thus, 31 becomes '31'}
var
    s:string;
    NextDigit:longint;
    NextDigitAsChar:char;
    m:longint; { Houses n div (10^p) for successively larger p.}
begin
    s := '';
    m := n;
    while m > 0 do
        begin
            NextDigit := m mod 10;
            NextDigitAsChar := Digit2Char(NextDigit);
            s := concat(NextDigitAsChar,s);
            m := m div 10;
        end;
    if n > 0 then
        Int2Str := s
    else
        Int2Str := '0'
end
```

Here is essentially what happens when this function is applied to n = 983:

```
program Example12_5;
{Scientific notation.}
var
   AnotherEg:char;
   x:real;
begin
   ShowText;
   AnotherEg := 'y';
   while AnotherEg <> 'n' do
      begin
      writeln('Specify a real number:  ');
      readln(x);
      writeln(' ', SciNot(x));
      writeln('Another example?  Enter y (yes) or n (no).');
      readln(AnotherEg);
   end
end.
```

The function SciNot is declared in the unit Chap12Codes. Sample output:

```
Specify a real number:
1234.56789
          .123457x10^4
Another example?  Enter y (yes) or n (no).
n
```

```
s := ''; { empty string}
NextDigit := n mod 10; { Isolate the 3 }
c:=Digit2Char(NextDigit);
s := concat(c,s); { s = '3'}
n := n div 10; { n is now 98 }

NextDigit := n mod 10; { Isolate the 8 }
c:=Digit2Char(NextDigit);
s := concat(c,s); { s = '83'}
n := n div 10; { n is now 9 }

NextDigit := n mod 10; { Isolate the 9 }
c:=Digit2Char(NextDigit);
s := concat(c,s); { s = '983'}
n := n div 10; { n is now 0 }
```

The combined use MantExp, Int2Str and SciNot is illustrated in Example12_5.

PROBLEM 12.11. Complete the following function

```
function InsertCommas (s:string):string;
```

```
{Post:Counting from the end, a comma is inserted after every 3rd character unless}
{that character is the first in the string.}
```

and use it to modify Example_4 so that it prints the total land area with commas.

PROBLEM 12.12. SciNot produces a string representation of x in 6-digit scientific notation. Generalize this function so that it produces a d-digit representation where $1 \leq d \leq 6$. In addition, your generalized SciNot should concatenate enough blanks to the front of the representation so that a length L string is returned. In particular, complete the function

```
function GenSciNot(x:real; d,L:integer):string;
{Pre:1<=d<=6, L>=d+10}
{Post:x in d-digit scientific notation.  Enough blanks are concatenated}
{to the front so that the returned string has length L. }
```

The condition on L ensures that a length L string is long enough to house the required representation.

We conclude this chapter with a discussion of the base-2 (binary) number system since it has an important role to play in digital computing. In the base-10 system, 10 symbols are used and their value depends upon their placement in the representation:

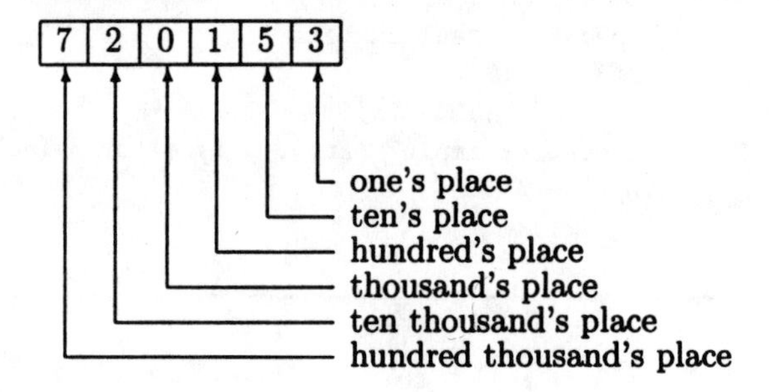

Thus,

$$720153 = 3 \cdot 10^0 + 5 \cdot 10^1 + 1 \cdot 10^2 + 0 \cdot 10^3 + 2 \cdot 10^4 + 7 \cdot 10^5.$$

In the binary system, two symbols bits are used: 0 and 1. These are called *bits* and the value that they designate also depends upon their place in the representation:

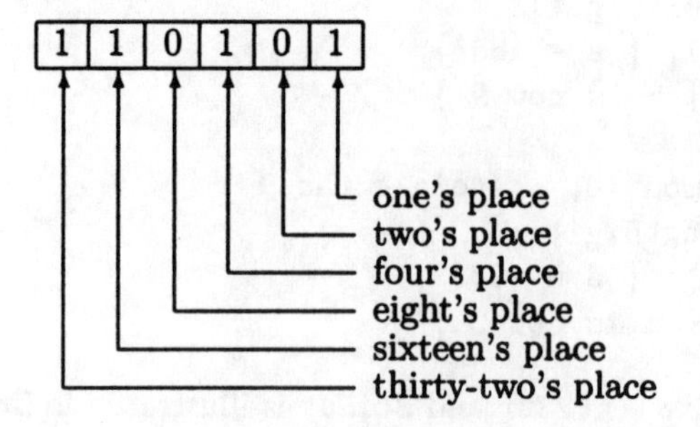

Analogously, the value represented in binary notation is a sum whose terms involve powers of two:

$$110101 \equiv 1 \cdot 2^0 + 0 \cdot 2^1 + 1 \cdot 2^2 + 0 \cdot 2^3 + 1 \cdot 2^4 + 1 \cdot 2^5 = 1 + 4 + 16 + 32 = 53.$$

To compute (as a string) the base-2 representation of an integer like 53 we pose a sequence of true/false questions:

Question	Answer	s	Take Away Calculation
How many 32's in 53?	1	'1'	53-32 = 21
How many 16's in 21?	1	'11'	21 - 16 = 5
How many 8's in 5?	0	'110'	5-0 = 5
How many 4's in 5?	1	'1101'	5-4 =1
How many 2's in 1?	0	'11010'	1-0=1
How many 1's in 1?	1	'110101'	1-1 = 0

The appropriate "starting" power of two to begin with is the largest power of two that is less than or equal to n. Handling separately the $n = 0$ case we obtain

```
function Int2Bin(n:longint):string;
{Pre:n>=0}
{Post:The base-2 representation of n.}
var
   m:longint; {A power of two.}
   s:string; {For building up the binary representation.}
begin
   if n = 0 then
      Int2Bin := '0'
   else
      begin
         m := 1;
         while 2 * m <= n do
            m := 2*m;
         {m is the largest power of two <= n}
         s := '';
         while m >= 1 do
            begin
               if m <= n  then
                  begin
                     {We need a "1" in the m's place.}
                     s := concat(s, '1');
                     n := n - m
                  end
               else
                  begin
                     {We need a "0" in the m's place.}
                     s := concat(s,'0');
                     m := m div 2;
                  end
            end;
         Int2Bin := s;
      end;
end
```

Example12_6 uses this function to print the binary representation of the integers 0 through 31.

```
program Example12_6;
{Binary notation.}
uses
   Chap12Codes;
var
   k:integer;
begin
   ShowText;
   writeln(' Base-10 Base-2 ');
   writeln('----------------------------');
   for k := 0 to 31 do
      writeln(k:8, ' ', Int2Bin(k,5):8);
end.
```

The function Int2Bin is in the unit Chap12Codes. Output:

```
 Base-10              Base-2
----------------------------
       0                   0
       1                   1
       2                  10
       3                  11
       4                 100
       5                 101
       6                 110
       7                 111
       8                1000
                 ⋮
      29               11101
      30               11110
      31               11111
```

PROBLEM 12.13. Complete the following functions:

```
function Reverse(s:string):string;
{Post:s in reverse order.}

function Bin2Int(s:string):longint;
{Pre:  s is a string of zeros and ones.}
{Post:The value of s as a binary number.}
```

Modify Example12_6 so that the k-th line of output has four items: k, k as a 5 bit binary number, the reversal of that string, its value.

PROBLEM 12.14. Complete the following function:

```
function Int2BaseP(n:longint; p:integer):string;
{Pre:0<=n}
{Post:The string 'equivalent' of n expressed in base-p notation.}
```

Chapter 13

Visualization

In Chapter 2 we developed the notion of numerical exploration, the main idea being that we could get a handle on difficult mathematical questions through computer experimentation. This is one of the most important aspects of computational science. To dramatize further this point, we enlist the services of computer graphics. Our geometric intuition and our ability to visualize go hand-in-hand. Both are essential in many problem-solving domains and graphics can lend a real helping hand.

We start by developing a handful of graphical tools that permit the construction of simple exploratory environments. Sometimes the computer visualization of a problem or a task is an end in itself. On other occasions, it merely sets the stage for analytical work. Regardless of how it is used, a "visualization system" is driven by many behind-the-scenes computations that permit the suppression of mundane detail. Typical among these are the coordinate transformations that take us from "problem space" to "screen space." The intelligent handling of these transformations leads to some important software issues.

13.1 Exploratory Environments

Here is a hard problem:

How many radius r disks are required to cover an s-by-s square?

We'll call the answer to this question the *cover number*. The cover number is a very complicated function of s and r and we have no intention of specifying it exactly. Our goal instead is to qualitatively explore its properties learning what we can through graphics. The "cast of characters" is displayed in FIGURE 13.1. Think of the square as a"target" at which we throw darts that leave a radius r disk upon impact. Two procedures declared in the unit `Chap13Codes` will be convenient:

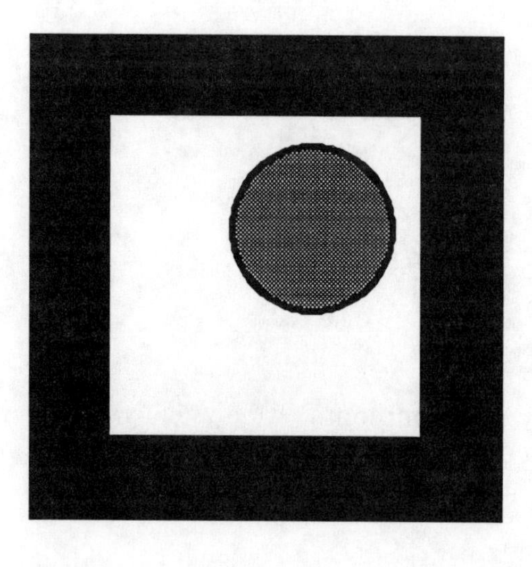

FIGURE 13.1 *The "Target" and a Disk*

```
procedure PaintDisk(h,v,r:integer);
{Post:Draws a bordered gray disk with radius r and center (h,v).}

procedure DrawTarget(s:string; vT,hL,vB,hR,r:integer);
{Post:Draws a border of width r around a white square defined by}
{hL<=h<=hR and vT<=v<=vB. The string s is displayed.}
```

We start our investigation by writing a program that allows us to use the mouse to click in the position of each disk. This is much simpler than having to solicit (h, v) locations using `readln` and `writeln`. To that end we make use of the following procedure:

```
procedure GetPosition(var h,v:integer);
{Post:(h,v) is the location of the next mouseclick.}
```

This procedure is declared in the unit `DDCodes`: The implementation of will not be discussed because it involves the use of records, a concept not discussed until Chapter 18. However, the use `GetPosition` is straight forward. The fragment

```
GetPosition(h,v);
MoveTo(h,v);
PaintDisk(h,v,3));
```

draws a small disk at the point of a mouse click. The processing of repeated mouse clicks is illustrated in `Example13_1`. Notice that the `while`-loop "forces" you click on the inner square area because the guard is true as long as the coordinates of a mouseclick are *not* in the rectangle defined by the constants `hL`, `hR`, `vT` and `vB`.

```
program Example13_1;
{Disks in a square.}
uses
   DDCodes, Chap13Codes;
const
   {Target rectangle defined by hL<=h<=hR and vT<=v<=vB}
   hL=150; hR=300; vT=80; vB=230;
   r = 40; {Disk radius}
var
   k:integer;
   h,v:integer; {Center of the dot being drawn.}
begin
   ShowDrawing;
   {Set up target and initialize the dot counter.}
   ClearScreen;
   DrawTarget('Click in 10 points.',vT,hL,vB,hR,r);
   for k:=1 to 10 do
      begin
         GetPosition(h,v);
         while (h < hL) or (hR < h) or (v < vT) or (vB < v) do
            GetPosition(h,v);
         PaintDisk(h,v,r);;
      end
end.
```

The procedures PaintDisk and DrawTarget are declared in the unit Chap13Codes. The function/procedures ClearScreen and GetPosition are declared in the unit DDCodes. Sample output: See FIGURE 13.2.

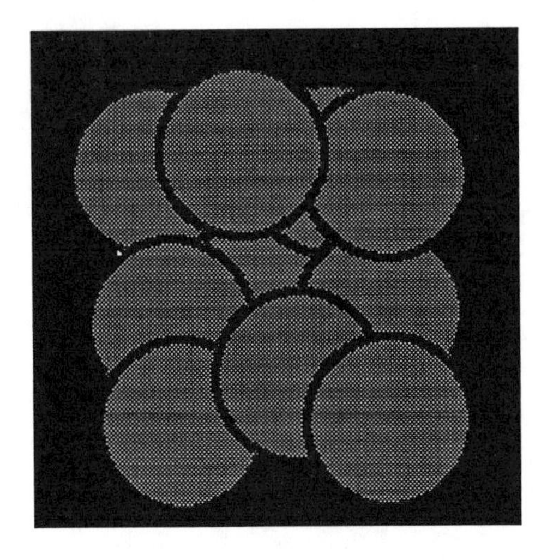

FIGURE 13.2 *Output from* Example13_1

FIGURE 13.3 *The* YesNo *Alternative*

Instead of just clicking in ten points, we may wish to click in an arbitrary number of points. In a situation like this, it is handy to be able to solicit boolean information by using the mouse. One way to do this is to have a boolean-valued function whose value depends upon which of two boxes is clicked:

```
function YesNo(Message:string):boolean;
{Post:  Displays two boxes at the top of the screen.}
{True if the displayed ''yes box'' is clicked.}
{False if the displayed ''no box '' is clicked.}
{In the latter case, both boxes are erased.}
var
   InYes,InNo:boolean; {False until one of the boxes clicked.}
   h,v:integer; {(h,v) = mouse click location}
begin
   {Clear message/box area.}
      ⟨:⟩
   InYes := false;
   InNo := false;
   while (not InYes) and (not InNo) do
      begin
         GetPosition(h,v);
         InYes := (20 <= h) and (h <= 30) and (15 <= v) and (v <= 25);
         InNo := (70 <= h) and (h <= 80) and (15 <= v) and (v <= 25);
      end;
   if InNo then
      EraseRect(0,0,25,100);
      YesNo := InYes;
end;
```

The `while` loop terminates only after one of the two boxes is clicked. When called, `YesNo` displays a two box alternative as illustrated in FIGURE 13.3. Using `YesNo` you can set the value of a boolean variable,

```
program Example13_2;
{Illustrate YesNo}
uses
   DDCodes,Chap13Codes;
const
   hL=150; hR=300; vT=80; vB=230;
   {Target rectangle defined by hL<=h<=hR and vT<=v<=vB}
   r=20; {Dot radius}
var
   rn:real; {For random numbers}
   h,v:integer; {Center of the dot being drawn.}
   count:longint; {Number of dots drawn.}
begin
   ShowDrawing;
   {Set up target and initialize the dot counter.}
   ClearScreen;
   DrawTarget(' ',vT,hL,vB,hR,r);
   count := 0;
   while YesNo('Another Dot') do
      {Generate dots until mouse is clicked.}
      begin
         GetPosition(h,v);
         PaintDisk(h,v,r);
         {Display the dot counter.}
         count:=count+1;
         EraseRect(200,400,240,500);
         MoveTo(410,220);
         writedraw('Count = ', count:4);
      end;
end.
```

The procedures `DrawTarget` and `PaintDisk` are declared in the unit `Chap13Codes`. The function/procedures `YesNo`, `GetPosition`, and `ClearScreen` are declared in the unit `DDCodes`.

```
BoolVar := YesNo('Some Message')
```

steer an alternative

```
if YesNo('Some Message') then
   ⟨:⟩
else
   ⟨:⟩
```

or control a `while`-loop as in `Example13_2`.

PROBLEM 13.1. Modify `Example13_2` so that it terminates only after the mouse is clicked outside of the target square.

Now let's modify the underlying question:

How many randomly placed, radius r disks are required to cover an s-by-s square?

We could continue to solicit disks by Yes/No clicks. However, the one-disk-at-a-time strategy is too slow when lots of disks are involved. The controlling while loop can be "turned loose" by using the built-in ThinkPascal function `Button`. This is a boolean-valued function with no arguments. It returns true if the mouse button is depressed and false otherwise. In `Example13_3` the while loop continues until the mouse is clicked.

```
program Example13_3;
{Illustrate Button}
uses
   DDCodes,Chap13Codes;
const
   hL=150; hR=300; vT=80; vB=230; {Target rectangle defined by hL<=h<=hR and vT<=v<=vB}
   seed = 0.123456;
   r = 8; {Dot radius}
var
   rn:real; {For random numbers}
   h,v:integer; {Center of the dot being drawn.}
   count:longint; {Number of dots drawn.}
begin
   ShowDrawing;
   rn:=seed;
   {Set up target and initialize the dot counter.}
   ClearScreen;
   DrawTarget('Click mouse to stop.',vT,hR,vB,hR,r);
   count:=0;
   while  not button do
      {Generate random dots until mouse is clicked.}
      begin
         rn:=rand(rn); h:=hL+trunc((hR-hL+1)*rn);
         rn:=rand(rn); v:=vT+trunc((vB-vT+1)*rn);
         PaintDisk(h,v,r);
         {Display the dot counter.}
         count := count + 1;
         EraseRect(200,400,240,500);
         MoveTo(410,220);
         writedraw('Count = ', count:4);
      end;
end.
```

The procedures `DrawTarget` and `PaintDisk` are declared in the unit `Chap13Codes`. The function/procedures `rand`, `YesNo`, `GetPosition`, and `ClearScreen` are declared in the unit `DDCodes`.

PROBLEM 13.2. Modify `Example13_3` so that it pauses after every m disks are drawn where m is a positive integer constant. At each pause, a yes/no alternative should be displayed that permits either the continuation or the termination of the program.

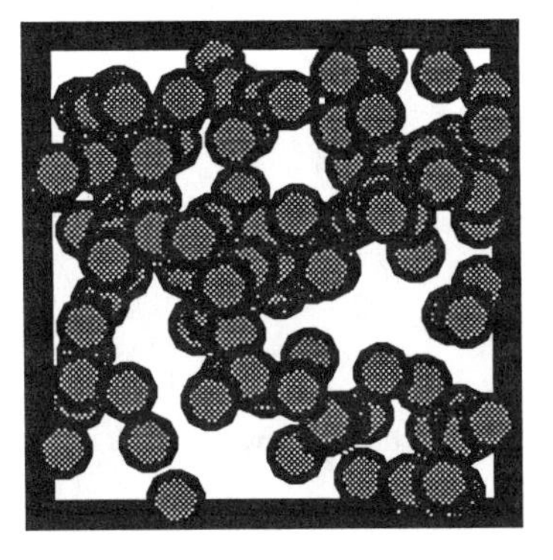

FIGURE 13.4 *The* ReviseInteger *Alternatives*

Suppose we wish to explore how the "covering number" varies with radius. We certainly have the tools to do this through **readln writeln**, and **WriteDraw**. However, it is sometimes handy to introduce numerical values using mouse clicks instead of the keyboard. A very simple function has been written to illustrate this:

```
function ReviseInteger(s:string; OldValue,L,R,Num:integer):integer;
{Pre:  L <= OldValue <= R and Num > 0.}
{Post:  An interactively determined value between L and R is returned.}
{The string s should name the parameter to be varied.}
{The new value is chosen from a finite subset of values in [L,R].}
{The spacing between possible values is (R-L) div Num.}
```

ReviseInteger is declared in the unit **DDCodes** and can be used to vary an integer parameter like the radius. When called, it presents a three box menu as illustrated in FIGURE 13.4. **Example13_4** uses **ReviseInteger** and provides a framework for us to explore the connection between the radius and the covering number.

PROBLEM 13.3. Modify **Example13_4** so that it leaves a trace of all the "experiments" in the DrawWindow. In particular, after each experiment, it should print the radius and the experimentally determined cover number. These results should be assembled in a table and remain on the screen. Thus, you'll have to replace the call to **ClearScreen** with a suitable call to **EraseRect** that leaves the table area alone.

PROBLEM 13.4. Develop an interactive environment that can be used to find the largest circle that fits inside a given triangle.

```
program Example13_4;
{Illustrate ReviseInteger}
uses
   DDCodes;
const
   hL=150;hR=300;vT= 80;vB= 230; {Target rectangle defined by hL<=h<=hR and vT<=v<=vB}
   seed = 0.123456;
var
   rn:real; {For random numbers}
   r:integer; {Radius of dots.}
   h,v:integer; {Center of the dot being drawn.}
   count:longint; {Number of dots drawn.}
begin
   ShowDrawing;
   r:=5; {Initial disk radius}
   rn := seed;
   while YesNo('Another Experiment?')  do
      begin
         {Set the value of the dot radius.}
         r := ReviseInteger('Dot radius',r,1,40,39);
         {Set up target and initialize the dot counter.}
         ClearScreen;
         DrawTarget('Click mouse to stop.',vT,hL,vB,hR,r);
         count:=0;
         while not button do
            {Generate random dots until mouse is clicked.}
            begin
               rn:=rand(rn);
               h:= hL+trunc((hR-hL+1)*rn);
               rn := rand(rn);
               v := vT+trunc((vB-vT+1)*rn);
               PaintDisk(h,v,r);
               {Display the dot counter.}
               count:=count+1;
               EraseRect(200,400,240,500);
               MoveTo(410,220);
               writedraw('Count = ',count:4);
            end;
      end;
end.
```

The function/procedures `rand`, `YesNo`, `GetPosition`, and `ClearScreen` are in the unit `DDCodes`. The procedures `DrawTarget` and `PaintDisk` are declared in the unit `Chap13Codes`.

FIGURE 13.5 *A Discrete Logarithmic Spiral*

Some visualization problems involve real parameters and for that purpose we can solicit numerical values via the following procedure:

```
function ReviseReal(s:string; OldValue,L,R:real; Num:integer):real;
{Pre:  L <= OldValue <= R and Num > 0.}
{Post:  An interactively determined value between L and R is returned.}
{The string s should name the parameter to be varied.}
{The new value is chosen from a finite subset of values in [L,R].}
{The spacing between possible values is (R-L)/Num.}
```

This procedure is declared in the unit DDCodes and works just like ReviseInteger. Let's use it to explore how the pattern produced by

```
procedure DrawSpiral(n,turn,hc,vc:integer; d:real);
{Post:Draws a polygonal line with n "legs".  }
{The k-th leg has length k*d and turns left turn degrees from leg k-1.}
{The first leg starts at (hc,vc) and ends at (hc+d*cos(turn),vc-d*sin(turn)).}
```

depends upon the values passed. This procedure draws something called a *discrete logarithmic spiral* and is declared in the unit Chap13Codes. See FIGURE 13.5. Example13_5 sets up a framework that permits easy variation of the three parameters n, *turn*, and d. Notice how a sequence of calls to ReviseInteger and ReviseReal can be used in situations where there is more than one parameter to vary.

```
program Example13_5;
{Draws a discrete logarithmic spiral.}
uses
   DDCodes, Chap13Codes;
var
   n,turn:integer;
   d:real;
begin
   ShowDrawing;
   n:=500;
   turn:= 60;
   d := 0.5;
   DrawSpiral(n,turn,230,140,d);
   while YesNo('More Revisions?')  do
      begin
         n := ReviseInteger('n',n,100, 2000,19);
         turn := ReviseInteger('turn',turn,0,180,180);
         d := ReviseReal('d', d, 0.01, 1.00, 99);
         ClearScreen;
         DrawSpiral(n,turn,230,140,d);
      end;
end.
```

The procedure **DrawSpiral** is declared in the unit **Chap13Codes**. The functions/procedures **ClearScreen**, **YesNo**, **ReviseInteger** and **ReviseReal** are declared in the unit **Chap13Codes**. For sample output, see FIGURE 13.5.

PROBLEM 13.5. Complete the following procedure:

```
procedure DrawOlympic(hc,vc,r:integer; hFactor,vFactor,dFactor:real);
{Pre:  r > 0}
{Post:Olympic symbol with center ring centered at (hc,vc).  Rings have radius r.}
{Center-to-center horizontal spacing between rings is hFactor*r.}
{Vertical spacing between centerlines of two rows is vFactor*r.}
{Ring thickness is dFactor*r.}
```

A call to DrawOlympic should produce a 5-ring design of the form

Notice that the first three parameters deal with position and size while the last three parameters deal with proportion. The act of finding the most pleasing 5-ring design is the act of working out these three proportionality factors. Set up an environment that enables you to determine the proportionality values that produce a pleasing Olympic symbol.

PROBLEM 13.6. Write a procedure that can draw a backgammon board:

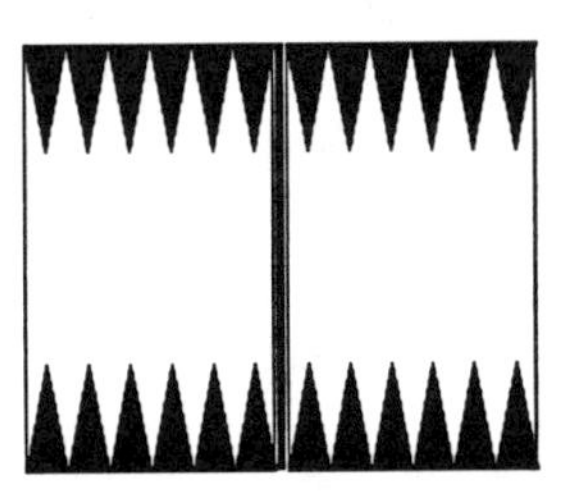

Proceed by developing the following hierarchy of procedures:

```
procedure DrawSpike(hc,vc,s:integer; SpikeFactor:real; up:boolean);
{Post:Draws a shaded triangle with base vertices at (hc-s,vc) and (hc+s,vc).}
{If up is true, third "tip" vertex is at (hc,vc- round(SpikeFactor*s)).  }
{Otherwise, it is at (hc,vc+ round(SpikeFactor*s))}

procedure DrawSpikeRow(n,hc,vc,s:integer; SpikeFactor:real; up:boolean);
{Post:Draws a row of n spikes.}
{The k-th spike is defined by the five parameters}
{(hc+2(k-1)s, s, vc, SpikeFactor, and up.  }

procedure DrawPanel(vT,hL,n,s:integer; SpikeFactor,PanelFactor:real);
{Post:Draws a rectangle defined by (vT,hL,vT+vHgt,hL+2ns) where}
{ vHgt =round(PanelFactor*s).  Inward pointing spike rows of length n }
{are drawn across the top and bottom of the rectangle.  }
{The spike dimensions are defined by s and SpikeFactor.}

procedure DrawBoard(vT,hL,n,s:integer; SpikeFactor,PanelFactor:real);
{Post:Draws a rectangle defined by (vT,hL,vT+vHgt,hL+2ns) where}
{vHgt =round(PanelFactor*s).  Inward pointing spike rows of length n }
{are drawn across the top and bottom of the rectangle.  }
{The spike dimensions are defined by s and SpikeFactor.}
```

Develop an environment that facilitates the design of a "nice looking" board.

PROBLEM 13.7. Write a procedure

```
procedure DrawClock(hc,vc,r,hour,min:integer);
{Pre:0<=min<=59, 1<=hour<=12}
{Post:  draws a circular clock, centered at (hc,vc) with radius r.}
{The clock shows the time specified by hour and min.}
```

There should be 60 hash marks that designate minutes. Determine a suitable length for these, making every fifth one longer. Hour and minute hands (of your own design) should also be drawn. Develop an environment that permits the easy selection of the hour and minute and displays the clock with the specified time.

PROBLEM 13.8. Complete the following procedure:

```
procedure DrawSpokes(n,r1,r2,p,hc,vc:integer);
{Pre:r1 > r2.}
{Post:Draws circles of radius r1 and r2 centered at (hc,vc).}
{Define t = 2*pi/n.}
{For k= 1 to n, draws a line between}
{ (r1*cos(kt),r1*sin(kt)) and (r2*cos((k-p)t),r2*sin((k-p)t)}
{ (r1*cos(kt),r1*sin(kt)) and (r2*cos((k+p)t),r2*sin((k+p)t)}
```

Develop an environment that can be used to explore how the values of p and n effect the design. Arrange the calls to ReviseInteger so that if the current value is changed, then it is either doubled or halved. Set $p = 8$ and $n = 128$ initially.

PROBLEM 13.9. Implement the following procedure two ways

```
procedure PaintSemiCircle(hc,vc,r,theta:longint);
{Pre:0<=theta<=180}
{Post:  Draws a semicircle with center (hc,vc) and radius r.}
{The curved side of the semicircle is defined by the parametric}
{equations h(a) = hc+r*cos(a) and v(a) = vc - r*sin(a) where a ranges}
{from theta degrees to theta + 180 degrees.}
```

Let PaintSemiCircle1 achieve the shading by drawing a suitable number of radial lines and let PaintSemiCircle2 shade by drawing every possible horizontal line segment within the semicircle. Set up an environment that permits the easy variation of the two parameters r and theta. The semicircles produced by the two methods should be displayed side by side along with benchmark information so that efficiency can be compared. Note: do not expect the radial shading approach to be perfect.

13.2 Coordinate Systems

So far in the text, all the geometric objects that we have displayed have been specified in terms of screen coordinates. For example, a call to

```
procedure DrawTriangle(h1,v1,h2,v2,h3,v3:integer);
{Post:Draws the triangle with vertices (h1,v1), (h2,v2), and (h3,v3).}
begin
   MoveTo(h1,v1);
   LineTo(h2,v2);
   LineTo(h3,v3);
   LineTo(h1,v1);
end
```

assumes that the three vertices are screen coordinates. What if we are "working in the continuous xy plane" and want to display the triangle depicted in FIGURE 13.6? Conceivably, we could work with a procedure of the form

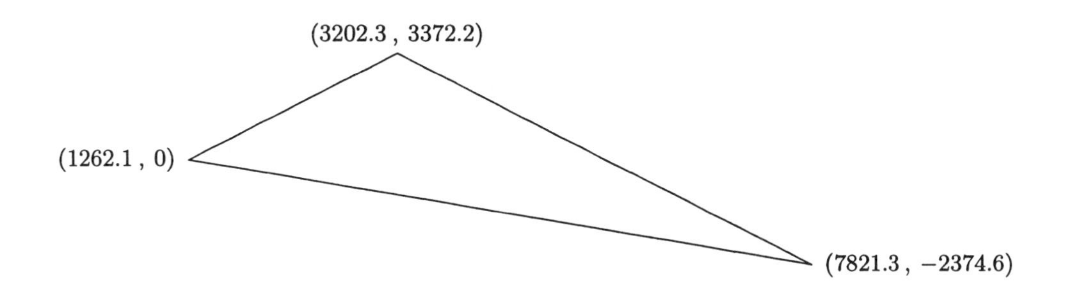

FIGURE 13.6 *A Big Triangle*

```
procedure DrawTriangle(x1,y1,x2,y2,x3,y3:real);
{Post:Draws the triangle with vertices (x1,y1), (x2,y2), and (x3,y3).}
begin
   MoveTo(round(x1),round(y1));
   LineTo(round(x2),round(y2));
   LineTo(round(x3),round(y3));
   LineTo(round(x1),round(y1));
end
```

but there are a number of drawbacks. The y and v axes have opposite orientation that would induce a "mirror effect." Moreover, a vertex will "fall off the screen" if its x or y coordinate is too big or negative.

The solution to these difficulties requires a *coordinate transformation*, from the xy plane to the screen. There are a number of ways that such a transformation can be defined. Our approach is to specify

- a point (x_0, y_0) that is to correspond to the screen center (h_{center}, v_{center}).
- the number of pixels p_{scale} that are to correspond to a unit of xy distance[1].

With these transformation parameters in place, we map point (x, y) to pixel (h, v) where

$$h = round(h_{center} + p_{scale}(x - x_0))$$
$$v = round(v_{center} - p_{scale}(y - y_0)).$$

To implement this idea, suppose `h_center`, `v_center`, `x_0`, `y_0`, and `p_scale` are main program constants that contain the transformation parameters. This allows us to write the following "continuous analogs" of `MoveTo` and `LineTo`:

[1]For greater generality we could scale differently in the x and y directions, but this will not be pursued.

```
procedure cMoveTo(x,y:real);
{Pre:(x_0,y_0) has screen location (h_center,v_center).}
{p_scale = pixels per xy unit.}
{Post:Moves mouse to a point on the screen that corresponds to (x,y).}
var
   h,v:integer; {(h,v) = screen location of (x,y).}
begin
   h := round(h_center + p_scale * (x - x_0));
   v := round(v_center - p_scale * (y - y_0));
   MoveTo(h,v);
end;

procedure cLineTo(x,y:real);
{Pre:(x_0,y_0) has screen location (h_center,v_center).}
{ p_scale = pixels per xy unit.}
{Post:Draws a line from the current mouse position to the point on the }
{screen that corresponds to (x,y).  }
var
   h,v:integer; {(h,v) = screen location of (x,y).}
begin
   h := round(h_center + p_scale * (x - x_0));
   v := round(v_center - p_scale * (y - y_0));
   LineTo(h,v);
end;
```

`Example13_6` shows how to design a triangle-drawing procedure that accepts xy plane vertices. To understand how this program works we need to understand how a procedure like `cLineTo` can seemingly circumvent the usual parameter-list channel and get hold of non-local data, e.g., the main program constant `p_scale`. Up until now, all of the procedures (and functions) that we have written "interact" with the calling program through their parameter list. Thus, the procedures `cMoveto` and `cLineTo` have a novel feature that requires an explanation. We start with a picture of the situation after `cDrawTriangle` calls `cMoveTo`:

main
h_center 280
v_center 150
x_0 4000
y_0 500
p_scale .03

cDrawTriangle
x1 1262.1
y1 0.0
x2 3202.3
y2 3372.2
x3 7821.3
y3 -2374.2

cMoveTo
x 1262.1
y 0.0
h ____
v ____

The current context is `cMoveTo` and the time has come to execute the assignment

```
h := round(h_center + p_scale * (x - x_0));
```

Three of the four ingredients in the arithmetic expression are not to be found inside the `cMoveTo` context. These are `h_center`, `p_scale`, and `x_0`. Instead complaining that these names are

```
program Example13_6;
{Draws a triangle}
const
   h_center = 280;
   v_center = 150; {(h_center,v_center) = screen center.}
   x_0 = 4000;
   y_0 = 500; {(x_0,y_0) corresponds to (h_center,v_center).}
   p_scale = 0.03; {Number of pixels per xy unit.}

procedure cMoveTo(x,y:real);
      ⟨:⟩
procedure cLineTo(x,y:real);
      ⟨:⟩
procedure cDrawTriangle(x1,y1,x2,y2,x3,y3:real);
{Post:Draws a triangle with vertices (x1,y1), (x2,y2), and (x3,y3).}
begin
   cMoveTo(x1,y1);
   cLineTo(x2,y2);
   cLineTo(x3,y3);
   cLineTo(x1,y1);
end;
begin
   ShowDrawing;
   cDrawTriangle(1262.1, 0.0, 3202.3, 3372.2, 7821.3, -2374.2);
end.
```

Output: See FIGURE 13.6.

undefined, the compiler attempts to locate the required values by searching the next larger context, in this case the main program. In this case, h_center, p_scale, and x_0 are found and used in the evaluation:

```
h:= round(280 + 0.03*(1261.2 - 4000))
```

This coordinate transformation framework is useful as long as we have no need to vary p_scale, x_0 or y_0. Unfortunately, this is not typically the case. During a graphical exploration, there is usually a need to change scale (zoom in / zoom out) and to vary the point that is "center stage", i.e., (x_0, y_0). For example, suppose that we want to display an arbitrary triangle whose vertices are obtained via readln's. Because we do not know the range of x and y values in advance, we cannot establish p_scale, x_0, and y_0 as constants and expect them to be satisfactory. Ideally, we would like to *compute* these quantities to ensure that the displayed triangle is appropriately sized and situated (for example) in the middle of the screen. A simple way to achieve this goal without having to rewrite the procedures cMoveTo and cLineTo is to make p_scale, x_0, and y_0 main program *variables*. After the three vertices are solicited, we can place the triangle's centroid at the center of the screen and set p_scale so that the point furthest from the centroid 100 pixels away. See Example13_7.

The main program variables p_scale, x_0, and y_0 are accessed by cMoveTo and cLineTo in exactly the manner as the main program constants p_scale, x_0, and y_0 are accessed in Example13_6. Main program variables used in this fashion are called *global variables.*

We could have chosen to rewrite cMoveTo, cLineTo, and cDrawTriangle so that x_0, y_0, and p_scale are parameters, e.g.,

```
procedure cMoveTo(x,y,x_0,y_0,p_scale:real)
```

It would then be business as usual with just parameter list interaction between the caller and the called. However, you can see why passing values via the global variable mechanism is handy. It shortens parameter lists, leaving just the "essential" parameters x and y. This has appeal, especially when there is a very involved hierarchy of procedures and functions each needing access to x_0, y_0, and p_scale.

The danger with global variables is the difficulty in keeping track of the changes that they undergo. They can be altered by any procedure or function that takes part in the program and this invites error and loss of clarity. For this reason *global variables should be used very sparingly.* If you find it essential to use a global variable, then clarify its role in the precondition of any procedure or function that relies upon it.

To facilitate xy plane visualization, we have incorporated a number of tools in the unit DDCodes. See FIGURE 13.7. If a main program begins with the **uses** DDCodes declaration, then it can freely reference these constants, variables, and procedures. We call particular attention to the procedure SetMap which we use to assign values to the global variables:

```
procedure SetMap(x0,y0,p:real);
{Post:Sets the global ''mapping'' variables x_0,y_0,p_scale.}
begin
   x_0 := x0;
   y_0 := y0;
   p_scale := p;
end
```

Modifying x_0, y_0, and p_scale only through this procedure forces us (in a small but important way) to pay extra attention to the status of these global variables.

```
program Example13_7;
{Draws a triangle}
uses
   Chap9Codes;
const
   h_center = 280; v_center = 150; {(h_center,v_center) = screen center.}
var
   x_0,y_0:real; {Global variables.  (x_0,y_0) maps to (h_center,v_center)}
   p_scale:real; {Global variable.  Number of pixels per xy unit.}
   x1,y1,x2,y2,x3,y3:real; {Triangle has vertices at (x1,y1), (x2,y2), (x3,y3).}
   d1,d2,d3:real; {Distances from (x_0,y_0) to (x1,y1), (x2,y2), (x3,y3) resp.}

procedure cMoveTo(x,y:real);
      ⟨:⟩
procedure cLineTo(x,y:real);
      ⟨:⟩
procedure cDrawTriangle(x1,y1,x2,y2,x3,y3:real);
      ⟨:⟩
begin
   ShowText;
   ShowDrawing;
   {Obtain the three vertices.}
   writeln('Enter x1 and y1'); readln(x1,y1);
   writeln('Enter x2 and y2'); readln(x2,y2);
   writeln('Enter x3 and y3'); readln(x3,y3);
   {Place centroid at screen center.}
   x_0 := (x1 + x2 + x3)/3; y_0 := (y1 + y2 + y3)/3;
   {Place the point furthest from centroid 100 pixels away from centroid.}
   d1 := Distance(x_0,y_0,x1,y1);
   d2 := Distance(x_0,y_0,x2,y2);
   d3 := Distance(x_0,y_0,x3,y3);
   if (d1 >= d2) and (d1 >= d3) then
      p_scale := 100/d1
   else
      {Either d2 or d3 is maximum}
      if d2 >= d3 then
         p_scale:=100/d2
      else
         p_scale:=100/d3;
   cDrawTriangle(x1,y1,x2,y2,x3,y3);
end.
```

The function `Distance` is declared in the unit `Chap9Codes`.

```
const
   h_center = 280; v_center = 150; {(h_center,v_center) = screen center.}
   ScreenWidth = 700; ScreenHeight = 340;

var
   x_0, y_0:  real; {Global variables.  (x_0,y_0) maps to (h_center,v_center)}
   p_scale:  real; {Global variable.  Number of pixels per xy unit.}

procedure SetMap(x0,y0,p:real);
{Post:Sets the global ''mapping'' variables x_0,y_0,p_scale.}

procedure cMoveTo(x,y:real);
{Pre:(x_0,y_0) has screen location (h_center,v_center).  p_scale = pixels per xy unit.}
{Post:Moves mouse to a point on the screen that corresponds to (x,y).}

procedure cLineTo(x,y:real);
{Pre:(x_0,y_0) has screen location (h_center,v_center).  p_scale = pixels per xy unit.}
{Post:Draws a line from the current mouse position to the point on the}
{screen that corresponds to (x,y).}

procedure cGetPosition (var x,y:real);
{Pre:(x_0,y_0) has screen location (h_center,v_center).  p_scale = pixels per xy unit.}
{Post:The xy coordinates of the next mouseclick.}

procedure cDrawDot(x,y:real);
{Pre:(x_0,y_0) has screen location (h_center,v_center).  p_scale = pixels per xy unit.}
{Post:Draws a dot at the screen location corresponding to (x,y).}

procedure cDrawBigDot(x,y:real);
{Pre:(x_0,y_0) has screen location (h_center,v_center).  p_scale = pixels per xy unit.}
{Post:Draws a big dot at the screen location corresponding to (x,y).}

procedure cDrawCircle(xc,yc,r:real);
{Pre:(x_0,y_0) has screen location (h_center,v_center).  p_scale = pixels per xy unit.}
{Post:Draws a circle with center (xc,yc) and radius r.}

procedure cFrameRect(yT,xL,yB,xR:real);
{Pre:(x_0,y_0) has screen location (h_center,v_center).  p_scale = pixels per xy unit.}
{Post:Draws the rectangle with vertices (xL,yB), (xR,yB), (xL,yT), and (xR,yT).}

procedure DrawAxes;
{Pre:(x_0,y_0) has screen location (h_center,v_center).  p_scale = pixels per xy unit.}
{Post:Draws the x and y axes.}
```

FIGURE 13.7 *xy Plane Display Tools in* DDCodes

The procedures specified in FIGURE 13.7 enable us to think in xy terms, hiding from our view the details associated with conversion to screen coordinates. For example, here is a procedure that produces a discrete plot of a "tilted" ellipse.

```
procedure cDrawEllipse(xc,yc,rx,ry,phi:real);
{Pre:(x_0,y_0) has screen location (h_center,v_center).}
{ p_scale = pixels per xy unit.}
{Post:Draws the ellipse obtained by rotating phi radians about (xc,yc)}
{the ellipse ((x-xc)/rx)^2 + ((y-yc)/ry)^2 = 1 }
const
   n = 80; {Number of dots used to approximate the ellipse.}
var
   k:integer;
   x,y:real;
begin
   for k := 0 to n  do
      {Draw the k-th point}
      begin
         x := xc+(cos(phi)*rx*cos(2*k*pi/n)-sin(phi)*ry*sin(2*k*pi/n));
         y := yc+(sin(phi)*rx*cos(2*k*pi/n)+cos(phi)*ry*sin(2*k*pi/n));
         cDrawDot(x,y);
      end;
end
```

The underlying parametric equations are discussed in §6.2. `Example13_8` uses `cDrawEllipse` and some of the procedures in `DDCodes` that facilitate the screen display of objects in the xy plane. The program draws a user-specified ellipse and (a) a big dot at the xy origin, (b) reference axes, and (c) a pair of circles that are centered at the origin and have radii `rx` and `ry` respectively.

PROBLEM 13.10. Write an xy analog of `PaintOval`, i.e.,

```
procedure cPaintOval(yT,xL,yB,xR:real);
```

PROBLEM 13.11. Complete the procedure

```
procedure DrawPoly(xc,yc,rx,ry:real; n:integer);
{Pre:(x_0,y_0) has screen location (h_center,v_center).}
{ p_scale = pixels per xy unit.}
{Post:Draws an n-gon with vertices at}
{(xc+rx*cos(t),yc+ry*sin(t)) where t=2pi*k/n, k=0,...,n-1.}
```

and use it to display a 3-row, 4-column array of regular n-gons with n ranging from 3 to 14.

PROBLEM 13.12. Write a procedure `DrawTiltedSquare(x0,y0,r,theta:real)` that draws the square obtained by rotating the square with vertices `(x0,y0)`, `(x0+r,y0)`, `(x0+r,y0+r)` and `(x0,y0+r)` `theta` radians counterclockwise about `(x0,y0)`.

As a final exercise in coordinate transformations, we develop a very simple environment that plots functions and enables us to "zoom in" on intervals of interest. The key idea is to recognize that if `f(x:real):real` is a declared function, then a loop of the form

```
program Example13_8;
{Tilted Ellipse.}
uses
   DDcodes;
var
   rx,ry real; {The semiaxes}
   phi:real; {Tilt angle in radians.}
procedure cDrawEllipse(xc,yc,rx,ry,phi:real);
      ⟨:⟩
begin
   ShowDrawing;
   rx := 5;
   ry := 2;
   phi := pi/4;
   while YesNo('Another Example?')    do
      begin
         rx := ReviseReal('rx',rx,1,10,9);
         ry := ReviseReal('ry',ry,1,10,9);
         phi:= ReviseReal('phi',phi,-pi/2, pi/2,20);
         ClearScreen;
         {Scale so longer semi axis is 100 pixels long}
         if rx > ry  then
            SetMap(0,0,100/rx)
         else
            SetMap(0,0,100/ry);
         cDrawEllipse(0,0,rx,ry,phi);
         cDrawCircle(0,0,rx);
         cDrawCircle(0,0,ry);
         DrawAxes;
         cDrawBigDot(0,0);
      end;
end.
```

The functions/procedures `YesNo`, `ClearScreen`, `ReviseReal`, `cDrawDot`, `cDrawBigDot`, `cDrawCircle`, and `SetMap` are in the unit `DDCodes`. For sample output, see FIGURE 13.8.

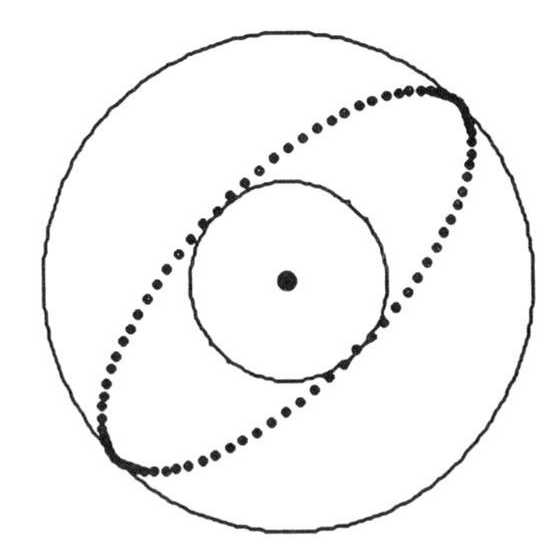

FIGURE 13.8 *Output from* Example13_8

```
{a < b and n a positive integer}
cMoveTo(a,f(a));
h:=(b-a)/n;
for k:=1 to n do
   cLineTo(a+k*h,f(a+k*h))
```

produces a plot of n line segments that connect equally-spaced points on the graph of f on the interval $[a, b]$. See FIGURE 13.9. The quality of the approximation improves as the value of n increases. The best value depends upon screen granularity and what the plot is used for. In the following encapsulation we have set n=200.

```
procedure SimplePlot(a,b:real; function f(x:real):real);
{Pre:(x_0,y_0) has screen location (h_center,v_center).}
{ p_scale = pixels per xy unit.  a<b, f defined on [a,b]}
{Post:Plots f(x) on [a,b] with axes though (x_0,y_0).}
const
   n = 200;
var
   h:real; {Space between sampling points}
   k:integer;
begin
   h := (b-a)/n;
   cMoveTo(a,f(a));
   for k := 1 to n do
      cLineTo(a+ k*h,f(a+ k*h));
   {Draw x-axis, report x_0 and scale.}
   MoveTo(0,v_center); LineTo(ScreenWidth, v_center);
   cDrawDot(x_0,y_0);
   EraseRect(0,0,25,ScreenWidth);
   MoveTo(250,20);
   WriteDraw('x_0 = ',x_0:10:6, ' Dot Diameter = ', (4/p_scale):10:6);
end
```

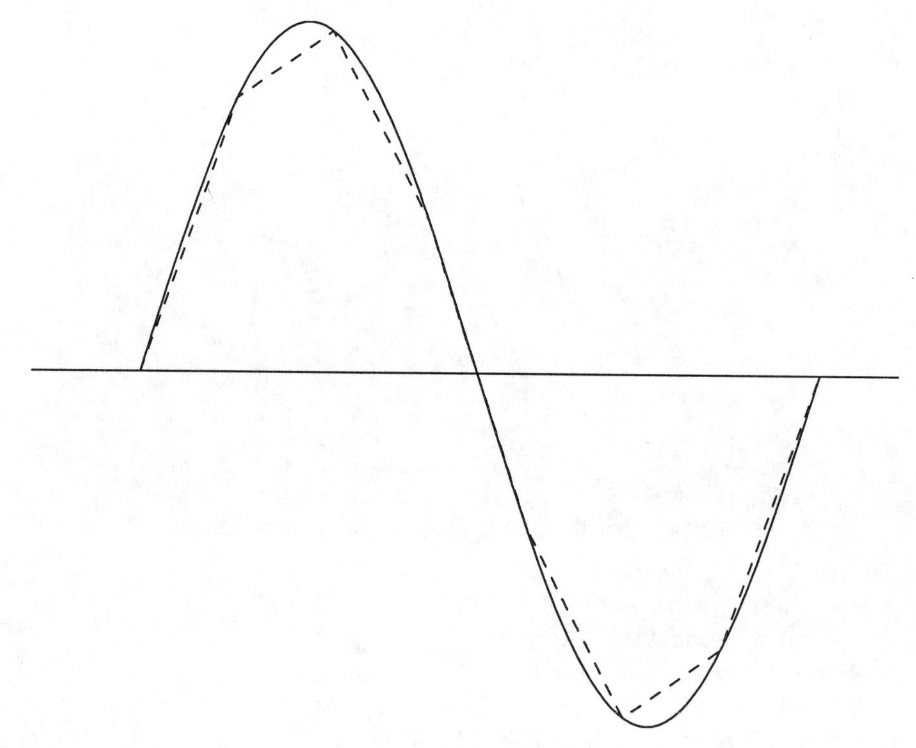

FIGURE 13.9 *Plotting a Piecewise Linear Approximation to a Function*

In addition to plotting the function, `SimplePlot` helps us interpret what we see by displaying the x-axis and information about the underlying scaling.

After a plot has been made, we can use `cGetPosition` to get the location of "a point of interest". By mapping this point to the screen center and increasing the value of `p_scale`, we can re-plot the function showing greater detail around the clicked point. See `Example13_9`.

PROBLEM 13.13. Modify `Example13_9` so that it gives you the choice to zoom in *or* zoom out. The latter can be achieved by *dividing* the current `p_scale` by the constant `ZoomFactor`.

PROBLEM 13.14. Using `SimplePlot`, modify the procedure `Bisection` so that the function `f` is plotted across the current bracketing after every iteration. Note, `Bisection` is declared in the unit `Chap10Codes`.

```
program Example13_9;
{Roots by Zooming}
uses
   DDCodes,Chap13Codes;
const
   a = 0; b = 6; {[a,b] = original interval}
   ZoomFactor = 2; {p_scale increased this factor for each plot.}
var
   x0,y0:real; {xy coordinate of mouse click}
function g(x:real):real;
begin
   g := sin(10*x)
end;
begin
   ShowDrawing;
   {Produce original plot using full width of screen.}
   SetMap((a+b)/2,0, ScreenWidth/(b-a));
   SimplePlot(a,b,g);
   while YesNo('Zoom In?') do
      begin
         EraseRect(0,0,25,200);
         MoveTo(10,20);
         WriteDraw('Click in new focal point.');
         cGetPosition(x0,y0);
         ClearScreen;
         {Move (x0,0) to screen center and increase p_scale for higher resolution.}
         SetMap(x0,0,p_scale*ZoomFactor);
         SimplePlot(x0-ScreenWidth/p_scale, x0 + ScreenWidth/p_scale,g);
      end;
end.
```

The functions/procedures `SetMap`, `ClearScreen`, `YesNo`, and `cGetPosition` are declared in the unit `DDCodes`. The procedure `SimplePlot` is declared in the unit `Chap13Codes`. For sample output, see FIGURE 13.10.

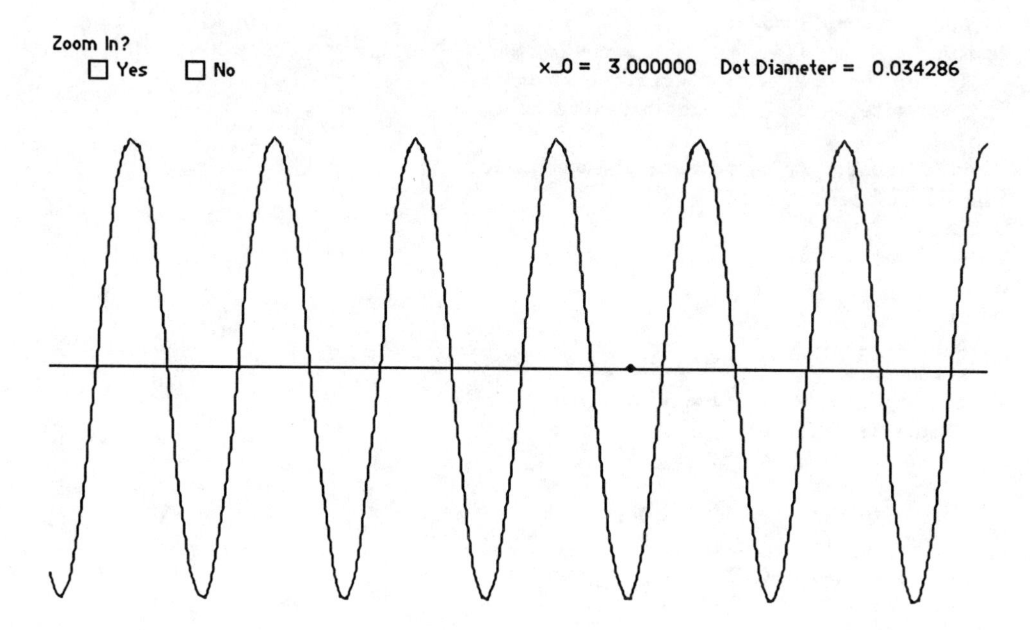

FIGURE 13.10 *Zooming In With* `SimplePlot`

Chapter 14

Points in the Plane

All of the programs that we have considered so far involve relatively few variables. We have seen problems that involve a lot of data, but there was never any need to store it "all at once." This will now change. Tools will be developed that enable us to store a large amount of data that can be accessed during program execution. We introduce this new framework by considering various problems that involve sets of points in the plane. If these points are given by $(x_1, y_1), \ldots, (x_n, y_n)$, then we may ask:

- What is their centroid?
- What two points are furthest apart?
- What point is closest to the origin (0,0)?
- What is the smallest rectangle that contains all the points?

The usual `readln/writeln` methods for input and output are not convenient for problems like this. The amount of data is too large and too geometric.

In this chapter we'll be making extensive use of `cGetPosition`, `cDrawDot`, `cDrawBigDot`, `DrawAxes`, `cMoveTo`, and `cLinetO`. These procedures are declared in `DDCodes` and are developed in §13.2. Throughout this chapter the underlying coordinate transformation is a non-issue and will be fixed with the xy origin at the screen center and 10 pixels per unit xy distance

14.1 Centroids

Suppose we are given n given points in the plane $(x_1, y_1), \ldots, (x_n, y_n)$. Collectively, they define a finite *point set.* Their *centroid* $(\bar{x}, \bar{y})$ of is defined by

$$\bar{x} = \frac{1}{n}\sum_{i=1}^{n} x_i \qquad \bar{y} = \frac{1}{n}\sum_{i=1}^{n} y_i.$$

See FIGURE 14.1. Notice that $\bar{x}$ and $\bar{y}$ are the averages of the x and y coordinates. The program

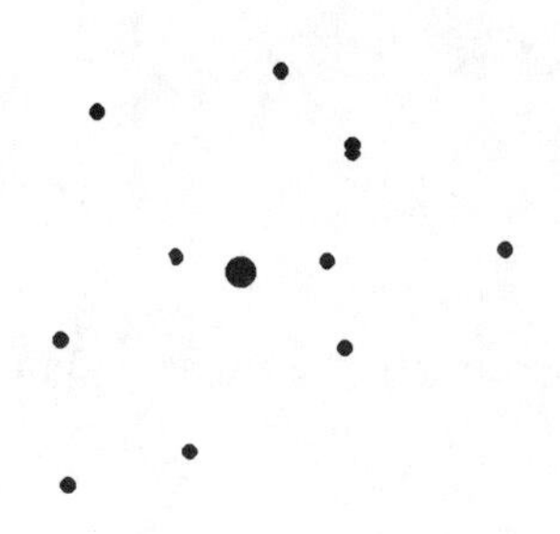

FIGURE 14.1 *Ten Points and Their Centroid*

`Example14_1` calculates the centroid of ten user-specified points. The xy values that define ten points are obtained by clicking the mouse. The summations that are required for the centroid computation are assembled as the data is acquired. The centroid is displayed using `cDrawBigDot` while the ten points that define the point set are displayed with `cDrawDot`.

Now let us augment `Example14_1` so that it draws a line from each of the ten points to the centroid as depicted in FIGURE 14.2. If we try to do this, then we immediately run into a problem: *the x and y values have not been saved.* If n is small, say 3, then we can solve this problem using existing techniques with a fragment like this:

```
sx := 0; sy := 0;
cGetPosition(x1,y1); cDrawDot(x1,y1);
sx := sx+x1; sy := sy+y1;
cGetPosition(x2,y2); cDrawDot(x2,y2,);
sx := sx+x2; sy := sy+y2;
cGetPosition(x3,y3); cDrawDot(x3,y3);
sx := sx+x3; sy := sy+y3;
xbar := sx/3; ybar := sy/3;
cDrawDot(xbar,ybar,4);
cMoveTo(xbar,ybar); cLineTo(x1,y1);
cMoveTo(xbar,ybar); cLineTo(x2,y2);
cMoveTo(xbar,ybar); cLineTo(x3,y3);
```

However, the feasibility of this approach diminishes rapidly as n gets large because approximately $2n$ variables have to be declared and approximately $6n$ statements are required to carry out the computation.

```
program Example14_1;
{Centroid Computation}
uses
   DDCodes;
const
   n = 10; {Number of points to click in.}
var
   k:integer;
   xk,yk:real; {(xk,yk) = kth point}
   sx,sy:real; {sum of first k x values and first k y values}
   xbar,ybar:real; {(xbar,ybar) = centroid}
begin
   ShowDrawing;
   SetMap(0,0,10);
   MoveTo(5,15);
   writeDraw('Click in ', n:2, ' points.');
   sx := 0;
   sy := 0;
   for k := 1 to n do
      {Enter the k-th point.}
      begin
         cGetPosition(xk,yk);
         cDrawDot(xk,yk);
         sx := sx+xk;
         sy := sy+yk;
      end;
   {Compute and display the centroid.}
   xbar := sx/n;
   ybar := sy/n;
   cDrawBigDot(xbar,ybar);
   EraseRect(0,0,20,100);
   MoveTo(5,15);
   WriteDraw('The centroid is displayed.');
end.
```

The procedures `SetMap`, `DrawAxes`, `cGetPosition`, `cDrawDot`, and `cDrawBigDot` are declared in the unit `DDCodes`. For sample output, see FIGURE 14.1.

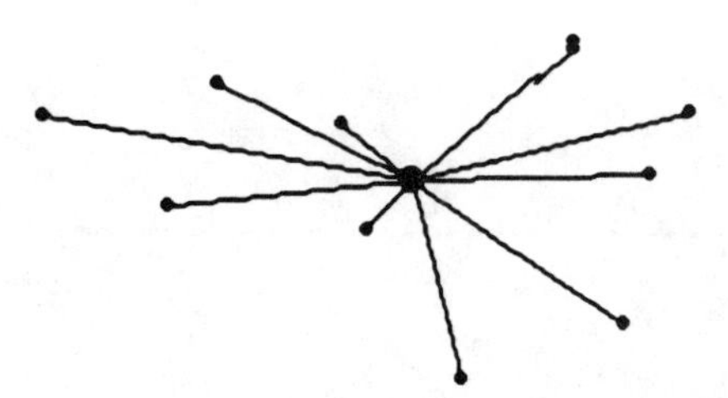

FIGURE 14.2 *Connecting the Centroid*

To solve this problem conveniently, we need the concept of the *array*. `Example14_2` introduces this all-important construction. The program has two *array variables* `x` and `y` which may be visualized as follows:

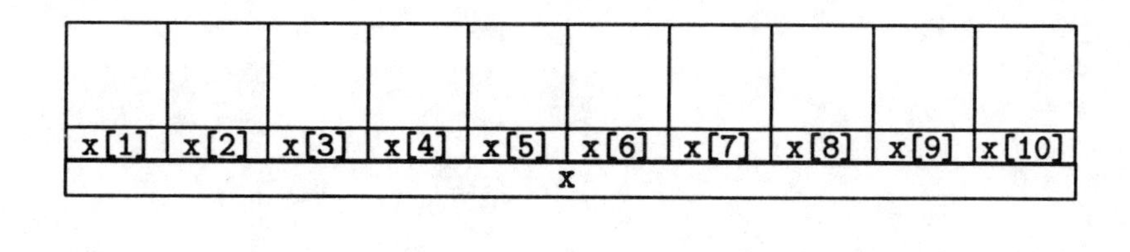

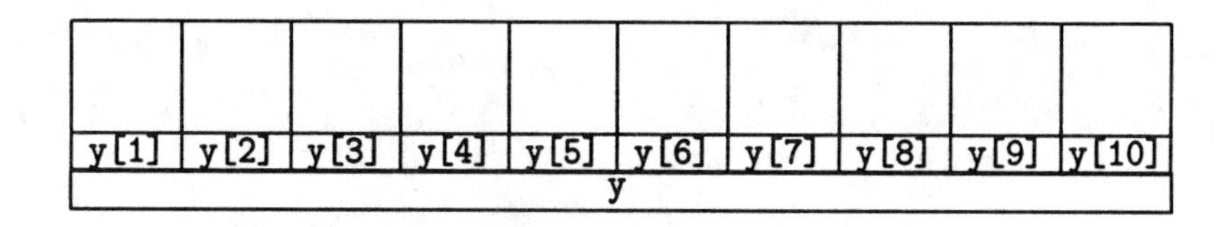

These arrays are each able to store up to 10 real values, one value per array *component*. The program uses the `x` and `y` arrays to store the coordinates of the points whose centroid is required. To understand `Example14_2` fully, we need to discuss how arrays are declared and how their components can participate in the calculations.

Beginning with declaration issues, we notice that

```
x,y:RealList
```

looks like a typical variable declaration except that the indicated type (`RealList`) is not one of the built-in types (`real`, `integer`, `boolean`, `char`, or `string`). Instead, it is a *programmer-defined* type that has been set up in the `type` declaration section of the program. Type declarations come before variable declarations. The line

```
RealList = array[1..10] of real
```

defines `RealList` to be an array type. The individual components of the array have type `real` and the "`1..10`" means that the components are indexed from 1 to 10.

To illustrate how values can be assigned to an array, consider the fragment

```
program Example14_2;
{Centroid Computation}
uses
   DDCodes;
const
   n = 10; {Number of points to click in.}
type
   RealList = array[1..10] of real;
var
   k:integer;
   x,y:RealList; {for the points}
   sx,sy:real; {sum of first k x values and first k y values}
   xbar,ybar:real; {(xbar,ybar) = centroid}
begin
   ShowDrawing;
   SetMap(0,0,10);
   MoveTo(5,15);
   writeDraw('Click in ', n:2, ' points.');
   sx := 0; sy := 0;
   for k := 1 to n  do
      {Enter the k-th point.}
      begin
         cGetPosition(x[k],y[k]);
         cDrawDot(x[k],y[k]);
         sx := sx+x[k]; sy := sy+y[k];
      end;
   {Compute and display the centroid.}
   xbar := sx/n;
   ybar := sy/n;
   cDrawBigDot(xbar, ybar);
   EraseRect(0,0,20,100);
   {Draw the connecting line segments.}
   for k := 1 to n  do
      {Connect the k-th point.}
      begin
         cMoveTo(xbar,ybar);
         cLineTo(x[k],y[k])
      end;
end.
```

The procedures **SetMap**, **cDrawDot**, **cDrawBigDot**, **cMoveTo**, and **cLineTo** are declared in the unit **DDCodes**. For sample output, see FIGURE 14.2.

```
x[1]:=3;
y[1]:=4;
x[2]:=6;
y[2]:=2;
x[3]:=1;
y[3]:=7;
```

This results in the following situation:

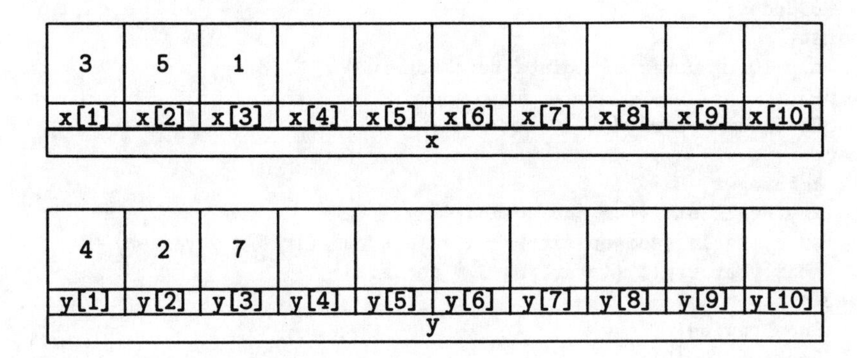

The key is to recognize that the subscript is part of the component's name. For example, `x[2]` is the name of a real variable that happens to be the second component of the array `x`. The assignment `x[2]:=6` is merely the assignment of a real value to a real variable.

A component of a real array like `x` or `y` can "show up" any place a simple real variable can "show up". Thus, the fragment

```
sx:=0; sy:=0;
cGetPosition(x[1],y[1]);
sx:=sx+x[1]; sy:=sy+y[1];
cGetPosition(x[2],y[2]);
sx:=sx+x[2]; sy:=sy+y[2];
cGetPosition(x[3],y[3]);
sx:=sx+x[3]; sy:=sy+y[3];
```

obtains the data for three points and stores the acquired x and y values in `x[1]`, `x[2]` and `x[3]` and `y[1]`, `y[2]` and `y[3]` respectively. But what makes arrays so powerful is that *array subscripts can be computed.* The preceding fragment is equivalent to

```
for i:=1 to 3 do
   begin
      cGetPosition(x[i],y[i]);
      sx:=sx+x[i];
      sy:=sy+y[i];
   end
```

When the loop index `i` has the value of 1, the effective loop body is

```
cGetPosition(x[1],y[1]);
sx:=sx+x[1];
sy:=sy+y[1];
```

This solicits the coordinates of the first point, puts the values in the first component of x and y, and updates the running sums sx and sy. During the second and third passes through the loop we have

```
cGetPosition(x[2],y[2]);
sx:=sx+x[2];
sy:=sy+y[2];
```

and

```
cGetPosition(x[3],y[3]);
sx:=sx+x[3];
sy:=sy+y[3];
```

In general, array references have the form

⟨*Array name*⟩ [⟨*integer-valued expression*⟩]

The value enclosed within the square bracket is the *subscript* and it must be integer-valued. Moreover, the value must be in the subscript range as established in the typing of the array. If x has type RealList, then references of the form x[0] or x[11] are illegal and result in program termination. Reasoning about subscript range violations gets complicated when the subscript values are the result of genuine expressions. For example, if x has type RealList and i is an integer, then x[2*i-1] is legal if the value of i is either 1, 2, 3, 4 or 5.

Another aspect of array computing that does not arise in simple variable situations concerns initialization. Arrays can be *partially* initialized as in the above examples, where just the first three components have been assigned values. In that setting, a reference to x[4] is legal but that component is not initialized.

When referring to parts of arrays, it is useful to use the "dot-dot" notation. Thus, to say that x[2..5] is initialized is to say that x[2], x[3], x[4] and x[5] have been assigned values. We refer to x[2..5] as a *subarray*.

Procedures can have array parameters. For example, the centroid computation can be encapsulated as follows:

```
procedure Centroid( n:integer; x,y:RealList; var xbar,ybar:real);
{Pre:x[1..n] and y[1..n] are initialized.}
{Post:(xbar,ybar) is the centroid of (x[1],y[1]),...,(x[n],y[n]).}
var
   k:integer;
   sx,sy:real; {running sums}
begin
   sx:=0; sy:=0;
   for i:=1 to n do
      {Incorporate the i-th point.}
      begin
         sx:=sx+x[k]; sy:=sy+y[k];
      end
   xbar:=sx/n; ybar:=sy/n;
end
```

Example14_3 illustrates the use of this procedure and is equivalent to **Example 14_2**. It is not necessary to redefine the type **RealList** inside the procedure because that type is established in the calling program.

```
program Example14_3;
{Centroid Computation}
uses
   DDCodes;
const
   n = 10; {Number of points to click in.}
type
   RealList = array[1..10] of real;
var
   k:integer;
   x,y:RealList; {for the points}
   xbar,ybar:real; {(xbar,ybar) = centroid}

procedure Centroid(n:integer; x,y:RealList; var xbar,ybar:real);
   ⟨:⟩
begin
   ShowDrawing;
   SetMap(0,0,10);
   MoveTo(5,15);
   writeDraw('Click in ', n:2, ' points.');
   DrawAxes;
   for k := 1 to  n  do
      {Get the k-th point.}
      begin
         cGetPosition(x[k],y[k]);
         cDrawDot(x[k],y[k]);
      end;
   Centroid(n,x,y,xbar,ybar);
   {Draw the centroid and the connecting line segments.}
   cDrawBigDot(xbar,ybar);
   EraseRect(0,0,20,100);
   MoveTo(5,15);
   WriteDraw('The centroid is displayed.');
   {Draw the connecting line segments.}
   for k := 1 to n do
      {Connect the k-th point.}
      begin
         cMoveTo(xbar,ybar);
         cLineTo(x[k],y[k])
      end;
end.
```

The procedures `SetMap`, `cDrawDot`, `cDrawBigDot`, `cMoveTo`, and `cLineTo` are declared in the unit `DDCodes`. For sample output, see FIGURE 14.2.

It is important to understand why the parameter list includes the integer n. Even though the x and y arrays can hold 10 values, we may be interested in computing the centroid of fewer than 10 points. Thus, if in a 5-point centroid problem, a call of the form Centroid(5,x,y,xbar,ybar) is appropriate. Why then did we make RealList arrays so long? The answer is that in a given application, we may have to call Centroid many times with varying numbers of points. Thus, the length of RealList arrays should be big enough to handle the largest anticipated centroid problem.

While on the topic of array bounds, it is good practice when typing arrays to use constants when specifying the upper subscript limit:

```
const
   nmax = 10;
type
   RealList = array[1..nmax] of real;
```

This is synonymous with

```
type
   RealList = array[1..10] of real;
```

By encapsulating the subscript bound in this way, it is easy to make future adjustments to its value should larger arrays be required.

PROBLEM 14.1. A line drawn from a vertex of a triangle to the midpoint of the opposite side is called a *median*. It can be proven that all three medians intersect at

$$(\bar{x},\bar{y}) = ((x_1+x_2+x_3)/3, (y_1+y_2+y_3)/3).$$

where the triangle's vertices are (x_1,y_1), (x_2,y_2), and (x_3,y_3). The point $(\bar{x},\bar{y})$ is the triangle's centroid and corresponds to its "center of mass." Write a program using arrays that solicits three points and then draws (a) the triangle that they define, (b) displays the triangle's centroid, and (c) draws the three medians.

PROBLEM 14.2. Complete the following procedure:

```
procedure LeftCentroid(n:integer; x,y:RealList; var xbar,ybar:real);
{Pre:n>=1 and x[1..n] and y[1..n] are initialized.}
{At least one of the points (x[1],y[1]),...,(x[n],y[n]) is strictly}
{to the left of the y-axis.  }
{Post:  (xbar,ybar) is the centroid of the points that are strictly }
{to the left of the y-axis.}
```

Using this procedure, write a program that (a) obtains ten points via cGetPosition, (b) uses cDrawBigDot to display the centroid of the points in the left half plane, and (c) draws a line segment from that point to each point *not* in the left half plane.

To create a useful environment for solving points-in-the-plane problems, we have set up a number of graphics procedures in the unit Chap14Codes. Their specification is given in FIGURE 14.3. Notice that the unit Chap14Tools uses the unit DDCodes. This means that when building a project that requires these two units, DDCodes should be added first.

By including the declaration of the RealList type in Chap14Codes, any program that uses this unit can freely reference this type. The program Example14_4 illustrates this feature. Here is how it works. Points are clicked in until the displayed "no box" is clicked. FIGURE 14.4 shows what the screen typically looks during this phase. The acquisition of this point data takes place inside the procedure GetPoints. That is why this procedure has a pair of array var parameters.

```
unit Chap14Codes;
interface
   uses
      DDCodes;
   const
      nmax = 500;
   type
      RealList = array[1..nmax] of real;

   procedure GetPoints(var n:integer; var x,y:RealList);
   {Post:n points are clicked and x[1..n] and y[1..n] contain }
   {the associated xy values.}

   procedure GetRandomPoints(n:integer; xL,xR,yB,yT:real;
               var rn:real; var x,y:RealList);
   {Pre:rn is the seed and satisfies 0<rn<1.  xL<=xR, yB<=yT}
   {Post:  x[1..n] houses random real numbers from the interval [xL,xR].}
   {y[1..n] houses random real numbers from the interval [yB,yT].}
   {rn contains a value obtained by applying rand 2n times to its }
   {original value.}

   procedure ShowPoints(n:integer; x,y:RealList);
   {Pre:x[1..n] and y[1..n] initialized.}
   {Post:  Draws the points (x[1],y[1]),...,(x[n],y[n]).}

   procedure ConnectPoints(n:integer; x,y:RealList);
   {Pre:x[1..n] and y[1..n] are initialized}
   {Post:Draws the polygon defined by the points }
   {(x[1],y[1]),...,(x[n],y[n]).}
```

FIGURE 14.3 *Tools for Point-In-the-Plane Problems from the unit* DDCodes

```
program Example14_4;
{Centroids}
uses
   DDCodes, Chap14Codes;
var
   n:integer; {Number of points.}
   x,y:RealList; {Coordinates of the points.}
   xbar,ybar:real; {Centroid of the points.}
begin
   ShowDrawing;
   while YesNo('Another Example?') do
      begin
         ClearScreen;
         GetPoints(n,x,y);
         Centroid(n,x,y,xbar,ybar);
         ConnectPoints(n,x,y);
         cDrawBigDot(xbar,ybar);
      end;
end.
```

The functions YesNo and ClearScreen. The procedures GetPoints, Centroid, and ConnectPoints and the type RealList are declared in the unit Chap14Codes. For sample output, see FIGURE 14.5.

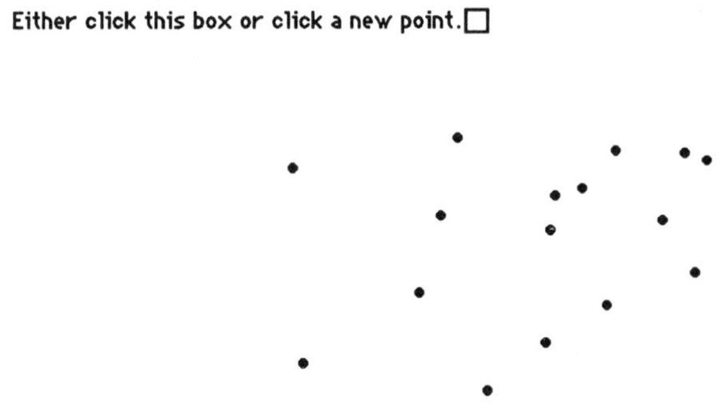

Either click this box or click a new point.☐

FIGURE 14.4 *The* GetPoints *Environment*

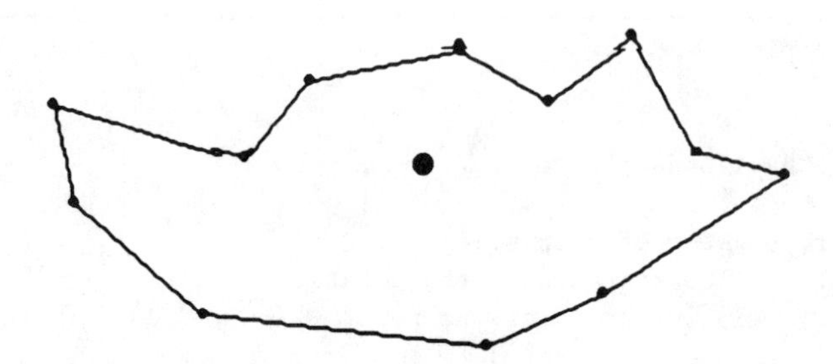

FIGURE 14.5 *Sample Output for* **Example14_4**

However, n is also a var parameter since the number of points that make up the final point set is determined interactively during the procedure call. After termination, the construction of the point set is completed, the points are connected and their centroid displayed as shown in FIGURE 14.5.

The body of the graphics procedure ConnectPoints is worth looking at since it is a typical for-loop computation that involve arrays:

```
cMoveTo(x[1],y[1]);
for i:= 2 to  n do
   cLineTo(x[i],y[i]);
cLineTo(x[1],y[1]);
```

The cMoveTo sends the mouse to the first point. The loop oversees the drawing of lines to each point in succession. Upon "arrival" at the n-th point, the loop terminates and the last leg is drawn back to the first point. Here is a slightly shorter fragment that does the same thing:

```
cMoveTo(x[n],y[n]);
for i:= 1 to  n do
   cLineTo(x[i],y[i]);
```

From the standpoint of the finished product, the order in which the points are connected is immaterial.

It is critical to understand a certain *unimportant* aspect of the subscript business. Assume that i and k are integer variables. The fragments

```
cMoveTo(x[1],y[1]);
for i:= 2 to  n do
   cLineTo(x[i],y[i]);
cLineTo(x[1],y[1]);
for i:=1 to n do
   begin
      cMoveTo(0,0);
      cLineTo(x[i],y[i]);
   end
```

and

```
cMoveTo(x[1],y[1]);
for i:= 2 to  n do
   cLineTo(x[i],y[i]);
cLineTo(x[1],y[1]);
for k:=1 to n do
   begin
      cMoveTo(0,0);
      cLineTo(x[k],y[k]);
   end
```

are equivalent. They both draw the polygon defined by the points and they both connect each point to the origin. The lesson to be learned is that you aren't compelled to use the same index when accessing components in an array.

We now turn to another example that illustrates the procedures **GetRandomPoints** and **ShowPoints**. The former can be used to set up random point sets in a specified rectangle as shown in **Example14_5**. The points are chosen from the rectangle R defined by

```
program Example14_5;
{Centroids of Random Sets of Points}
uses
   DDCodes, Chap14Codes;
 bf var
   rn:real; {For random numbers.}
   n:integer; {Number of random points.}
   x,y:RealList; {Coordinates of the random points.}
   xbar,ybar:real; {Centroid of the random points.}
begin
   ShowDrawing;
   rn := 0.123456;
   n := 5;
   while YesNo('Another Example?')     do
      begin
         n := ReviseInteger('n',n,5,100,19);
         ClearScreen;
         GetRandomPoints(n,-15,15,-10,10,rn,x,y);
         DrawAxes;
         ShowPoints(n,x,y);
         Centroid(n,x,y,xbar,ybar);
         cDrawBigDot(xbar,ybar);
      end;
end.
```

The procedures **GetRandomPoints**, **Centroid**, and **ShowPoints** are in the unit **Chap14Codes**. For sample output, see FIGURE 14.6.

$$R = \{(x,y) : -15 \leq x \leq 15,\ -10 \leq y \leq 10\}.$$

The number of points n is obtained using the function **ReviseInteger** discussed in §13.1. By running the program you will find that as n increases, the computed centroid tends to be closer to the origin. See FIGURE 14.6.

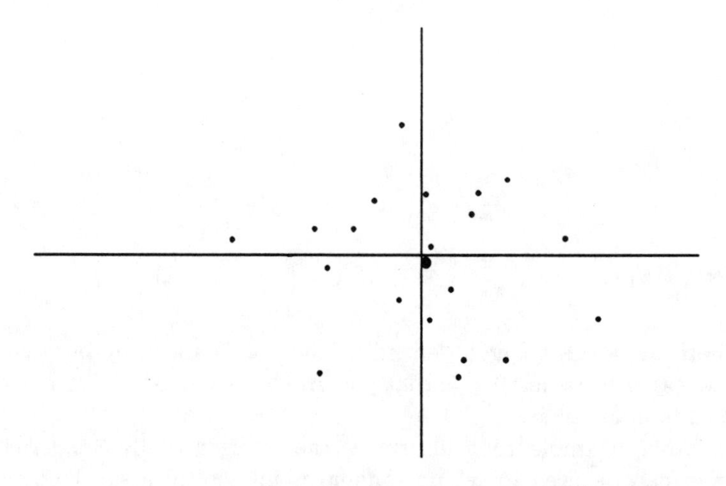

FIGURE 14.6 *Centroids of Random Point Sets*

PROBLEM 14.3. Define the rectangles R_1 and R_2 as follows:

$$R_1 = \{(x, y) : 4 \le x \le 10, \ -5 \le y \le 5\}$$
$$R_2 = \{(x, y) : -10 \le x \le -4, \ -5 \le y \le 5\}$$

Write a program that displays 10 random points $(a_1, b_1), \ldots, (a_{10}, b_{10})$ from R_1 and 10 random points $(c_1, d_1), \ldots, (c_{10}, d_{10})$ from R_2. The line segments that connect (a_i, b_i) to (c_i, d_i) for $i = 1, \ldots, 10$ should also be displayed.

PROBLEM 14.4. Complete the following procedure:

```
procedure SmoothPoints(n:integer; var x,y:RealList);
{Pre:n>=1 and x[1..n] and y[1..n] are initialized}
{Post:  Replace (x[i],y[i]) with midpoint of the line segment}
{that connects this point with its successor (x[i+1],y[i+1]).}
{(Think of (x[1],y[1]) as the successor of (x[n],y[n]).)}
```

14.2 Max's and Mins

Consider the problem of finding the smallest rectangle that encloses the points $(x_1, y_1), \ldots, (x_n, y_n)$. We assume that the sides of the sought-after rectangle are parallel to the coordinate axes as depicted in FIGURE 14.7. From the picture we see that the left and right edges of the rectangle are situated at

$$x_L = \min\{x_1, \ldots, x_n\}$$
$$x_R = \max\{x_1, \ldots, x_n\}$$

while the bottom and top edges are specified by

$$y_B = \min\{y_1, \ldots, y_n\}$$
$$y_T = \max\{y_1, \ldots, y_n\}$$

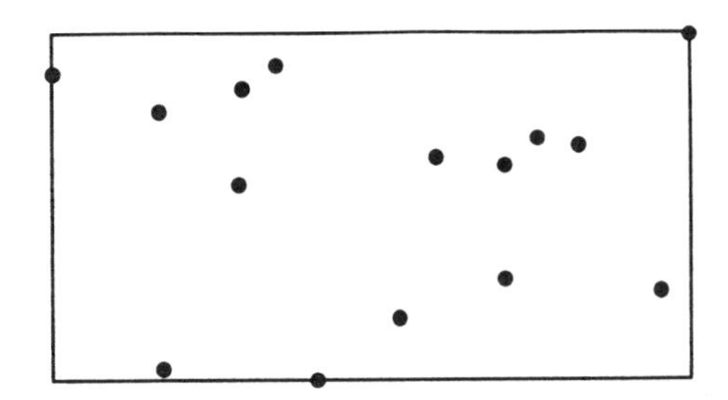

FIGURE 14.7 *Minimum Enclosing Rectangle*

The problem that confronts us is clearly how to find the minimum and maximum value in a given array.

Suppose x[1..4] is initialized[2]. To obtain the maximum value in this array, we could proceed as follows:

```
s:=x[1];
if x[2] > s then
   s:=x[2];
if x[3] > s then
   s:=x[3];
if x[4] > s then
   s:=x[4];
```

The idea behind this fragment is (a) to start the search by assigning the value of x[1] to s and then (b) to scan x[2..4] for a larger value. This is done by "visiting" x[2], x[3], and x[4] comparing the value found with s. The mission of s is to house the largest value "seen so far." A loop is required for general n:

```
s:=x[1];
for i:=2 to n do
   if x[i]>s then
      s:=x[i];
```

Note that after the i-th pass through the loop, the value of s is the largest value in x[1..i]. Thus, if

x[1..6] = | 3 | 2 | 5 | 2 | 7 | 5 |

then the value of s changes as follows during the search:

i	Value of s after i-th pass.
Start	3
2	3
3	5
4	5
5	7
6	7

[2]The dot-dot notation reads as "on up to." Thus, x[1..4] means x[1] on up to x[4].

Packaging these find-the-max ideas we get

```
function MaxInList(n:integer; x:RealList):real;
{Pre:n>=1 and x[1..n] is initialized}
{Post:The largest value in x[1..n].}
var
   k:integer;
   s:real;
begin
   s:=x[1];
   for k:=2 to n do
      {s = largest value in x[1..k].}
      if x[k] > s then
         s := x[k];
   MaxInList:= s;
end;
```

This shows that a function can have array parameters.

Searching for the minimum value in an array is entirely analogous. We merely replace the conditional

```
if x[k] > s then
   s:=x[k];
```

with

```
if x[k] < s then
   s:=x[k];
```

so that s is revised downwards anytime a new smallest value is encountered.

```
function MinInList(n:integer; x:RealList):real;
{Pre:n>=1 and x[1..n] is initialized}
{Post:The smallest value in x[1..n].}
var
   k:integer;
   s:real;
begin
   s:=x[1];
   for k:=2 to n do
      {s = smallest value in x[1..k].}
      if x[k] < s then
         s := x[k];
   MinInList:= s;
end;
```

With `MaxInList` and `MinInList` available we can readily solve the smallest enclosing rectangle problem. `Example14_6` shows that two calls to each of these functions are required.

Max and min calculations abound in applications and as a further illustration of "max/min" thinking, we consider an important "nearest point" problem. Assume the availability of a distance function

```
program Example14_6;
{Smallest Enclosing Rectangle}
uses
   DDCodes,Chap14Codes;
var
   n:integer; {The number of points.}
   x,y:  RealList; {The n points are (x[1],y[1]),...,(x[n],y[n]).}
   xL,xR: real; {The min and the max x-value.}
   yB,yT: real; {The min and the max y-value.}
begin
   ShowDrawing;
   SetMap(0,0,10);
   DrawAxes;
   GetPoints(n,x,y);
   xL := MinInList(n,x);
   xR := MaxInList(n,x);
   {The x-values are contained in the interval [xL,xR].}
   yB := MinInList(n,y);
   yT := MaxInList(n,y);
   {The y-values are contained in the interval [yB,yT].}
   {Draw the enclosing rectangle:}
   cFrameRect(yT,xL,yB,xR);
end.
```

The functions/procedures `DrawAxes`, `cFrameRect`, and `SetMap` are declared in the unit `DDCodes`. The functions/procedures `MaxInList`, `MinInList` and `GetPoints` are declared in the unit `Chap14Codes`. For sample output, see FIGURE 14.7.

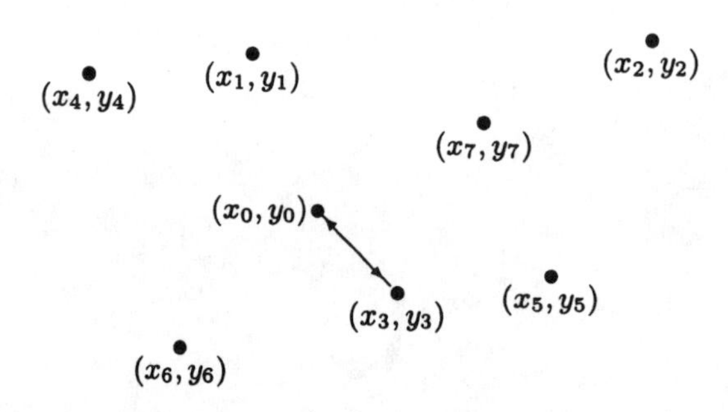

FIGURE 14.8 *Nearness to a Point Set*

```
function Distance(x1,y1,x2,y2:real):real;
{Post:the distance from (x1,y1) to (x2,y2)}
```

and that our goal is to find that point in the set $\{(x_1, y_1), \ldots, (x_n, y_n)\}$ that is closest to a given point (x_0, y_0). See FIGURE 14.8. There are actually *two* things for us to find: the index of the closest point and its distance from (x_0, y_0). Since two values are to be returned, a procedure is required with two var parameters:

```
procedure Nearest(x0,y0:real; n:integer; x,y:RealList;
          var p:integer; var dmin:real);
{Pre:n>=1 and x[1..n] and y[1..n] are initialized.}
{Post:1<=p<=n and (x[p],y[p]) is the closest point to (x0,y0)}
{dmin is the distance from (x0,y0) to (x[p],y[p]).}
var
   j:integer;
   dj:real; {The distance from (x0,y0) to (x[j],y[j])}
begin
   dmin := distance(x0,y0,x[1],y[1]);
   for j := 2  to n  do
      {Check the j-th point.}
      begin
         dj := distance(x0,y0,x[j],y[j]);
         if dj < dmin then
            {New closest point found.}
            begin
               dmin := dj;
               p := j;
            end;
      end;
end;
```

The "search philosophy" is identical to that used in the function MinInList. An initial value for the minimum is established and then a loop steps through all the other possibilities. Whenever a new minimum distance is found, two variables are revised. The new minimum value is assigned to dmin and the index of the new closest point is assigned to p.

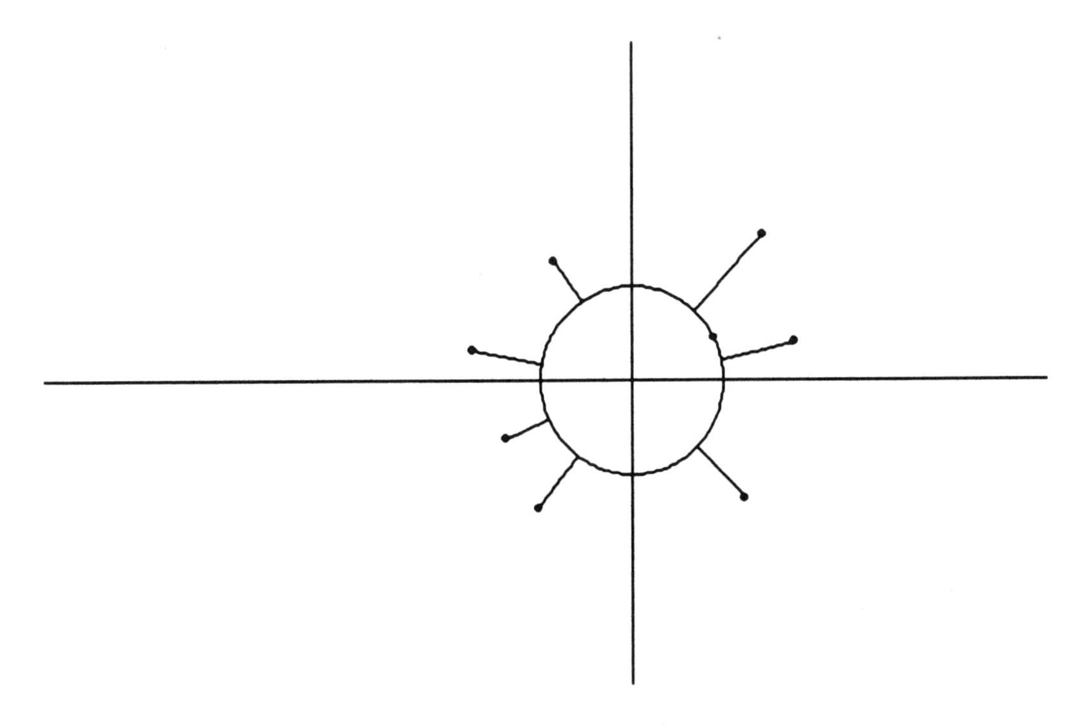

FIGURE 14.9 *Sample output from* Example14_7

Example14_7 illustrates the use of the procedure Nearest. It displays the largest point-free circle centered at the origin as shown in FIGURE 14.9.

We mention that if we are only interested in the index of the closest point, then a function of the form

```
function Nearest(x0,y0:real; n:integer; x,y:RealList):integer
```

could be used.

```
program Example14_7;
{Paths to Largest Point-Free Circle}
uses
   DDCodes,Chap14Codes;
var
   n:integer; {Number of points.}
   x,y:RealList; {The x and y coordinates of the points.}
   p:integer; {Index of the closest point to the origin.}
   rmin:real; {Distance to closest point to origin.}
   scale:real; {Scale factor.}
   i:integer;
begin
   ShowDrawing;
   SetMap(0,0,10);
   DrawAxes;
   GetPoints(n,x,y);
   Nearest(0,0,n,x,y,p,rmin);
   ClearScreen;
   DrawAxes;
   {Draw the largest point-free circle centered at origin.}
   cDrawCircle(0,0,rmin);
   for i := 1 to n do
      if i <> p then
         {The i-th point is not the closest point.}
         begin
            {Draw a radial line from i-th point to circle.}
            cDrawDot(x[i],y[i]);
            cMoveTo(x[i],y[i]);
            scale := rmin / Distance(0,0,x[i],y[i]);
            cLineTo(scale*x[i], scale * y[i])
         end
end.
```

The procedures **SetMap**, **DrawAxes**, **ClearScreen**, **cDrawCircle**, **cDrawDot**, **cMoveTo**, and **cLineTo** are declared in the unit **DDCodes**. The procedures **GetPoints** and **Nearest** and the type **RealList** are declared in the unit **Chap14Codes**. For sample output, see FIGURE 14.9.

PROBLEM 14.5. We say that the set

$$\{ (x_1 + h_x, y_1 + h_y), \ldots, (x_n + h_x, y_n + h_y) \}$$

is a *translation* of the set

$$\{ (x_1, y_1), \ldots, (x_n, y_n) \} .$$

Write a program that translates into the first quadrant, an arbitrary set of points obtained by GetPoints. Choose the translation factors h_x and h_y so that the maximum number of points in the translated set are on the x and y axes. Display the axes and the translated set only.

PROBLEM 14.6. Complete the following procedure

```
procedure TwoNearest(x0,y0:real; n:integer; x,y:RealList;
          var p,q:integer; var dp,dq:real);
{Pre:n>=2 x[1..n] and y[1..n] are initialized.}
{Post:  Points (x[p],yp]) and (x[q],y[q]) are the }
{closest and second closest point to (x0,y0).}
{dp and dq are their respective distances.}
```

Using TwoNearest, write a program that obtains an arbitrary set of points using Getpoints and draws a triangle defined by the origin and the two points nearest to it.

PROBLEM 14.7. Complete the following:

```
procedure SmoothPoints(n:integer; var x,y:RealList);
{Pre:n>=1 and x[1..n] and y[1..n] are initialized}
{Post:Replace (x[i],y[i]) with the midpoint between it and its successor.}
{(The successor of the i-th point is point (i+1) unless i = n.  The successor of}
{the n-th point is the 1st point)}

procedure TranslatePoints(n:integer; var x,y:RealList);
{Pre:x[1..n] and y[1..n] initialized.}
{Post:For i=1..n, x[i] and y[i] are replaced by x[i]-xbar and y[i]-ybar}
{where (xbar,ybar) is the centroid.}

procedure ScalePoints(n:integer; r0:real; var x,y:RealList);
{Pre:x[1..n] and y[1..n] initialized.}
{Post:The point set is scaled so that r0 is the distance of the furthest}
{point to the origin.}

function Variation(n:integer; x,y:RealList):real;
{Pre:x[1..n and y[1..n] initialized.}
{Post:The ratio of the longest to shortest edge of the polygon obtained by}
{connecting the points (x[1],y[1]),...,(x[n],y[n]) in order.}
```

Using these tools, build an environment that permits the easy exploration of the changes that a random point set undergoes with repeated smoothing. After each smoothing, the point set should be translated and scaled for otherwise it "collapses" and disappears from view. You should find that the polygon defined by the points takes on an increasingly regular appearance. The function Variation measures this and its value should be displayed.

Next, we look at another max/min setting that reveals something about *how long* it takes to perform a search calculation. The problem we consider is to find the minimum separation between any pair of points in the set $\{(x_1, y_1), \ldots, (x_n, y_n)\}$. This entails looking at all pairwise distances and will require a nesting of loops. To warm-up to the idea, let's write a fragment that connects all possible pairs of points as depicted in FIGURE 14.10 Taking the top down approach, we start with the following high-level formulation:

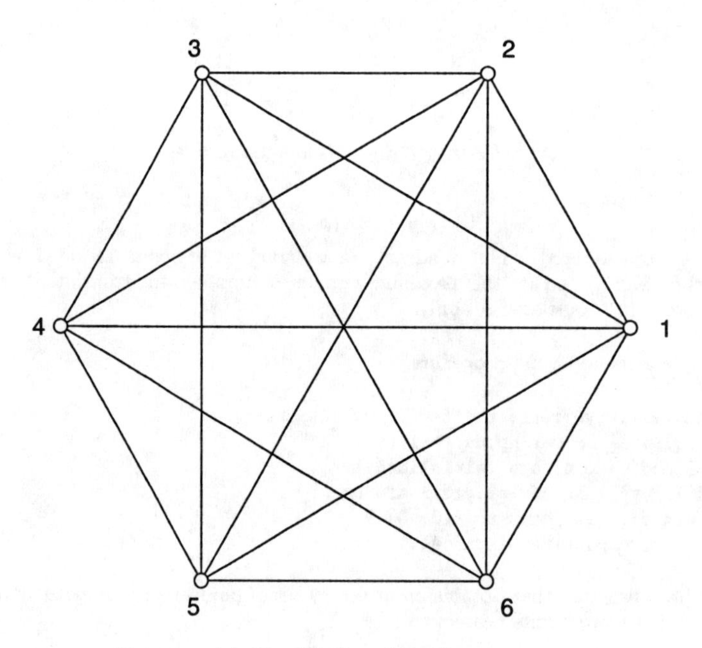

FIGURE 14.10 *All Possible Connections*

```
for i:=1 to n do
   {Draw a line from (x[i],y[i]) to every other point.}
```

If i is 1, then the fragment

```
for j:=2 to n do
   begin
      cMoveTo(x[1],y[1]);
      cLineTo(x[j],y[j]);
   end
```

Connects (x_1, y_1) to $(x_2, y_2), \ldots, (x_n, y_n)$. See FIGURE 14.11. After this we can establish the rest of the second point connections with

```
for j:=3 to n do
   begin
      cMoveTo(x[2],y[2]);
      cLineTo(x[j],y[j]);
   end
```

Notice that the loop begins at 3 because the line connecting the first and second point has already been drawn. The interconnect status at this stage is displayed in FIGURE 14.12.

Picking up the pattern from these examples, we see that a loop of the form

```
for j:=i+1 to n do
   begin
      cMoveTo(x[i],y[i]);
      cLineTo(x[j],y[j]);
   end
```

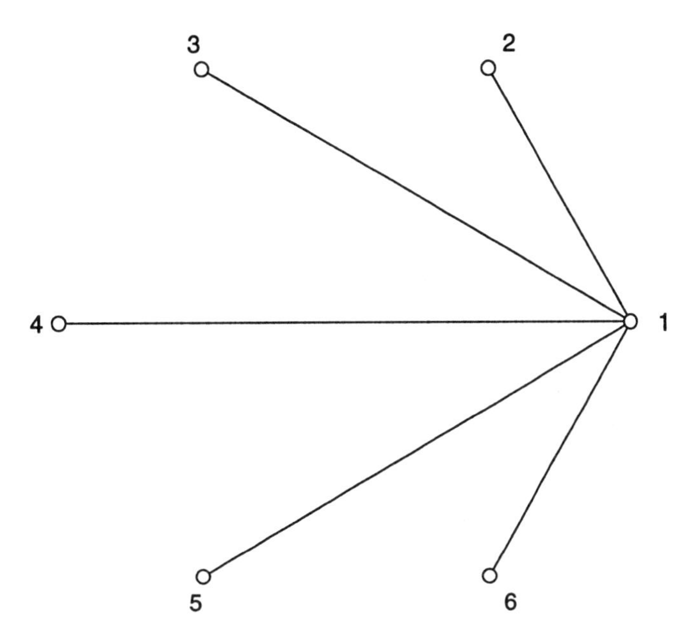

FIGURE 14.11 *Connections that Involve the First Point*

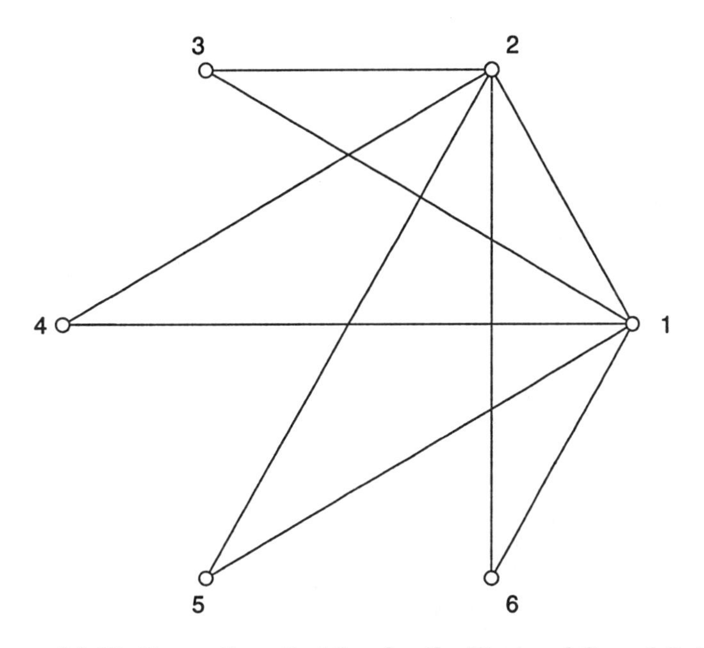

FIGURE 14.12 *Connections that Involve the First and Second Point*

```
for i:=1 to n do
   for j:=i+1 to n do
      begin
         cMoveTo(x[i],y[i]);
         cLineTo(x[j],y[j]);
      end
```

constructs all the required line segments. Note that the upper bound in the outer loop can be replaced by n-1.

The minimum separation problem requires identical looping. However, instead of drawing every possible connecting line segment, we need to compute its length and produce the smallest such value obtained:

```
dmin := Distance(x[1],y[1],x[2],y[2]);
for i:=1 to n-1 do
   for j:=i+1 to n do
      begin
         dij := Distance(x[i],y[i],x[j],y[j]);
         if dij < dmin then
            dmin := dij;
      end
```

This could be encapsulated as a real-valued function. However, if we also wish to return the indices of the closest two points, then a procedure is required:

```
procedure MinSep(n:integer; x,y:RealList;
         var p,q:integer; var dmin:real);
{Pre:n>=1 x[1..n] and y[1..n] are initialized.}
{Post:Among the the points (x[1],y[1]),...,(x[n],y[n]), point p }
{and point q have the minimum separation and the distance}
{ between them is dmin.}
var
   i,j:integer;
   dij:real; {The distance from (x[i],y[i]) to (x[j],y[j])}
begin
   dmin := distance(x[1],y[1],x[2],y[2]);
   p:=1; q:=2;
   for i:=1 to n-1 do
      for j:=i+1 to n do
         {Examine the distance between the i-th and j-th points.}
         begin
            dij := Distance(x[i],y[i],x[j],y[j]);
            if dij < dmin then
               {A new minimum is found.}
               begin
                  dmin := dij;
                  p:=i; q := j
               end;
         end;
end;
```

```
program Example14_8;
{Uniform point-free circles.}
uses
   DDCodes,Chap14Codes;
var
   n:integer; {Number of points.}
   x,y:RealList; {The x and y values of the points.}
   i:integer;
   p,q:integer; {Indices of the minimally separated points.}
   dmin:real; {The distance between points p and q.}
begin
   ShowDrawing;
   GetPoints(n,x,y);
   MinSep(n,x,y,p,q,dmin);
   for i := 1 to n do
      cDrawCircle(x[i],y[i],dmin/2);
end.
```

The procedures `GetPoints` and `MinSep` and the type `RealList` are declared in the unit `Chap14Codes`. For sample output see FIGURE 14.13

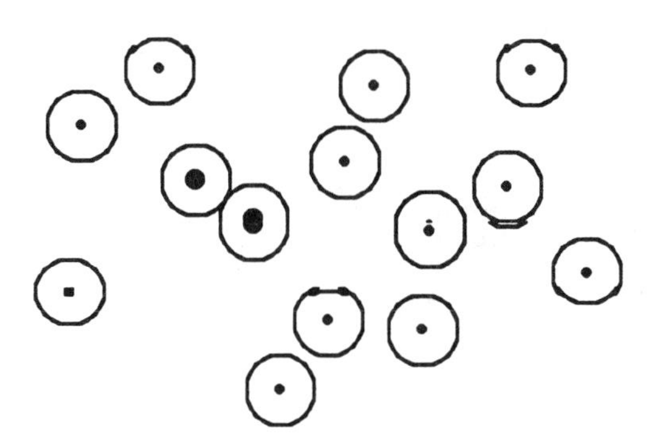

FIGURE 14.13 *Sample Output from* `Example14_8`.

This procedure is used in `Example14_8`. Circles with radius equal to half of the minimum separation are drawn around each point.

Let's compare the times it takes to execute the procedures `Nearest` and `MinSep` for a given n. Intuitively we suspect that `MinSep` should take much longer because all pairwise distances must be checked whereas with `Nearest`, there is only one distance per point to compute. `Example14_9` confirms this conjecture. The actual times reported are less important than how they vary with n:

- If n is doubled, then the time required by `Nearest` is roughly doubled.
- If n is doubled, then the time required by `MinSep` increases by roughly 4.

The relationships are easily explained. The amount of arithmetic required by `Nearest` increases linearly with n. Indeed, there are n distances to compute. On the other hand, `MinSep` requires the calculation of about $n^2/2$ distances. That's why doubling n increases the amount of work by a factor of 4.

As discussed in §6.2, the "big O" notation is used to describe qualitatively the running time of programs like `Nearest` and `MinSep`. `Nearest` is $O(n)$ because its running time depends linearly upon n, an integer that effectively quantifies the amount of input. `MinSep` is $O(n^2)$ because its execution time is essentially a quadratic function of n.

PROBLEM 14.8. Write a program that prints a table of all pairwise distances between 5 randomly chosen points from a unit square. The table should not have any redundant values.

PROBLEM 14.9. Write a program that generates 10 random points from the square $-10 \le x \le 10$, $-10 \le y \le 10$ and connects each point with all the other points that are in a different quadrant.

PROBLEM 14.10. Complete the following procedure:

```
procedure ConnectNeighbors(n:integer; x,y:RealList; p:integer);
{Pre:p>=1, n>=1 and x[1..n] and y[1..n] initialized.}
{Post:Draws a line from (x[i],y[i]) to (x[j],y[j]) if |i-j| <= p or}
{if |n+i-j| <= p or if |i-j-n|<=p.}
```

PROBLEM 14.11. Complete the following procedure:

```
procedure Diameter(n:integer; x,y:RealList; var p,q:integer; var dmax:real);
{Pre:n>=1 and x[1..n] and y[1..n] are initialized.}
{Post:The maximum distance between any pair of the points}
{(x[1],y[1]),...,(x[n],y[n]) occurs between the p-th and q-th points}
{and dmax is their separation.}
```

Write a program that uses `Diameter` to draw a circle whose diameter is the diameter of a point set obtained using `GetPoints`. The two maximally separated points should be on the circle.

PROBLEM 14.12. Suppose $r > 0$ and that the points $(x_1, y_1), \ldots, (x_n, y_n)$ are given. Our goal is to compute (x_c, y_c) so that the circle

$$(x - x_c)^2 + (y - y_c)^2 = r^2$$

encloses the largest possible number of points. The actual "best" circle is very hard to compute. However, here are two methods that render an approximate solution to the problem:

Method 1. Let C_i be the circle of radius r that is centered at $(x[i], y[i])$. Choose the C_i that encloses the largest number of points.

*Method 2.*Let $\{(x, y) : a \le x \le b,\ c \le y \le d\}$ be the smallest rectangle that encloses all the points. Choose positive integers n_x and n_y and let C_{ij} be the circle of radius r that is centered at $(a+i(b-a)/n_x, c+j(d-c)/n_y)$. Choose the C_{ij} that encloses the largest number of points.

Implement the procedure

```
procedure FindCenter(r:real; n:integer; x,y:RealList; var xc,yc:real);
{Pre:n>=1 and points (x[1],y[1]) ....(x[n],y[n]).}
{Post:The radius r circle centered at (xc,yc) contains a large number of the points.}
```

in each of these ways. Write a program that illustrates both methods on a random distribution of points. For the second method, set $n_x = 10$ and $n_y = 5$.

PROBLEM 14.13. Write a program that (a) reads in n points using `GetPoints`, (b) draws the odd-indexed points with `cDrawBigDot` and the even-indexed points with `cDrawDot`, and (c) draws a line that connects the two points that are as far apart as possible subject to the constraint that one point has an even index and one point has an odd index.

```
program Example14_9;
uses
   DDCodes,Chap14Codes;
const
   MyComputer = 'Macintosh Powerbook 170, System 7.1.';
   maxN = 160;
var
   rn:real; {Random number.}
   i:integer;
   n:integer; {Number of points.}
   x,y:RealList; {x and y coordinates of the points.}
   p,q:integer;
   dmin: real;
   StartTime, StopTime: real; {Clock snapshots.}
   NearTime,MinSepTime: real; {Time for call to Nearest and Minsep.}
begin
   ShowText;
   rn := 0.123456;
   GetRandomPoints(maxN,-10,10,-10,10,rn,x,y);
   writeln(' n Nearest MinSep');
   writeln('--------------------------');
   n := 20;
   while n <= 160 do
      begin
         StartTime := TickCount;
         for i := 1 to n do
            Nearest(0,0,n,x,y,p,dmin);
         StopTime := TickCount;
         NearTime := (StopTime - StartTime) / (n * 60);
         StartTime := TickCount;
         MinSep(n,x,y,p,q,dmin);
         StopTime := TickCount;
         MinSepTime := (StopTime - StartTime) / 60;
         writeln(n:5, NearTime:10:3, MinSepTime:10:3);
         n := 2*n;
      end;
   writeln;
   writeln(MyComputer);
end.
```

The procedures `Nearest`, `MinSep`, and `GetRandomPoints` are declared in the unit `Chap14Codes`. Sample output:

```
    n  Nearest       MinSep
 --------------------------
   20    0.011        0.083
   40    0.020        0.350
   80    0.038        1.450
  160    0.074        5.833
```

Chapter 15

Tables

Suppose it costs one dollar to evaluate a function $f(x)$ and that a given fragment calls f 1000 times. If each function call involves a different value of x, then \$1000 must be spent on f evaluations. However, if only 10 different x values are involved, then there is a \$10 solution to the f-evaluation problem:

(a) "precompute" the 10 necessary f evaluations and store them in an array. (This costs \$10.)

(b) extract the necessary f-values from the array during the execution of the fragment.

Storing x-values and f-values in a pair of arrays is just a method for representing a table in the computer.

A plotting environment is developed that allows us to display in a window the values in a table. Although the plotting tools that we offer are crude, they are good enough to build an appreciation for plotting as a vehicle that builds intuition about a function's behavior.

The setting up of a table is an occasion to discuss several efficiency issues that have to do with function evaluation. A sine/cosine example is used to show how to exploit recursive relations that may exist between table entries. The "parallel " construction of the entries in a table using array-level operations is also discussed

Once a table is set up, there is the issue of looking up values that it contains. The methods of linear search and binary search are discussed. The "missing" data problem is handled by linear interpolation.

15.1 Set-Up

In table problems, it is handy to work with arrays that are subscripted from zero:

```
const
   fEvalMax = 200;
type
   RealList0 = array[0..fEvalMax] of real
```

Components 0 through 8 of an array x that has this type may be visualized as follows:

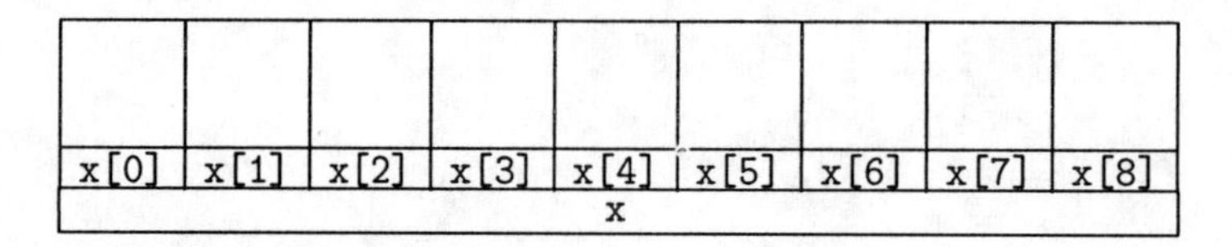

To represent a table of the form

x_0	x_1	$\cdots$	x_n
$f(x_0)$	$f(x_1)$	$\cdots$	$f(x_n)$

we use a pair of arrays xval and yval, agreeing to store x_k in xval[k] and $f(x_k)$ in yval[k]. Here is a function that can be used to set up yval[0..n], assuming that the x-values are already established in xval[0..n]:

```
function SetUp(n:integer; xval:RealList0; function f(x:real):real):RealList0;
{Pre:0<=n<=fEvalMax.  f(x) defined at x = xval[0],...,xval[n].}
{Post:SetUp[k] = f(xval[k]), k=0,...,n.}
var
   k:integer;
begin
   for k := 0 to  n  do
      SetUp[k]:=f(xval[k]);
end
```

This is an example of a function that returns a programmer-defined type[1]. The declaration of such a function is no different than if it returned a built-in type:

```
   function ⟨Name⟩(⟨parameter list⟩):⟨type⟩
```

As an application of SetUp, suppose

```
function g(x:real):real;
begin
   g := x*x - 5*x + 6;
end
```

is declared, n = 8, and

xval[0..8] = | 1.0 | 1.5 | 2.0 | 2.5 | 3.0 | 3.5 | 4.0 | 4.5 | 5.0 |.

The assignment

[1]Standard Pascal does not permit this.

```
yval := SetUp(n,g,xval);
```

produces:

yval[0..8] =

2.00	0.75	0.00	−0.25	0.00	0.75	2.00	3.75	6.00

It is frequently the case that the values at which f is sampled in a table are equally spaced across an interval $[L, R]$. To handle this situation, typical in plotting applications, we have

```
function xEqual(n:integer; L,R:real):RealList0;
{Pre:1<= n<=fEvalMax}
{Post:xEqual[k] = L + k(R-L)/n, k=0,...,n.}
var
   h:real; {Spacing between x-values}
   k:integer;
begin
   h := (R - L)/n;
   for k := 0 to n do
      xEqual[k]:= L + k*h;
end
```

The program Example15_1 shows how xEqual and SetUp can be used to set up a table.

PROBLEM 15.1. Complete the following function:

```
function xLog(n:integer;L,R:real):RealList0;
{Pre:0<L<R}
{Post:xLog[0]=L, xLog[n]=R, and for k=1..n, log(xLog[k])-log(xLog[k-1]) = (log(R)-log(L))/n.}
```

Thus, if xval:=xLog(5,0.01,1000), then xval[0..5] =

−2	−1	0	1	2	3

PROBLEM 15.2. A function $f(x)$ is *even* if $f(-x) = f(x)$ for all x. The functions $\cos(x)$ and $x^4 - 2x^2 + 1$ are examples. Complete the function

```
function SetUpEven(n:integer;R:real; function f(x:real):RealList0
{Pre:f even and defined on [-R,R]}
{Post:SetUpEven[k] = f(-R+kh) for k = 0 to n where h = 2R/n.}
```

Take steps to minimize the number of f evaluations.

The function passed to SetUp must be a real-valued function with a single real argument. However, many functions of a single variable involve extra parameters that define the value produced. For example, the cubic $q(x) = ax^3 + bx^2 + cx + d$ is defined by the four coefficients a, b, c, and d. In a situation where lots of different cubic polynomials are involved, it is natural to work with a function of the following form:

```
function Cubic(a,b,c,d,x:real):real;
{Post:ax^3 + bx^2 + cx +d}
begin
   Cubic := ((a*x + b)*x + c)*x + d;
end;
```

```
program Example15_1;
{Table of values for g(x) = (x-1)(x-2)(x-3)}
uses
   Chap15Codes;
const
   n = 100; {Length of Table}
var
   k:integer;
   xval,yval:RealList0;

function g(x:real):real;
begin
   g := (x - 1)*(x - 2)*(x - 3);
end;

begin
   ShowText;
   xval := xEqual(n,0,4);
   yval := Setup(n, xval, g);
   writeln(' x g(x)');
   writeln('------------------');
   for k := 0 to n do
      writeln(xval[k]:5:3, yval[k]:10:3)
end.
```

The functions xEqual and SetUp and the type RealList0 are declared in the unit Chap15Codes.
Output:

```
    x          g(x)
------------------
0.880        -0.285
0.920        -0.180
0.960        -0.085
1.000         0.000
1.040         0.075
1.080         0.141
```

But now there is a problem if you want to apply `SetUp` to obtain a table of cubic function values. A reference of the form `SetUp(n,xval,cubic)` is illegal because `Cubic` is *not* a function of the required type. One solution is to rewrite `SetUp`, replacing each reference to `f` with an appropriate reference to `Cubic`, e.g.,

```
function SetUpCubic(n:integer; xval:RealList0; a,b,c,d:real):RealList0;
{Pre:0<=n<=fEvalMax.  f(x) defined at x = xval[0],...,xval[n].}
{Post:SetUp[k] = q(xval[k]), k=0,...,n where q(x) = ax^3 + bx^2 + cx +d.}
var
   k:integer;
begin
   for k := 0 to  n  do
      SetUpCubic[k]:= Cubic(a,b,c,d,x[k])
end
```

However, making changes to an existing piece of software is a risky business. Of course, `SetUp` is very simple and so this strategy is quite feasible. But it is not hard to imagine a setting where the required modifications are extensive and it is just not worth jeopardizing the integrity of a working piece of software.

We could also rewrite `cubic` so that the coefficients a, b, c, and d are obtained by referencing non-local variables:

```
function NiceCubic(x:real):real;
{Pre:a,b,c,d contain the values of the cubic's coefficients.}
{Post:ax^3 + bx^2 + cx +d}
begin
   NiceCubic := ((a*x+b)*x+c)*x+d;
end
```

A call of the form `SetUp(n,xval,NiceCubic)` is permissible because `NiceCubic` has the required typing. But, going the global variable route with `NiceCubic` is also fraught with danger. The main program may already use variables with names `a`, `b`, `c`, and `d`. If the changes to these variables are not carefully monitored, then errors will result.

Fortunately, there is a way out of these "software-use" dilemmas. The idea is to implement `SetUpCubic` with a properly typed, nested function for cubic polynomial evaluation:

```
procedure SetUpCubic(n:integer; xval:RealList0;a,b,c,d:real):RealList0;
{Pre:0<=n<=fEvalMax.  f(x) defined at x = xval[0],...,xval[n].}
{Post:SetUp[k] = q(xval[k]), k=0,...,n where q(x) = ax^3 + bx^2 + cx +d.}
   function NiceCubic(x:real):real;
   {Post:ax^3 + bx^2 + cx +d}
   begin
      NiceCubic := cubic(a,b,c,d,x);
   end;
begin
   SetUpCubic := SetUp(n,xval,NiceCubic);
end
```

Notice how attractive this solution is. It is short, simple, and most importantly, *neither the function* `cubic` *nor* `SetUp` *are modified.* Tracing a call to `SetUpCubic` requires an understanding

```
program Example15_2;
{Table of values for general cubic}
uses
   Chap15Codes;
const
   n=3; {Length of Table}
   L=10; R=13; {Table ranges for x=L to x=R.}
var
   k:integer;
   a0,a1,a2,a3:real; {The cubic's coefficients.}
   xval,yval:RealList0;
function cubic(a,b,c,d,x:real):real;
      ⟨:⟩
function SetUpCubic(n:integer; xval:RealList0; a,b,c,d:real):RealList0;
      ⟨:⟩
begin
   ShowText;
   writeln('Enter the coefficients of the cubic:');
   readln(a3,a2,a1,a0);
   xval := xEqual(n,L,R);
   yval := SetUpCubic(n,xval,a3,a2,a1,a0);
   writeln;
   writeln(' k xval[k] yval[k]');
   writeln('------------------------');
   for k := 0  to  n  do
      writeln(k:3, xval[k]:10:3, yval[k]:10:3);
end.
```

The functions xEqual and SetUp and the type RealList0 are declared in the unit Chap15Codes. Sample output:

```
Enter the coefficients of the cubic:
1 2 3 4

 k   xval[k]     yval[k]
------------------------
 0    10.000    1234.000
 1    11.000    1610.000
 2    12.000    2056.000
 3    13.000    2578.000
```

of the nested function concept and the passing of values through nonlocal variables[2]. We discuss these points by considering `Example15_2`. The state of the program just after the main program call to `SetUpCubic` is as follows:

Example15_2
n 3
a 10
b 13
a3 1
a2 2
a1 3
a0 4
xval[0] 10
xval[1] 11
xval[2] 12
xval[3] 13
yval[0] ___
yval[1] ___
yval[2] ___
yval[3] ___

SetUpCubic
n 3
xval[0] 10
xval[1] 11
xval[2] 12
xval[3] 13
a 1
b 2
c 3
d 4
NiceCubic

(To save space, only the first four entries of the array variables are displayed.) The body of `SetUpCubic` consists of just the call to `SetUp` with `NiceCubic` passed as the function:

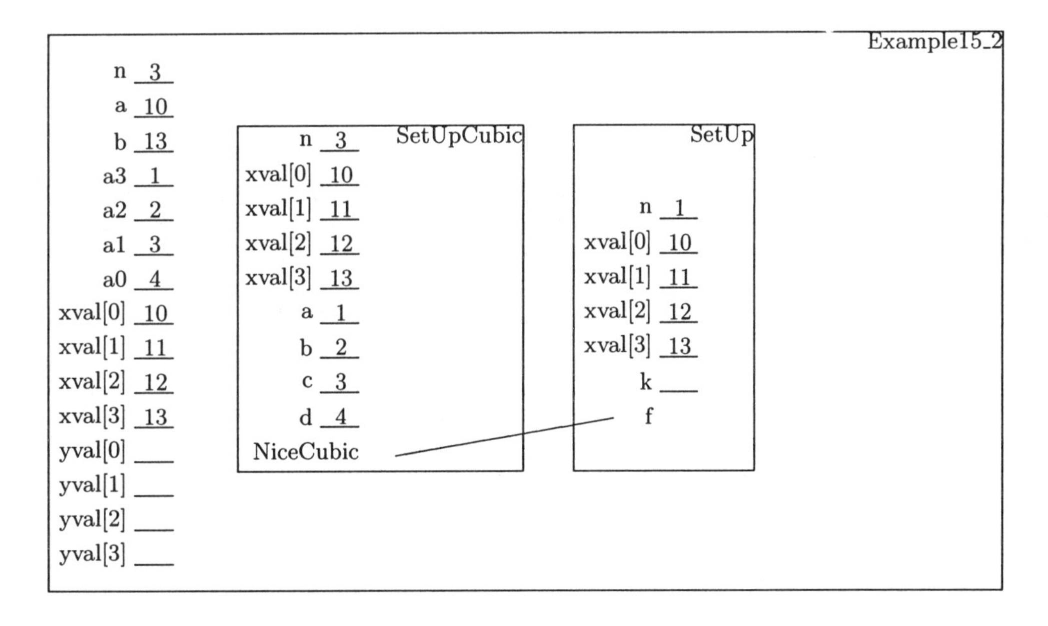

`SetUp` proceeds to apply `NiceCubic` to each value in `xval[0..3]`. For example, during the first pass through the loop, `k = 0` and in effect we execute

[2]Nested functions are introduced in §10.3. The use of non local variables is discussed in §13.2.

```
SetUp[0] := NiceCubic(xval[0])
```

Here is the context box description:

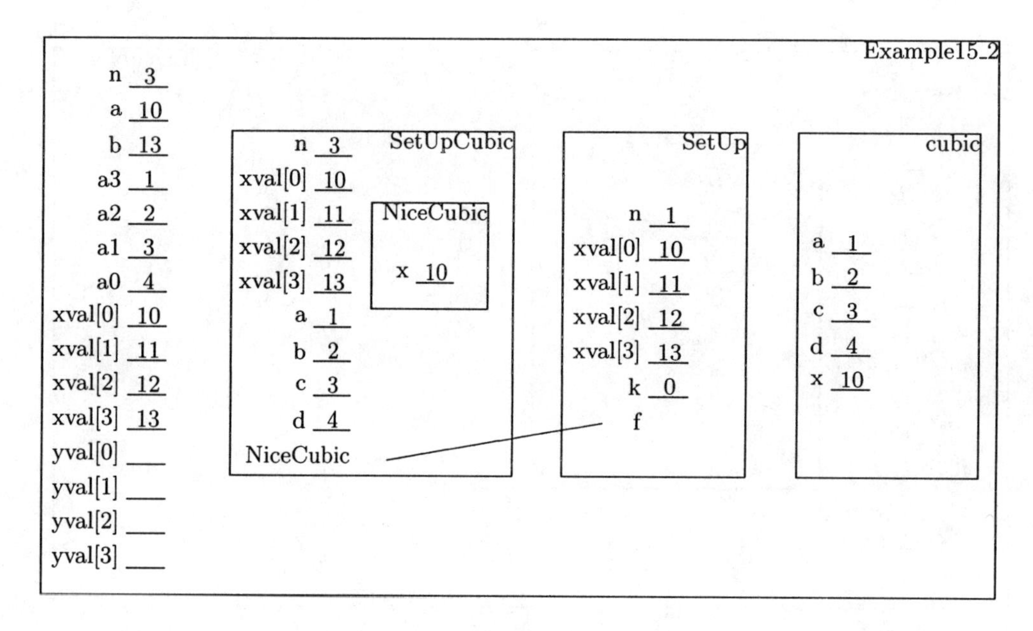

Inside `NiceCubic` a call is made to `Cubic` and this involves references to the nonlocal variables `a`, `b`, `c`, and `d`. These values are found in `SetUpCubic`, the "box" that surrounds `NiceCubic`. Notice that the `a` and `b` values defined in the main program context are *not* used.

PROBLEM 15.3. Define the function $f(x,k)$ by

$$f(x,k) = \frac{x(x-1)\cdots(x-k+1)}{1\cdot 2\cdots k}$$

where x is real and k is nonnegative. (Assume that $f(x,k)$ is 1 whenever $k=0$.) Recognize f as a generalized binomial coefficient. (See §7.2.) Complete the following function:

```
function RealBC(x:real; k:integer):real;
{Pre:k>=0}
{Post:x(x-1)(x-2)*...*(x-k+1)/k!}
```

Write a program that prints a table whose k-th line reports the value of $f(x,k)$ for $x = 1, 1.5, 2.0, 2.5, \cdots, 4.5, 5.0$. The integer k should range from 0 to 5. Make effective use of `xEqual` and `SetUp`, using the latter to produce the function values that are printed on each line of the table.

15.2 Plotting

By connecting the points

$$(1, g(1)),\ (1.5, g(1.5)),\ (2, g(2)), \ldots,\ (5, g(5))$$

we obtain a plot of the function g on the interval [1,5]. See FIGURE 15.1. One way to accomplish this is to use the procedures `cMoveto` and `cLineTo`:

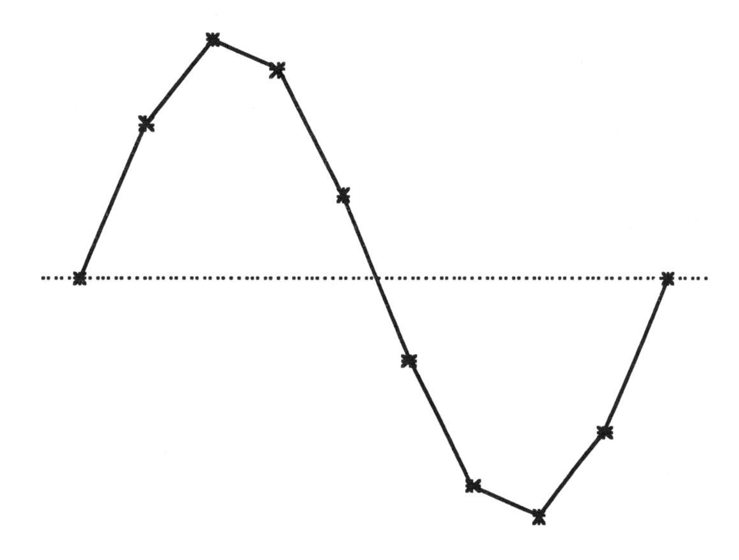

FIGURE 15.1 *Connecting Dots*

```
cMoveTo(xval[0],yval[0]));
for k:=1 to 8 do
   cLineTo(xval[k],yval[k])
```

We discussed this strategy in §13.2 and developed the procedure SimplePlot. But with arrays we can handle the scaling issue a little more effectively. Our plan is to place the plot in a "window" defined by four global variables, v_Top, h_Left, v_Bottom, and h_right:

```
procedure Plot(n:integer; xval,yval:RealList0; xmin,xmax,ymin,ymax:real);
{Pre:  0<=n<=nmax}
{Post:  Plots yval[0..n] vs xval[0..n] in the rectangular window}
{ ( v_Top,h_Left,v_Bottom,h_right)}
{Across the window horizontally x ranges from xmin to xmax.}
{Across the window vertically, y ranges from ymin to ymax.}
```

The details of plot are not particularly instructive. It produces the graph by "connecting the dots." The tabulations along the x and y axes are crude and could be refined. It is assumed that the values in xval[0..n] are between xmin and xmax. Portions of the plot that fall outside the window are "whited out." Example15_3 illustrates the use of Plot. Notice how several plots can be superimposed in the same window and that different scales prevail in the x and y directions. The location and size of the plotting window is defined by global variables v_Top, h_Left, v_Bottom, and h_Right that are declared in the unit DDCodes. The procedure StandardPlot assigns values to these variables so that a single, reasonably sized window is placed in the middle of the screen.

```
program Example15_3;
{Plot some simple functions.}
uses
   Chap15Codes;
const
   n = fEvalMax; {Number of function values used in the plot.}
   xmin = -1; xmax = 10;
   ymin = -2; ymax = 2;
var
   xval:RealList0;
   fpval,fmval,gval:RealList0;

function fplus(x:real):real;
{Post:  exp(-x/5)}
begin
   fplus := exp(-0.2*x);
end;

function fminus(x:real):real;
{Post:  -exp(-x/5)}
begin
   fminus := -exp(-0.2*x);
end;

function g(x:real):real;
{Post:  exp(-x/5)sin(x ^2)}
begin
   g := exp(-0.2*x)*sin(sqr(x));
end;

begin
   ShowDrawing;
   StandardPlot;
   xval := xEqual(n,xmin,xmax);
   fpval := SetUp(n,xval,fplus);
   fmval := SetUp(n,xval,fminus);
   gval := SetUp(n,g,xval);
   Plot(n,xval,fpval,xmin,xmax,ymin,ymax);
   Plot(n,xval,fmval,xmin,xmax,ymin,ymax);
   Plot(n,xval,gval,xmin,xmax,ymin,ymax);
end.
```

The functions/procedures `StandardPlot`, `xEqual`, `SetUp`, and `Plot` and the type `RealList0` are declared in the unit `Chap15Codes`. For output, see FIGURE 15.2

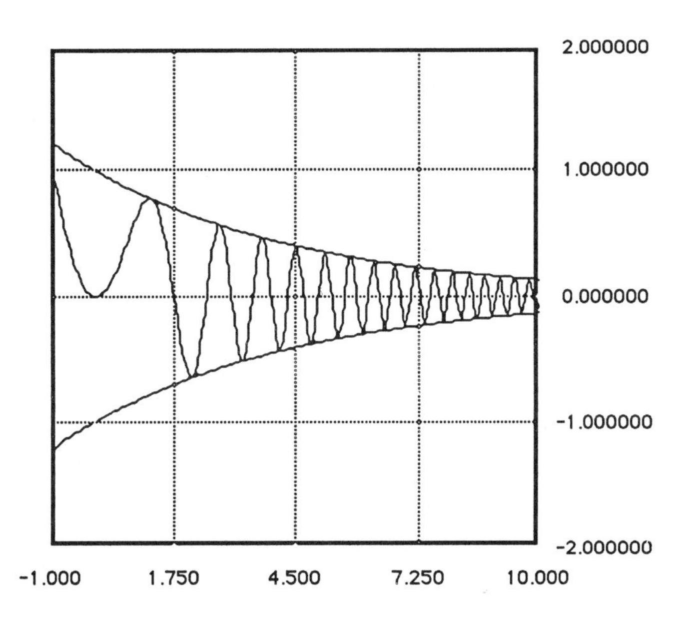

FIGURE 15.2 *Output from* Example15_3.

PROBLEM 15.4. Plot the functions $\sin(x)$ and $x - x^3/6$ on the interval $[0, \pi]$. [3]

PROBLEM 15.5. Complete the following procedure:

```
procedure TangentPlot(a:real; function f(x:real):real;
      function fp(x:real):real; xmin,xmax,ymin,ymax:real);
{Pre:f and fp are defined at x = a.}
{Post:Plots the tangent line to f(x) at x = a.  }
{The left(right) edge of the graph window corresponds to x = xmin(xmax).}
{The bottom(top) edge of the graph window corresponds to y = ymin(ymax).}
```

PROBLEM 15.6. What is the probability that a quintic polynomial of the form

$$q(x) = x^5 + a_4x^4 + a_3x^3 + a_2x^2 + a_1x + a_0$$

has three real roots in the interval [-3,3] assuming that each a_i is randomly selected from [-1,1]? Set up an environment that uses `Plot` to shed light upon this question.

A problem with `Plot` is that it requires the specification of the x and y ranges. For a simple functions this is not difficult. But in more complicated settings we may not have good estimates for `xmin`, `xmax`, `ymin`, or `ymax`. A mistake in choosing the scale can result in a plot that does not display any valuable information. For example, if we plot $g(x) = (x - 2)(x - 3)$ on the interval [0,6], then we would hope to discover the two roots. But if the y range extends from -1 to +100, then this would not be the case because the vertical compression would be too great. Thus,

[3]It is illegal for built-in functions to be passed as parameters. Thus, to pass the sine function as a parameter you have to write a function `MySine` with the one-line body `MySine := sin(x)`. A reference of the form `SetUp(n,MySine,x)` is valid.

it would be handy to have a plot procedure that automatically takes care of the fit-to-window calculations. One way to ensure that the complete plot is displayed in the window is to associate the

min value in `xval[0..n]`	with	`h = h_Left`, the left edge of the plot window
max value in `xval[0..n]`		`h = h_Right`, the right edge of the plot window
min value in `yval[0..n]`		`v = v_Bottom`, the bottom edge of the plot window
max value in `yval[0..n]`		`v = v_Top`, the top edge of the plot window

To compute the max's and min's we use

```
function MaxInList0(n:integer; x:RealList0):real;
{Pre:0<=n<=nmax.}
{Post:The largest value in x[0..n].}

function MinInList0(n:integer; x:RealList0):real;
{Pre:0<=n<=nmax.}
{Post:The smallest value in x[0..n].}
```

both of which are declared in the unit `Chap15Codes`. (These functions are just subscript-from-zero versions of the max/min functions developed in Chapter 14.) A fragment of the form

```
xmin := MinInList0(n,xval);
xmax := MaxInList0(n,xval);
ymin := MinInList0(n,yval);
ymax := MaxInList0(n,yval);
Plot(n,xval,yval,xmin,xmax,xmin,xmax);
```

fits the plot to the window. However, the x and y scales would generally be different. Although this is sometimes appropriate, we have chosen to implement our automatic plot procedure with the same scaling along both axes:

```
procedure AutoPlot(n:integer; xval,yval:RealList0);
{Pre:0<=n<=fEvalMax}
{Post:Plots yval[0..n] vs xval[0..n] with auto scaling.}
var
   xmin,xmax,ymin,ymax:real;
begin
   xmin := MinInList0(n,xval);
   xmax := MaxInList0(n,xval);
   ymin := MinInList0(n,yval);
   ymax := MaxInList0(n,yval);
   if (xmax - xmin) > (ymax - ymin) then
      Plot(n,xval,yval,xmin,xmax,xmin,xmax)
   else
      Plot(n,xval,yval,ymin,ymax,ymin,ymax)
end
```

`Example15_4` illustrates the use of this procedure by plotting the " tilted" ellipse defined by

$$(x(t), y(t)) \;=\; (5\cos(t) + 3\sin(t), 7\cos(t) - 4\sin(t)) \qquad 0 \le t \le 2\pi.$$

```
program Example15_4;
{Plot Ellipse}
uses
   Chap15Codes;
const
   n = nmax;
var
   tval,xval,yval:RealList0;

function fx(t:real):real;
begin
   fx := 5*cos(t) + 3*sin(t);
end

function fy(t:real):real;
begin
   fy := 7*cos(t) - 4*sin(t);
end

begin
   ShowDrawing;
   tval := xEqual(n,0,2*pi);
   xval := SetUp(n,tval,fx);
   yval := SetUp(n,tval,fy);
   AutoPlot(n,xval,yval);
end.
```

The functions/procedures `xEqual`, `SetUp`, and `AutoPlot` are declared in the unit `Chap15Codes`. For sample output, see FIGURE 15.3.

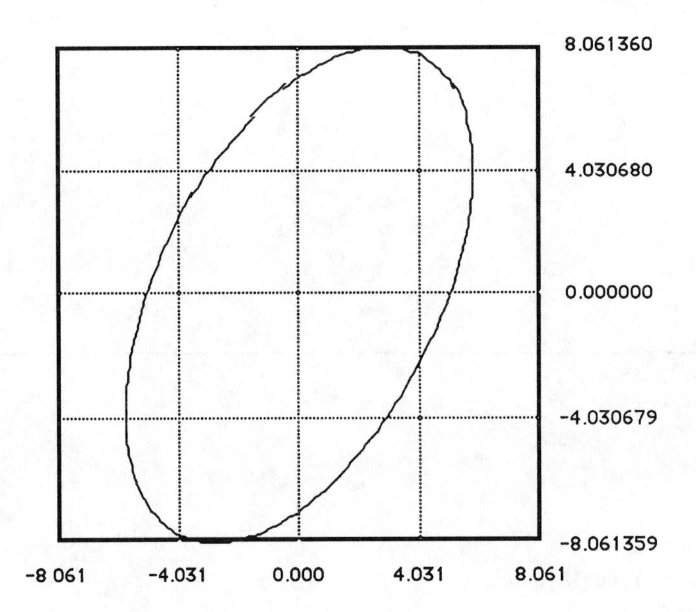

FIGURE 15.3 *Output from* Example15_4.

This requires the production of a set of points $(x(t_i), y(t_i))$ where $0 = t_0 < t_1 < \cdots < t_n = 2\pi$. In Example15_4, the t-values are equally spaced. Without the automatic scaling, we would have to analyze the parametric equations ourselves to determine a good fit-to-window scaling.

PROBLEM 15.7. Complete the following procedure:

```
procedure fPlot(function f(x:real):real; L,R:real);
{Pre:f(x) defined on [L,R].}
{Post:Plots f(x) on [L,R]. Scales so that the displayed plot fits in the graph window.}
```

Base the plot on fEvalMax function evaluations.

PROBLEM 15.8. Complete the following procedure:

```
procedure fLogPlot(function f(x:real):real; L,R:real);
{Pre:f(x) is defined and positive on [L,R]. 0<L<=R. }
{Post:Plots log(f(x)) versus log(x) for x in [L,R]. }
{Scales so that the displayed plot fits in the graph window.}
```

Base the plot on fEvalMax function evaluations where nmax is the constant that defines the length of the RealList0 array type.

PROBLEM 15.9. Observe that the plot produced by Example15_4 is actually tangential to the boundary of the graph window. Modify AutoPlot so that the x and y ranges are expanded by 10 percent, thereby "pulling in" the plot from the graph window boundary.

15.3 Efficiency Issues

The function SetUp generates each function value independently. However, there are occasions when this may not be the most efficient course of action. Consider the generation of sines and cosines at equally spaced points across the interval $[0, \pi]$. The straightforward approach is simply to call sin and cos at each required x value:

```
procedure SinCos1(n:integer; s,c:RealList0);
{Pre:n>0}
{Post:s[k] = sin(k*pi/n), c[k] = cos(k*pi/n), k=0,...,n.}
var
   k:integer;
   theta:real; {pi/n}
   angle:real; {k*theta}
begin
   theta := pi/n;
   for k:=0 to n do
      begin
         angle:=k*theta;
         s[k]:=sin(angle);
         c[k]:=cos(angle);
      end
end
```

However, by using various trigonometric identities we can reduce the number of sin and cos evaluations by a factor of 4 without any additional floating point arithmetic. To illustrate, assume that n is a multiple of 4 and start by generating the sines and cosines for angles in the interval $[0, \pi/4]$:

```
   p:= n div 4;
   theta := pi/n;
   for k:=0 to p do
      begin
         angle:=k*theta;
         s[k]:=sin(angle);
         c[k]:=cos(angle);
      end
```

To get the values over $(\pi/4, \pi/2]$, we use the following identities:

$$\begin{aligned} \sin(\pi/2 - \phi) &= \cos(\phi) \\ \cos(\pi/2 - \phi) &= \sin(\phi) \end{aligned}$$

Setting $\phi = k\theta$ and $m = n/2$ we obtain

$$\begin{aligned} \sin((m-k)\theta) &= \cos(k\theta) \\ \cos((m-k)\theta) &= \sin(k\theta) \end{aligned}$$

where $m = n/2$. The computation of s[p+1..m] and c[p+1..m] may therefore proceed as follows:

```
m := 2*p;
for k := 0 to p-1 do
   begin
      s[m-k]:=c[k]; c[m-k]:=s[k];
   end;
```

Note that this fragment merely involves copying from one array to another. There is no floating point arithmetic and there are no calls to `sin` and `cos`. To appreciate the subscripting, assume that $p = 3$ and notice that the loop is equivalent to

```
s[6]:=c[0]; c[6]:=s[0];
s[5]:=c[1]; c[5]:=s[1];
s[4]:=c[2]; c[4]:=s[2];
```

To complete the computation we exploit the identities

$$\begin{aligned} \sin(\pi/2 + \phi) &= \cos(\phi) \\ \cos(\pi/2 + \phi) &= -\sin(\phi). \end{aligned}$$

once again setting $\phi = k\theta$. This gives

```
for k:=1 to m do
   begin
      s[m+k]:=c[k]; c[m+k]:=-s[k];
   end
```

Putting it all together we obtain:

```
procedure SinCos2(n:integer; var s,c:RealList0);
{Pre:n>0, n mod 4 = 0}
{Post:s[k] = sin(k*pi/n), c[k] = cos(k*pi/n), k=0,...,n.}
var
   k:integer;
   p,m:integer; {n/4 and n/2}
   theta:real; {pi/n}
   angle:real; {k*theta}
begin
   p:=n div 4; m:=n div 2; theta:=pi/n;
   for k:=0 to p do
      begin
         angle:=k*theta; s[k]:=sin(angle); c[k]:=cos(angle);
      end;
   for k:=0 to p-1 do
      {Entries for (pi/4,pi/2].}
      begin
         s[m-k]:=c[k]; c[m-k]:=s[k];
      end;
   for k:=1 to m do
      {Entries for (pi/2,pi]}
      begin
         s[m+k]:=c[k]; c[m+k]:=-s[k];
      end;
end;
```

```
program Example15_5;
{Sine/Cosine Table Construction.}
uses
   Chap15Codes;
const
   n = nmax;
   MyComputer = 'Macintosh Powerbook 170, System 7.1.';
   RepeatFactor = 20;
var
   k:integer;
   cos1,sin1,cos2,sin2:RealList0;
   StartTime,StopTime:real; {Clock Snapshots}
   T1,T2:real; {Time for SinCos1 and SinCos2}
begin
   ShowText;
   StartTime:=tickcount;
   for k:=1 to RepeatFactor do
      SinCos1(nmax,sin1,cos1);
   StopTime:=tickcount;
   T1 := StopTime - StartTime;
   StartTime:=tickcount;
   for k:=1 to RepeatFactor do
      SinCos2(nmax,sin1,cos1);
   StopTime:=tickcount;
   T2 := StopTime - StartTime;
   writeln('SinCos2 Time/SinCos1 Time = ', (T2/T1):6:3);
end.
```

The functions `SinCos1` and `SinCos2` are in the unit `Chap15Codes`. Output:

Macintosh Powerbook 170, System 7.1.
SinCos2 Time / SinCos1 Time = 0.312

`Example15_5` benchmarks `SinCos1` and `SinCos2`.

The timings confirm that it usually pays to avoid unnecessary function evaluations.

PROBLEM 15.10. Produce yet another implementation of `SinCos1`, basing it upon the following identities:

$$\begin{aligned}\sin((k+1)\theta) &= \sin(k\theta)\cos(\theta) + \cos(k\theta)\sin(\theta)\\ \cos((k+1)\theta) &= \cos(k\theta)\cos(\theta) - \sin(k\theta)\sin(\theta)\end{aligned}$$

In particular, use the identities to generate `s[2..n]` and `c[2..n]` after computing

```
s[0]:=0; c[0]:=1;
s[1]:=sin(pi/n); c[1]:=pi/n)
```

Modify `Example15_3` so that it benchmarks this implementation and compares it to `SinCos1`.

PROBLEM 15.11. Complete the following procedure:

```
procedure SinCos2Pi(n:integer; var s,c:RealList0);
{Pre:n is a positive even integer}
{Post:s[k] = sin(k*2*pi/n), c[k] = cos(k*2*pi/n), k=0,...,n.}
```

Use sin and cos to get the sines and cosines of angles in the interval $[0, \pi/4]$. Obtain the remaining sines and cosines without floating point arithmetic.

PROBLEM 15.12. Complete the following procedure:

```
procedure GenSinCos2(n,L:integer; var s,c:LongRealList0);
{Pre:L> 0,n>0, n mod 4 = 0, }
{Post:s[k] = sin(k*pi/n), c[k] = cos(k*pi/n), k=0,...,L.}
```

Use sin and cos to get the sines and cosines of angles in the interval $[0, \pi/4]$. Obtain the remaining sines and cosines without floating point arithmetic.

PROBLEM 15.13. Recall that a function $f(x)$ that is defined everywhere has period p if $f(x+p) = f(x)$ for all x. Complete the following procedure:

```
procedure PeriodicfPlot (function f(x:real):real; L,p:real; q:integer);
{Pre:f(x) has period p.  q>=1.}
{Post:  Plots f(x) on [L,L+q*p].  Scales so plot fits in graph window.]}
```

Your implementation should exploit the fact that it is only necessary to evaluate the function over the interval $[L, L+p]$.

Another efficiency issue associated with table set-up concerns the process of *vectorization*. The act of vectorizing a program is the act in making it rich in array-level operations. To motivate this important idea, we consider the following fragment which plots the functions x, $x^2, \ldots, x^{15}$ on the interval [0,1].

```
n := fEvalMax;
x := xEqual(n,0,1);
for k:=0 to n do
   y[k]:=x[k];
for p:=1 to 15 do
   {y[k] = x[k]^p,k=0..n}
   begin
      Plot(n,x,y,0,1,0,1);
      for k:=0 to n do;
         y[k] := x[k]*y[k];
   end;
```

Notice that the two k-loops effectively carry out an array-level assignment and an array-level multiplication. Thus, if the functions

```
function Copy(n:integer; a:RealList0):RealList0;
{Pre:  0<=n<=fEvalMax.}
{Post:  Copy[k] = a[k], k=0,...,n.}
var
   k:integer;
begin
   for k:=0 to n do
      Copy[k]:=a[k] ;
end
```

and

```
function Mult(n:integer; a,b:RealList0):RealList0;
{Pre:0<=n<=fEvalMax.}
{Post:Mult[k] = a[k]*b[k], k=0,...,n.}
var
   k:integer;
begin
   for k:=0 to n do
      Mult[k]:=a[k]*b[k];
end
```

are available, then these two loops can be replaced by function calls. See Example15_6. Many

```
program Example15_6;
{Plot y = x, x^2,...,x^15}
uses
   Chap15Codes;
const
   n=nmax;
var
   p:integer;
   xval,yval:RealList0;
begin
   ShowDrawing;
   StandardPlot;
   xval := xEqual(n,0,1);
   yval := Copy(n,xval);
   for p:=1 to 15 do
      {yval[k] = xval[k]^p,k=0..n}
      begin
         Plot(n,xval,yval,0,1,0,1);
         yval := Mult(n,xval,yval);
      end;
end.
```

The type RealList0 and the functions/procedures xEqual, Copy, Plot and Mult are declared in the unit Chap15Codes. For output, see FIGURE 15.4.

computers are designed so that they can perform certain array-level operations very quickly. The ability to vectorize a program and to think at the array level is very important in computational science.

To get an idea about what it is like to vectorize a computation, we have established a "library" of vector operations. See FIGURE 15.5. The vectorization that leads to Example 15_6 is obvious. Less so is the vectorization of the fragment

```
for k:=0 to n do
   Error[k] := abs((1 - xval[k]/2)/(1 + xval[k]/2) - exp(-x));
```

which evaluates the function

$$err(x) = \left| \frac{1 - \frac{x}{2}}{1 + \frac{x}{2}} - e^{-x} \right|$$

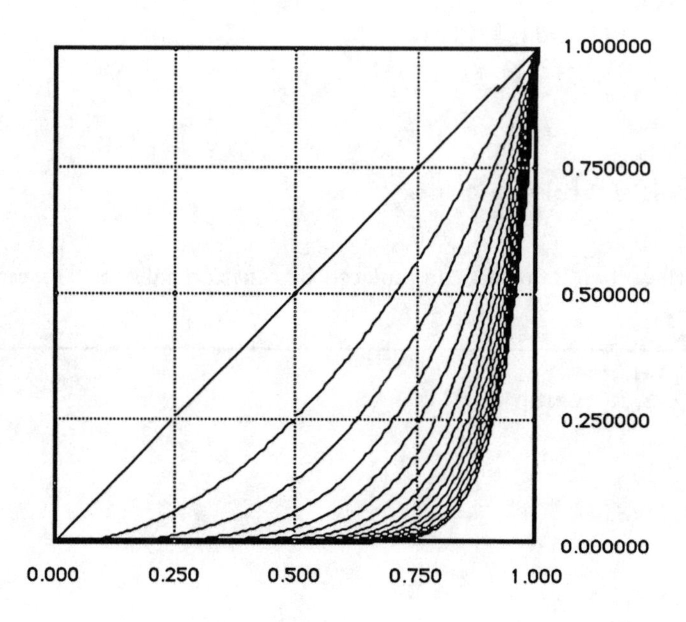

FIGURE 15.4 *Output from* Example15_6

at each of the values in xval[0..n]. The components of Error[0..n] are computed sequentially. Alternatively, we can develop these values "in parallel" by using our vectorization library. To derive the "vector version", we insert a number of temporary real variables in the original fragment so that there is no more than one operation per statement:

```
for k:=0 to n do
   begin
      xHalf:= xval[k]/2;
      Num:=1;
      Num:=1-xHalf;
      Denom:=1;
      Denom:=Denom+xHalf;
      Approx:=Num/Denom;
      Exact:=exp(-xval[k]);
      Diff:=Approx-Exact;
      Error[k] := abs(Diff);
   end
```

We then do something that looks very extravagant with respect to memory: we turn these temporary simple variables into arrays so that all the intermediate quantities are saved:

```
function Add(n:integer;a,b:RealList0):RealList0;
{Pre:0<=n<=fEvalMax.}
{Post:Add[k] = a[k]+b[k], k=0,...,n.}

function Sub(n:integer; a,b:RealList0):RealList0;
{Pre:0<=n<=fEvalMax.}
{Post:Sub[k] = a[k]-b[k], k=0,...,n.}

function Mult(n:integer; a,b:RealList0):RealList0;
{Pre:0<=n<=fEvalMax.}
{Post:Mult[k] = a[k]*b[k], k=0,...,n.}

function Divide(n:integer; a,b:RealList0):RealList0;
{Pre:0<=n<=fEvalMax, b[k] <>0, k=0,...,n}
{Post:Divide[k] = a[k]+b[k], k=0,...,n.}

function Scale(n:integer; a:real; b:RealList0):RealList0;
{Pre:0<=n<=fEvalMax.}
{Post:Scale[k] = a*b[k], k=0,...,n.}

function Translate(n:integer; a:real; b:RealList0):RealList0;
{Pre:0<=n<=fEvalMax.}
{Post:Translate[k] = a+b[k], k=0,...,n.}

function Constant(n:integer; a:real):RealList0;
{Pre:0<=n<=fEvalMax.}
{Post:Constant[k] = a, k=0,...,n.}

function CopyV(n:integer; a:RealList0):RealList0;
{Pre:0<=n<=fEvalMax.}
{Post:CopyV[k] = a[k], k=0,...,n.}

function AbsV(n:integer; a:RealList0):  RealList0;
{Pre:0<=n<=fEvalMax.}
{Post:AbsV[k] = |a[k]|, k=0,...,n.}
```

FIGURE 15.5 *Vectorization Functions in* Chap15Codes

```
for k:=0 to n do
   begin
      xHalf[k]:= xval[k]/2;
      Num[k]:=1;
      Num[k]:=Num[k]-xHalf[k];
      Denom[k]:=1;
      Denom[k]:=Denom[k]+xHalf[k];
      Approx[k]:=Num[k]/Denom[k];
      Exact[k]:=exp(-xval[k]);
      Diff[k]:=Approx[k]-Exact[k];
      Error[k] := abs(Diff[k]);
   end
```

Observe that this single loop is the merge of nine separate loops:

```
for k:=0 to n do
   xHalf[k]:= xval[k]/2;
for k:=0 to n do
   Num[k]:=1;
for k:=0 to n do
   Num[k]:=Num[k]-xHalf[k];
for k:=0 to n do
   Denom[k]:=1;
for k:=0 to n do
   Denom[k]:=Denom[k]+xHalf[k];
for k:=0 to n do
   Approx[k]:=Num[k]/Denom[k];
for k:=0 to n do
   Exact[k]:=exp(-xval[k]);
for k:=0 to n do
   Diff[k]:=Approx[k]-Exact[k];
for k:=0 to n do
   Error[k] := abs(Diff[k]);
```

Each of these loops can be handled by one of the array-valued functions in `Chap15Codes`. See `Example15_7`. A *vectorizing* compiler is able to perform this kind of reorganization automatically.

PROBLEM 15.14. Plot the function

$$f(x) = \frac{1}{(x-.3)^2+.01} + \frac{1}{(x-.9)^2+.04}$$

on the interval [0.2]. Make effective use of the vectorization functions in `Chap15Codes`. when setting up the array of function values.

PROBLEM 15.15. Vectorize the following function which evaluates the polynomial

$$q(z) = \sum_{j=0}^{p}(-1)^p \frac{z^{2p}}{(2p)!}$$

at $z = x_0, \ldots, x_n$:

```
program Example15_7;
{Plot exp(-x) and (1-x/2)/(1+x/2)}
uses
   DDCodes,Chap15Codes;
const
   n = fEvalMax; {Number of function evaluations}
var
   xval:RealList0; {x values}
   xHalf:RealList0; {x/2}
   Num:RealList0; {1-(x/2)}
   Denom:RealList0; {1+(x/2)}
   Approx:RealList0; {(1-(x/2))/(1+(x/2)}
   Exact:RealList0; {exp(-x)}
   Diff:RealList0; {Approx - Exact}
   Error:RealList0; {Exact - Approx}
   R:real; {Examine error on [0,R]}

function f(x:real):real;
begin
   f := exp(-x);
end;

begin
   ShowDrawing;
   StandardPlot;
   R := 8;
   while YesNo('Another Plot?') do
      begin
         {Error on [0,R] plotted.}
         R := ReviseReal('R',R,R/2,R,1);
         xval := xEqual(n,0,R);
         xHalf := Scale(n,0.5,xval);
         Num := Constant(n,1);
         Num := Sub(n, Num,xHalf);
         Denom := Constant(n,1);
         Denom := Add(n,Denom,xHalf);
         Approx := Divide(n,Num,Denom);
         Exact := SetUp(n,xval,f);
         Diff := sub(n,Exact,Approx);
         Error := AbsV(n,Diff);
         ClearScreen;
         Plot(n,xval,Error,0,R,0,MaxInList(n,Error));
      end;
end.
```

The constant `fEvalMax`, the type `RealList` and the functions/procedures `xEqual`, `Scale`, `Constant`, `Add`, `Sub`, `Divide`, `AbsV`, `SetUp`, and `Plot` are declared in the unit `Chap15Codes`. The functions/procedures `YesNo`, `ReviseReal` and `ClearScreen` are declared in the unit `DDCodes`. For sample output, see FIGURE 15.6.

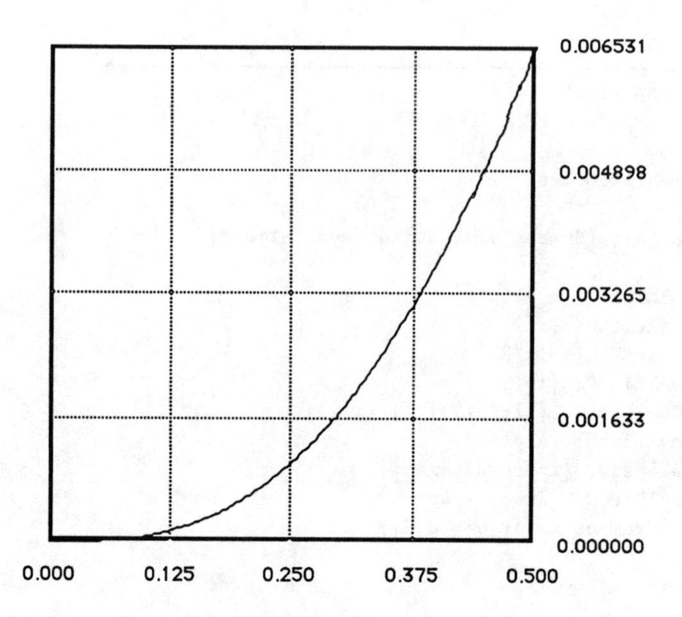

FIGURE 15.6 *Output from* Example15_7

```
function MyCos(n:integer; x:RealList0; p:integer):RealList0;
{Pre:1<=n<=nmax, 0<=p}
{Post:x - x^2/2!  + x^4/4!  - x^6/6!  +...+ (-1)^p x^(2p)/(2p)!  }
var
   j,k:integer;
   y,xsquare,term:RealList0;
begin
   for k := 0 to n  do
      begin
         y[k] := 1;
         xsquare[k] := sqr(x[k]);
         term[k] := -xsquare[k]/2;
      end;
   for j := 1 to  p do
      begin
         for k := 0 to  n  do
            begin
               y[k] := y[k] + term[k];
               if j < p then
                  term[k] := -xsquare[k]*term[k]/((2*j+1)*(2*j+2))
               else
                  MyCos[k] := y[k];
            end;
      end;
end;
```

PROBLEM 15.16. Develop a vectorized implementation of the following function:

```
function MyExp(n:integer; x:RealList0; p:integer):RealList0;
{Pre:1<=n<=nmax, 0<=p}
{Post:  MyExp[k] = q(x[k]) for k=0..n where}
{q(z) = 1 - z + z^2/2!  - z^3/3!  +...  = (-1)^p z^p / p!}
```

15.4 Look-Up

Suppose that the execution of a given program fragment is dominated by a series of calls to a costly-to-evaluate function `f(x:real):real`. Assume that the number of `f`-calls is large compared to the number of different x values that are passed. If there are m function calls and n distinct arguments passed, then the time it takes to execute the fragment can be approximately reduced by a factor of n/m by adopting the principle of *table look-up*. The idea is simply to precompute the necessary function values, store them in an array, and to "look-up" the f-values in the array as needed. The reason for the n/m reduction is that the number of function evaluations is reduced from m to n.

The applicability of the look-up principle can be extended if we are willing to *interpolate* between the precomputed function values. Suppose we need the value of f at a point z that is in between two table entries:

$\vdots$	$\vdots$
x_k	$f(x_k)$
z	$f(z)$
x_{k+1}	$f(x_{k+1})$
$\vdots$	$\vdots$

The idea behind *linear interpolation* is to approximate $f(z)$ with the value at z of the line that passes through $(x_k, f(x_k))$ and $(x_{k+1}, f(x_{k+1}))$:

$$f(z) \approx f(x_k) + \frac{f(x_{k+1}) - f(x_k)}{x_{k+1} - x_k}(z - x_k).$$

Of course, before this recipe can be invoked, we must determine k. This is particularly easy if the table is based upon equally spaced x-values. Suppose `L`, `R`, `n` and `y[0..n]` are initialized with the understanding that `y[i]` houses the value of a function $f(x)$ at $x_i = L + ih$ where $h = (R - L)/n$. Assume that $x_0 \le z \le x_n$. To locate the subinterval $[x_k, x_{k+1}]$ that contains z, we determine how far z is beyond x_0 in "h-units" and truncate the result: `k := trunc((x-x0)/h)`. Combining this with linear interpolation we obtain the following procedure:

```
function LookUpE(z:real; n:integer; L,R: real; y:RealList0):real;
{Pre:L<=z<=R and for k=0,...,n y[k], is the value of a function at L + kh }
{where h = (R-L)/n.}
{Post:The interpolated value of the function at z.}
var
   h:real; {Table spacing.}
   k:integer; {Index of subinterval that contains z.}
begin
   h := (R - L)/n;
   k := trunc((z - L) / h);
   if k = n  then
      {z=R}
      LookUpE := y[n]
   else
      {L+kh <= z <= L+(k+1)h.}
      LookUpE := y[k] + ((y[k + 1] - y[k])/h)*(z - (L+ k*h));
end;
```

The program Example15_8 explores the accuracy of linear interpolation applied to the function $g(x) = \sin(4x)$. For $n = 10,20,40,80$, and 160, a table of function values is set up based on the uniform sampling across the interval $[0, \pi]$. On each of the n subintervals, the discrepancy between $f(x)$ the linear interpolant is computed at 9 different places. The absolute value of the largest such number is reported. Notice that if the spacing between table entries is halved, then the error is approximately reduced by a factor of 4.

Let us examine the table look-up/interpolation problem when the entries are not equally spaced. Given z in the interval $[x_0, x_n]$, our task is to compute k so that $x_k \leq z \leq x_{k+1}$. Here is a solution that uses a for-loop:

```
for j:=0 to n-1 do
   if (x[j] <= z) and (z<=x[j+1]) then
      k:=j
```

The problem with this is that the search continues even after the bracketing interval is found. The situation is more efficiently handled with a while-loop:

```
function LinearSearch(z:real; n:integer; xval:RealList0):integer;
{Pre:n>=1, xval[0]<xval[1]<...<xval[[n] and xval[0]<=z<=xval[n]}
{Post:If k = LinearSearch(z,n,xval), then 0<=k<n and xval[k]<=z<=xval[k+1].}
var
   k:integer;
   FoundInterval:boolean;
begin
   k:=0;
   FoundInterval := (xval[k]<=z) and (z<=xval[k+1]);
   while not FoundInterval do
      begin
         k:=k+1;
         FoundInterval := (xval[k] <= z) and (z <= xval[k+1]);
      end;
   LinearSearch := k;
end
```

The boolean variable FoundInterval "turns true" as soon as the interval that contains z is discovered.

An even faster method of search involves the "divide and conquer" idea, encountered earlier in Chapter 10 with the method of bisection. We begin with an $n = 7$ example in which z and xval[0..7] are arranged as follows:

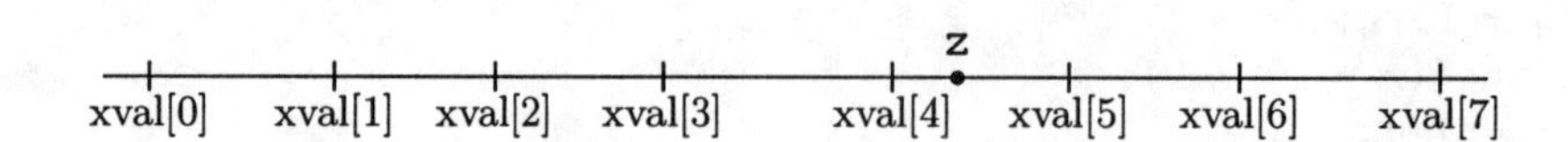

The idea is to maintain a pair of integers L and R with the property that at all times the inequalities L < R, xval[L] <= z, and z <= xval[R] are true. We start off with L:=0 and R:=7. The value of z is then compared to xval[m] where m:=(L+R) div 2:

```
program Example15_8;
{Error in Linear Interpolation.}
uses
   DDCodes,Chap15Codes;
const
   L = 0; {Left endpoint.}
   R = pi; {Right endpoint.}
   p = 10; {p-1 = number of interpolated values between table entries.}

var
   n:integer; {Length of table.}
   yval:RealList0; {Table values.}
   k:integer;
   del:real; {(R-L)/pn}
   z:real; {Value at which to check error of interpolant.}
   err:real; {Error at z.}
   MaxErr:real; {Maximum error encountered.}

function g(x:real):real;
{Post:sin(4x)}
begin
   g := sin(4*x)
end;

begin
   ShowText;
   n := 10;
   writeln(' n MaxErr');
   writeln('----------------');
   while n <= nmax do
      begin
         yval := SetUpE(n,L,R,g);
         MaxErr := 0;
         del:=(R-L)/(p*n);
         for k:=0 to p*n do
            begin
               z := L+del*k;
               err := abs(g(z) - LookUpE(z,n,L,R,yval));
               if err > MaxErr  then
                  MaxErr := err;
            end;
         writeln(n:4, MaxErr:10:6);
         n := 2*n;
      end;
end.
```

The functions SetUpE and LookUpE are declared in the unit Chap15Codes. Output:

```
   n    MaxErr
----------------
  10  0.181636
  20  0.048944
  40  0.012160
  80  0.003073
 160  0.000770
```

`L=0, R=7, m=3`

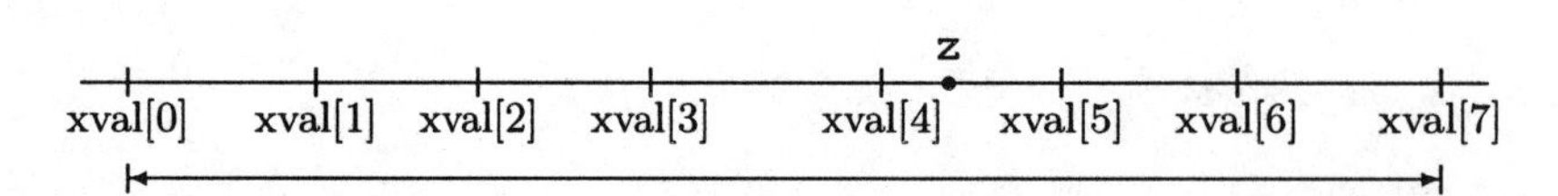

This gives us the chance to halve the current bracketing interval which is indicated by the double arrow:

```
if xval[m] < z then
    L:=m
else
    R:=m
```

Regardless of the alternative chosen, `z` is still in between `xval[L]` and `xval[R]`. In our example, `L` is updated and we repeat the process:

`L=3, R=7, m=5`

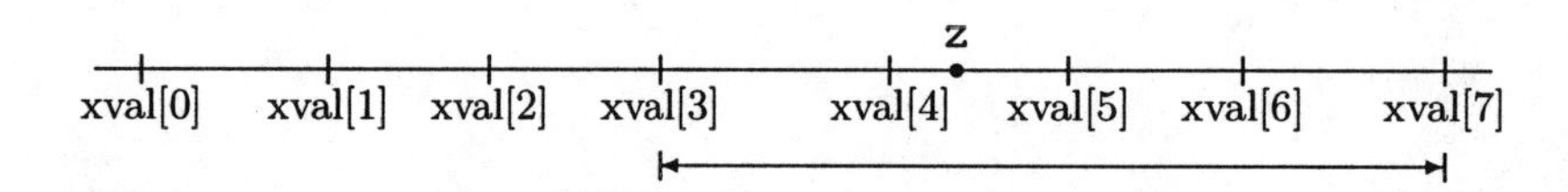

Since `z < xval[5]`, we set `R=5` and proceed with the third step:

`L=3, R=5, m=4`

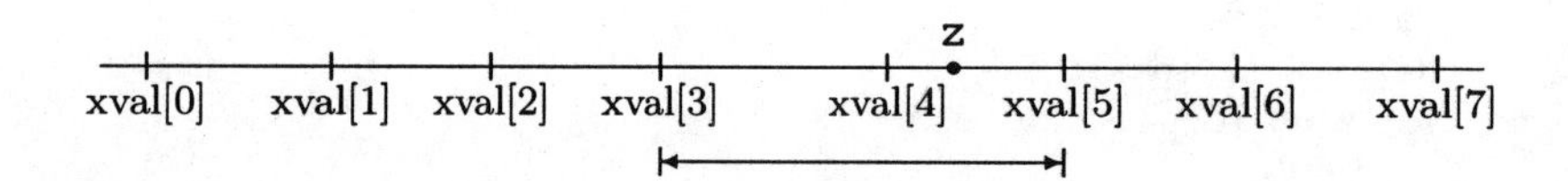

Since `xval[4] < z` , we set `L=4`. But now the interval that contains `z` is known:

`L=4, R=5`

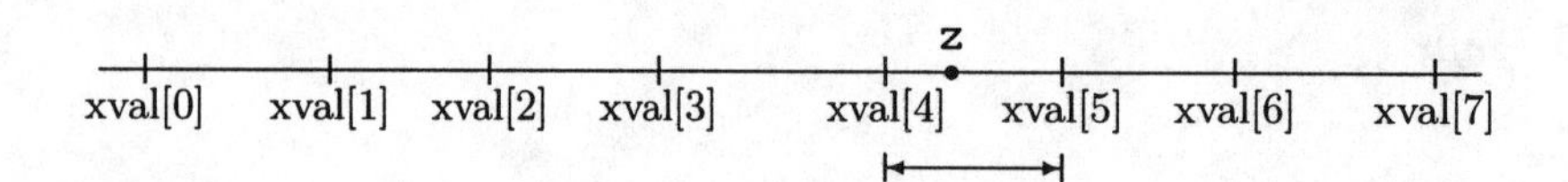

The search is terminated with the value of `L` being the index of the sought-after interval.

The overall process is clear. The value of `R-L` is roughly halved at each step[5]. Eventually, `R = L+1` and the iteration is brought to a close. This is the method of binary search which we

[5] Recall that in the method of bisection, the interval length is halved at each step.

encapsulate as follows:

```
function BinarySearch(z:real; n:integer; xval:RealList0):integer;
{Pre:n>=1, xval[0]<xval[1]<...<xval[[n] and xval[0]<=z<=xval[n]}
{Post:If k = BinarySearch(z,n,xval), then 0<=k<n and xval[k]<=z<=xval[k+1].}
var
   L,R,m:integer;
begin
   L := 0;
   R := n;
   while R>L+1 do
      {xval[L] <= z<= xval[R]}
      begin
         m := (R+L) div 2; {L<m<R}
         if z > xval[m] then
            {xval[m] <= z <= xval[R]}
            L := m
         else
            {xval[L] <= z <= xval[m]}
            R := m;
      end;
   BinarySearch := L;
end;
```

The assertion about `m` ensures that we never have `L = R`. Approximately $\log_2 n$ steps are required. On average, linear search requires about $n/2$ steps. The two methods are compared in `Example15_9` by benchmarking the following look-up-and-interpolate functions:

```
function LookUpULin(z:real; n:integer; xval,yval:RealList0):real;
{Pre:xval[0]<...xval[n] and xval[0]<=z<=xval[n].  For k=0,.. ,n}
{yval[k], is the value of a function at xval[k].}
{Post:The interpolated value of the function at z.}

function LookUpUBin(z:real; n:integer; xval,yval:RealList0):real;
{Pre:xval[0]<...xval[n] and xval[0]<=z<=xval[n].  For k=0,...,n}
{yval[k], is the value of a function at xval[k].}
{Post:The interpolated value of the function at z}
```

These are based on linear search and binary search respectively. A table of cube roots is used in the program. The y_k are equally spaced across [0,10] and $x_k = y_k^3$. Thus, the x-values are not equally spaced. The results show the disadvantage of linear search, for as z increases, it takes the linear search method longer to "reach" the bracketing interval.

Problem 15.17. Generalize `LinearSearch` so that if `z < xval[0]` it returns -1 and if `z > xval[n]` it returns `n`.

Problem 15.18. Modify `LinearSearch` so that it requires only one comparison each time through the loop.

Problem 15.19. Include print statements in `BinarySearch` so that it prints the bracketing interval after each pass through the loop.

```
program Example15_9;
{Arctan}
uses
   DDCodes, Chap15Codes;
const
   trials = 100; {Repeat factor for benchmark}
   n = fEvalMax; {Length of Table}
var
   xval,yval:RealList0; {For the table of arctangent values.}
      ⟨:⟩
function cube(x:real):  real;
      ⟨:⟩
begin
   ShowText;
   {Set up table.}
   xval := xEqual(n,0,89);
   yval := SetUp(n,xval,tan);
   z := 0;
   while z<=3 do
      begin
         {Benchmark binary search method.}
            ⟨:⟩
         OurArcTan := LookUpUBin(z,n,yval,xval);
            ⟨:⟩
         {Benchmark linear search method.}
            ⟨:⟩
         OurArcTan := LookUpULin(z,n,yval,yval);
            ⟨:⟩
         z := z + 0.5;
      end
end.
```

The functions LookUpULin and LookUpUBin are declared in the unit Chap15Codes. Output:

z	Cube Root	Interpolated Value	LinTime/BinTime
1.0	1.00000	1.00000	1.83
2.0	1.25992	1.25961	2.45
4.0	1.58740	1.58710	2.82
8.0	2.00000	2.00000	3.17
16.0	2.51984	2.51961	4.70
32.0	3.17480	3.17461	5.09
64.0	4.00000	4.00000	5.67
128.0	5.03968	5.03960	7.64
256.0	6.34960	6.34960	8.67
512.0	8.00000	8.00000	11.82

Chapter 16

Divisors

A division problem need not have a "happy ending." Quotients like $1 \div 0$ are not defined. Ratios like $1/3$ have no finite base-10 expansion. Real numbers like $\sqrt{2}$ cannot be obtained by dividing one integer into another. Integers like 7 have no proper divisors. Etc, Etc.

Division *is hard.* That's why we learn it last in grade school. That's why the IRS permits rounding to the nearest dollar. That's why base-60 systems were favored by the Maya and the Babylonians.[1]

Yes, division is by far the most interesting of the four arithmetic operations. But the idea of division transcends the purely numerical. Geometry and combinatorics are filled with partitioning problems. How can a polygon be divided into a set of triangles? How many ways can a set of m objects be divided into n non-empty subsets?

In this chapter we consider a pair of representative division problems. One is purely arithmetic and involves the prime numbers. A prime number is an integer that has no divisors except itself and one. They figure in many important applications. Our treatment of the prime numbers in §16.1 is designed to build intuition about integer divisibility.

The second division problem we consider also involves the integers, but it is essentially a partitioning problem with social constraints. This is the problem of apportionment, which in its most familiar form is this: how can 435 Congressional districts be divided among 50 states? Few division problems have such far-reaching ramifications and that alone is reason enough to study the computation. the algorithms that solve the apportionment But the apportionment problem is a good place to show how reasonable methods may differ in the results that they produce, a fact of life in computational science.

16.1 Prime Numbers

If there is no remainder when an integer p is divided by an integer q, then q is a *divisor* of p. Thus, the divisors of 24 are 1, 2, 3, 4, 6, 8,12, and 24. Let's write a procedure that produces a list of divisors for a given integer. For that purpose we assume the following declarations:

[1] More numbers divide 60 than 10, and this permits a simpler arithmetic life.

```
const
   pmax = 1000;
type
   LongIntList = array[1..pmax] of longint;
```

Suppose p is an integer and that d has type LongIntList. Using the mod function to check every integer $\leq p$ for divisibility we obtain

```
n:=0;
for k:=1 to p do
   if p mod k = 0 then
      begin
         {k divides p}
         n := n+1;
         d[n] := k;
      end
```

The fragment terminates with d[1..n] housing all of p's divisors. Thus, if $p = 30$, then

$$d[1..8] = \boxed{1 \mid 2 \mid 3 \mid 5 \mid 6 \mid 10 \mid 15 \mid 30}.$$

A more efficient construction results by noticing that we need only test the integers in $[1, \sqrt{p}]$ for divisibility. The divisors in $[\sqrt{p}, p]$ can then be obtained by division. For example, if $p = 30$, then

$$d[1..4] = \boxed{1 \mid 2 \mid 3 \mid 5}$$

an

$$d[5..8] = \boxed{30/5 \mid 30/3 \mid 30/2 \mid 30/1}$$

Taking care to handle correctly the case when p is a perfect square we obtain

```
procedure DivisorList(p:longint;var n:integer; var d:LongIntList);
{Pre:p>=1}
{Post:n is the number of divisors of p and they are contained }
{in d[1..n], ranging from smallest to largest.}
var
   k,m:longint;
   bound:longint;{Largest integer <=sqrt(p)}
begin
   m := 0;
   bound:=trunc(sqrt(p));
   for k := 1 to bound do
      if p mod k = 0 then
         begin
            m := m + 1;
            d[m] := k;
         end;
   {m is the number of divisors <= sqrt(p)}
   if p div d[m] = d[m] then
      {p is a perfect square}
      begin
         n := 2*m-1;
```

```
          for k:=1 to m-1 do
             d[m+k]:= p div d[m-k]
       end
    else
       {p is not a perfect square}
       begin
          n:=2*m;
          for k := 1 to m do
             d[m+k]:=p div d[m-k+1]
       end
 end
```

This implementation is quicker by a factor of $\sqrt{p}$ than the first solution that checks every integer in $[1,p]$. Example16_1 solicits integers and uses DivisorList to display their divisors.

PROBLEM 16.1. Let n be a positive integer and let $s(n)$ be the sum of its divisors excluding the number itself. We say that n is a *perfect number* if $s(n) = n$ The first five perfect numbers are 6, 28, 496, 8128, and 33550336. Define an integer as near-perfect if $|s(n) - n| <= 2$. Write a program that prints a table of all near perfect numbers less than or equal to 10000. The table should also include the value of $|s(n) - n|$ and the divisors of n that are less than n.

PROBLEM 16.2. Complete the following procedure:

```
procedure CommonDivisors(p,q:longint; var n; var d:LongIntList);
{Pre:p,q are positive.}
{Post:The common divisors of p and q are in d[1..n].}
```

Do the problem by getting a list of p's divisors and then forming the required list by checking if those integers divide q.

A divisor of an integer p is a *proper divisor* if it is strictly in between 1 and p. Thus, the proper divisors of 12 are 2,3,4, and 6. If p has no proper divisors, then it is *prime*. The first 10 prime numbers are

$$1 \quad 2 \quad 3 \quad 5 \quad 7 \quad 11 \quad 13 \quad 17 \quad 19 \quad 23\,.$$

Prime numbers figure heavily in a branch of mathematics called number theory and they are involved in a wide range of signal processing applications.

The procedure DivisiorList can be used to check if a specific integer p is prime or not. If it is, then it has most two divisors and

```
DivisorList(p,n,d);
IsPrime := (n <= 2);
```

assigns true to IsPrime. However, this is inefficient method. The production of divisors that goes on inside DivisorList should halt as soon as the first proper divisor is found, for then we know that p is not a prime. In particular, we would like to terminate the loop

```
m := 0;
bound:=trunc(sqrt(p));
for k := 1 to bound do
   if p mod k = 0 then
      begin
         m := m + 1;
```

```
program Example16_1;
{Divisors and Greatest Common Divisors}
uses
   Chap16Codes;
var
   AnotherEg:char; {Continuation indicator.}
   p:longint; {The integer being tested.}
   n:integer; {Number of p's divisors.}
   d:LongIntList; {List of p's divisors.}
   k:integer;
begin
   ShowText;
   AnotherEg := 'y';
   while AnotherEg <> 'n' do
      begin
         writeln('Enter a positive integer.');
         readln(p);
         DivisorList(p,n,d);
         writeln;
         writeln('The divisors of', p:9, ':');
         for k:=1 to n do
            begin
               if k mod 8 = 0 then
                  writeln;
               write(d[k]:9);
            end;
         writeln;
         writeln('Another example?  Enter y (yes) or n (no).');
         readln(AnotherEg);
      end;
end.
```

The type LongIntList and the procedure DivisorList are defined in Chap16Codes. Sample output:

```
Enter a positive integer.
720
The divisors of 720:
    1     2     3     4     5     6     8     9
   10    12    15    16    18    20    24    30
   36    40    45    48    60    72    80    90
  120   144   180   240   360   720
Another example?  Enter y (yes) or n (no).
n
```

```
            d[m] := k;
          end;
```

should `m>1` be true. This requires a `while`-loop implementation:

```
m:=0;
bound:=trunc(sqrt(p));
k:=1;
while (m<=1) and (k<=bound) do
   begin
      if p mod k = 0 then
         begin
            m:=m+1;
            d[m]:=k;
         end;
      k:=k+1;
   end;
IsPrime:=(m=1);
```

Note that since the loop only checks for divisors in the interval $[1, \sqrt{p}]$, the number of proper divisors that it finds is given by $m - 1$. Recognizing also that there is no need to assemble p's divisors in an array merely to answer the question "Is p a prime?." We therefore obtain

```
function IsPrime(p:longint):boolean;
{Pre:p>=1}
{Post:p is a prime}
var
   m:integer; {Counts divisors}
   k:integer; {Possible divisor}
   bound:  integer; {Largest integer <= sqrt(p).}
begin
   m:=0;
   bound:=trunc(sqrt(p));
   k:=1;
   while (m<=1) and (k<=bound) do
      begin
         if p mod k = 0 then
            m:=m+1;
         k:=k+1;
      end;
   IsPrime:=(m=1);
```

The program `Example16_2` computes the ten largest prime numbers that can be represented in a `longint` variable.

PROBLEM 16.3. Let p be the quadratic $p(x) = x^2+x+41$. Write a program that prints the values $p(0), p(1), \ldots, p(n)$ where n is the largest integer so that $p(0), p(1), \ldots, p(n)$ are all prime.

Let's compute a list of all the prime numbers that are less than or equal to a given integer x. The function `IsPrime` can certainly be used for this purpose:

```
program Example16_2;
{Big primes.}
uses
   Chap16Codes;
var
   count:integer;
   p:longint; {The integer being tested for primality}
begin
   ShowText;
   p := maxlongint;
   count := 0;
   writeln('The ten biggest primes <= 2^31 -1 = maxlongint:');
   while count < 10 do
      begin
         if IsPrime(p) then
            begin
               writeln(p:60);
               count:=count+1
            end;
         p:= p-1;
      end
end.
```

The function `IsPrime` is declared in the unit `Chap16Codes`. Output:

```
The ten biggest primes <= 2^31 -1 = maxlongint:
                                      2147483647
                                      2147483629
                                      2147483587
                                      2147483579
                                      2147483563
                                      2147483549
                                      2147483543
                                      2147483497
                                      2147483489
                                      2147483477
```

```
procedure PrimeList(x:longint; var n:integer; var p:LongIntList);
{Pre:x>=1}
{Post:n is the number of prime numbers <=x and they are contained}
{in p[1.. n], ranging from smallest to largest.}
var
   k:integer;
begin
   n:=0;
   for k:=1 to x do
      if IsPrime(k) then
         begin
            n:=n+1;
            p[n]:=k
         endend
```

Thus, if $x = 25$, then $n = 10$ and

p[1..10] =

1	2	3	5	7	11	13	17	19	23

`Example16_3` uses this procedure to print lists of primes.

PROBLEM 16.4. The positive integers p and q are *twin primes* if they are both prime and $|p - q| = 2$. Write a program that prints a list of all the twin prime pairs in $[1, 1000]$.

PROBLEM 16.5. The sequence $i..j$ is *prime-free* if there are no primes k that satisfy $i \leq k \leq j$. Write a program that prints the longest prime-free sequence with $1 \leq i \leq j \leq 1000$. Next to each entry in the list should be printed its proper divisors.

PROBLEM 16.6. Let p_k be the k-th prime and define $f_k = k/p_k$. Note that f_k is the fraction of integers in the interval $[1, p_k]$ that are prime. Write a program that explores the value of f_k as k grows. Proceed by plotting a sufficient number of the f_k.

PROBLEM 16.7. If p and q are primes, then there exist positive integers m and n so that $mp = nq + 1$. Verify this for all distinct pairs of primes that are less than or equal to 13.

PROBLEM 16.8. The *Goldbach conjecture* states that every even integer is the sum of two primes. Assuming that the conjecture is true, complete the following procedure:

```
procedure TwoPrime(n:integer; p:LongIntList; i:integer; var p1,p2:integer);
{Pre:p[1..n] contains the first n primes in order and 2<=i<=p[n].}
{Post:i = p1+p2 where both p1 and p2 are contained in p[1..n].}
```

PROBLEM 16.9. Complete the following procedure:

```
procedure PrimeDivisorList(x:longint; var n:integer; var d:LongIntList);
{Pre:x>=1}
{Post:n is the number of prime divisors of x and they are contained }
{in d[1..n], ranging from smallest to largest.}
```

PROBLEM 16.10. Complete the following function:

```
function FirstPrimes(n:integer):LongIntList;
{Pre:n>=1}
{Post:  FirstPrimes[1..n] contains the first n primes.}
```

```
program Example16_3;
{The primes in [1,300].}
uses
   Chap16Codes;
const
   x = 300;
var
   k:integer;
   n:integer; {Number of primes <= x}
   p:LongIntList; {p[1..n] the primes <= x}
begin
   ShowText;
   PrimeList(x,n,p);
   writeln('The primes <= ', x :  4, ':');
   writeln;
   for k := 1 to  n do
      begin
         write(p[k]:5);
         if k mod 10 = 0 then
            writeln
      end
end.
```

The type LongIntList and the function PrimeList are defined in Chap16Codes. Output:

```
    1    2    3    5    7   11   13   17   19   23
   29   31   37   41   43   47   53   59   61   67
   71   73   79   83   89   97  101  103  107  109
  113  127  131  137  139  149  151  157  163  167
  173  179  181  191  193  197  199  211  223  227
  229  233  239  241  251  257  263  269  271  277
  281  283  293
```

16.2 Apportionment

The apportionment of Congressional districts among the 50 states is essentially a 50-fold division problem with rules about the divisors. Let d_i be the number of districts allocated to the i-th state and let that state have population p_i. The ideal of equal representation states that the quotients p_i/d_i be the same, i.e., from state to state, the number of people per district should be equal. The goal of perfectly equal representation cannot be achieved because the total number of districts is limited

$$435 = d_1 + d_2 + \cdots + d_{50}$$

and because each state is guaranteed to have at least one district,

$$d_i \geq 1 \qquad i = 1, 2, \cdots, 50.$$

Thus, in the 1990 apportionment Montana had the most people per district (799,000) and Wyoming the fewest (453,000). Ohio, with approximately 570,000 people per district, was closest to the "ideal" ratio $(p_1 + \cdots + p_{50})/435$.

The method used by the Federal government to construct an apportionment can be likened to a card deal. The deck has $C = 435$ cards (districts) and there are $n = 50$ players (states). The dealer always gives the next card (district) to the most "the most needy" player (state).

Our goal is to describe this process in detail by writing an apportionment procedure.. To that end we make the following declarations:

```
const
   nState := 51;
type
IntegerList = array[1..nState] of longInt;
RealList = array[1..nState] of real;
```

Let `C` and `n` be constants that contain the number of districts and the number states that are involved. The process starts by assigning one district to each state:

```
for i := 1 to n do
   d[i] := 1;
```

Here, `d` is an `IntegerList` array. When we're done, —tt d[1..n] will contain the apportionment.

At this stage, `C-n` districts have yet to be allocated. Since equal representation means that the quotients $p_1/d_1, \ldots, p_n/d_n$ should have small variation, it is reasonable to "deal" the next district to the state with the highest population-to-districts quotient. We represent these quotients in a `RealList` array `PV` and use the function

```
function MaxIndex(n:integer; x:RealList):integer;
{Pre:x[1..n] initialized.}
{Post:x[k] >= every component of x[1..n].}
```

to identify the state most deserving of the district to be allocated next. Thus, after the initial assignment of one district to each state, `PV[1..n]` is initialized

```
for i:=1 to n do
   PV[i] := p[i];
```

and the allocation of the remaining districts proceeds as follows:

```
DistrictsLeft := C - n;
while DistrictsLeft > 0 do
   {Award the next district to the state that maximizes p[i]/d[i]}
      begin
         k := MaxIndex(n,PV);
         {The k-th state gets the district.}
         d[k] := d[k] + 1;
         PV[k] := p[k]/d[k];
         DistrictsLeft := DistrictsLeft - 1;
      end
end
```

Each time through the loop, `PriorityVals[1..n]` is searched for its largest value and `k` is assigned the index where the max occurs. The values in `d[k]` and `PriorityVals[k]` are updated to reflect the allocation of the district. The number of districts that have yet to be assigned is stored in `DistrictsLeft` and when the value of this counter is zero, the loop terminates.

The method of apportionment just illustrated is referred to as the method of *Smallest Divisors.* It is one of five "modern" methods of apportionment. Each method is identified with a priority function $f(p_i, d_i)$ that quantifies "the need" of the i-th state after it has been allocated d_i districts. They are all combinations of the current people-per-district ratio p_i/d_i and the next people-per-district ratio $p_i/(d_i+1)$:

Method	$f(p_i, d_i)$
Smallest Divisors	$\frac{p_i}{d_i}$.
Largest Divisors	$\frac{p_i}{d_i+1}$
Major Fractions	$\frac{1}{2}\left(\frac{p_i}{d_i} + \frac{p_i}{d_i+1}\right)$
Equal Proportions	$\sqrt{\frac{p_i}{d_i}\frac{p_i}{d_i+1}}$
Harmonic Mean	$\frac{2}{\frac{1}{p_i/d_i} + \frac{1}{p_i/(d_i+1)}}$

It can be shown that the method of smallest divisors tends to favor the smaller states while the method of greatest divisors tends to favor the larger states. The other three methods take the mean of these two extremes. The arithmetic, geometric, and harmonic mean are used in the methods of Major Fractions, Equal Proportions, Harmonic Means.

The unit `Chap16Codes` includes the functions

```
function SmallestDivisors(pop:longint; di:integer):real;
function GreatestDivisors(pop:longint; di:integer):real;
function MajorFractions(pop:longint; di:integer):real;
function EqualProportions(pop:longint; di:integer):real;
function HarmonicMean(pop:longint; di:integer):real;
```

By parameterizing the priority function, we obtain the following procedure that can be used to compute an apportionment:

```
procedure Apportion(C,n:integer; p:IntegerList;
        function f(pop:longint;di:integer):real;
        var d:IntegerList; var PV:RealList);
{Pre:C = number of districts, n = number of states, p[1..n] their populations,}
{f is the priority function.}
{Post:For i=1..n, d[i] is the number of districts awarded to state i}
{and PV[i] = f(p[i],d[i])}
var
   i:integer;
   DistrictsLeft:integer; {The number of districts that remain to be allocated.}
   k:integer; {Index of the next state picked.}
begin
   {Initialize d and priority values}
   for i := 1 to  n  do
      begin
         d[i]:=1;
         PV[i]:=f(p[i],d[i]);
      end;
   DistrictsLeft:= C - n;
   while DistrictsLeft > 0 do
      {Award the next district to the state that maximizes f(p[i],d[i])}
      begin
         k := MaxIndex(n,PV);
         {State[k] gets the district.}
         d[k]:=d[k]+1;
         PV[k]:=f(p[k],d[k]);
         DistrictsLeft:= DistrictsLeft-1;
      end
end;
```

To facilitate experimentation with this procedure, we have set up data files pop1980 and pop1990. They include population data and state names. The information in these files can be placed into a pair of arrays with a call to

```
procedure GetPopulationData(year:string; var state:stringlist; var p:IntegerList);
{Pre:The relevant census year.  Either '1980' or '1990'}
{Post:  Define Washington DC as the 51st state.}
{The names of the 51 states are in state[1..51] and their populations in p[1..51].}
```

For this to work, it is essential that the project be opened in the same folder as these data files. The type `stringlist` is defined by

```
stringlist = array[1..nState] of string[20]
```

Example16_4 compares the five methods of apportionment. Observe that the different priority functions can produce different apportionments.

PROBLEM 16.11. Write a program that obtains census data for a specified year and computes the apportionments based on the methods of Smallest Divisors (SD) and Greatest Divisors (GD). A list of all those states that receive more districts with the GD method should be printed. A second list of all states that receive more districts via the SD method should also be printed. The results should confirm that the GD method favors large states and the SD method favors small states.

```
program Example16_4;
{Prints state population and congressional district information.}
uses
   Chap16Codes;
const
   C = 435; {Total number of districts.}
   n = 50; {Apportionment size.}
var
   k:integer;year:string; {The census year.}
   state:StringList; {The state names.}
   p:IntegerList; {The state population (in millions).}
   dEP,dMF,dHM,dSD,dGD:IntegerList; {Apportionments by the various methods.}
   PV:RealList; {Priority values array.}
begin
   ShowText;
   writeln('Enter census year.  (1980 or 1990)');
   readln(year);
   GetPopulationData(year,state,p);
   Apportion(C,n,p,EqualProportions,dEP,PV);
   Apportion(C,n,p,MajorFractions,dMF,PV);
   Apportion(C,n,p,HarmonicMean,dHM,PV);
   Apportion(C,n,p,SmallestDivisors,dSD,PV);
   Apportion(C,n,p,GreatestDivisors,dGD,PV);
   writeln;
   writeln(' ', year, ' Apportionment ');
   writeln;
   writeln(' State Population Equal Major Harm.  Small.  Great.');
   writeln(' Propor.  Fract Mean Divisors Divisors');
   writeln;
   for k := 1  to n do
      writeln(state[k]:18,p[k]:12,dEP[k]:10,dMF[k]:8,dHM[k]:8,dSD[k]:8,dGD[k]:8);
end.
```

The types StringList, IntegerList, and RealList and the functions/procedures GetPopulationdata, Apportionment, EqualProportions, MajorFractions, HarmonicMean, SmallestDivisors, and GreatestDivisors are declared in the unit Chap16Codes. Sample output:

```
Enter census year.  (1980 or 1990)
1980

                        1980 Apportionment

           State  Population    Equal  Major  Harm.    Small.    Great.
                               Propor. Fract   Mean  Divisors  Divisors

         Alabama     3893888        7      7      7         8         7
          Alaska      401851        1      1      1         1         1
         Arizona     2718215        5      5      5         5         5
            ⋮           ⋮           ⋮      ⋮      ⋮         ⋮         ⋮
   West Virginia     1949644        4      4      4         4         3
       Wisconsin     4705767        9      9      9         9         9
         Wyoming      469557        1      1      1         1         1
```

PROBLEM 16.12. Write a program that prints out the 50-state, Equal Proportions apportionment. The state that would lose a district if Washington DC is treated as a 51st state should be highlighted. Hint: Compute the 51-state, Equal Proportions apportionment with the 50-state, Equal Proportions apportionment.

PROBLEM 16.13. Suppose from 1990 onwards that the population of California, Florida, and Texas grows at 10 percent per decade and that the population of the other 47 states grows at 5 percent per decade. After what census will these three states have a majority of the 435 districts? Write a program that answers this question. It should print a table showing the districts awarded to these states after each decade until the sum exceeds 218. Use the method of equal proportions. For your information, California, Florida, and Texas have index 5, 9, and 43 respectively.

PROBLEM 16.14. If `C = 435`, then after the procedure call `Apportion(C-1,n,p,EqualProportions,dEP,PV)` it is possible to determine the number of people each state requires for it to "win" the 435th district. Print a table showing what this number is for each state.

PROBLEM 16.15. Define a small state to be a state that has either one or two districts. Write a program that solicits the number of districts, prints the list of small states and districts. Equal proportion Use interactive framework.

PROBLEM 16.16. For an n state apportionment $d_1, \ldots, d_n$ involving states that have populations $p_1, \ldots, p_n$, we define the following quantities:

$$
\begin{aligned}
P &= \sum_{k=1}^{n} p_k \\
S &= \sum_{k=1}^{n} p_k \left(\frac{p_k}{d_k}\right) \\
A &= S/P \\
\sigma &= \frac{1}{n}\sqrt{\sum_{k=1}^{n}\left(\frac{p_k}{d_k} - A\right)^2}
\end{aligned}
$$

Write a procedure of the form

```
procedure Deviation(n:integer; p,d:IntegerList; var Ave,Dev:real);
```

that returns the quantities A and σ in `Ave` and `Dev`.

Chapter 17

The Second Dimension

As we have said before, the ability to think at the array level is very important in computational science. This is challenging enough when the arrays involved are linear, i.e., one-dimensional. Now we consider the two-dimensional array using this chapter to set the stage for more involved applications that involve this structure. Two-dimensional array thinking is essential in application areas that involve image processing. (A digitized picture is a 2-dimensional array.) Moreover, many 3-dimensional problems are solved by solving a sequence of 2-dimensional, "cross-section" problems.

We start by considering some array set-up computations in §17.1. The idea is to develop an intuition about the parts of a 2-dimensional array: its rows, its columns, and its subarrays.

Once an array is set up, it can be searched and its entries manipulated. Things are not too different from the 1-dimensional array setting, but we get additional row/column practice in §17.2 by considering a look-for-the-max problem and also a mean/standard deviation calculation typical in data analysis. Computations that involve both 1- and 2-dimensional arrays at the same time are explored through a cost/purchase order/inventory application. Using a 2-dimensional array to store a finite snapshot of a 2-dimensional continuous function $f(x, y)$ is examined in §17.3.

In the last section we present the 2-dimensional boolean array as a vehicle for representing some familiar patterns of "yes-no" data.

17.1 "ij" Thinking

If $f(x)$ is a function of a single variable, then an array can be used to represent a table of its values. Various aspects of this are discussed in Chapter 15. In that chapter and until now, we have only dealt with *1-dimensional* arrays. A single subscript is sufficient to specify the location of a value in a 1-dimensional array.

Many applications involve a function of *two* variables and a *2-dimensional array* is often handy for the representation of its values. Our experience with 2-dimensional tables begins in grade school with the times table:

1	2	3	4	5	6	7	8	9
2	4	6	8	10	12	14	16	18
3	6	9	12	15	18	21	24	27
4	8	12	16	20	24	28	32	36
5	10	15	20	25	30	35	40	45
6	12	18	24	30	36	42	48	54
7	14	21	28	35	42	49	56	63
8	16	24	32	40	48	56	64	72
9	18	27	36	45	54	63	72	81

Times table construction is shown in `Example17_1` which illustrates the declaration, setting up, and printing of a 2-dimensional array. It prints a designated *subarray* of a 12-by-10 times table. The integers `i1` and `i2` define the range of the involved rows while `j1` and `j2` specify the column range. Extending our "dot-dot" notation in the obvious way, we see that `Example17_1` displays the 4-by-6 subarray `T[6..9,5..10]`.

There are many things to discuss. The 2-dimensional real array type `Real2DArray` is declared as follows:

```
const
   rowmax = 12;
   colmax = 10;
type
   Real2DArray = array[1..rowmax,1..colmax] of real;
```

If `T` has this type, then it amounts to a 12-row, 10-column array of real variables:

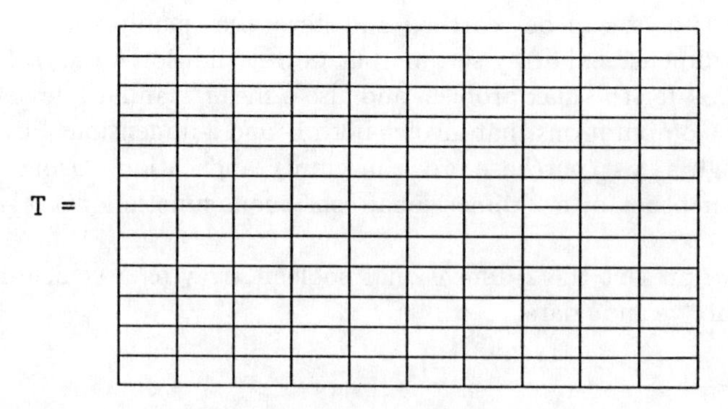

In general, the range of allowable row and column indices is established in a type declaration of the following form:

⟨*Name of type*⟩ = **array**[⟨*row range*⟩,⟨*column range*⟩] **of** ⟨*type*⟩

```
program Example17_1;
{Times Tables}
const
   rowmax = 12;
   colmax = 10;
type
   Real2DArray = array[1..rowmax, 1..colmax] of real;
var
   T:Real2DArray;
   i,j:integer;
   AnotherEg:  char;
   i1,i2:integer;
   j1,j2:integer;
begin
   ShowText;
   for i:=1 to rowmax do
      {Set up the i-th row}
      for j:=1 to colmax do
         T[i,j]:= i*j;
   AnotherEg := 'y';
   while AnotherEg <> 'n' do
      begin
         writeln('Enter j1 and j2, the x range.  Must have 1<=j1<=j2<=',colmax:2);
         readln(j1,j2);
         writeln('Enter i1 and i2, th3 y range.  Must have 1<=i1<=i2<=',rowmax:2);
         readln(i1,i2);
         {Print T[i1..i2,j1..j2]}
         for i:=i1 to i2 do
            {Print T[i,j1..j2]}
            begin
               writeln;
               for j := j1 to j2 do
                  write(T[i,j]:4:0);
            end;
         writeln;
         writeln;
         writeln('Another example?  Enter y (yes) or n (no).');
         readln(AnotherEg);
      end
end.
```

Sample output:

```
Enter j1 and j2, the x range.  Must have 1<=j1<=j2<=10
5      10
Enter i1 and i2, th3 y range.  Must have 1<=i1<=i2<=12
6      9
  30  36  42  48  54  60
  35  42  49  56  63  70
  40  48  56  64  72  80
  45  54  63  72  81  90
Another example?  Enter y (yes) or n (no).
n
```

Notice how constants are used to hold the values that define the row and column ranges. The declaration of the `Real2DArray` variable `T` sets aside a portion of memory that is capable of storing `rowmax*colmax` real variables. These parameters do not have to be particularly large before the capacity of the underlying memory system becomes a constraint. Effective 2-dimensional array computing requires a sensitivity to this issue.

With respect to the setting up of a 2-dimensional array, each element in `T` is capable of storing a real value and a double subscript notation is used to indicate its location in the array. A fragment of the form

```
T[1,1]:=1*1;
T[1,2]:=1*2;
     ⟨:⟩
T[1,10]:=1*10
```

sets up the first row of `T`:

T =

1	2	3	4	5	6	7	8	9	10

In general, a reference to an entry in a 2-dimensional array has the form

⟨*Name*⟩`[`⟨*row index*⟩`,`⟨*column index*⟩`]`

Successful subscript-level "thinking" in 2-dimensions requires being able to keep track of both the row and column indices. *Remember that the row index comes first.* In `Example17_1`, the array `T` is filled row-by-row. The inner loop in the "set up" portion of the program establishes the i-th row of the table. If `i=2`, then

```
for j:=1 to colmax do
   T[i,j]:= i*j;
```

is equivalent to

```
T[2,1]:=2*1;
T[2,2]:=2*2;
     ⟨:⟩
T[2,10]:=2*10
```

and the array `T` now looks like

T =

1	2	3	4	5	6	7	8	9	10
2	4	6	8	10	12	14	16	18	20

Double loops abound in 2-dimensional array settings and the mastery of the subscript concept is essential. Here is what T looks like upon completion of the double loop:

T =

1	2	3	4	5	6	7	8	9	10
2	4	6	8	10	12	14	16	18	20
3	6	9	12	15	18	21	24	27	30
4	8	12	16	20	24	28	32	36	40
5	10	15	20	25	30	35	40	45	50
6	12	18	24	30	36	42	48	54	60
7	14	21	28	35	42	49	56	63	70
8	16	24	32	40	48	56	64	72	80
9	18	27	36	45	54	63	72	81	90
10	20	30	40	50	60	70	80	90	100
11	22	33	44	55	66	77	88	99	110
12	24	36	48	60	72	84	96	108	120

If we reverse the order of the i and j loop, then T is filled column by column. Thus, after two passes through the outer loop of

```
for j:=1 to colmax do
   {Set up the j-th column}
   for i:=1 to rowmax do
      T[i,j]:= i*j;
```

we have

T =

1	2								
2	4								
3	6								
4	8								
5	10								
6	12								
7	14								
8	16								
9	18								
10	20								
11	22								
12	24								

We also mention that it is perfectly legal to use only a portion of a declared 2-dimensional array. Thus,

```
for i:=1 to 4 do
   {Set up the i-th row}
   for j:=1 to 3 do
      T[i,j]:= i*j;
```

produces

T =

1	2	3							
2	4	6							
3	6	9							
4	8	12							

We have established a number of tools in the unit `Chap17Codes` that facilitate the setting up and display of 2-dimensional arrays. See FIGURE 17.1. Note that the unit must be linked to `DDCodes`, `Chap14Codes`, and `Chap15Codes`. To refine further our double index thinking abilities, let's consider the following `Chap17Codes` function that returns a 2-dimensional array type:

```
function SubArray(A:Real2DArray; i1,i2,j1,j2:integer):Real2DArray;
{Pre:1<=i1<=i2, 1<=j1<=j2 and A[i1..i2,j1..j2] initialized.}
{Post:The array A[i1..i2,j1..j2]}
var
   i,j:integer;
begin
   for j := j1 to j2 do
      for i:= i1 to i2 do
         SubArray[i-i1+1,j-j1+1] := A[i,j]
end
```

The required subarray happens to be assembled column by column. A modest amount of subscript analysis is required. Elements from column `j` in `A` are "mapped" into column `j-j1+1` in the subarray, assuming `j1<=j<=j2`. Likewise, row `i` in `A` is associated with row `i-i1+1` in the subarray. As a check on these relationships, note that `A[i1,j1]` is "mapped" on to `SubArray[1,1]`. In the program `Example 17_2` extracts a subarray from a times table and displays the result using `Show2DArray` and `HighLightSubArray`.

Sometimes the entries in a 2-dimensional array are defined in terms of other entries. As an example of such a recursive specification, consider the following definition of the n-by-n *Pascal array*, whose (i, j) entry is defined by

$$p_{ij} = \begin{cases} 1 & \text{if } i = 1 \text{ or } j = 1 \\ p_{i,j-1} + p_{i-1,j} & \text{otherwise} \end{cases}$$

```
unit Chap17Codes;
interface
uses
   DDCodes, Chap14,Codes, Chap15Codes;

const
   rowmax = 30;
   colmax = 20;

type
   Real2DArray = array[1..rowmax,1..colmax] of real;
   Real2DArray0 = array[0..rowmax, 0..colmax] of real;
   Int2DArray = array[1..rowmax,1..colmax] of integer;
   LongInt2DArray = array[1..rowmax,1..colmax] of longint;

procedure ShowElement(hL,vT,p,size:integer; x:real);
{Pre:0<=p<=3}
{Post:Prints the value of x to p decimal places in a size-by-size square with}
{upper left corner at (hL,vT).}

procedure Show2DArray(hL,vT,p,size:integer; A:Real2DArray; m,n:integer);
{Pre:0<=p<=3, A[1..m,1..n] initialized.}
{Post:Displays the array A[1..m,1..n], with upper left corner at }
{ (hL,vT). The value of each}
{entry is printed to p decimal places in a size-by-size square.}

procedure HighLightSubArray(hL,vT,size,i1,i2,j1,j2:integer);
{Post:  Draws a thick line rectangle defined by hL+(j1-1)size <= h <= hL+(j2-1)size }
{and vT+(i1-1)size <= v <= vT+(i2-1)size}

function SetUp2D(m,n:integer; xval,yval:RealList0;
             function f(x,y:real):real):  Real2DArray0;
{Pre:xval[0..m] and yval[0..n] and f defined at all (xval[i],yval[j]).}
{Post:SetUp2D[i,j] = f(xval[i],yval[j]), i=0..m, j=0..n}

function RandM(m,n:integer; L,R:real; var rn:real):Real2DArray;
{Pre:0<rn<1, L < R}
{Post:An m-by-n array with entries chosen randomly from the interval [L,R].}
{The value returned in rn is the value obtained by executing m*n times the}
{assignment rn:=rand(rn).}
```

FIGURE 17.1 *Some Declarations in* Chap17Codes.

```
program Example17_2;
{Times Tables}
uses
   DDCodes,Chap14Codes,Chap15Codes,Chap17Codes;
const
   {m-by-n times table displayed with size-by-size boxes, with}
   {upper left corner = (hL,vT).}
   size=25; hL=5; vT=30; m=9; n=7;
   {Row range = i1..i2, column range = j1..j2}
   i1=3; i2=5; j1=3; j2=6;
var
   i,j:integer;
   T:Real2dArray; {The ''master'' times table.}
   B:Real2DArray; {T[i1..i2,j1..j2]}
begin
   ShowDrawing;
   {Set up and display the master times table.}
   for i:=1 to m do
      for j:=1 to n do
         T[i,j] := i*j;
   Show2DArray(hL,vT,0,size,T,m,n);
   {Set up and display the designated subarray.}
   B := SubArray(T,i1,i2,j1,j2);
   Show2DArray(hL+(n+2)*size,vT,0,size,A,i2-i1+1,j2-j1+1);
   HighLightSubArray(hL,vT,size,i1,i2,j1,j2);
   MoveTo(hL+(n+2)*size,20);
   WriteDraw('T[',i1:2,':',i2:2,',',j1:2,':',j2:2,']');
end.
```

The type Real2DArray and the functions/ procedures SubArray, Show2DArray, and HighLightSubArray are declared in Chap17Codes. See FIGURE 17.2 for output.

1	2	3	4	5	6	7
2	4	6	8	10	12	14
3	6	9	12	15	18	21
4	8	12	16	20	24	28
5	10	15	20	25	30	35
6	12	18	24	30	36	42
7	14	21	28	35	42	49
8	16	24	32	40	48	56
9	18	27	36	45	54	63

T[3:5, 3:6]

9	12	15	18
12	16	20	24
15	20	25	30

FIGURE 17.2 Example17_2 *Output*

Thus, the first row and column of the array are made up of ones. Elsewhere, the (i, j) entry is the sum of the "previous" entry in its row, $p_{i,j-1}$, and the previous entry in its column, $p_{i-1,j}$. Encapsulating this we obtain

```
function Pascal(n:integer):LongInt2DArray;
{Post:The n-by-n Pascal array.}
var
   i,j:integer;
   A:LongInt2DArray; {For building the array.}
begin
   for i:=1 to n do
      {Set up the ith row.}
      for j := 1 to n do
         if (i = 1) or (j = 1) then
            A[i,j] := 1
         else
            A[i,j] := A[i,j-1] + A[i-1,j];
   Pascal:=A;
end
```

Note that it is necessary to have the local array A because

```
Pascal[i,j] := Pascal[i,j-1]+Pascal[i-1,j]
```

would be illegal. We also mention that instead of having a pair of "1-to-n" loops with an **or**, we could set up A as follows:

```
for i:=1 to n do
   begin
      A[i,1]:=1;
      A[1,i]:=1;
   end;
for i:=2 to n do
   for j:=2 to n do
      A[i,j]:=A[i,j-1]+A[i-1,j];
```

Here, row 1 and column 1 are established first, and then A[2..n,2..n]. Example17_3 illustrates the use of Pascal. Note that each row is printed with a sequence of write statements under the control of a for-loop.

PROBLEM 17.1. Modify Example17_1 so that it repeatedly solicits values for i1, i2, j1, and j2 using ReviseInteger. Display the master times table and the designated subarray at the same time using Show2DArray. Highlight the subarray using HighLightSubarray.

PROBLEM 17.2. Complete the following procedure

```
procedure SinCosArray(n:integer; theta:real; var S,C: Real2DArray);
{Post:S[i,j] = sin(ij*theta) and C[i,j] = cos(ij*theta), 1<=i<=n, 1<=j<=n}
```

Implement the simplest approach which is to call the functions sin and cos for each entry. The number of function evaluations can be reduced by exploiting the identities

$$\begin{aligned}\sin(ij\theta) &= \sin((i-1)j\theta)\cos(j\theta) + \cos((i-1)j\theta)\sin(j\theta)\\ \cos(ij\theta) &= \cos((i-1)j\theta)\cos(j\theta) - \sin((i-1)j\theta)\sin(j\theta)\end{aligned}$$

```
program Example17_3;
{Pascal Array}
uses
   Chap14Codes, Chap15Codes, Chap17Codes;
const
   n = 10; {Size of problem.}
var
   i,j:integer;
   A:LongInt2DArray;
begin
   ShowText;
   A := Pascal(n);
   writeln('The Pascal array with n =', n :  2, ':');
   for i:=1 to n do
      begin
         writeln;
         for j:=1 to n do
            write(A[i,j]:8);
      end;
end.
```

The type LongInt2D and the function Pascal is declared in Chap17Codes. Output:

```
The Pascal array with n = 6:

1  1   1   1    1    1
1  2   3   4    5    6
1  3   6  10   15   21
1  4  10  20   35   56
1  5  15  35   70  126
1  6  21  56  126  252
```

In particular, the identities can be used to generate the first row and column of S and C from S[1,1] and C[1,1]. Once the first rows and columns are available, the recursions can be used to generate S[2..n,2..n] and C[2..n,2..n]. Using these ideas, complete the following procedure:

```
procedure SinCosArray(n:integer; theta:real; var S,C: Real2DArray);
{Post:S[i,j] = sin(ij*theta) and C[i,j] = cos(ij*theta), 1<=i<=n, 1<=j<=n}
```

In order to check for correctness, implement and use the following function:

```
function Diff(m,n:integer; A,B:Real2DArray):real;
{Pre:A[1..m,1..n] and B[1..m,1..n] initialized.}
{Post:The maximum value of |A[i,j] - B[i,j]|, i=1..m, j=1..n}
```

PROBLEM 17.3. The (i, j) entry of the n-by-n Pascal array P is also specified by

$$p_{ij} = \sum_{k=1}^{\min\{i,j\}} \binom{i-1}{k-1}\binom{j-1}{k-1}$$

Write an alternative Pascal1 to the function Pascal that returns the n-by-n Pascal array using this recipe. Make use of the fact that $p_{ij} = p_{ji}$.

PROBLEM 17.4. A *magic square* of size n has the property is an n-by-n array with the following properties:

- It is "made up" of the integers $1, 2, \ldots, n^2$.
- The numbers in every row, column, and diagonal sum to $n(n^2 + 1)/2$. (There are two diagonals, one from the upper left corner to the lower right corner, and one from the lower left corner to the upper right corner.)

Here are the magic squares of size 3, 5, and 7:

2	7	6
9	5	1
4	3	8

9	3	22	16	15
2	21	20	14	8
25	19	13	7	1
18	12	6	5	24
11	10	4	23	17

20	12	4	45	37	29	28
11	3	44	36	35	27	19
2	43	42	34	26	18	10
49	41	33	25	17	9	1
40	32	24	16	8	7	48
31	23	15	14	6	47	39
22	21	13	5	46	38	30

Pick up the pattern and complete the following function:

```
function Magic(n:integer):Int2DArray;
{Pre:n is odd.}
{Post:Magic is an n-by-n magic square.}
```

For checking purposes, implement the following boolean-valued function:

```
function IsMagic(n:integer;A:Int2DArray):boolean;
{Pre:n is odd}
{Post:A[1..n,1..n] is a magic square.}
```

PROBLEM 17.5. Complete the following function:

```
function VanderMonde(n:integer;x:RealList):Real2DArray;
{Pre:x[0..n-1] initialized}
{Post:VanderMonde[i,j] = x[i]^(j-1), 1<=i<=n, 1<=j<=n}
```

PROBLEM 17.6. Complete the following function:

```
function Toeplitz(n:integer; d:RealList0):Real2DArray;
{Pre:d[0..n-] initialized.}
{Post:Toeplitz[i,j] = d[|i-j|), 1<=i<=n, 1<=j<=n}
```

17.2 Operations

Searching for a maximum or minimum value in a 2-dimensional array is very similar to searching a 1-dimensional array. The only difference is that a double loop is required, one for stepping through the columns and one for the rows. Here's a procedure that computes the row and column index of the maximum entry:

```
procedure MaxM(m,n:integer; A:Real2DArray; var imax,jmax:integer);
{Pre:A[1..m,1..n] is initialized.}
{Post:A[imax,jmax] is the largest entry in A[1..m,1..n].}
var
   i,j:integer;
   MaxSoFar:real;
begin
   imax:=1;
   jmax:=1;
   MaxSoFar:=A[1,1];
   for i:=1 to m do
      {Scan the i-th row.}
      for j:=1 to  n do
         if A[i,j] > MaxSoFar then
            {A new max has been found.}
            begin
               MaxSoFar:= A[i,j];
               imax:=i;
               jmax:=j;
            end
end;
```

The search through the array is row-by-row. Within each row, the array entries are checked left to right. Any time a new largest value is encountered, its row and column indices are stored in imax and jmax respectively. Example17_4 illustrates the use of the procedure.

PROBLEM 17.7. Complete the following procedure:

```
procedure MaxMcol(m,n:integer; A:Real2DArray; var imax:IntegerList);
{Pre:A[1..m,1..n] is initialized.}
{Post:imax[j] is the row index of the largest entry in column j.}
```

PROBLEM 17.8. We say that A[i,j] is a *saddle point* of the array A[1..m,1..n] if it is at least as large as any other entry in its row and no bigger than any other entry in its column. Complete the functions

```
function MinInCol(A:Real2DArray; m,j:integer):integer;
{Pre:A[1..m,j] initialized.}
{Post:the index of the smallest entry in A[1..m,j].}

function MaxInRow(A:Real2DArray; n,i:integer):integer;
{Pre:A[i,1..n] initialized.}
{Post:  the index of the largest entry in A[i,1..n]).}
```

and use them to implement the following procedure:

```
procedure SaddlePoint(m,n:integer; A:Real2DArray; var Saddle:boolean; var i,j:integer);
{Pre:A[1..m,1..n] is initialized.}
{Post:If Saddle = true, A[i,j] is a saddle point.  }
{If Saddle = false, A[1..m,1..n] has no saddle point.}
```

PROBLEM 17.9. Complete the function

```
function EdgeOk(A:Real2DArray; i,j,r:integer):boolean;
{Pre:r>=0,A[i-r..i+r,j-r..j+r] initialized.}
{Post:A[i,j] >= A[i-r,j+p], A[i+r,j+p], A[i+p,j-r], and A[i+p,j+r] for p=-r..r}
```

and use it effectively to implement the following:

```
function View(m,n:integer; A:Real2DArray; i,j:integer):Integer;
{Pre:A[1..m,1..n] initialized.}
{Post:Largest r so that A[i-r..i+r,j-r..j+r] is a subarray of A[1..m,1..n] and so that}
{no entry in A[i-r..i+r,j-r..j+r] is bigger than A[i,j].}
```

```
program Example17_4;
{Maximum entry.}
uses
   DDCodes,Chap14Codes, Chap15Codes, Chap17Codes;
const
   seed = 0.123456;
   m=6;
   n=8; {m-by-n problem}
   LBound = -10;
   UBound = 10; {Array entries randomly selected from [LBound,UBound].}
   places=1; size = 35; hL = 40;vT = 40; {Display Parameters}
var
   A:Real2DArray;
   i,j:integer;
   r:real;
begin
   ShowDrawing;
   r:=seed;
   while YesNo('Another Example?') do
      begin
         ClearScreen;
         A := RandM(m,n,LBound,UBound,r);
         Show2DArray(hL,vT,places,size,A,m,n);
         MaxM(m,n,A,i,j);
         HighLightSubArray(hL,vT,size,i,i,j,j);
      end
end.
```

The function YesNo and procedure ClearScreen are declared in DDCodes. RandM, MaxM, HighLightSubArray, Show2DArray and the type Real2DArray are declared in Chap17Codes. For sample output see FIGURE 17.3.

3.4	-8.9	8.4	7.1	7.3	6.9	-8.5	-7.5
7.3	9.6	-2.6	-9.4	-4.4	-7.0	-2.0	-7.9
8.2	9.6	-1.5	-3.5	4.9	-9.1	-1.9	-8.2
9.2	6.8	9.8	-4.8	8.3	6.3	0.1	-7.7
-6.7	-3.2	9.4	-2.4	-1.0	4.8	-8.0	-5.8
3.8	-4.8	-4.5	-5.9	10.0	-8.0	1.8	-1.9

FIGURE 17.3 *Output from* Example17_4.

A broad family of 2-dimensional array operations amount to a sequence of 1-dimensional array operations in which the same calculations are performed on each column (or row). For example, a common enterprise in data analysis is to "normalize" the values in a 1-dimensional array by their mean μ and standard deviation σ. Thus, if

x = | 10 | 40 | 20 | 30 |,

then $\mu = 25$, $\sigma = \sqrt{(10-25)^2 + (40-25)^2 + (20-25)^2 + (30-25)^2} = 10\sqrt{5}$, and

z = | $-3/(2\sqrt{5})$ | $3/(2\sqrt{5})$ | $-1/(2\sqrt{5})$ | $1/(2\sqrt{5})$ |

is the normalization of x. The following fragment performs this task:

```
mu:=0;
for i:=1 to m do
   mu:=mu+x[i];
mu:=mu/m;
sigma:=0;
for i:=1 to m do
   sigma:= sigma+sqr(x[i]-mu);
sigma:=sqrt(sigma/m);
for i:=1 to m do
   z[i]:=(x[i]-mu)/sigma;
```

It is easy to check that z has zero mean and unit standard deviation, hence the term "normalize". Here is a procedure that normalizes each of the columns in an array A[1..m,1..n]:

```
function Normalize(m,n:integer; A:Real2DArray):Real2DArray;
{Pre:A[1..m,1..n] initialized}
{Post:Each column of Normalize[1..m,1..n] is the normalization of the}
{corresponding column in A.}
```

```
var
   i,j:integer;
   mu,sigma:real;
begin
   for j:=1 to n do
      {Normalize the j-th column}
      begin
         {Compute the mean}
         mu:=0;
         for i:=1 to m do
            mu:=mu+A[i,j];
         mu:= mu/m;
         {Compute the standard deviation.}
         sigma:=0;
         for i:=1 to m do
            sigma:=sigma + sqr(A[i,j]-mu);
         sigma := sqrt(sigma/m);
         {Normalize}
         for i:=1 to m do
            Normalize[i,j] := (A[i,j]-mu)/sigma;
      end;
end;
```

See Example17_5.

```
program Example17_5;
{Normalizing Data}
uses
   DDCodes, Chap14Codes, Chap15Codes, Chap17Codes;
const
   seed = 0.123456;
   L=0; R=10; {Random numbers chosen from [L,R].}
   m=6; n=4; {Problem size = m-by-n}
   size=45; hL=20; vT=30; {Display parameters}
var
   rn:real;
   A:Real2DArray;
begin
   ShowDrawing;
   rn := seed;
   while YesNo('Another Example?') do
      begin
         ClearScreen;
         A := RandM(m,n,L,R,rn);
         Show2DArray(hL,vT,2,size,A,m,n);
         A:=Normalize(m,n,A);
         Show2DArray(hL+(n+2)*size,vT,2,size,A,m,n);
      end
end.
```

PROBLEM 17.10. Complete the following procedure

```
function Smooth(m,n:integer; A:Real2DArray):Real2DArray;
{Pre:A[1..m,1..n] initialized.}
{Post:An m-by-n array where Smooth[i,j] = A[i,j] if i=1 or i=m or j=1 or j=n.}
{Otherwise, Smooth[i,j] is the average of A[i-1,j],A[i+1,j],A[i,j-1],A[i,j+1]}
```

Computations that involve 1- and 2-dimensional arrays "at the same time" arise in numerous applications. Consider a situation where a company has m factories, each of which can produce any of n products. An integer array `Cost[1..m,1..n]` can be used to store cost-of-production information, the value of `Cost[i,j]` being the cost to factory i of producing product j. Here is a sample cost array:

`Cost[1..3,1..5]` =

10	32	21	73	5
12	27	25	67	6
9	30	26	73	4

Thus, Factory 2 can produce product 4 at a cost of $67 per unit.

A customer wishes to purchase a certain number of each product. A 1-dimensional integer array `w[1..n]` can be used to represent this request, e.g.,

`w[1..5]` =

100	50	80	10	30

We wish to determine the factory that can fill the purchase order most cheaply. Note that the cost to factory i is a summation that involves numbers from the i-th row of `Cost` multiplied by the corresponding numbers from `w`:

$$\begin{array}{lcrcrcrcrcrcr} \text{Cost to Factory 1} & = & 10\cdot 100 & + & 32\cdot 50 & + & 21\cdot 80 & + & 73\cdot 10 & + & 5\cdot 30 & = & 5160 \\ \text{Cost to Factory 2} & = & 12\cdot 100 & + & 27\cdot 50 & + & 25\cdot 80 & + & 67\cdot 10 & + & 6\cdot 30 & = & 5400 \\ \text{Cost to Factory 3} & = & 9\cdot 100 & + & 30\cdot 50 & + & 26\cdot 80 & + & 73\cdot 10 & + & 4\cdot 30 & = & 5330 \end{array}$$

In terms of the arrays, the three production costs are given by

```
C1:= Cost[1,1]*w[1] + Cost[1,2]*w[2] + Cost[1,3]*w[3] + Cost[1,4]*w[4] + Cost[1,5]*w[5];
C2:= Cost[2,1]*w[1] + Cost[2,2]*w[2] + Cost[2,3]*w[3] + Cost[2,4]*w[4] + Cost[2,5]*w[5];
C3:= Cost[3,1]*w[1] + Cost[3,2]*w[2] + Cost[3,3]*w[3] + Cost[3,4]*w[4] + Cost[3,5]*w[5];
```

Picking up the pattern and assuming that n contains the number of products, we see that

```
s:=0;
for j:=1 to n do
   s:=s+Cost[i,j]*w[j]
```

assigns to `s` the i-th factory's cost of production. Our goal is to determine the value of `i` that makes this summation as small as possible. To that end we make the declarations

```
const
   nFact = 20;
   nProd = 50;
type
   IntTable = array[1..nFact,1..nProd];
   PurchaseOrder = array[1..nProd] of integer;
```

and define

```
function iCost(i,n:integer; Cost:IntTable; w:PurchaseOrder):integer;
{Pre:Cost[i..i,1..n] and w[1..n] initialized.}
{Post:Cost[i,1]w[1] +...+ Cost[i,n]w[n]}
var
   j:integer;
   s:integer;
begin
   s:=0;
   for j:=1 to n do
      s:=s+Cost[i,j]*w[j];
   iCost:=s;
end;
```

The search for the factory with the cheapest production cost is a straightforward look-for-the-min calculation:

```
procedure Cheapest(m,n:integer; Cost:IntTable; w:PurchaseOrder;
          var imin,MinCost:integer);
{Pre:Cost[1..m,1..n] and w[1..n] initialized}
{Post:i=imin minimizes Cost[i,1]w[1] + ...  + Cost[i,n]w[n] and }
{MinCost is the associated value.}
var
   i:integer;
   c:integer;
begin
   imin:=1;
   MinCost := iCost(1,n,Cost,w);
   for i:=2 to m do
      begin
         c := iCost(i,n,Cost,w);
         if c < MinCost then
            begin
               imin := i;
               MinCost := c
            end;
      end;
end
```

Some interesting boolean computations arise with the introduction of an "inventory array." Assume that Inventory[1..m,1..n] is initialized with the understanding that Inventory[i,j] contains the number of product j units on inventory at factory i. Factory i can process the purchase order if Inventory[i,j]>=w[j] for $j = 1$ to n. Thus if

Inventory[1..3,1..5] =

150	200	100	120	110
200	40	130	20	40
300	50	100	12	80

and

w[1..5] =

100	50	80	10	30

then factories 1 and 3 have sufficient inventory but factory 2 does not. Here is a boolean-valued function that tests for sufficient inventory:

```
function iCanDo(i,n:integer; Inv:IntTable; w:PurchaseOrder):boolean;
{Pre:Inv[i,1],...,Inv[i,n] initialized}
{Post:Inv[i,j]>=w[j] for j=1..n}
var
   OK:boolean;
   j:integer;
begin
   j := 1;
   OK := w[1] <= Inv[i, 1];
   while (j < n) and OK do
      begin
         j:=j+1;
         OK := w[j] <= Inv[i,j];
      end;
   iCanDo := OK;
end;
```

Example17_6 provides a menu-driven environment for experimenting with the iCost and iCanDo. functions. See FIGURE 17.4.

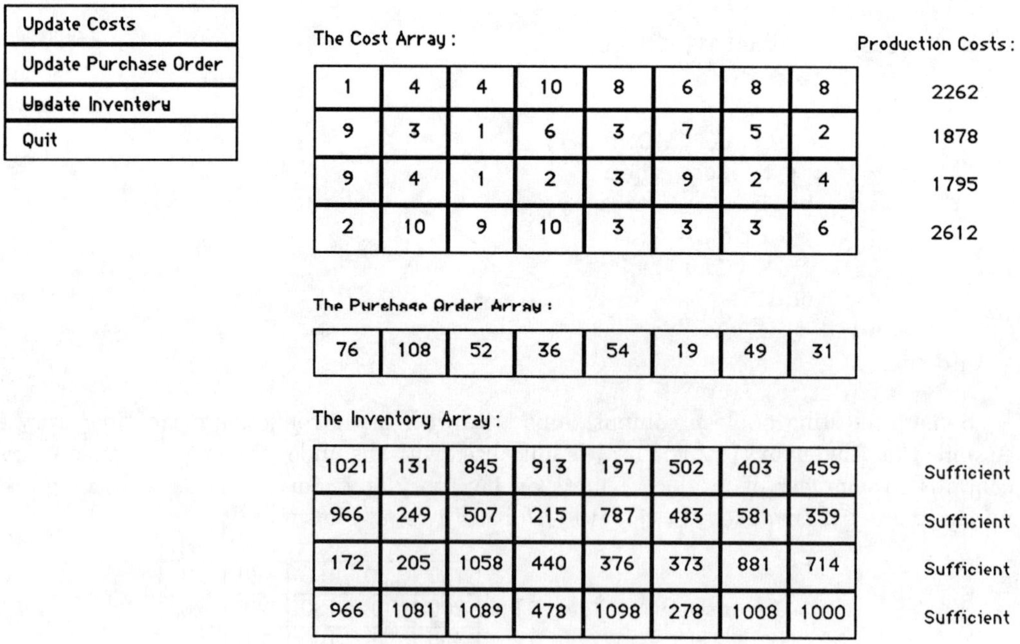

Update Costs
Update Purchase Order
Update Inventory
Quit

The Cost Array :

								Production Costs :
1	4	4	10	8	6	8	8	2262
9	3	1	6	3	7	5	2	1878
9	4	1	2	3	9	2	4	1795
2	10	9	10	3	3	3	6	2612

The Purchase Order Array :

76	108	52	36	54	19	49	31

The Inventory Array :

1021	131	845	913	197	502	403	459	Sufficient
966	249	507	215	787	483	581	359	Sufficient
172	205	1058	440	376	373	881	714	Sufficient
966	1081	1089	478	1098	278	1008	1000	Sufficient

FIGURE 17.4 *A Production Cost Environment*

By clicking the appropriate "update box" and an individual entry in the designated array, it becomes possible to modify its value. The production costs and inventory status are updated automatically. After the modification is complete, the inventory of the "cheapest factory" is depleted by amounts stored in the purchase order.

PROBLEM 17.11. Complete the following:

```
function TotalValue(m,n:integer; Cost,Inv:IntTable):longint;
{Pre:Cost[1..m,1..n] and Inv[1..m,1..n] are initialized.}
{Post:The sum of the products C[i,j]*Inv[i,j], i=1..m, j=1..n}

procedure CheapestPossible(m,n:integer; Cost,Inv:IntTable; w:PurchaseOrder;
      var possible:boolean; var imin,MinCost:integer);
{Pre:Cost[1..m,1..n], Inv[1..m,1..n] and w[1..n] initialized}
{Post:Possible is true if there is at least one factory with sufficient inventory.}
{In that case, i=imin minimizes Cost[i,1]w[1] + ...  + Cost[i,n]w[n] }
{over all factories with sufficient inventory}
{and MinCost is the associated value. }
```

17.3 Tables in Two Dimensions

We learned in Chapter 15 that it is sometimes efficient to pre-compute function values and store them in a (1-dimensional) table for "future use." Two-dimensional arrays can be used to do the same thing if the underlying function depends upon a pair of variables. To illustrate, consider the problem of finding the triangle with the longest perimeter assuming that the three vertices are chosen from a given finite point set. See FIGURE 17.5. Assume that x[1..n] and y[1..n] contain the coordinates of the points. The most obvious solution is simply to check every possible triangle:

```
Pmax:=0;
for i:=1 to n do
   for j:=1 to n do
      for k:=1 to n do
         begin
            Pij:=Distance(x[i],y[i],x[j],y[j]);
            Pjk:=Distance(x[j],y[j],x[k],y[k]);
            Pki:=Distance(x[k],y[k],x[i],y[i]);
            Pijk:=Pij+Pjk+Pki;
            if Pijk>Pmax then
               begin
                  Pmax:=Pijk; imax:=i; jmax:=j; kmax:=k;
               end
         end
```

Here, Distance(a,b,c,d) returns $\sqrt{(a-c)^2+(b-d)^2}$, the distance between (a,b) and (c,d). The triply nested loop steps through all the possibilities, *but it is redundant.* Let the notation (i,j,k) stand for the triangle obtained by connecting points i, j, and k. Note that the triangles (1,2,3), (1,3,2), (2,1,3), (2,3,1), (3,1,2), and (3,2,1) are all the same. This 6-fold repetition can be avoided by abbreviating the loop ranges as follows:

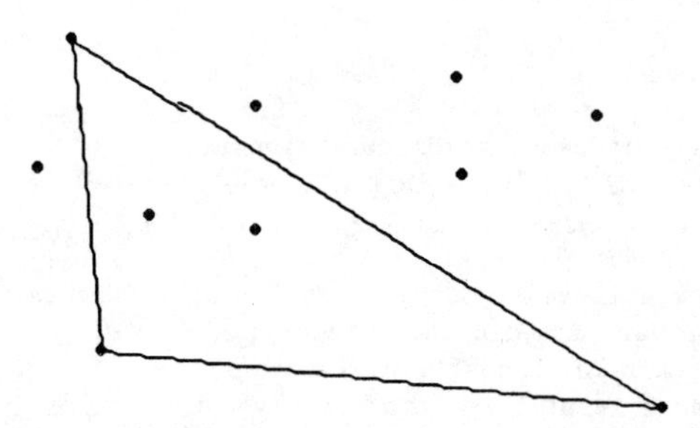

FIGURE 17.5 *Maximum Triangle*

```
for i:=1 to n do
   for j:=i to n do
      for k:=j to n do
```

Think of the i-th point as the one selected first, the j-th point as the one selected second, and the k-th point as the one selected third. A little geometric thinking indicates that the optimum triangle will involve three *distinct* vertices and so the lower bounds in the inner two loops can be increased by 1:

```
for i:=1 to n do
   for j:=i+1 to n do
      for k:=j+1 to n do
```

Here is an enumeration of all the triangles that are checked in the $n = 6$ case:

$(1,2,3)$	$(1,2,4)$	$(1,2,5)$	$(1,2,6)$
$(1,3,4)$	$(1,3,5)$	$(1,3,6)$	
$(1,4,5)$	$(1,4,6)$		
$(1,5,6)$			
$(2,3,4)$	$(2,3,5)$	$(2,3,6)$	
$(2,4,5)$	$(2,4,6)$		
$(2,5,6)$			
$(3,4,5)$	$(3,4,6)$		
$(3,5,6)$			
$(4,5,6)$			

There are four groups corresponding to i=1,2,3, and 4. Across each line we list all the triangles that involve a particular pair of points, avoiding the mention of triangles that have already been listed or that do not involve three distinct points.

Despite these modifications, there remains an even bigger redundancy. Each pairwise distance is computed about n times because each line segment that connects two points participates in about n triangles. This suggests that we pre-compute all the pairwise distances and store them in an array D. Incorporating this idea and casting the final solution in the form of a procedure we obtain:

```
procedure MaxTriangle(n:integer; x,y:RealList;
      var imax,jmax,kmax:integer; var Pmax:real);
{Pre:x[1..n] and y[1..n] contain the xy coordinates of n points and n>=3.}
{Post:Of all the triangles defined by selecting three distinct points,}
{the one with vertices (x[imax],y[imax]), (x[jmax],y[jmax]), and }
{(x[kmax],y[kmax]) has the longest perimeter.}
{Pmax returns the maximum perimeter.}

var
   i,j,k:integer;
   Pijk:real; {Perimeter of triangle defined by points with indices i,j, and k.}
   D:Real2DArray; {Array of pairwise distances}

begin
   {Set up a pairwise distance array.}
   for i:=1 to n do
      for j:=i+1 to n do
         D[i,j] := Distance(x[i],y[i],x[j],y[j]);
   {Search for the maximum perimeter triangle.}
   Pmax:=0;
   for i:=1 to n-2 do
      {First vertex = (x[i],y[i]).}
      for j:=i+1 to n-1 do
         {Second vertex = (x[j],y[j]).}
         for k:=j+1 to n do
            {Third vertex = (x[k],y[k]).}
            begin
               Pijk := D[i,j] + D[j,k] + D[i,k];
               if Pijk > Pmax then
                  begin
                     Pmax:=Pijk;
                     imax:=i; jmax:=j; kmax:=k;
                  end;
            end;
end;
```

Example17_7 solicits a finite point set and then uses MaxTriangle to compute the triangle with maximum perimeter.

Notice that only the upper triangular portion of the array is used. By setting up and using the D array, the number of calls to Distance is reduced by a factor of about n. Of course, Distance is not a particularly expensive function to execute, so this example of "trading space for time" is not very dramatic. But in other settings, the use of a 2-dimensional array to store function values can be crucial.

PROBLEM 17.12. Complete the following function exploiting symmetry as much as possible:

```
function GridDist1(n:integer):Real2DArray0;
{Pre:n>=0}
{Post:GridDist1[i,j] is the distance from (0,0) to (i,j)}
```

```
program Example17_7;
{Maximum triangle.}
uses
   DDCodes, Chap14Codes, Chap15Codes, Chap17Codes;
var
   n:integer;
   x,y:RealList; {The coordinates of the n points.}
   imax,jmax,kmax:integer; {Vertex indices of the maximum perimeter triangle.}
   Pmax:real;
begin
   ShowDrawing;
   SetMap(0,0,10);
   GetPoints(n,x,y);
   MaxTriangle(n,x,y,imax,jmax,kmax,Pmax);
   cMoveTo(x[imax],y[imax]);
   cLineTo(x[jmax],y[jmax]);
   cLineTo(x[kmax],y[kmax]);
   cLineTo(x[imax],y[imax]);
end.
```

The procedures `SetMap`, `cMoveTo`, and `cLineTo` are declared in `DDCodes`. The type `RealList` and the procedure `GetPoints` are declared in `Chap14Codes`. For sample output, see FIGURE 17.5.

PROBLEM 17.13. Complete the following function exploiting symmetry as much as possible:

```
function GridDist2(n:integer):Real2DArray0;
{Pre:n>=0}
{Post:GridDist1[i,j] is the distance from (0,n) to (i,j).}
```

PROBLEM 17.14. Complete the following function exploiting symmetry as much as possible:

```
function GridDist3(n:integer):Real2DArray0;
{Pre:n>=0 and is even }
{Post:GridDist1[i,j] is the distance from (n/2,n/2) to (i,j).}
```

17.4 Bit Maps

For many output devices, characters are displayed using a 7-by-5 *dot matrix*. For the digits we have

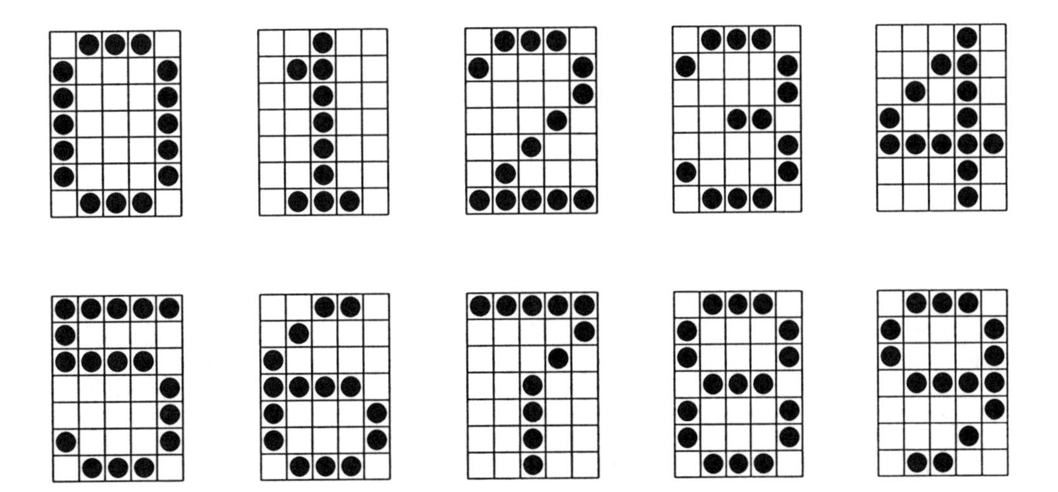

A 7-by-5 boolean array can be used to encode each digit. In particular, if B has the type

```
DotMatrix = array[1..7, 1..5] of boolean;
```

then we can set B[i,j] to true if and only if the (i,j) "tile" is "inked". Here is a procedure that graphically displays such a representation:

```
procedure Show7x5(hL,vT,d:integer; B:DotMatrix);
{Post:Displays B as a 7-by-5 array of d-by-d tiles.}
{The upper left corner of the upper left tile is at (hL,vT ). }
{The (i,j) tile is ''inked'' if B[i,j] is true.}
var
   h,v:integer;
   i,j:integer;
begin
   for i:=1 to 7 do
      for j:=1 to 5 do
         begin
            h := hL + d*(j-1);
            v := vT + d*(i-1);
            if B[i,j] then
               PaintOval(v,h,v+d,h+d);
            FrameRect(v,h,v+d, h+d);
         end;
end;
```

The entries of B are boolean, and so may appear in any context where such variables are allowed.

An environment for "designing" dot matrix representations is provided in Example17_8. It makes use of the function

```
function Build7x5(hL,vT,d:integer):DotMatrix;
{Pre:Interactively set's up a 7-by-5 boolean array.}
{The displayed 7-by-5 array is made up of d-by-d tiles}
{and the upper left corner is at (hL,vT).}
```

which is declared in Chap17Codes.

```
program Example17_8;
{Design 7-by-5 bit array.}
uses
   DDCodes, Chap14Codes, Chap15Codes, Chap17Codes;
const
   {Grid location parameters:}
      hL = 240;
      vT = 40;
      d = 20;
var
   B:DotMatrix;
begin
   ShowDrawing;
   while YesNo('Another Example?')  do
      begin
         B := Build7x5(hL,vT,d);
         ClearScreen;
      end;
   Show7x5(hL,vT,d,B);
end.
```

The functions/procedures YesNo and ClearScreen are declared in DDCodes. The function Build7x5 and the type DotMatrix are declared in Chap17Codes.

A length 35 "bit string" can also be used to represent the information in a 7-by-5 dot matrix. Using 1 and 0 for true and false, we see that the string

```
seven := '11111000010001000100001000010000100';
```

encodes in row-by-row fashion the information present in the dot matrix representation of the digit seven:

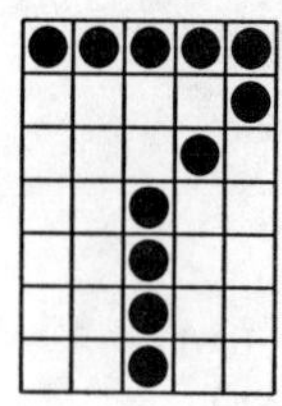

In particular,

$$\underbrace{11111}_{\text{Row 1}}\,\underbrace{00001}_{\text{Row 2}}\,\underbrace{00010}_{\text{Row 3}}\,\underbrace{00100}_{\text{Row 4}}\,\underbrace{00100}_{\text{Row 5}}\,\underbrace{00100}_{\text{Row 6}}\,\underbrace{00100}_{\text{Row 7}}$$

using these ideas we obtain

```
function Digit7by5(n:integer):DotMatrix;
{Pre:  0<=n<=9}
{Post:  The 7-by-5 dot matrix encoding of n.}
type
   ShortStrList = array[0..9] of string;
var
      p,i,j:integer;
      s:ShortstrList; {String encodings of the 7-by-5 representations}

begin
   {Row-by-row encodings for each digit.  1=true, 0=false.}
   s[0] := '01110100011000110001100011000101110';
   s[1] := '00100011000010000100001000010001110';
   s[2] := '01110100010000100010001000100011111';
   s[3] := '01110100010000100110000011000101110';
   s[4] := '00010001100101010010111110001000010';
   s[5] := '11111100001111000001000011000101110';
   s[6] := '00110010001000011110100011000101110';
   s[7] := '11111000010001000100001000010000100';
   s[8] := '01110100011000101110100011000101110';
   s[9] := '01110100011000101111000100010001000';

   p:=1;
   for i:=1 to 7 do
      for j := 1 to 5 do
         {Establish the value for the dot in row i, column j.}
         begin
            if s[n][p] = '1' then
               Digit7by5[i,j] := true
            else
               Digit7by5[i,j] := false;
            p:= p + 1;
         end;
end;
```

Notice the use of the *local type* `ShortStrList`.

A "dictionary" of these representations can be established in a variable `D` with the following type:

```
   DotMatrixDigits = array[0..9] of DotMatrix;
```

This is an array whose entries are arrays, a perfectly legal declaration. It illustrates how one programmer defined type can be defined in terms of another. Here is a fragment that sets up `D`:

```
for k := 0 to 9 do
   D[k] := Digit7by5(k);
```

`Example17_9` illustrates the "granularity" of the dot matrix idea by displaying the ten digits over a range of scales. It makes use of the following procedure:

```
procedure DrawBigDigit(k,hL,vB,p:integer; D:DotMatrixDigits);
{Pre:  0<=k<=9, D houses the 7-by-5 dot matrix representations.}
{Post:  Draws the digit k as a 7-by-5 dot array with radius p dots.  The lower}
{left corner of the digit area is at (hL,vB).}
```

```
program Example17_9;
{Dot Matrix Numbers}
uses
   DDCodes, Chap14Codes, Chap15Codes, Chap17Codes;
var
   k:integer;
   m:integer;
   n:integer;
   v:integer; {Vertical coordinate of current row.}
   D:DotMatrixDigits; {For the 7-by-5 encodings.}
begin
   ShowDrawing;
   for k := 0 to 9 do
      D[k] := Digit7by5(k);
   v:=20;
   for m:=1 to 6 do
      {Show the m pixels per dot case.}
      begin
         for n := 0 to 9 do
            {Display the digit n.}
            DrawBigDigit(n,10+40*n,v,m,D);
      end;
         v:=v+10*m;
      end;
end.
```

The function/procedures `Digit7by5` and `DrawBigDigit` are declared in `Chap17Codes`. For output, see FIGURE 17.6.

PROBLEM 17.15. Modify `Example17_9` so that it stores each completed dot matrix in a list. After that it should "play them back" one by one.

PROBLEM 17.16. Write a program that prints out a 7-by-5 array of integers with the property that the j-th integer in the i-th row is the number of times that dot matrix entry (i, j) goes on and off as the sequence $0, 1, 2, 3, \ldots, 9$ is displayed.

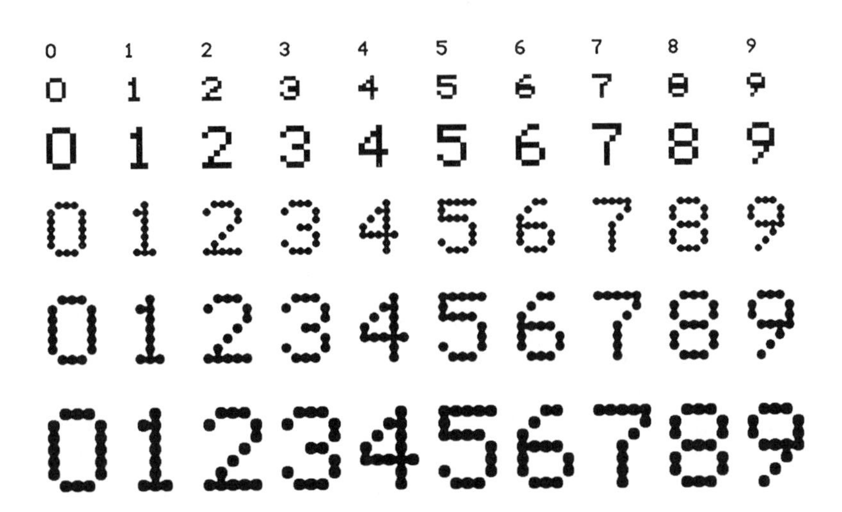

FIGURE 17.6 *Dot Matrix Displays*

Chapter 18

Polygons

Plane geometry is filled with hierarchies. For example, each side of a polygon is a line segment. In turn, each line segment is defined by two points, and each point is defined by two real numbers. Problem solving in this domain is made easier by using *records*. With records, the data that defines a problem can be "packaged" in a way that facilitates our geometric thinking.

18.1 Points

Consider the problem of computing the midpoint M of the line segment that connects the points $P_1 = (x_1, y_1)$ and $P_2 = (x_2, y_2)$. We begin the solution process with a sketch. See Figure 18.1.

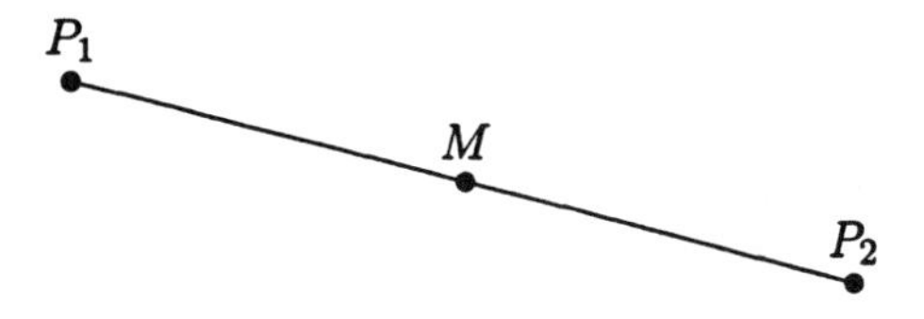

FIGURE 18.1 *Midpoint computation*

Our intuition is bolstered by looking at the picture and thinking at the point level. The unknown is a single geometric object: a point. However, two numbers are required to specify this point and ultimately we must think at the xy-coordinate level and work out the required recipes:

$$M = \left(\frac{x_1 + x_2}{2}, \frac{y_1 + y_2}{2}\right).$$

Encapsulating this we obtain

```
procedure MidPoint(x1,y1,x2,y2:real; var a,b:real);
{Post:  (a,b) is the midpoint of the line segment that extends from}
{(x1,y1) to (x2,y2).}
begin
   a:= (x1+x2)/2;
   b:= (x2+y2)/2;
end
```

This packaging of the midpoint computation is not satisfying. We have hidden the computation but not the coordinates. *To use* `Midpoint` *we are compelled to think at the coordinate level even though it is concerned with points.* Ideally we'd like to have a point *type* and to be able to write function `Midpoint` of the form depicted in FIGURE 18.2. This turns out to be possible in

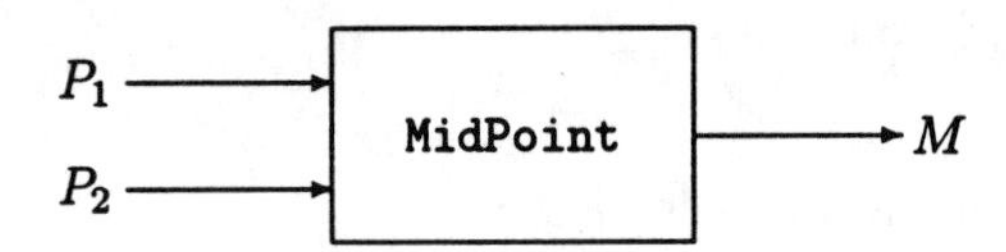

FIGURE 18.2 *A Point-Valued Function with a Point Argument*

ThinkPascal through the use of *records*. Records are used to organize data in a way that reflects the logical structure of the data. Properly chosen records can simplify the problem-solving process because they conceal detail and free us to think at more natural levels. In this regard, records have the same clarifying influence as procedures and functions.

The midpoint problem is a good way introduce the elementary use of records. Subsequent examples will be more dramatic. We start with the program `Example18_1` that reads in the coordinates that define points P_1 and P_2 and the prints the coordinates of the connecting line segment's midpoint. The first new feature that we see in this program is the declaration of a record type. It has the name `RealPoint` and is comprised of two *fields* called `x` and `y`. The general framework for declaring a record is

⟨*Name of Record*⟩ = **record**
  ⟨*Name of First Field*⟩ : ⟨*type*⟩ ;
  ⟨*Name of Second Field*⟩ : ⟨*type*⟩ ;
  ⋮
  ⟨*Name of Last Field*⟩ : ⟨*type*⟩ ;
end

As in **var** declarations, similarly typed fields can be grouped. Thus,

```
program Example18_1;
{Midpoint computation.}
type
   RealPoint = record
         x:real;
         y:real
      end;
var
   P1,P2,Mid:RealPoint;
begin
   writeln('Enter the coordinates of P1:');
   readln(P1.x,P1.y);
   writeln('Enter the coordinates of P2:');
   readln(P2.x,P2.y);
   Mid.x:=(P1.x+P2.x)/2;
   Mid.y:=(P1.y+P2.y)/2;
   writeln('Midpoint = (',Mid.x,',',Mid.y,')');
end.
```

Sample output:

```
Enter the coordinates of P1:
1      7
Enter the coordinates of P2:
9      15
Midpoint = ( 5.000000, 11.000000)
```

```
type
   RealPoint = record
         x,y:real
      end;
```

is equivalent to

```
type
   RealPoint = record
         x:real;
         y:real
      end
```

Since the type **RealPoint** has two fields, a variable P of this type should be visualized as a box with two compartments as shown in FIGURE 18.3. Each compartment (field) is capable of

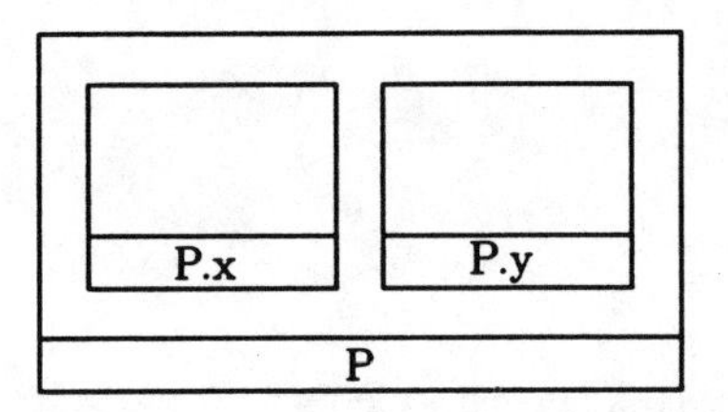

FIGURE 18.3 *A Two-Field Record*

storing a real number. A "dot" notation system is used to access the data that is housed within these "inner boxes." Consider the three point variables P1, P2 and Mid in **Example18_1**. Thus, P1.x names the x-box and is used to house the x-coordinate of the point. Indeed, P1.x *is* a real variable. The statement

```
readln(P1.x,P1.y)
```

says "read the next two values and place them in P1.x and P2.y respectively." It is not permissible to write **readln(P1)**, even though such a statement makes perfect sense to us.

The assignment

```
Mid.x := (P1.x + P2.x)/2
```

says " add the contents of P1.x and P2.x and place the result in Mid.x." Thus, if P1 and P2 are initialized as follows

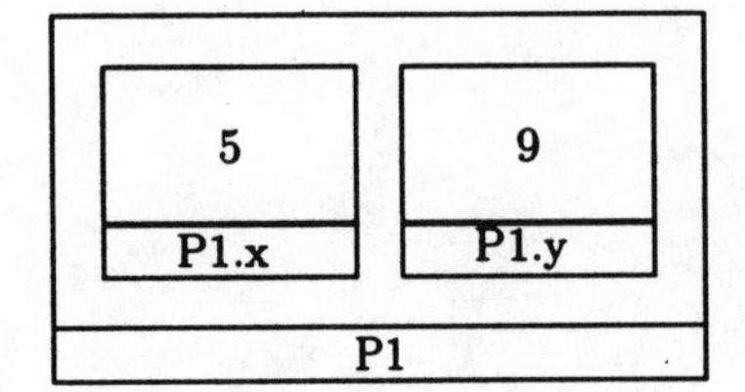

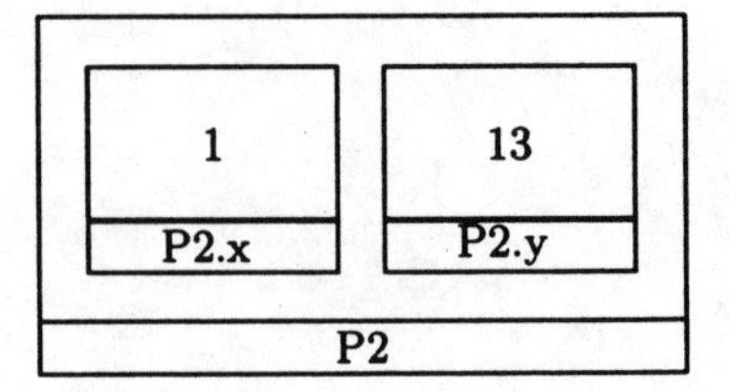

then after the assignments

```
Mid.x := (P1.x + P2.x)/2;
Mid.y := (P1.y + P2.y)/2;
```

the RealPoint variable Mid looks like this:

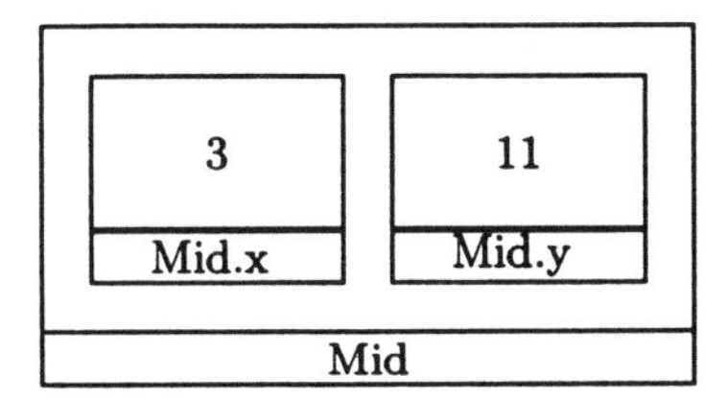

It is *not* legal to compute the midpoint via `Mid := (P1+P2)/2`.

To assign the value of Mid to a fourth RealPoint variable Q, we write

```
Q := Mid
```

This is equivalent to the following pair of real assignments:

```
Q.x:=Mid.x;
Q.y:=Mid.y
```

In general, an assignment of the form

⟨*Record Variable 1*⟩ := ⟨*Record Variable 2*⟩

is legal if the two record variables have *exactly* the same type. This is a very handy feature, especially if the records are complicated and involve a lot of data.

It is not yet apparent why the record idea is useful. The x/y aspect of the midpoint computation is still visible. However, in ThinkPascal a function can return a record type and therein lies the key. We can hide the x/y details inside a function that returns the midpoint:

```
function MidPoint(P1,P2:RealPoint):RealPoint;
{Post:The midpoint of the line segment that extends from}
{P1 to P2.}
begin
   MidPoint.x:= (P1.x+P2.x)/2;
   MidPoint.y:= (P1.y+p2.y)/2
end
```

Example18_2 illustrates its use.

We could also hide the x/y details associated with the readln and writeln with the procedures

```
procedure ReadPoint(p:RealPoint);
begin
   readln(p.x,p.y)
end;
procedure WritePoint(p:RealPoint);
begin
   writeln(p.x,p.y)
end
```

```
program Example18_2;
{Midpoint computation.}
type
   RealPoint = record
         x:real;
         y:real
      end;
function MidPoint(P1,P2:RealPoint):RealPoint;
      ⟨:⟩
var
   P1,P2,Mid:RealPoint;
begin
   writeln('Enter the coordinates of P1:');
   readln(P1.x,P1.y);
   writeln('Enter the coordinates of P2:');
   readln(P2.x,P2.y);
   Mid := MidPoint(P1,P2);
   writeln('Midpoint = (',Mid.x,',',Mid.y,')');
end.
```

Sample output:

```
Enter the coordinates of P1:
10      20
Enter the coordinates of P2:
-4      30
Midpoint = ( 3.000000, 25.000000)
```

```
uses
   DDCodes;
type
   RealPoint = record
         x,y:real;
      end;
      {If p is a RealPoint, then it represents the point (p.x,p.y).}

function MakePoint(x,y:real):RealPoint;
{Post:The point (x,y).}

procedure ShowPoint(p:RealPoint);
{Post:Draws a dot at the screen location corresponding to p.}

procedure ShowBigPoint(p:RealPoint);
{Post:Draws a big dot at the screen location corresponding to p.}

function GetPoint(Message:string):RealPoint;
{Post:Displays a message and returns the next clicked point.}

procedure MoveToPoint(p:RealPoint);
{Post:Moves mouse to the screen coordinate corresponding to p.}

procedure LineToPoint(p:RealPoint);
{Post:Draws a line from the current mouse position to the}
{screen coordinate corresponding to p.}

function MidPoint(P1,P2:RealPoint):RealPoint;
{Post:The midpoint of the line segment extending from P1 to P2.}

function Distance(p1,p2:RealPoint):real;
{Post:The distance from p1 to p2.}
```

FIGURE 18.4 *Declarations from* Chap18Code

However, in anticipation of more intricate point/line/polygon problems, we have chosen to repackage using records some of the graphical i/o tools that are in the unit DDCodes. See FIGURE 18.4. For example, corresponding to

```
procedure LineToPoint(x,y:real);
{Post:Draws a line from the current mouse position to the}
{screen coordinate corresponding to (x,y).}
var
   h,v:integer;
begin
   h:= round(hc+p*x);
   v:=round(vc-p*y);
   LineTo(h,v)
end
```

we have

```
procedure LineToPoint(p:RealPoint);
{Post:Draws a line from the current mouse position to the}
{screen coordinate corresponding to p.}
begin
   cLineTo(p.x p.y)
end
```

Since some of the procedures in this unit build on procedures in DDCodes, it begins with the declaration uses DDCodes. This means that any program that uses Chap18Codes must begin with the declaration

```
uses
   DDCodes,Chap18Codes
```

Example18_3 shows how we can tap into the Chap18Codes resource. The program repeatedly

```
program Example18_3;
{MidPoint}
uses
   DDCodes,Chap18Codes;
var
   P1,P2,Mid:  RealPoint;
begin
   ShowDrawing;
   while YesNo('Another Midpoint Computation?')     do
      begin
         ClearScreen;
         {Acquire the two points that define the line segment.}
         P1 := GetPoint('Enter First Endpoint');
         P2 := GetPoint('Enter Second Endpoint');
         {make a point out of the x and y coordinate that define the midpoint.}
         Mid := MidPoint(P1,P2);
         {Draw a line segment from P1 to P2.}
         MoveToPoint(P1);
         LineToPoint(P2);
         {Display the midpoint as a big dot.}
         ShowBigPoint(mid);
      end;
end.
```

YesNo and ClearScreen are in the unit DDCodes. The type RealPoint and the function/procedures GetPoint, Midpoint, MoveToPoint, LineToPoint and ShowBigPoint are in the unit Chap18Codes.

solicits two points using the mouse. It then computes the midpoint using MidPoint and displays the result using ShowBigPoint.

PROBLEM 18.1. Modify Example 18_3 so that it draws large dots at p and q and nine other points equally spaced points along the connecting line segment.

Once a record is defined, we can declare arrays whose components are of that type. Thus, the declarations

```
const
   nmax=50;
type
   RealPoint = record
         x:real;
         y:real
      end;
   RealPointList = array[1..nmax] of RealPoint;
var
   z:point;
   a:RealPointList;
```

set up a point variable z and an array a that is capable of storing 50 points. Here is how you should visualize a[1..3]:

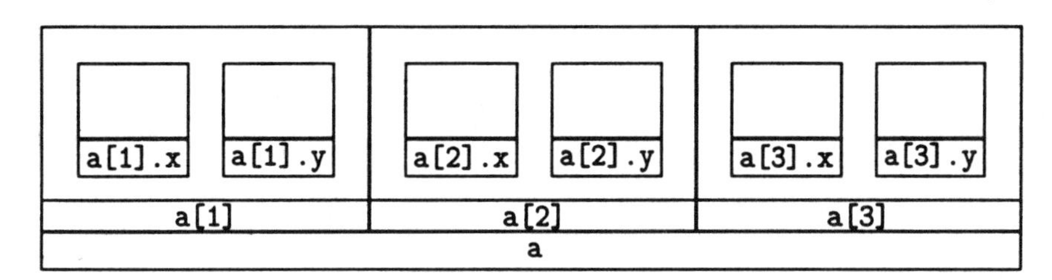

To illustrate the manipulations with this data type, here is a fragment that sets up the vertices of a triangle in a[1..3] and then computes the perimeter:

```
a[1].x:=9; a[1].y := 7;
a[2].x:=3; a[2].y := 8;
a[3].x:=5; a[3].y := 2;
perimeter:= Distance(a[1],a[2]) + Distance(a[2],a[3]) + Distance(a[3],a[1]);
```

The six assignments place the values in a as follows:

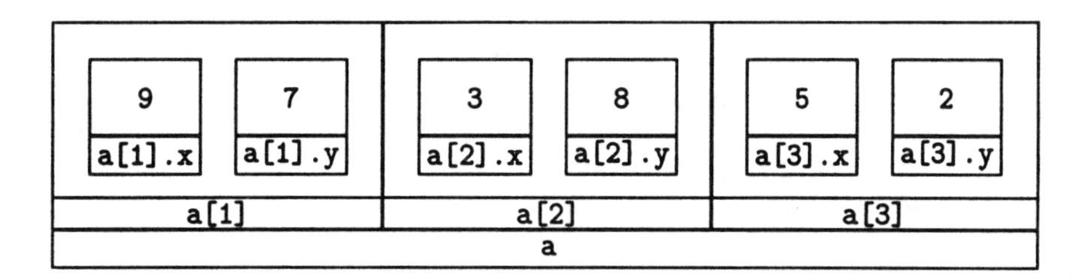

To understand a reference like a[1].y, it is necessary to read from left to right, carefully applying the rules of subscripts and fields:

- a is an array of points.
- a[1] is a component in the array and has type point.
- a[1].y is the y-coordinate of the first point represented in the array.

Think of "a[1].y" as a top-down address. We are placing the value "7" somewhere in the array a. But a is an array and there are lots of components. To nail down the correct component, we need a subscript: a[1]. But the address is still incomplete because a[1] is a variable with two fields. We resolve that last bit of ambiguity by pinpointing the exact field that is to house the "7": a[1].y. If you do not reason in this top-down fashion, then a mistake of the form

```
a.x[1] := 7
```

is likely to emerge. This assignment does not make sense for two reasons. The variable a is not a point and therefore has no x coordinate field and a.x is not an array and therefore cannot be subscripted.

As an application of the RealPointList type, let us revisit the centroid computation. In §14.1 we developed the following procedure:

```
procedure Centroid( n:integer; x,y:RealList; var xbar,ybar:real);
{Pre:x[1..n] and y[1..n] are initialized.}
{Post:(xbar,ybar) is the centroid of (x[1],y[1]),...,(x[n],y[n]).}
var
   k:integer;
   sx,sy:real; {running sums}
begin
   sx:=0; sy:=0;
   for k:=1 to n do
      begin
         sx:=sx+x[k];
         sy:=sy+y[k];
      end
   xbar:=sx/n;
   ybar:=sy/n;
end
```

Let's convert this to an equivalent function using our newly developed record types. If p[1..n] is a RealPointList and contains the point coordinates, then the summation becomes

```
for k:=1 to n do
   begin
      sx:= sx + p[k].x
      sy := sy + p[k].y
   end
```

Going a step further, if s is a RealPoint and we make use of

```
function MakePoint(x,y:real):RealPoint;
{Post:The point (x,y).}
begin
   MakePoint.x:=x;
   MakePoint.y:=y
end
```

then we can build the required running sums as follows:

```
s:= MakePoint(0,0);
for k:=1 to n do
   s := MakePoint(s.x + p[k].x,sy + p[k].y)
```

With respect to the two **var** parameters `xbar` and `ybar` used to return the coordinates of the centroid, we could replace them with a single `RealPoint` parameter `pbar`. We need only replace the assignments

```
xbar:=sx/n;
ybar:=sy/n;
```

with

```
pbar := MakePoint(s.x/n,s.y/n)
```

Moreover, with just one point value to return, it is preferable to package the centroid computation as a function:

```
function Centroid(n:integer; p:RealPointList):RealPoint;
{Post:The centroid of p[1],...,p[n].}
var
   k:integer;
   s:RealPoint;
begin
   s := MakePoint(0,0);
   for k := 1  to n  do
      s := MakePoint(s.x + p[k].x, s.y + p[k].y);
   Centroid := MakePoint(s.x/n, s.y/n);
end
```

A number of procedures have been included in `Chap18Codes` that facilitate calculations when many points are involved. See FIGURE 18.5. The implementations are quite straight forward.

```
procedure GetManyPoints(var n:integer; var p:  RealPointList);
{Post:n points are clicked and p[1..n] contains the associated points.  }

procedure ShowManyPoints(n:integer; p:RealPointList);
{Post:Draws the points p[1]..p[n].}

procedure ConnectPoints(n:integer; p:RealPointList);
{Post:Connects the points p[1],...,p[n],p[1] in order.}
```

FIGURE 18.5 *More Declarations from* `Chap18Codes`

For example, `ConnectPoints` requires just the loop

```
MoveToPoint(p[n]);
for k:=1 to n do
   LineToPoint(p[k])
```

As an illustration of these graphical tools, `Example18_4` solicits points and displays their centroid.

```
program Example18_4;
{Centroids}
uses
   DDCodes, Chap18Codes;
var
   n:integer; {Number of points.}
   p:RealPointList; {Coordinates of the points.}
   pbar:RealPoint; {Centroid of the points.}
begin
   ShowDrawing;
   while YesNo('Another Example?') do
      begin
         ClearScreen;
         GetManyPoints(n,p);
         ConnectPoints(n,p);
         pbar := Centroid(n, p);
         ShowBigPoint(pbar);
      end
end.
```

ClearScreen and **YesNo** are in the unit **DDCodes**. The function/procedures **GetManyPoints**, **ConnectPoints**, **Centroid**, and **ShowBigPoint** and the types **RealPoint** and **RealPointList** are in the unit **Chap18Codes**.

PROBLEM 18.2. Convert the **Chap14Codes** procedure

```
procedure Nearest(x0,y0:real; n:integer; x,y:RealList; var q:integer; var dmin:real);
{Pre:n>=1 and x[1..n] and y[1..n] are initialized.}
{Post:1<=q<=n and (x[q],y[q]) is the closest point to (x0,y0)}
{dmin is the distance from (x0,y0) to (x[q],y[q]).}
```

to

```
procedure Nearest(p0:RealPoint; n:integer; p:RealPointList;
      var q:integer; var dmin:real);
{Pre:n>=1 p[1..n] initialized.}
{Post:  1<=q<=n and p[q] is the closest point to p0.}
{dmin is the distance from p0 to p[q]}
```

PROBLEM 18.3. Complete the following procedure:

```
procedure TwoFurthest(n:integer; p:RealPointList; p0:RealPoint; var Q1,Q2:RealPoint);
{Pre:  p[1..n] initialized}
{Post:Q1,Q2 are furthest and second furthest points in p[1..n] to p0.}
```

18.2 Line Segments

As we mentioned at the beginning of this chapter, there is a natural hierarchy of objects in geometry and this gives us an opportunity to work with records whose fields are other records. For example, a line segment is defined by two distinct points. In an application that is concerned with the manipulation of line segments, it is handy to have a type `LineSegment` that is tailored to that purpose:

```
type
   RealPoint = record
         x,y:real;
      end
      {If p is a point it represents the ordered pair (p.x,p.y).}
   LineSegment = record
         p,q:RealPoint;
      end;
      {If L is a line segment, then it represents the line segment from}
      {point Lp to point L.q.}
```

If one record type is defined in terms of another, then the order of the type declarations is important. Thus, `RealPoint` must be declared before `LineSegment` because the latter is defined in terms of the former. Notice the comments that attend the typing and how they briefly describe the idea behind the record.

If L is a `LineSegment`, then it should be visualized as follows:

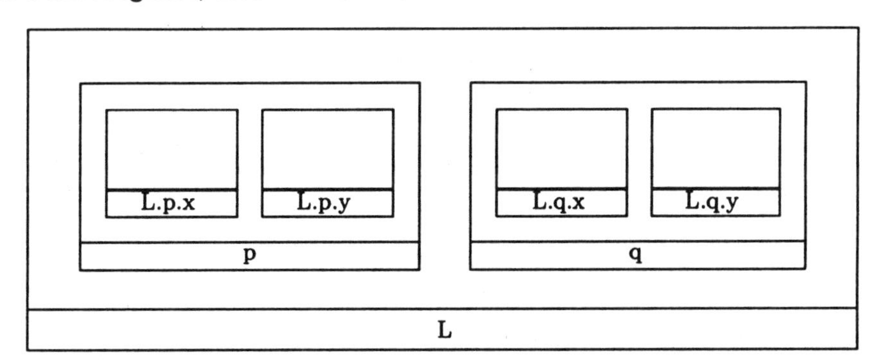

Notice how the dot notation is used to name the innermost boxes. An assignment of the form `L.p.x := 1` results in the storage of a "1" in the x-field of the p-field of L:

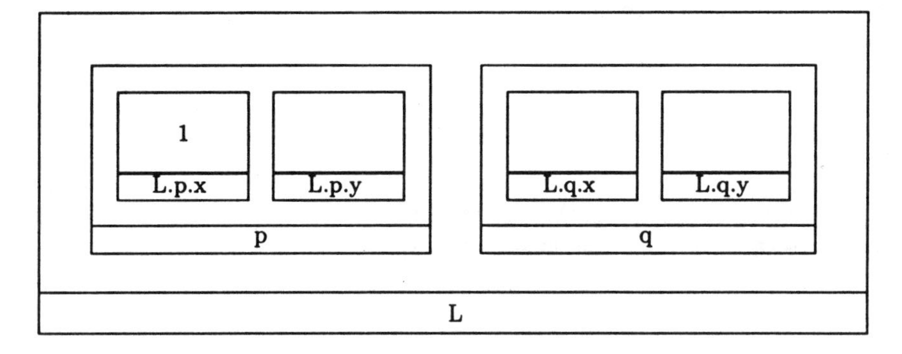

Suppose P1 and P2 are `RealPoint` variables initialized as follows:

```
P1 := MakePoint(1,2);
P2 := MakePoint(3,4)
```

If L is a LineSegment variable, then

```
L.p := P1;
L.q := P2
```

assigns the line segment defined by these two points to L. An assignment like `L.p := P1` is legal because `L.p` and `P1` have the same type. The more verbose fragment

```
L.p.x := P1.x;
L.p.y := P1.y;
L.q.x := P2.x;
L.q.y := P2.y;
```

is equivalent but forces us to think at the x/y level. As an application of RealPoint and LineSegment, let's rewrite the procedure

```
procedure PointToLineSegment(x0,y0,x1,y1,x2,y2:real; var a,b:real);
{Pre:(x1,y1) and (x2,y2) are distinct points.}
{Post:(a,b) is the closest point to (x0,y0) on the line segment from}
        { (x1,y1) to (x2,y2).}
```

that we developed in Chapter 8. Recalling the details of the "coordinate-level" implementation, we obtain the following function:

```
function PointToLineSegment(p0:RealPoint; L:LineSegment):RealPoint;
{Post:  The closest point to p0 on L}
var
tStar:real; {The ''time'' associated with the closest point on the}
        { line defined by L.}
begin
   tStar := ((p0.x-L.p.x)*(L.q.x-L.p.x)+(p0.y-L.p.y)*(L.q.y-L.p.y))/
         (sqr(L.q.x-L.p.x)+sqr(L.q.y-L.p.y));
   if tstar <= 0 then
      PointToLineSegment := L.p
   else if tstar >= 1 then
      PointToLineSegment := L.q
   else
      PointToLineSegment :=
         MakePoint(L.p.x+tStar*(L.q.x-L.p.x), L.p.y+tStar*(L.q.y-L.p.y));
end
```

The program Example18_5 illustrates this function, graphically depicting what it does. It also makes use of

```
function MakeLine(p,q:RealPoint):LineSegment;
{Post:The line segment through points p and q.}
begin
   MakeLine.p:=p;
   MakeLine.q:=q
end
```

```
program Example18_5;
{Point To Line.}
uses
   DDCodes, Chap18Codes;
var
   L:LineSegment;
   p,q:RealPoint; {The points that define L.}
   p0,z:RealPoint; {z is the closest point on L to p0.}
begin
   ShowDrawing;
   while YesNo('Another Example?') do
      begin
         ClearScreen;
         {Set up and show L:}
         p := GetPoint('Enter first endpoint.');
         q := GetPoint('Enter second endpoint.');
         L := MakeLineSegment(p,q);
         ShowLineSegment(L);
         {Obtain p0 and the closest point to it on L.}
         p0 := GetPoint('Enter any point.');
         z := PointToLineSegment(p0,L);
         {Connect p0 and z and highlight z.}
         ShowLineSegment(MakeLineSegment(p0,z));
         ShowBigPoint(z);
      end
end.
```

YesNo and ClearScreen are in the unit DDCodes. The functions/procedures GetPoint, MakeLineSegment, PointToLineSegment, ShowLineSegment, and ShowBigPoint and the types RealPoint and LineSegment are in the unit Chap18Codes.

and

```
procedure ShowLine(L:LineSegment);
{Post:Draws the line segment L.}
begin
   MoveToPoint(L.p);
   LineToPoint(L.q)
end
```

PROBLEM 18.4. Complete the following procedure:

```
procedure Thirds(L:LineSegment; var L1,L2:LineSegment);
{Post:L1 and L2 are the first and last third of L.}
```

Two points P and Q are on the same side of a line L if the line segment that connects them does not intersect L. See FIGURE 18.6. To deduce which of the two situations applies, we

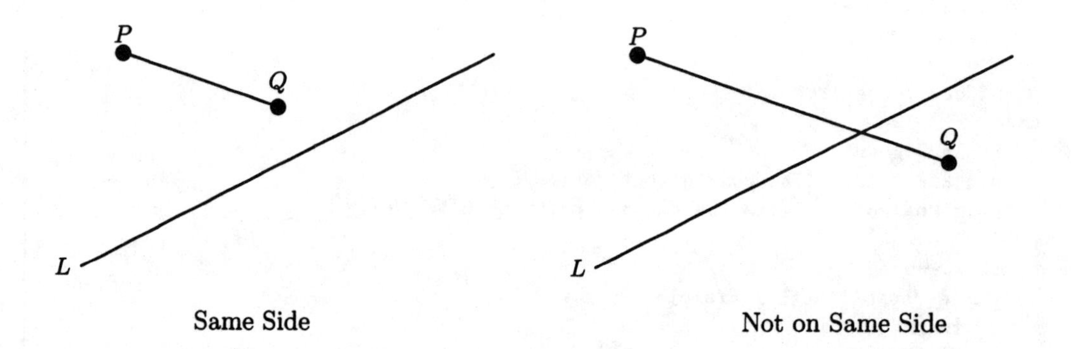

FIGURE 18.6 *Two Points and a Line*

consider the parametric equations

$$\begin{aligned} x(t) &= a + (c - a)t \\ y(t) &= b + (d - b)t \end{aligned}$$

of the line segment that connects $p = (a, b)$ and $q = (c, d)$ and the corresponding ones that define the line L through (x_1, y_1) and (x_2, y_2):

$$\begin{aligned} x(t) &= x_1 + (x_2 - x_1)t \\ y(t) &= y_1 + (y_2 - y_1)t \end{aligned}$$

Note that if we can find "times" t_{pq} and t_L so that

$$\begin{aligned} a + (c - a)t_{pq} &= x_1 + (x_2 - x_1)t_L \\ b + (d - b)t_{pq} &= y_1 + (y_2 - y_1)t_L \end{aligned}$$

then the line that passes through p and q intersects L. If t_{pq} satisfies $0 \leq t_{pq} \leq 1$, then the line *segment* defined by p and q intersects L and p and q are *not* on the same side. These observations are the basis of the following boolean-valued function:

```
function SameSide(p,q:RealPoint; L:Line):boolean;
{Post:  the points p,q are on the same side of line L}
```

that is declared in **Chap18Codes**. See **Example18_6**.

PROBLEM 18.6. Modify **Example18_6** so that it solicits a line segment and n points, and then prints the number of points on either side of the line.

```
program Example18_6;
{Same side.}
uses
   DDCodes,Chap18Codes;
var
   L:LineSegment;
   p,q:RealPoint;
begin
   ShowDrawing;
   while YesNo('Another Example?')  do
      begin
         ClearScreen;
         {Set up and show L:}
         p := GetPoint('Enter first endpoint.');
         q := GetPoint('Enter second endpoint.');
         L := MakeLineSegment(p,q);
         ShowLineSegment(L);
         {Enter two points and see if they are on the same side.}
         p := GetPoint('Enter first point.');
         q := GetPoint('Enter second point.');
         MoveTo(200,25);
         if SameSide(p,q,L) then
            writedraw('Same side')
         else
            writeDraw('Not same side');
      end
end.
```

YesNo and **ClearScreen** are in the unit **DDCodes**. The functions/procedures **GetPoint**, **MakeLineSegment**, **PointToLineSegment**, **ShowLineSegment**, **ShowBigPoint**, and **SameSide** and the types **RealPoint** and **LineSegment** are in the unit **Chap18Codes**.

18.3 Triangles and Rectangles

We have almost completed the journey from point to polygon. But before we tackle general n-gon problems, we look at the representation of triangles and rectangles. One way to define a triangle is to specify its three vertices and that's the idea behind the following type:

```
type
   Triangle = record
      p,q,r:RealPoint
   end;
   {If T is a triangle, then it represents the triangle with vertices}
   { T.p, T.q, and T.r.}
```

Example18_7 illustrates the use of this type which is declared in Chap18Codes along with

```
function MakeTriangle(p,q,r:RealPoint):Triangle;
{Post:The triangle with vertices p, q, and r.}

procedure ShowTriangle(T:Triangle);
{Post:Draws T.}
```

```
program Example18_7;
{Triangles}
uses
   DDCodes,Chap18Codes;
var
   p,q,r:RealPoint; {The vertices of the triangle.}
   T:Triangle;
begin
   ShowDrawing;
   while YesNo('Another Triangle?') do
      begin
         ClearScreen;
         p := GetPoint('Enter first triangle vertex.');
         q := GetPoint('Enter second triangle vertex.');
         r := GetPoint('Enter third triangle vertex.');
         T := MakeTriangle(p,q,r);
         ShowTriangle(T);
      end;
end.
```

YesNo and ClearScreen are in the unit DDCodes. The function/procedures GetPoint, MakeTriangle, and ShowTriangle and the types RealPoint and Triangle are in the unit Chap18Codes.

PROBLEM 18.6. Complete the following functions

```
function Area(T:Triangle):real;
{Post:The area of T.}
```

```
function CenterOfT(T:triangle):RealPoint;
```

Modify Example18_7 so that for each triangle, the centroid is displayed with a large dot and the area is printed.

PROBLEM 18.7. Complete the following procedure:

```
procedure BisectTriangle(T:Triangle; var T1,T2:Triangle);
{Post:T1 and T2 are the triangles that are formed by connecting}
{the midpoint of T's longest side with the opposite vertex.}
```

PROBLEM 18.8. One way to draw a shaded triangle is to draw lots of lines from a vertex to equally spaced points along the opposite side. With that goal in mind, complete the procedure:

```
procedure LotsOfLines(n:integer; p,q,r:RealPoint);
{Post:Draws n+1 lines from p to equally spaced points on line segment from q to r.}
```

and apply it three times to produce

```
procedure ShadeTriangle(n:integer; T:triangle);
{Pre:n>=1}
{Post:Draws T with n+1 shading lines to each opposite side.}
```

PROBLEM 18.9. Using SameSide, complete the following function:

```
function InTriangle(p:RealPoint; T:Triangle):boolean;
{Post:p is inside T.}
```

PROBLEM 18.10. Write a program that prints a list of all Pythagorean right triangles whose legs are 100 or less in length. The no two triangles on the list should be similar. Thus, if the 3-4-5 triangle is on the list, then the 4-3-5 and 6-8-10 triangles cannot be. Hint: set up a type PythTri for representing Pythagorean triangles and build up the required list using an array type PythTriList.

There are usually several reasonable ways to represent a geometric object. Consider the rectangle whose sides are parallel to the coordinate axes. The typing can be based upon range:

```
Rectangle1 = record
      x0,x1:real;
      y0,y1:real
   end;
   {If R is a rectangle, then it represents the set of all points (x,y)}
   {that satisfy R.x0 <= x <= R.x1 and R.y0 <= y <= R.y1.}
```

Or, by opposite vertices:

```
Rectangle2 = record
      p,q:RealPoint
   end;
   {If R is a rectangle, then it represents it has vertices (p.x,p.y),}
   {(p.x,r.y), (r.x,r.), and (r.x,p.y).}
```

Or, by center, length, and width:

```
Rectangle3 = record
      c:RealPoint;
      L,W:real
   end;
   {If R is a rectangle, then it has vertices (c.x-L/2,c.y-W/2),}
   {(c.x+L/2,c.y-W/2), (c.x+L/2,c.y+W/2), and (c.x-L/2,y+W/2).}
```

It is important to recognize that the choice of type "sets the tone" for computation. For example, if R is a rectangle, then how we compute its area depends upon its representation. For the three type possibilities above we have

```
A := (R.x1 - R.x0)*(R.y1 - R.y0);
A := abs(R.p.x - R.q.x)*abs(R.p.y - R.q.y);
A := R.L*R.w
```

Clearly, the choice of representation should be colored by the application in question. Efficiency and whatever facilitates thinking are the key issues. In Chap18Codes we have declared the first of the three typing possibilities

```
Rectangle = record
     x0,x1:real;
     y0,y1:real
  end;
  {If R is a rectangle, then it represents the set of all points (x,y)}
  {that satisfy R.x0 <= x <= R.x1 and R.y0 <= y <= R.y1.}
```

along with

```
function MakeRectangle(p,q:RealPoint):Rectangle;
{Post:The rectangle that has the line segment from p to q as a diagonal.}

procedure ShowRectangle(R:Rectangle);
{Post:Draws rectangle R.}
```

The parameters for MakeRectangle were taken to be opposite vertices for convenience. Example18_8 illustrates their use.

PROBLEM 18.11. Define a type Square that can be used to represent squares and complete the following:

```
function LargestSquare(R:Rectangle):square;
{Post:The largest square in area that fits inside R and has}
{the same upper right corner}

procedure ShowSquare(A:square);
{Post:Draws A.}
```

PROBLEM 18.12. Suppose we are given two rectangles A and B and that we wish to determine the smallest rectangle C that contains them both. We make four observations:

1. The top edge of C should line up with the top edge of A or the top edge of B, whichever is higher.
2. The left edge of C should line up with the left edge of A or the left edge of B, whichever is further to the left.
3. The bottom edge of C should line up with the bottom edge of A or the bottom edge of B, whichever is lower.
4. The right edge of C should line up with the right edge of A or the right edge of B, whichever is further to the right.

Complete the following function:

```
function Enclosing(R1,R2:Rectangle):Rectangle;
{Post:The smallest rectangle that contains R1 and R2.}
```

Note that if we wrote this as a procedure with simple variables, then there would be twelve parameters, four for each of the three rectangles that participate in the problem.

PROBLEM 18.13. Complete the following procedure:

```
procedure Intersect(R1,R2:Rectangle; var Empty:boolean; var R:Rectangle);
{Post:If R1 and R2 do not intersect, then Empty is true.  Otherwise, Empty is}
{false and R is their intersection.}
```

```
program Example18_8;
{Rectangles}
uses
   DDCodes,Chap18Codes;
var
   p1,p2:RealPoint; {Opposite vertices of rectangle.}
   R:rectangle;
begin
   ShowDrawing;
   while YesNo('Another Rectangle?')  do
      begin
         ClearScreen;
         p1 := GetPoint('Enter first vertex.');
         p2 := GetPoint('Enter opposite vertex.');
         R := MakeRectangle(p1,p2);
         ShowRectangle(R)
      end
end.
```

YesNo and ClearScreen are in the unit DDCodes. The function/procedures GetPoint, MakeRectangle, and ShowRectangle and the types RealPoint and Rectangle are in the unit Chap18Codes.

18.4 N-gons

There are a number of ways to represent a polygon in a record. Perhaps the simplest is to use an integer field to house the number of sides and a RealPointList field for the vertices:

```
type
   Polygon = record
         n:integer;
         v:RealPointList
   end;
   {If P is a polygon, then it obtained by connecting the points}
   { P.v[1], P.v[2],..., P.v[P.n], P.v[1] in turn.}
```

This is our first example of a record that has an array field and here is how to visualize a variable of this type:

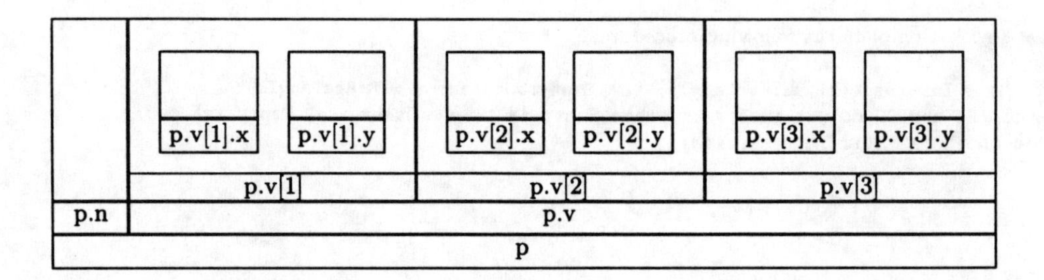

If P is a polygon, then the fragment

```
P.n:=5;
P.v[1].x := 4; P.v[1].y := 48;
P.v[2].x := 50; P.v[2].y := 25;
P.v[3].x := 130; P.v[3].y := 45;
P.v[4].x := 130; P.v[4].y := 135;
P.v[5].x := 70; P.v[5].y := 150;
```

sets up the pentagon depicted in FIGURE 18.7. Again, be careful with the interpretation of

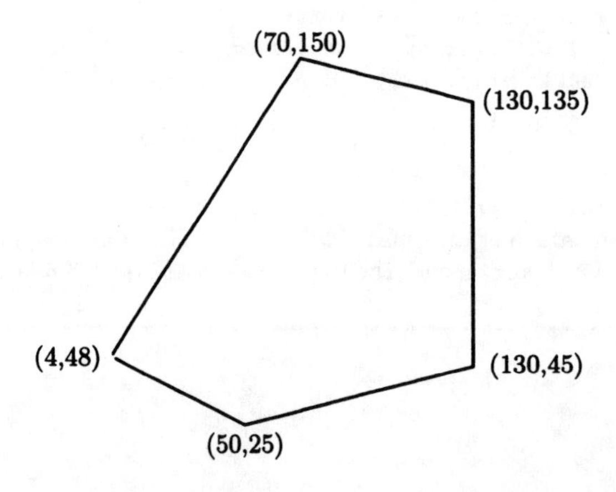

FIGURE 18.7 *A Pentagon*

references like `P.v[1].y := 1`.

- P is polygon.
- `P.v` is an array of points that houses the polygon's vertices.
- `P.v[1]` is a point that houses the coordinates of the first vertex.
- `P.v[1].y` is a real variable that houses the y-coordinate of the first vertex.

The perimeter of the pentagon can be computed as follows:

```
perimeter:= Distance(P.v[1],P.v[5]);
for k:=1 to 4 do
   perimeter := perimeter + Distance(P.v[k],P.v[k+1]);
```

As is our habit, we define a build function

```
function MakePolygon(n:integer; v:RealPointList):Polygon;
{Pre:v[1..n] initialized}
{Post:The polygon obtained by connecting vertices v[1],..,v[n].}
```

and a show procedure

```
procedure ShowPolygon(P:polygon);
{Post:Draws P.}
```

to facilitate polygon computation. These are declared in Chap18Codes. The implementation of MakePolygon shows reminds us that it is legal to assign the value of one record variable to another that has the same type:

```
MakePolygon.n:=n;
for k:=1 to n do
   MakePolygon.v[k]:=v[k]
```

Example19_9 solicits the vertices and then displays the resulting polygon.

```
program Example18_9;
{Polygons.}
uses
   DDCodes,Chap18Codes;
var
   n:integer;
   P:polygon;
   v:RealPointList;
begin
   ShowDrawing;
   while YesNo('Another Polygon?')  do
      begin
         ClearScreen;
         GetManyPoints(n,v);
         P := MakePolygon(n,v);
         ShowPolygon(P);
      end
end.
```

YesNo and ClearScreen are in the unit DDCodes. The function/procedures GetManyPoints, MakePolygon and ShowPolygon and the types RealPointList and Polygon are in the unit Chap18Codes.

A polygon P is *convex* if any pair of points in P can be connected with a line segment that is contained entirely in P. See FIGURE 18.8. Many problems with polygons are easier to solve

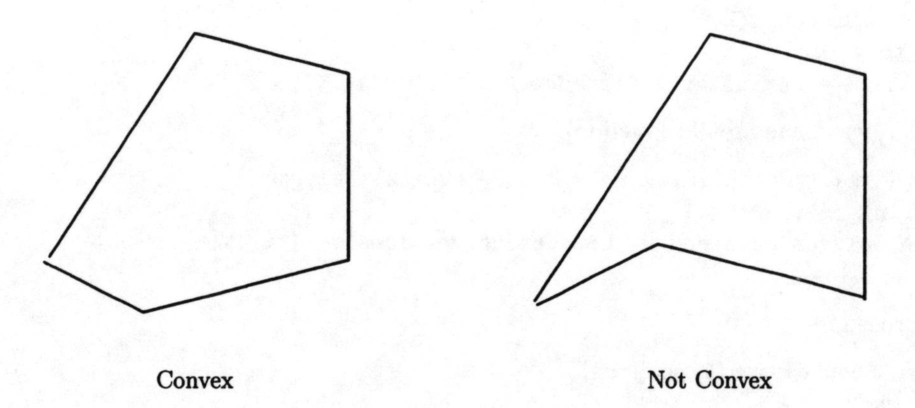

FIGURE 18.8 *Convexity*

if the polygons involved are convex. For example, the area of an n-gon can be computed as the sum of n-2 triangular areas involving the vertex points as suggested in FIGURE 18.9. Thus, if

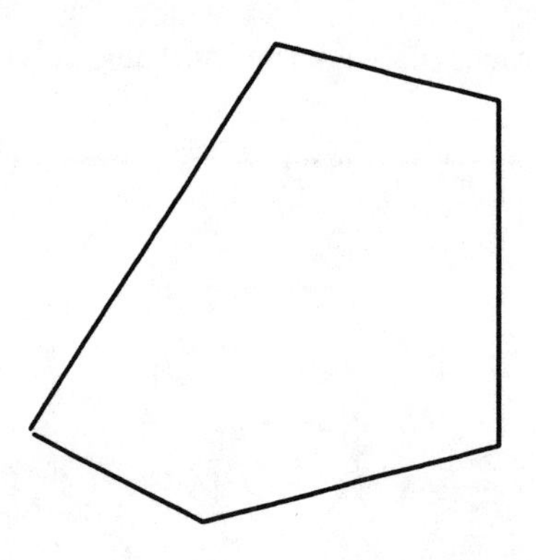

FIGURE 18.9 *Area of a Convex Polygon*

the function

```
function TriangleArea(T:triangle):real;
{Post:The area of T.}
```

is available, then

```
s:= 0;
for k:=2 to P.n -1 do
   s := s + TriangleArea(MakeTriangle(P.v[k-1],P.v[k],P.v[k+1]);
```

assigns the area of P to s.

PROBLEM 18.14. Complete the following function.

```
function ConvexPoly(P:polygon):boolean;
{Pre:P.n>=3}
{Post:P is convex}
```

PROBLEM 18.15. Complete the following function

```
function PointToPolygon(z:RealPoint; P:polygon):realPoint;
{Post:Nearest point to z on boundary of P.}
```

making effective use of `PointToLineSegment`.

PROBLEM 18.16. Suppose (a, b) and (c, d) are given points and that $0 \leq theta \leq 2\pi$. The point (a', b') defined by

$$\begin{aligned} a' &= a + (c-a)\cos(\theta) - (d-b)\sin(\theta) \\ b' &= b + (c-a)\sin(\theta) + (d-b)\cos(\theta) \end{aligned}$$

is the rotation of (a, b) about (c, d). The angle of the rotation is θ. Complete the following functions

```
function RotatePt(p,q:RealPoint; theta:real):RealPoint;
{Post:Rotate p theta radians about q.}

function RotatePolygon(P:Polygon; q:RealPoint; theta:real):Polygon;
{Post:Rotate P theta radians about the point q.}
```

PROBLEM 18.17. Suppose A, B, C and D are the vertices of a quadrilateral Q. Q is convex if A and C are not on the same side of the line through B and D and B and D are not on the same side of the line through A and C. Complete the function

```
function ConvexQuad(P:polygon):boolean;
{Pre:P.n=4}
{Post:P is convex.}
```

Chapter 19

Special Arithmetics

In this chapter we push out from the constraints of computer arithmetic. We have no way to represent exactly 100! or even 1/3 or $\sqrt{-1}$. To address these constraints we develop three *environments*. The first is for very long integer arithmetic and will permit us to compute very large integers. The idea is to use an array to represent an integer. Functions are developed that permit the manipulation of integers that are stored in this fashion and enable us to compute exactly things like

$$\begin{aligned} 100! = \; & 33262154439441526816992388562667004907159682643812146859296389521759999322991 \\ & 5608941463976156518286253697920827223758251185210916864000000000000000000000000 \end{aligned}$$

Quotients of integers give us the rational numbers. Unfortunately, even simple rational numbers like 1/3 have no exact (base-2) floating point representation. Clearly, the thing to do is to represent a rational number as a pair of integers in the computer. By doing this and developing arithmetic functions that can operate on rational numbers, we can compute exactly rational numbers like

$$1 + \frac{1}{2} + \frac{1}{3} + \cdots + \frac{1}{10} = \frac{7381}{2580}$$

Our third extended arithmetic system deals more with a shortcoming of the real numbers than with real floating point arithmetic. The problem is that the square root of a negative real number is not real, but complex. But if we let i stand for $\sqrt{-1}$, then many interesting doors are opened. For starters, square roots like $\sqrt{-36}$ have a complex representation, e.g., $6i$. General complex numbers have the form $a + bi$ where a and b are real. We develop an environment that supports their representation and manipulation. The display of complex numbers in the complex plane enables us to acquire a geometric intuition about their behavior in certain computational settings.

19.1 The Very Long Integer

Suppose we have a task that requires the manipulation of very big integers, say 100 digits in length. Since the largest value that can be stored in a **longint** variable is a few billion, an alternative representation must be sought.

One approach is to use an array. Indexing from zero, we use the k-th entry in the array to house the value of the 10^k place in the base-10 expansion. For example, the integer 123456789012 can be represented in a length 12 integer array:

2	1	0	9	8	7	6	5	4	3	2	1

To permit the storage of negative values and to indicate the "length" of the integer represented, we work with the following type:

```
const
   nmax = 100;
type
   digit = 0..9;
   VeryLongInt = record
         sign:boolean;
         L:integer;
         a:array[0..nmax] of digit
      end
```

Here is a schematic to help visualize this structure:

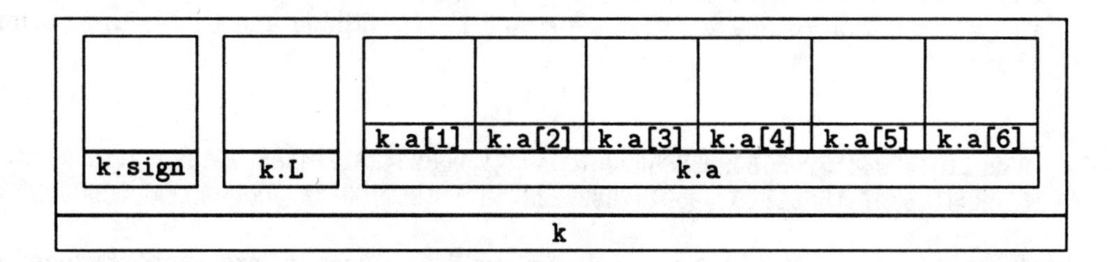

If k is an initialized **VeryLongInteger**, then we assume that

$$k = \begin{cases} +(k.a[0] + k.a[1]10^1 + k.a[2]10^2 + \cdots + k.a[k.L]10^{k.L}) & \text{if } \texttt{k.sign} \text{ is true} \\ -(k.a[0] + k.a[1]10^1 + k.a[2]10^2 + \cdots + k.a[k.L]10^{k.L}) & \text{if } \texttt{k.sign} \text{ is false} \end{cases}$$

Let's build an environment for **VeryLongInt** manipulation including functions that facilitate the initialization and display of **VeryLongInteger** variables. We start with the development of

```
function Int2VLInt(n:longint):VeryLongInt;
{Post:The representation of n as a VeryLongInt.}
```

(See §12.2 where integer-to-string conversion is discussed.) A small example that converts an integer n to its **VeryLongInteger** equivalent x illustrates the main idea. If $n = 638$, then the fragment

```
x.a[0] := n mod 10; {The one's place.}
n := n div 10;
x.a[1] := n mod 10; {The ten's place.}
n := n div 10;
x.a[2] := n mod 10; {The hundred's place.}
x.L := 2;
x.sign := true
```

produces

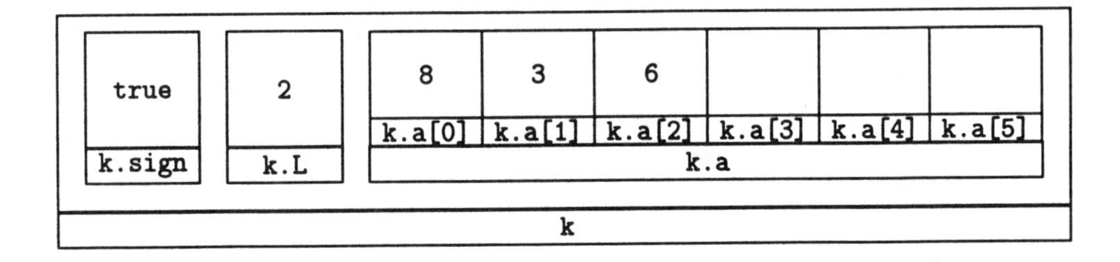

Note that the digits are remainders¡

8 is the remainder of $638 \div 10$.
3 is the remainder of $63 \div 10$.
6 is the remainder of $6 \div 10$.

and that the required dividends (638, 63, and 6) are obtained through a sequence of integer divides.

The idea of "pulling off" the digits one-by-one using **mod** and **div** was used in the integer-to-string conversion function `Int2Str` that we developed in §12.x. Extrapolating from the example and taking care to set up the sign and length fields we obtain

```
function Int2VLInt(n:longint):VeryLongInt;
{Post:The representation of n as a VeryLongInt.}
var
   k:integer;
   m:longint;
begin
   {Take care of the sign.}
   Int2VLint.sign := (n >= 0);
   m := abs(n);
   k := 0;
   Int2VLint.a[k] := m mod 10;
   m := m div 10;
   while m > 0  do
      {Isolate the k-th digit.}
      begin
         k := k + 1;
         Int2VLInt.a[k] := m mod 10;
         m := m div 10;
      end;
   Int2VLInt.L := k;
end
```

It is also convenient to have a function that converts the value of a VeryLongInt variable into an equivalent string so as to permit "pretty printing:"

```
function VLInt2Str(x:VeryLongInt; w:integer):string;
{Post:Converts x to equivalent string representation.  }
{If fewer than w characters are needed, then enough blanks are added to}
{to the front so that the returned string has length w.}
var
   j:integer;
   s:string;
begin
   s := '';
   for j := 0 to x.L do
      {Incorporate the digit in the jth place.}
      s := concat(copy('0123456789',x.a[j]+1,1),s);
   if not x.sign then
      s := concat('-',s);
   {Append blanks if necessary.}
   for j := length(s)+1 to w do
      s := concat(' ',s);
   VLInt2Str := s;
end
```

A repertoire of functions for performing arithmetic operations between VeryLongInt values is also needed. We work on the easiest of these:

```
function AddPosVLInt(x,y:VeryLongInt):VeryLongInt;
{Post:The sum of x and y.}
```

The addition $x + y$ of two whole numbers involves a sequence of three-term summations each of which involves a digit from x, the corresponding digit from y, and a carry from the previous position. For example, 168 + 917 proceeds as follows:

- 8 plus 7 is 15 carry 1.
- 6 plus 1 plus the previous carry is 8 carry 0.
- 1 plus 9 plus the previous carry is 10 carry 1.
- 0 plus 0 plus the previous carry is 1 carry 0.

The simplicity of this algorithm, learned in grade school, is the basis of the following implementation:

```
function AddPosVLInt(x,y:VeryLongInt):VeryLongInt;
{Pre:x>=0 and y>=0}
{Post:The sum of x and y.}
var
   i:integer;
   carry:integer; {The carry from the i-th place.}
   z:VeryLongInt; {x + y}
begin
   carry := 0;
   i:=-1;
   while (i < x.L) or (i < y.L) do
      {Addition complete through the 10^i place.}
      begin
         i:=i+1;
         z.a[i] := carry;
         if i <= x.L then
            z.a[i] := z.a[i] + x.a[i];
         if i <= y.L then
            z.a[i] := z.a[i] + y.a[i];
         carry := z.a[i] div 10;
         if carry > 0 then
            z.a[i] := z.a[i] mod 10;
      end;
   {Handle the last carry and determine the length.}
   if carry>0 then
      {There is a carry from the last place.}
      begin
         z.L := i+1;
         z.a[z.L] := carry;
      end
   else
      {No carry from the last place.}
      z.L := i;
   z.sign := true;
   AddPosVLInt := z;
end
```

Notice that so that there is uniformity in each position, we bring a carry of zero into the one's place. `Example19_1` uses `Int2VLInt` and `AddPosVLIInt` to print the first 100 powers of 2.

```
program Example19_1;
{First 100 powers of 2.}
uses
   Chap19Codes;
var
   j:integer;
   Power:VeryLongInt; {2^j}
begin
   ShowText;
   writeln(' 2^j j');
   writeln(' ---------------------------');
   Power := IntToVLInt(1);
   for j := 1 to 100 do
      {Print 2^j}
      begin
         Power := AddPosVeryLongInt(Power,Power);
         writeln(VLIntToStr(Power,40), j:5);
      end;
end.
```

The type VeryLongInt and the functions IntToVLint, AddPosVeryLogInt, and VLIntToStr are in the unit Chap19Codes. Output:

```
                           2^j        j
------------------------------------------
                             2        1
                             4        2
                             8        3
                            16        4
                            32        5
                             :        :
    39614081257132168796771975168    95
    79228162514264337593543950336    96
   158456325028528675187087900672    97
   316912650057057350374175801344    98
   633825300114114700748351602688    99
  1267650600228229401496703205376   100
```

PROBLEM 19.1. Modify **Example19_1** so that it prints all positive powers of three that are less than or equal to 10^{50}.

PROBLEM 19.2. The Fibonacci sequence $\{f_n\}$ is generated via the recursion $f_n = f_{n-1} + f_{n-2}$ with $f_0 = f_1 = 1$. Modify **Examplel19_1** so that it prints all the Fibonacci numbers $\{f_n\}$ that are less than or equal to 10^{50}.

PROBLEM 19.3. The "Up-and-Down" sequence $\{a_n\}$ is generated via the recursion

$$a_n = \begin{cases} 3a_{n-1} + 1 & \text{if } a_{n-1} \text{ is odd} \\ a_{n-1}/2 & \text{if } a_{n-1} \text{ is even} \end{cases}$$

where a_0 is an arbitrary positive integer. Complete the functions

```
function Up(a:VeryLInt):VeryLInt;
{Post:3a+1}

function DivBy2(x:VeryLInt):VeryLInt;
{Pre:x>=0}
{Post:x div 2}

function IsOne(x:VeryLInt):boolean;
{Post:x=1}
```

and use them to determine the smallest n such that $a_n = 1$ where a_0 = **maxlongint**.

PROBLEM 19.4. Suppose Digit = 0..9999. and that a[0..L] is initialized, We can then identify its contents with the integer

$$a[0] + a[1]10000^1 + a[2]10000^3 + \cdots + a[L]10000^L$$

Rewrite **AddPosVLInt** and Int2VLInt to work with this representation and use them effectively to compute and print 2^{1000}.

PROBLEM 19.5. Write a program that prints 100!. This requires the development of a function that can multiply a pair of VeryLongInt numbers. Start by completing the following functions:

```
function DigitProd(n:integer; x:VeryLongInt):VeryLongInt;
{Pre:  0 <= n <=9}
{Post:  The value of n*x.}

function TenProd(e:integer; x:VeryLongInt):VeryLongInt;
{Pre:e>=1}
{Post:  (10^e) times x}
```

In writing **DigitProd** you'll have to multiply each digit in **x** by **n**. The digits are housed in **x.a**. Of course, you'll have to propagate the carries. Study **AddPosVLInt**. **TenProd** involves multiplyingx by a power of 10. For this, you'll have to shift the digits in the array **x.a** *e* places to the right.

These two functions when combined with **AddPosVLInt** can be used to develop

```
function MultLongInt(x,y:VeryLongInt):VeryLongInt;
{Pre:x>=0, y>=0}
{Post:The product of x and y.}
```

A small example suggests how to proceed:

```
{ Compute 237 times 456.}
z := 0; (via IntToVLInt)
w := 237*6 = 1422; (via DigitProd)
z := z + w = 1422; (via AddLongInt)
w := 237*5 = 1185; (via DigitProd)
w := 10*w = 11850; (via TenProd)
z := z + w = 13272; (via AddLongInt)
w := 237*4 = 948; (via DigitProd)
w := 100*w = 94800 (via TenProd)
z := z + w = 108072; (via AddLongInt)
```

PROBLEM 19.6. The Roman Numeral system is a place value system in which the symbol used to represent a place value depends upon the place. For example, 4 is represented with 'IV' in the ones place, 'XL' in the tens place, and 'CD' in the hundreds place. The empty string stands for 0. Here is a table that fully describes what we mean:

Digit	Ones Place	Tens Place	Hundreds Place
0	empty	empty	empty
1	I	X	C
2	II	XX	CC
3	III	XXX	CCC
4	IV	XL	CD
5	V	L	D
6	VI	LX	DC
7	VII	LXX	DCC
8	VIII	LXXX	DCCC
9	IX	XC	CM

Any number between 0 and 999 is the concatenation of three strings, one from each column. Thus, 708 = `concat('DCC','','VIII')`. Write a program that prints a list of the Roman numeral representation of the integers 0 through 999.

19.2 The Rational

A number is *rational* if it is the quotient of two integers. Notice that the sum, difference, product, and quotient of two rational numbers is again rational. Here is a convenient type for working in this system:

```
rational = record
      num:longint;
      denom:longint
      {If p has type rational, then it represents the value p.num/p.denom}
   end
```

If `x` is an initialized `rational`, then

```
x.num := 22;
x.denom := 7
```

sets up a representation for the rational number 22/7. Note that if `z` is real, then the floating point representation of this quotient via `z := 22/7` is not exact.

An important feature of the rational numbers is that any number, even a number like π or $\sqrt{2}$, is arbitrarily close to a rational number. Here is a fragment that assigns to the rational variable `r`, the closest rational number to π subject to the constraint that both the numerator and denominator are less than or equal to 9999:

```
n:= 9999;
r.num:=3;
r.denom:=1;
Err := abs(pi-(r.num/r.denom));
for q=1 to n do
   for p=1 to n do
      if abs(((p/q) - pi) < Err then
         begin
            r.num:=p;
            r.denom:=q;
            Err:=abs((p/q)-pi);
         end;
```

The dimension of the search can be reduced by noting that once we "pick" the denominator q, the best numerator is the round of $q\pi$. See Example19_2.

PROBLEM 19.6. Complete the following function:

```
function ClosestRational(x:real; bound:integer):rational;
{Pre:x>=0}
{Post:The closest rational number p/q to x with the property that}
{p and q satisfy p<=bound and q<=bound}
```

PROBLEM 19.7. Complete the following two functions

```
function EqualRational(a,b:rational):boolean;
{Post:a=b}

function Belong(a:rational; z:RationalList; m:integer):boolean;
{Post:  a is in the list z[1..n]}
```

where RationalList = array[1..1000] of rational. Let R_n be the number of *distinct* positive rational numbers whose numerator and denominator are no bigger than n. (Thus, $R_2 = 3$ because there are three distinct values in the set $\{(1/1), (1/2), (2/1), (2,2)\}$.) Using the above functions, write a program that prints a 3-column table that prints n, R_n, and R_n/n^2 for $n = 1, \ldots, 40$. In organizing your solution, note that the calculation of R_n can exploit R_{n-1}.

Note that any rational number can be expressed in an infinite number of ways: 2/3, 4/6, 6/9, etc. For that reason, it is handy to work with quotients that are reduced to lowest terms meaning that the numerator and denominator have no common divisors greater than 1. The problem of dividing out the common divisors is easily handled through the greatest common divisor function

```
function gcd(p,q:integer):integer;
{Pre:p>=0,q>=0}
{Post:the greatest common divisor of p and q.  }
```

that we discussed in §16.1. Using gcd we obtain

```
program Example19_2;
{Rational approximations to pi of the form p/q where both p and q}
{are <= n.}
const
   n=9999;
type
   rational = record
         num:longint;
         denom:longint
   end
var
   r:rational;
   Err:real;
begin
   ShowText;
   writeln(' p q |(p/q) - pi| ');
   Err := abs(pi - (3/1));
   r.denom:=1;r.num:=3;
   writeln(p:5, q:5, ' ', Err:10:7);
   while r.num <= n do
      begin
         if abs((r.num/r.denom)-pi) < Err then
            begin
            Err := abs((r.num/r.denom) - pi);
            writeln(r.num:5, r.denom:5, ' ', Err:10:7);
            end;
         r.denom := r.denom+1;
         r.num := round(r.denom*pi);
      end
end.
```

The type rational is in the unit Chap19Codes. Output:

```
    p    q     |(p/q) - pi|
------------------------------
    3    1      0.1415927
   13    4      0.1084073
   16    5      0.0584073
   19    6      0.0250740
   22    7      0.0012645
  179   57      0.0012418
  201   64      0.0009677
  223   71      0.0007476
  245   78      0.0005670
  267   85      0.0004162
  289   92      0.0002883
  311   99      0.0001785
  333  106      0.0000832
  355  113      0.0000003
```

```
function reduce(r:rational):rational;
{Post:r reduced to lowest terms.}
var
   d:longint;
   q:rat;
begin
   if r.num = 0 then
      q.denom := 1
   else
      begin
         d := gcd(abs(r.num),abs(r.denom));
         q.num := r.num div d;
         q.denom := r.denom div d
      end
   reduce := q;
end
```

The rules for rational addition, subtraction, multiplication, and division are as follows:

$$
\begin{aligned}
\frac{p_1}{q_1} + \frac{p_2}{q_2} &= \frac{p_1 q_2 + p_2 q_1}{q_1 q_2} \\
\frac{p_1}{q_1} - \frac{p_2}{q_2} &= \frac{p_1 q_2 - p_2 q_1}{q_1 q_2} \\
\frac{p_1}{q_1}\frac{p_2}{q_2} &= \frac{p_1 p_2}{q_1 q_2} \\
\frac{p_1}{q_1} \div \frac{p_2}{q_2} &= \frac{p_1 q_2}{q_1 p_2} \quad p_2 \neq 0\,.
\end{aligned}
$$

Here is an implementation of the add operation:

```
function AddRational(a,b:rational):rational;
{Post:The sum of a and b reduced to lowest terms.}
var
   c:rational;
begin
   c.num := a.denom*b.num + a.num*b.denom;
   c.denom := a.denom*b.denom;
   AddRational := reduce(c)
end
```

We also have

```
function SubRational(a,b:rational):rational;
{Post:The difference a - b reduced to lowest terms.}
function MultRational(a,b:rational):rational;
{Post:The product of a and b reduced to lowest terms.}
function DivRational(a,b:rational):rational;
{Pre:b <> 0}
{Post:The quotient a/b reduced to lowest terms.}
```

Example19_3 illustrates the use of the rational arithmetic system made possible by these functions. It computes (in rational form) first ten *harmonic numbers*

```
program Example19_3;
{Harmonic Numbers}
uses
   Chap19Codes;
var
   h,term:rational;
   n:integer;
begin
   ShowText;
   h.num := 0;
   h.denom := 1;
   writeln(' n n-th Harmonic Number');
   writeln('----------------------------');
   for n:=1 to 10 do
      {h = 1 + 1/2 + 1/3 + ...  + 1/n}
      begin
         term.num := 1;
         term.denom := n;
         h := AddRational(h,term);
         writeln(n:3, h.num:10,' /', h.denom:6);
      end;
end.
```

The type `rational` and the function `AddRational` are in the unit `Chap19Codes`. Output:

```
 n       n-th Harmonic Number
-----------------------------
 1                        1/1
 2                        3/2
 3                       11/6
 4                      25/12
 5                     137/60
 6                      49/20
 7                    363/140
 8                    761/280
 9                  7129/2520
10                  7381/2520
```

$$H_n = 1 + \frac{1}{2} + \frac{1}{3} + \cdots + \frac{1}{n}.$$

Thus, for $n = 6$ it finds that

$$H_6 = 1 + \frac{1}{2} + \frac{1}{3} + \frac{1}{4} + \frac{1}{5} + \frac{1}{6} = \frac{60 + 30 + 20 + 15 + 12 + 10}{60} = \frac{147}{60} = \frac{49}{20}$$

PROBLEM 19.8. Bernoulli Numbers are defined by

$$B_n = -\frac{1}{n+1}\sum_{j=0}^{n-1}\binom{n+1}{j}B_j$$

where $B_0 = 1$. The coefficient of B_j in the summation is the binomial coefficient $n+1$-choose-j. Binomial coefficients are discussed in §7.2. Thus,

$$\begin{aligned} B_1 &= -\frac{1}{2}\left(\binom{2}{0}B_0\right) = -\frac{1}{2} \\ B_2 &= -\frac{1}{3}\left(\binom{3}{0}B_0 + \binom{3}{1}B_0\right) = -\frac{1}{3}\left(1 - 3\frac{1}{2}\right) = \frac{1}{6} \end{aligned}$$

Write a program that computes $B_1, \ldots, B_{20}$ in rational form and prints the results. Make us of the function `BinCoeff(n,k:integer):longint` that is available through **Chap7Codes**.

19.3 The Complex

A complex number has the form $a + bi$ where a and b are real and $i^2 = -1$. They can be represented with the following type:

```
complex = record
        r:real;
        i:real
    end
{If z is complex, then z.r and z.i are its real and imaginary parts.}
```

Thus, if z is complex, then the assignments

```
z.r := 3;
z.i := 4;
```

assigns the complex number $3 + 4i$ to z.

The operations of addition, multiplication, division, and absolute value are defined by

$$\begin{aligned} (a+bi)+(c+di) &= (a+c)+(b+d)i \\ (a+bi)(c+di) &= (ac-bd)+(ad+bc)i \\ (a+bi)/(c+di) &= \frac{ac+bd}{\sqrt{c^2+d^2}} + \frac{-ad+bc}{\sqrt{c^2+d^2}}i \\ |a+bi| &= \sqrt{a^2+b^2} \end{aligned}$$

To facilitate the manipulation of complex numbers, the following functions are available through Chap19Codes:

```
function AddComplex(u,v:complex):complex;
{Post:u+v}
```

```
function SubComplex(u,v:complex):complex;
{Post:u-v}

function MultComplex(u,v:complex):complex;
{Post:u*v}

function DivComplex(u,v:complex):complex;
{Pre:v<> 0}
{Post:u/v}

function ScaleComplex(u:complex; s:real):complex;
{Post:s*u}

function Cabs(z:complex):real;
{Post:The absolute value of z.}
```

The implementation of these functions is straightforward, e.g.,

```
function DivideComplex(u,v:complex):complex;
{Pre:v is not zero.}
{Post:The quotient u/v.}
var
   z:complex;
begin
   s := sqr(v.r) + sqr(v.i);
   z.r := (u.r*v.r +u.i*v.i)/s;
   z.i := (-u.r*v*i + u.i*v.r)/s;
   DivideComplex:=z;
end
```

To illustrate the use of these functions, let us explore the behavior of the summation

$$S_n = 1 + z + \frac{z^2}{2!} + \cdots + \frac{z^n}{n!} = \sum_{k=0}^{n} \frac{z^k}{k!}$$

for complex z. For real z this could be handled as follows:

```
term := 1;
k := 0;
s := term;
Converged := false;
whilenot Converged
   begin
      k := k + 1;
      term := z*term / k;
      Converged := (abs(term) <= (eps * abs(s))) or (k >= itmax);
      s := s + term;
   end
```

Termination occurs when the current term is small compared to the running sum or if 30 iterations have been completed–whichever comes first. To carry this out with complex z, we merely use

AddComplex instead of +, MultComplex and ScaleComplex instead of z*term/k, and Cabs for absolute value. This leads to

```
function ExpComplex(z:complex):complex;
{Post:  Define S(n) = 1+z + z^2/2!  +...+ z^n/n!  and let m be the smallest}
{value positive integer so that |z^m/m!| <= .00001*|S(-1)|.  If m<= 30, then }
{S(m) returned.  Otherwise, S(30) is returned.}
const
   eps = 0.00001; {Termination criteria}
   itmax = 30; {Iteration maximum.}
var
   k:integer;
   term:complex; {The k-th term z^k/k!}
   s:complex; {S(k)}
   Converged:boolean;
begin
   term.r := 1;
   term.i := 0;
   k := 0;
   s := term;
   Converged := false;
   while not Converged do
      begin
         k := k + 1;
         term := ScaleComplex(MultComplex(z,term),1/k);
         Converged := (Cabs(term) <= (eps*Cabs(s))) or (k >= itmax);
         s := AddComplex(s,term);
      end;
   ExpComplex := s;
end
```

PROBLEM 19.9. Modify Example19_4 so that it can be used to affirm *Euler's formula* $e_{ix} = \cos(x) + i \cdot \sin(x)$ for real x.

PROBLEM 19.10. Let a be a nonzero complex number and consider the sequence $z_0, z_1, \ldots$ defined by

$$z_{n+1} = \left(z_n + \frac{a}{z_n}\right)/2$$

where $z_0 = a$. As n gets large $z_n^2 \approx a$ and so the iteration can be used to generate an approximate square root. Develop a function SqrtComplex(z:complex):complex that does this. Terminate the iteration as soon as $|z_n^2 - a| \leq 10^{-5}$ or as soon as $n = 30$, whichever comes first.

PROBLEM 19.11. The n roots of unity are defined by $w^0, w^1, \ldots, w^{n-1}$ where $w = \cos(2\pi/n) + i \cdot \sin(2\pi/n)$. These complex numbers are distinct and have the property that they equal one when raised to the n-th power. Complete the following procedure

```
procedure RootsOfUnity(n:integer; var z:ComplexList0);
{Pre:n>=1}
{Post:  z[k] = w^k where w = cos(2pi/n) + i*sin(2pi/n) for k=0..n-1.}
```

Using this procedure, write an efficient implementation of

```
procedure DFT(n:integer; var A:ComplexArray0);
{Pre:n>=1}
{Post:A[p,q] = w^(pq) where w = cos(2pi/n) + i*sin(2pi/n) for k=0..n-1.}
```

```
program Example19_4;
{Complex Summations}
uses
   Chap19Codes;
var
   AnotherEg:char;
   z,w:complex;
begin
   ShowText;
   AnotherEg := 'y';
   while AnotherEg <> 'n'  do
      begin
         writeln('Enter the real and imaginary parts of a complex number:');
         readln(z.r,z.i);
         w := ExpComplex(z);
         writeln(' ExpComplex.r =', w.r :  13:6);
         writeln(' ExpComplex.i =', w.i :  13:6);
         writeln;
         writeln('Another Example?  (y/n)');
         readln(AnotherEg);
      end
end.
```

The type complex and the function ExpComplex are in the unit Chap19Codes. Sample output:

```
Enter the real and imaginary parts of a complex number:
3    4
                    ExpComplex.r = -13.128825
                    ExpComplex.i = -15.200783

Another Example?  Enter y (yes) or n (no).
n
```

Note that every entry in A is an n-th root of unity.

Complex numbers can be displayed graphically merely by identifying the x and y axes with the real and imaginary parts. Assume the availability of a graphics procedure ShowComplex(z:complex)[1] that displays (as a dot on the screen) the complex number z. Using ShowComplex we are able to explore visually the behavior of various functions that involve complex numbers. For example if we have the function,

```
function Circle(t:real):complex;
{Post:  cos(2pi*t) + i*sin(2pi*t)}
begin
   Circle.r := cos(2*pi*t);
   Circle.i := sin(2*pi*t);
end
```

then the loop

```
for j:=0 to n-1 do
   ShowComplex(Circle(j/n))
```

displays the unit circle.

Complex-valued functions of a complex variable are particularly important in mathematics, e.g.,

$$f(z) = z - \frac{z^4 - 1}{4z^3} = \frac{3z^4 + 1}{4z^3}.$$

Here is an implementation of this function:

```
function Newton(z:complex):complex;
{Post:(3z ^4 +1)/4z^3 }
var
   one:complex;
   zcube,Num,Denom:complex;
begin
   one.r := 1.0;
   one.i := 0.0;
   zcube := MultComplex(z,MultComplex(z,z));
   Num := AddComplex(ScaleComplex(MultComplex(z,zcube),3),one);
   Denom := ScaleComplex(zcube,4);
   Newton := DivComplex(Num, Denom);
end
```

An interesting thing to do with such a function is to plot its *orbit* starting at a particular point in the complex plane. To illustrate what we mean by this, here is a fragment that plots a length-m orbit of Newton starting at the complex number z0:

```
z:=z0;
ShowComplex(z);
for k:=1 to m do
   begin
      z:=Newton(z);
      ShowComplex(z)
   end
```

[1]Behind the scenes here would be the conversion of the real and imaginary parts of z to horizontal and vertical screen coordinates. Not much different than what we discuss in §13.2.

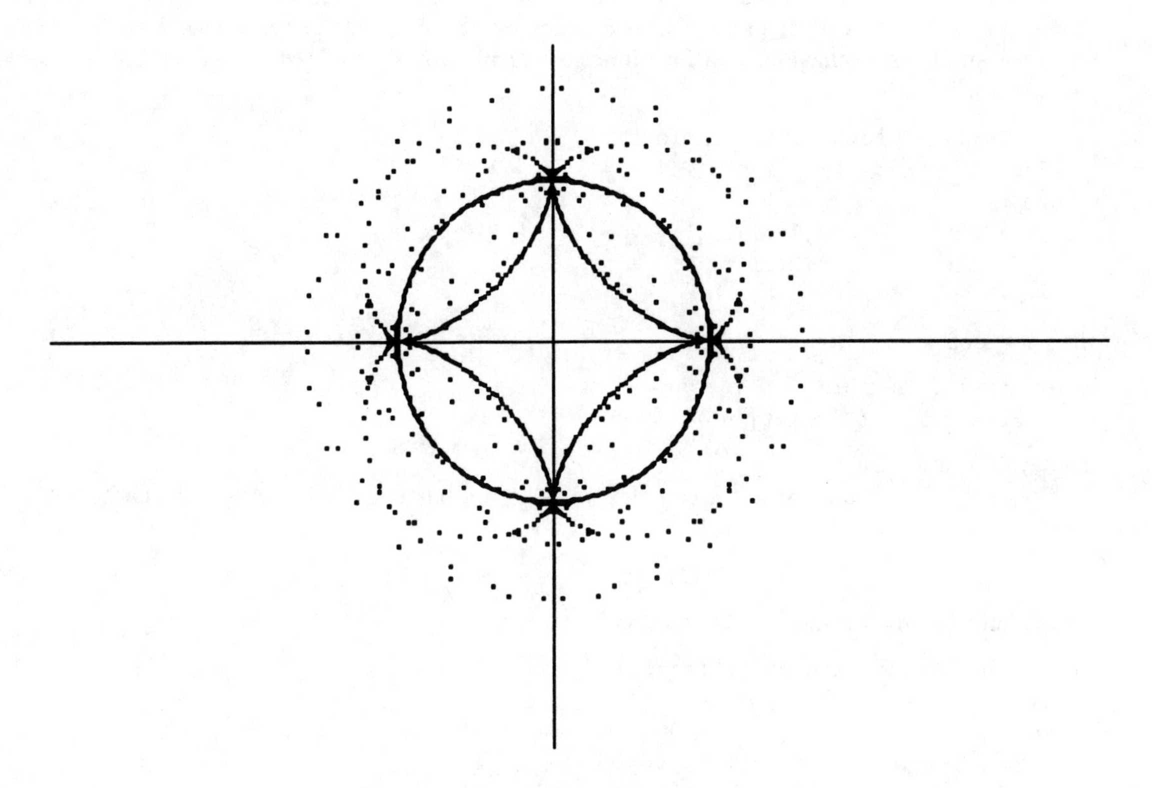

FIGURE 19.1 *Orbits of* `Newton`.

Fascinating patterns emerge when we plot sequences of orbits that start at points along a specified curve. For example, the following fragment plots `n` length `m` orbits of the function `Newton` with starting points that are equally spaced around the unit circle:

```
for j:=0 to n-1 do
    begin
        z:=Circle(j/n);
        ShowComplex(z);
        for k:=1 to m do
            begin
                z:=Newton(z);
                ShowComplex(z)
            end
    end
```

See FIGURE 19.1. The following procedure can be used to plot orbits of a complex function `f(z:complex):complex` with starting points that lie on a curve in the complex plane defined by `g(t:real):complex`.

```
procedure DisplayOrbits(function f(z:complex):complex; function g(t:real):complex;
            n,m:integer; a,b,p:real);
{Pre:  (a+bi) corresponds to the center of the screen.  p is the scale factor}
{in pixels per unit distance along both the real and imaginary axes.}
{f(z) is defined if z = g(t), 0<=t<=1.}
{Post:Plots the length-m orbits the points g(k/n), k=0..n under the action of f.}
const
   h0=260; v0=150; {(h0,v0) = screen center}
var
   k:integer;

   procedure ShowComplex(z:complex);
   {Post:Displays the complex number z}
   var
      h,v:integer; {(h,v) = screen location corresponding to z.}
   begin
      {Display real and imaginary axes.}
      MoveTo(h0,0); LineTo(h0,500);
      MoveTo(0,v0);LineTo(1000,v0);
      {Convert to screen coordinates.}
      h:=round(h0+(z.r-a)*p); v:=round(v0-(z.i-b)*p);
      PaintOval(v-1,h-1, v+1, h+1);
   end;

   procedure ShowOrbit(z0:complex; function f(z:complex):complex; m:integer);
      {Post:Define the complex number w(k) by w(k) = f(w(k-1)) where w(0) = z0.  }
      {Displays the length m orbit w(1),...,w(m).}
   var
      k:integer;
      z,zNext:complex;
   begin
      z:=z0;
      ShowComple(z);
      for k := 1 to m do
         begin
            z := f(z);
            ShowComplex(z);
         end;
   end;

begin
   for k := 0 to n do
      ShowOrbit(g(k/n),f,m)
end
```

Example19_5 illustrates the use of this plotting procedure.

```
program Example19_5;
{Orbits}
uses
   DDCodes,Chap19Codes;
var
   n:integer; {Number of orbits.}
   m:integer; {Orbit Length}
begin
   ShowDrawing;
   m := 2;
   n := 100;
   DisplayOrbits(Newton,Circle,n,m,0,0,80);
   while YesNo('Change m or n?') do
      begin
         n := ReviseInteger('Orbits',n,100,1000,9);
         m := ReviseInteger('Orbit Length',m,0,100,100);
         ClearScreen;
         DisplayOrbits(Newton,Circle,n,m,0,0,80);
      end;
end.
```

The functions/procedures **YesNo**, **ReviseInteger** and **ClearScreen** are declared in **DDCodes**. The functions/procedures **Newton** and **DisplayOrbit** and the type **Complex** are declared in **Chap19Codes**. Output: See FIGURE 19.1.

Chapter 20

Polynomials

A polynomial is a function of the following form:

$$p(x) = a_0 + a_1x + a_2x^2 + \cdots + a_nx^n.$$

If a_n is nonzero, then the *degree* of $p(x)$ is n. An n-degree polynomial has exactly n roots. Some of these roots may be complex, but if the coefficients are real, then the complex roots come in conjugate pairs. For quadratics (degree = 2), cubics (degree = 3) and quartics (degree = 4), there are closed formulae for the roots. A famous theorem by Galois states that no such recipes exist for polynomials having degree ≥ 5. Polynomials are widely used because

- They have a tractable algebra with many interesting and useful properties.
- They are easy to integrate and differentiate.
- They can be used to approximate more complicated functions.
- They are used to build rational functions.

In this chapter we build an appreciation for these things and generally develop an ability to work with this important family of functions.

20.1 Representation and Operations.

A polygon with n sides is defined by its n vertices and for its representation we chose a two-field record. One field was for the number of sides and the other, an array for holding the vertices.

A similar structure is useful for representing a polynomial which is defined by its coefficients. One field for its degree and the other, an array for holding its coefficients: The representation of polynomials is made convenient with the definitions

```
const
   DegreeMax = 20;
type
   CoeffList = array[0..DegreeMax] of real;
   polynomial = record
         deg:integer;
         a:CoeffList
      end
```

If p has type polynomial, then it represents

$$p.a[0] + p.a[1]x + p.a[2]x^2 + \cdots + p.a[p.deg]x^{p.deg} .$$

For example, the fragment

```
p.deg := 3;
p.a[0]:=1; p.a[1]:=0; p.a[2]:=3; p.a[3]:=-5
```

assigns the polynomial $1 + 3x^2 - 5x^3$ to p and may be visualized as follows:

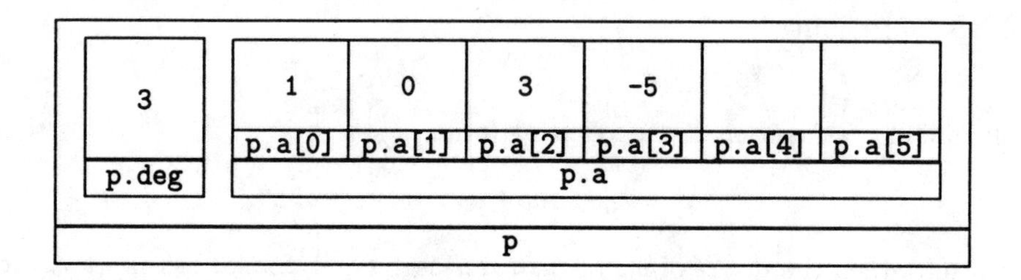

Notice that zero values for zero coefficients must be assigned. However, the values in p.a beyond p.a[3] are ignored since p.deg = 3.

A good way to become familiar with the polynomial type is to develop the following function:

```
function DerPoly(p:polynomial):polynomial;
{Post:The derivative of p.}
```

Recall that if $p(x) = a_0 + a_1x + \cdots + a_nx^n$, then

$$p'(x) = a_1 + 2a_2x + 3a_3x^2 + \cdots + na_nx^{n-1}.$$

Thus,

$$(1 + 3x^2 - 5x^3)' = 6x - 15x^2$$

and the DerPoly mission is as follows:

3	1 | 0 | 3 | −5 | ···	→	2	0 | 6 | −15 | ···
deg	a		deg	a

In general, the degree of $p'(x)$ is one less than the degree of $p(x)$ the coefficient of x^k in $p'(x)$ is $k + 1$ times the coefficient of x^{k+1} in $p(x)$. Noting the special situation when $p(x)$ is a constant polynomial we obtain

```
function DerPoly(p:polynomial):polynomial;
{Post:The derivative of p.}
var
   k:integer;
begin
   if p.deg > 0  then
      {p is not a constant.}
      begin
         for k := 0 to p.deg - 1 do
            {Compute the k-th coefficient of the derivative.}
            DerPoly.a[k]  := p.a[k+1]*(k+1);
         DerPoly.deg  := p.deg-1
      end
   else
      {p is a constant.}
      begin
         DerPoly.a[0]  := 0;
         DerPoly.deg  := 0
      end
end
```

Example20_1 computes the first 10 derivatives of the polynomial

$$p(x) = 1 + x + x^2 + x^3 + x^4 + x^5 + x^6 + x^7 + x^8 + x^9.$$

Notice the factorial-like coefficients that arise during the repeated differentiation.

PROBLEM 20.1. Recall that if $p(x) = a_0 + a_1x + \cdots + a_nx^n$, then

$$\int p(x)dx = a_0x + \frac{a_1}{2}x^2 + \frac{a_2}{3}x^3 + \cdots + \frac{a_n}{n+1}x^{n+1} + const$$

Complete the following function

```
function IntPoly(p:polynomial):polynomial;
{Pre:p.deg < MaxDegree.}
{Post:The indefinite integral of p.  The constant of}
{integration is set to zero.}
```

We now develop functions for polynomial addition and multiplication. To multiply a polynomial by a constant, multiply each coefficient by the constant:

$$2(1 + 2x + 7x^4) = 2 + 4x + 14x^4.$$

To add two polynomials, you add the corresponding coefficients. The degree of the result is the larger of the two degrees:

$$(1 + 2x + 7x^4) + (3x + 5x^3 - x^4 + x^5) = 1 + 5x + 5x^3 + 6x^4 + x^5.$$

Combining these two operations we have the following function:

```
program Example20_1;
{Polynomial differentiation.}
uses
   Chap20Codes;
var
   p:polynomial;
   k,j:integer;
begin
   ShowText;
   {Set up the polynomial 1 + x + x^2 + ...  + x^9 .}
   for k := 0 to 9 do
      p.a[k] := 1;
   p.deg := 9;
   for k := 0 to 10 do
      begin
         if k > 0 then
            p := DerPoly(p);
         write(k:2, ' ');
         for j := 0  to p.deg  do
            write(p.a[j]:7:0);
         writeln;
      end
end.
```

The type `polynomial` and the function `DerPoly` are in the unit `Chap20Codes`. Output:

k-th derivative of 1 + x + x^2 +...+ x^9

k	1	x	x^2	x^3	x^4	x^5	x^6	x^7	x^8	x^9
0	1	1	1	1	1	1	1	1	1	1
1	1	2	3	4	5	6	7	8	9	
2	2	6	12	20	30	42	56	72		
3	6	24	60	120	210	336	504			
4	24	120	360	840	1680	3024				
5	120	720	2520	6720	15120					
6	720	5040	20160	60480						
7	5040	40320	181440							
8	40320	362880								
9	362880									
10	0									

```
function AddPoly(alfa,beta:real; p,q:polynomial):polynomial;
{Post:The sum of the polynomials alfa*p and beta*q.}
var
   k:integer;
   r:polynomial;

begin
   k := 0;
   while  (k <= p.deg) or (k <= q.deg)  do
      begin
         if k<=p.deg then
            begin
               r.a[k] := alfa*p.a[k];
               if k<=q.deg then
                  r.a[k] := r.a[k] + beta*q.a[k]
            end
         else
            r.a[k] := beta*q.a[k];
         k:=k+1;
      end;
   r.deg := k-1;
   AddPoly := r;
end
```

No attempt is made to adjust the degree in the event of fortuitous cancellation. Thus, when $p(x) = x^2 + 2x - 1$ is added to $q(x) = -x^2 + 3x + 4$, AddPoly returns $5x + 3$, but the value assigned to the degree is 2.

The product of two polynomials is more involved and we consider products of the form $(b + cx)p(x)$ first. If $p(x) = a_0 + a_1x + a_2x^2 + a_3x^3$, then

$$(b + cx)p(x) = ba_0 + (ba_1 + ca_0)x + (ba_2 + ca_1)x^2 + (ba_3 + ca_2)x^3 + ca_3x^4.$$

In general,

$$(b + cx)\sum_{k=0}^{n} a_kx^k = ba_0 + \left(\sum_{k=1}^{n}(ba_k + ca_{k-1})x^k\right) + ca_nx^{n+1}$$

and so we have

```
function LinearTimesPoly(b,c:real; p:polynomial):polynomial;
{Pre:  1 + p.deg <= degmax.}
{Post:The polynomial (b+cx)p(x).}
var
   k:integer;
   q:polynomial;
begin
   q.a[0] := b*p.a[0];
   for k:=1 to p.deg do
      q.a[k] := b*p.a[k] + c*p.a[k-1];
   q.a[p.deg+1] := c*p.a[p.deg];
   q.deg := p.deg +1;
   LinearTimesPoly := q;
end
```

Many "structured" polynomials arise in applications and they give us an opportunity to practice with the functions AddPoly and LinearTimesPoly. For example, the *Chebyshev polynomials* $T_0(x), T_1(x), \ldots$ are defined as follows:

$$T_k(x) = \begin{cases} 1 & k = 0 \\ x & k = 1 \\ 2xT_{k-1}(x) - T_{k-2}(x) & k \geq 2 \end{cases}$$

Thus, $T_2(x) = 2x^2 - 1$, $T_3(x) = 2x(2x^2 - 1) - x = 4x^3 - 3x$, etc. Suppose p, q, and r are polynomials and that q and r contain representations of $T_{k-1}(x)$ and $T_{k-2}(x)$ respectively. It follows that

```
p := AddPoly(1,-1,LinearTimesPoly(0,2,q),r)
```

assigns T_k to p. Encapsulating these ideas we obtain

```
function Chebychev(n:integer):polynomial;
{Pre:n>=0}
{Post:The n-th Chebychev polynomial}
var
   k:integer;
   p,q,r:polynomial;
begin
   for k := 0 to  n  do
      if k = 0 then
         begin
            p.deg := 0;
            p.a[0] := 1;
         end
      else if k = 1 then
         begin
            q := p;
            p.deg := 1;
            p.a[0] := 0;
            p.a[1] := 1
         end
      else
         begin
            r := q;
            q := p;
            p := AddPoly(2,-1,LinearTimesPoly(0,1,q),r)
         end;
      Chebychev :=p;
end
```

Notice that the cases $k = 0$ and $k = 1$ are specially handled. After that, the general recursion $T_k = 2xT_{k-1} - T_{k-2}$ is in force.

Example21_2 prints a table of the coefficients of $T_0, T_1, \ldots, T_9$.

```
program Example20_2;
{Examines the Chebyshev polynomials.}
uses
   Chap20Codes;
var
   n,j:integer;
   p:polynomial;
begin
      ⟨:⟩
   for n := 0 to 9  do
      begin
         p := Chebychev(n);
         write(n :  2, ' ');
         for j := 0 to n do
            write(p.a[j]:7:0);
         writeln;
      end
end.
```

Output:

Coefficients of n-th Chebyshev Polynomial

n	1	x	x^2	x^3	x^4	x^5	x^6	x^7	x^8	x^9
0	1									
1	0	1								
2	-1	0	2							
3	-0	-3	0	4						
4	1	-0	-8	0	8					
5	0	5	-0	-20	0	16				
6	-1	0	18	-0	-48	0	32			
7	-0	-7	0	56	-0	-112	0	64		
8	1	-0	-32	0	160	-0	-256	0	128	
9	0	9	-0	-120	0	432	-0	-576	0	256

PROBLEM 20.2. The *Pochhammer polynomial* $P_n(x)$ is defined by

$$P_n(x) = x(x+1)(x+2)\cdots(x+n-1) \;=\; \prod_{j=0}^{n-1}(x+j)$$

Modify Example20_2 so that it print the coefficients of $P_n(x)$ for $n = 0$ to 10.

PROBLEM 20.3. The *Hermite polynomials* are defined as follows:

$$H_k(x) \;=\; \begin{cases} 1 & k=0 \\ 2x & k=1 \\ 2xH_{k-1}(x) - 2(k-1)H_{k-2}(x) & k \geq 2 \end{cases}$$

Modify Example20_2 so that it print the coefficients of $H_k(x)$ for $k = 0$ to 10.

PROBLEM 20.4. The *Legendre polynomials* are defined as follows:

$$L_k(x) \;=\; \begin{cases} 1 & k=0 \\ x & k=1 \\ \dfrac{(2k-1)xL_{k-1}(x) - (k-1)L_{k-2}(x)}{k} & k \geq 2 \end{cases}$$

Modify Example20_2 so that it print the coefficients of $L_k(x)$ for $k = 0$ to 10.

Consider the multiplication of the polynomial

$$p(x) = a_0 + a_1x + a_2x^2 + a_3x^3 + a_4x^4$$

by the monomial x^2:

$$x^2 \cdot p(x) = 0 + 0 \cdot x + a_0x^2 + a_1x^3 + a_2x^4 + a_3x^5 + a_4x^6.$$

In effect, the degree is raised by 2 and the coefficients are "bumped" two terms to the right. In general, if

$$p(x) = \sum_{k=0}^{n} a_kx^k,$$

then

$$x^s \cdot p(x) = \sum_{k=m}^{n+m} a_{k-m}x^k$$

and we have

```
function MonoTimesPoly(p:polynomial; s:integer):polynomial;
{Pre:s>=0}
{Post:p(x)*x^s}
var
   k:integer;
begin
   for k := 0 to s - 1  do
      MonoTimesPoly.a[k]  := 0;
   for k := s to  s + p.deg  do
      MonoTimesPoly.a[k]  := p.a[k - s];
   MonoTimesPoly.deg := s + p.deg;
end
```

The general product

$$r(x) = (a_0 + a_1x + \cdots a_nx^n)(b_0 + b_1x + \cdots + b_mx^m)$$

can be computed as a sum of scaled polynomial-monomial products:

$$\begin{aligned} r(x) &\leftarrow 0 \\ r(x) &\leftarrow r(x) + b_0x^0p(x) \\ r(x) &\leftarrow r(x) + b_1x^1p(x) \\ &\vdots \\ r(x) &\leftarrow r(x) + b_mp(x) \end{aligned}$$

This is the basis of the following function:

```
function MultPoly(p,q:polynomial):polynomial;
{Pre:(p.deg)*(q.deg) <=MaxDegree}
{Post:The product of p and q.}
var
   r:polynomial;
   k:integer;
begin
   r.a[0] := 0;
   r.deg := 0;
   for k := 0 to q.deg do
      r := AddPoly(1,q.a[k],r,MonoTimesPoly(p,k));
   MultPoly := r;
end
```

Example20_3 uses this function to compute powers of $1 + 2x + x^2$.

PROBLEM 20.5. Define the polynomial

$$p_{n,m}(x) = \left(\sum_{k=0}^{n} \frac{(x/m)^k}{k!} \right)^m$$

Write a function ExpPoly(n,m:integer):polynomial that computes this polynomial.

PROBLEM 20.6. It can be shown that

$$(1+x)^n = \sum_{k=0}^{n} (\ n//k \ ) x^k$$

where

$$(\ n//k \ ) = \frac{n!}{k!(n-k)!}$$

is the binomial coefficient n-choose-k. Complete the following function

```
function BinCoeff(n:integer):CoeffList;
{Pre:1<=n<=MaxDegree}
{Post:BinCoeff[k] is binomial coefficient n-choose-k, k=0 to n.}
```

```
program Example20_3;
{Polynomial Multiplication.}
uses
   Chap20Codes;
var
   p,q:polynomial;
   k,j:integer;
begin
   ShowText;
   q.a[0] := 1;
   q.a[1] := 2;
   q.a[2] := 1;
   q.deg := 2;
      ⟨:⟩
   for k := 1 to  5  do
      begin
         if k = 1  then
            p := q
      else
         p := MultPoly(p,q);
      write(k:2, ' ');
         for j := 0 to  p.deg do
            write(p.a[j]:7:0);
         writeln;
      end
end.
```

The type Polynomial and the function MultPoly are in the unit Chap20Codes.

The Powers (1 + 2x + x^2)^k

k	1	x	x^2	x^3	x^4	x^5	x^6	x^7	x^8	x^9	x^10
1	1	2	1								
2	1	4	6	4	1						
3	1	6	15	20	15	6	1				
4	1	8	28	56	70	56	28	8	1		
5	1	10	45	120	210	252	210	120	45	10	1

20.2 Evaluation

The evaluation of a polynomial represented in p at $x = t$ is a straightforward exercise in powers and summation. Indeed, upon termination of

```
val := p.a[0]; power := 1;
for k := 1 to p.deg do
   begin
      power := power*t; {power = t**k }
      val := val + p.a[k]*power;
      {val = the value of the polynomial through the x**k term.}
   end
```

the variable val houses $p(t)$. However, a more efficient scheme is based upon the following nested factorization, which we illustrate for the degree 3 case:

$$a_0 + a_1x + a_2x^2 + a_3x^3 \;=\; ((a_3x + a_2)x + a_1)x + a_0,$$

Thus, to evaluate $p(x)$ at $x = t$ we execute

```
val := p.a[3];
val := t*val + p.a[2];
val := t*val + p.a[1];
val := t*val + p.a[0]
```

In general we have

```
function EvalPoly(p:polynomial; t:real):real;
{Post:The value of p at t.}
var
   val:real;
   k:integer;
begin
   val := p.a[p.deg];
   for k:= p.deg -1 downto 0 do
      val := t*s + p.a[k];
   EvalPoly := val;
end
```

The plotting of a polynomial requires its evaluation at a sufficient number of points across the required range of x-values. Using EvalPoly we obtain

```
function SetUpPoly(n:integer; p:polynomial; xval:RealList0):RealList0;
{Pre:  0<=n<=nmax.  p(x) defined at x = xval[0],...,xval[n].}
{Post:  SetUpPoly[k] = p(x[k]), k=0,...,n.}
var
   k:integer;
begin
   for k := 0  to  n  do
      SetUpPoly[k] := EvalPoly(p, xval[k]);
end
```

```
program Example20_4;
{Plot a Chebychev polynomial}
uses
   DDCodes,Chap20Codes;
var
   n:integer;
   p:polynomial;
   xvals, pvals:RealList0;
begin
   ShowDrawing;
   xvals := xEqual(nmax,-1,1);
   n := 0;
   while YesNo('Try next n?')    do
      begin
         ClearScreen;
         n := n + 1;
         p := Chebychev(n);
         pvals := SetUpPoly(nmax,p,xvals);
         AutoPlot(nmax, xvals, pvals);
         MoveTo(20,150);
         writeDraw('n = ', n :  2);
      end;
end.
```

The type RealList0 and the functions/procedures YesNo, ClearScreen, xEqual, and AutoPlot are defined in the unit DDCodes. The function SetUpPoly is defined in Chap20Codes.

Example20_4 plots a sequence of Chebychev polynomials across the interval [-1,1].

PROBLEM 20.7. Complete the following function:

```
function GivenRootsPoly(n:integer; r:CoeffList):polynomial;
{Pre:n>=1}
{Post:  The polynomial (x-r[1])*(x-r[2])*...*(x-r[n])}
```

PROBLEM 20.8. The *composition* of the polynomial $p(x) = a_0 + a_1x + \cdots + a_nx^n$ with $ax + b$ is is the polynomial

$$p(ax + b) = a_0 + a_1(ax + b) + a_2(ax + b)^2 + \cdots + a_n(ax + b)^n$$

Complete the following function:

```
function ShiftNScalePoly(a,b:real; p:polynomial):polynomial;
{Post:  p(ax+b)}
```

PROBLEM 20.9. Complete the following functions:

```
function IntPoly(p:polynomial):polynomial;
{Pre:p.deg < MaxDegree}
{Post:The indefinite integral of p with zero constant term.}

function DefIntPoly(p:polynomial; a,b:real):real;
{Pre:p.deg < MaxDegree}
{Post:  The integral of p from a to b.}
```

PROBLEM 20.10. Complete the following function:

```
procedure OddEven(p:polynomial; var pe,po:polynomial);
{Pre:p.deg>=1}
{Post:  pe(x) = (p(x)+p(-x))/2 and po(x) = (p(x)-p(-x))/2}
```

PROBLEM 20.11. Modify the procedure **Bisection** in **Chap10Codes** to produce:

```
procedure BisectionPoly(p:polynomial; L,R:real; eps:real; var root,proot:real);
{Pre:p(L)p(R)<0, and eps > 0}
{Post:root is is within eps of a true root and proot = p(root).}
```

PROBLEM 20.12. Modify the procedure **Newton** in **Chap10Codes** to produce:

```
procedure NewtonPoly(p:polynomial; InitialGuess:real; ItMax:integer; eps:real; L,R:real;
        var root,proot:  real; var its:integer);
{Pre:  ItMax > 0, L <= R and eps > 0.}
{ InitialGuess is in [L,R] and is an estimate of a root.}
{Post:  froot = f(root).  Either|froot| <= eps or:}
{ Its = ItMax indicating that too many iterations required.}
{ Its < ItMax indicating that another Newton step would jumps out of [L,R].}
```

20.3 Quotients

The quotient of two polynomials is called a *rational function*, e.g.,

$$r(x) = \frac{1 + 3x^2 - 5x^3}{6 + 5x + x^2}.$$

The zeros of the numerator polynomial are the zeros of the rational, and the zeros of the denominator polynomial are its *poles*. The manipulation of rational functions is made convenient through the definitions

```
CoeffList = array[0..MaxDegree] of real;
polynomial = record
   deg:integer;
   a:CoeffList
end;
RatFunc = record
   num:polynomial;
   denom:polynomial
end
```

where MaxDegree is a constant. As an illustration, here is a fragment that assigns to r the above rational function:

```
r.num.deg:=3;
r.num.a[0]:=1; r.num.a[1]:=0; r.num.a[2]:=3; r.num[3]:=-5;
r.denom.deg:=2;
r.denom.a[0]:= 6; r.denom.a[1]:=5; r.denom[2]:=1;
```

Here is how r can be visualized:

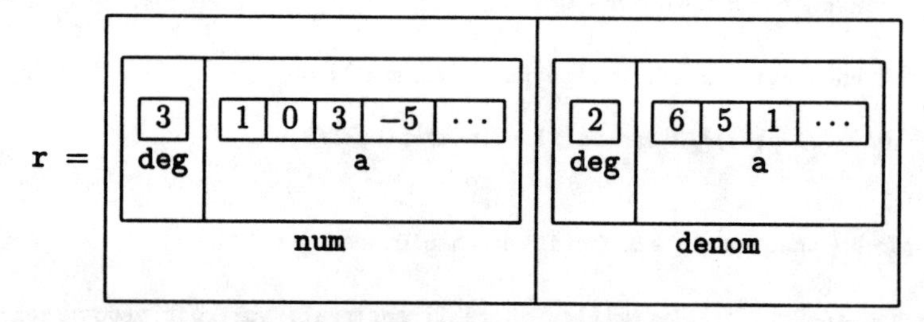

The (p, q) *Pade approximation* to e^x is given by

$$R_{p,q}(x) = \frac{\sum_{j=0}^{p} c_j x^j}{\sum_{j=0}^{q} d_j (-x)^j}$$

where

$$c_j = \frac{(p+q-j)!p!}{(p+q)!j!(p-j)!}$$

$$d_j = \frac{(p+q-j)!q!}{(p+q)!j!(q-j)!}$$

Noting that

$$c_j = \begin{cases} 1 & \text{if } j = 0 \\ \dfrac{j(p-j+1)}{p+q-j+1} c_{j-1} & \text{if } j > 0 \end{cases}$$

and

$$d_j = \begin{cases} 1 & \text{if } j = 0 \\ \dfrac{-j(q-j+1)}{p+q-j+1} d_{j-1} & \text{if } j > 0 \end{cases}$$

Using these recursions, we can encapsulate the setting up a Pade approximate as follows:

```
function Pade(n,m:integer):RatFunc;
{Pre:n,m>=0}
{Post:The (n,m) Pade approximate of exp(x).}
var
   j:integer;
   TopPoly,BottomPoly:polynomial;
begin
   TopPoly.a[0] := 1;
   for j:=1 to n do
      TopPoly.a[j] := (j*(n-j+1)/(n+m-j+1))*TopPoly.a[j-1];
   TopPoly.deg := n;
   BottomPoly.a[0] := 1;
   for j:=1 to m do
      BottomPoly.a[j] := -(j*(m-j+1)/(n+m-j+1))*BottomPoly.a[j-1];
   BottomPoly.deg := m;
   Pade.Num := TopPoly;
   Pade.Denom := BottomPoly;
end;
```

To plot a rational function $r(x)$ we need to set up an array of$r$-values. One approach is to apply EvalPoly to the numerator and denominator:

```
rval := EvalPoly(r.num,t)/EvalPoly(r.denom,t)
```

This assigns the value of $r(t)$ to rval. On the otherhand, we can build upon SetUpPoly as follows:

```
function SetUpRatFunc(n:integer; r:RatFunc; xval:RealList0):RealList0;
{Pre:0<=n<=nmax.  p(x) defined at x = xval[0],...,xval[n].}
{Post:SetUpPoly[k] = p(x[k]), k=0,...,n.}
var
   k:integer;
   TopPolyVals, BottomPolyVals:RealList0;
begin
      TopPolyVals := SetUpPoly(n,r.num,xval);
      BottomPolyVals := SetUpPoly(n,r.denom,xval);
      for k:=0 to n do
         SetUpRatFunc[k] := TopPolyVals[k]/BottomPolyVals[k];
end;
```

```
program Example20_5;
{Pade Plots}
uses
   DDCodes, Chap20Codes;
const
   nF = 100; {Plots based on this many function evaluations.}
   xL = -3;
   xR = 1; {x range for plots is [xL,xR].}
   yB = -1;
   yT = 3; {y range for plots is [yB,yT].}
var
   k:integer;
   More:Boolean; {Continuation indicator.}
   n,m:integer;
   r:RatFunc; {The (n,m) Pade approximate.}
   xvals:RealList0; {Plotting abscissae.}
   rvals,ExpVals:RealList0; {The Pade and exp values.}
begin
   ShowDrawing;
   xvals := xEqual(nF,xL,xR);
   for k := 0 to nF do
      ExpVals[k] := exp(xvals[k]);
   More := true;
   while More do
      begin
         n := ReviseInteger('n',n,0,8,8);
         m := ReviseInteger('m',m,0,8,8);
         Eraserect(0,0,1000,1000);
         r := Pade(n,m);
         rvals := SetUpRatFunc(nF,r,xvals);
         Plot(nF,xvals,rvals,xL,xR,yB,yT);
         PenSize(2,2);
         Plot(nF,xvals,Expvals,xL,xR,yB,yT);
         PenSize(1,1);
         MoveTo(440,160);
         WriteDraw('(',n:1, ',', m:1, ') Approximation');
         More := YesNo('Another Example?');
      end;
end.
```

PROBLEM 20.13. The addition of rational functions is defined by:

$$\frac{p_1(x)}{q_1(x)} + \frac{p_2(x)}{q_2(x)} = \frac{p_1(x)q_2(x) + p_2(x)q_1(x)}{q_1(x)q_2(x)}$$

Complete the function

function `AddRatFunc(r:RatFunc):RatFunc`

Note that the degree of the numerator and denominator of the sum may exceed `MaxDegree`. The precondition must guard against this possibility.

PROBLEM 20.14. The derivative of a rational function is defined by

$$\frac{d}{dx}\left(\frac{p(x)}{q(x)}\right) = \frac{q(x)p'(x) - p(x)q'(x)}{q(x)^2}$$

Complete the function

function `DerRatFunc(r:RatFunc):RatFunc`

Note that the degree of the numerator and denominator of the derivative may exceed `MaxDegree`. The precondition must guard against this possibility.

Chapter 21

Permutations

The when data is re-ordered it undergoes a *permutation*. An ability to compute with permutations and to reason about them is very important in computational science. Unfortunately, it's an activity that is very prone to error because it often involves intricate subscripting. So we start gently by discussing two very straightforward but important permutations: the shift and the perfect shuffle. These operations play a key role in many signal processing applications.

Sorting is by far the most important permutation that arises in applications and three elementary methods are discussed in §21.2: bubble sort, insertion sort, and merge sort. These methods are developed and compared in the context of real arrays. When arrays or records are to be sorted, other issues come to the fore and these are discussed in §21.3.

An important undercurrent throughout the chapter is the concept of data motion. On most advanced machines, program efficiency is a function of how much data flows back and forth between the computer's memory and processing units. The volume of arithmetic is almost a side issue. Because programs that implement permutations deal almost exclusively with moving data, they are good for rounding out our intuition about efficiency.

21.1 Shifts and Shuffles

An array `x[1..n]` is a *permutation* of an array `y[1..n]` if it can be obtained from `y[1..n]` by reordering its values. Thus,

x[1..4] = | 20 | 40 | 10 | 30 |

is a permutation of

y[1..4] = | 40 | 10 | 20 | 30 |.

Note, however, that `x[1..4]` is *not* a permutation of

z[1..4] = | 20 | 40 | 10 | 20 |.

Even though every value in `x[1..4]` appears in `z[1..4]`, it is not a reordering of the values in `z[1..4]`.

We begin our study of permutations with the *left shift* permutation. To fix the discussion, we work with real arrays that are indexed from 1:

```
const
   nmax = 1024;
type
   RealList = array[1..nmax] of real;
```

The *left shift* of a given array `x[1..n]` is obtained by moving each of its values one position left with the value in `x[1]` going to `x[n]`:

| 2 | 1 | 5 | 8 | → | 1 | 5 | 8 | 2 |

Writing this as a procedure we obtain

```
procedure LeftShift(n:integer;var x:RealList);
{Pre:x[1..n] initialized.}
{Post:Its values are permuted so x[k] contains the value originally}
{in x[k+1], x=1..n-1 and x[n] contains the value originally in x[1].}
var
   k:integer;
   t:real;
begin
   t:=x[1];
   for k:=1 to n-1 do
      x[k]:=x[k+1];
   x[n]:=t;
end
```

The procedure works by saving `x[1]`, "bumping" the values in `x[2..n]` one to the left, and then copying `x[1]` into the "freed up" position in `x[n]`[1].

Permutations can also be implemented as functions:

---

[1]Of course, we could work with arrays that are subscripted from zero, and for some permutations like the left shift, this subscript range makes specifications a little easier. Thus, `y[0..n-1]` is a left shift of `x[0..n-1]` if `y[k] = x[(k+1) mod n]`, for $k = 0..n-1$.

```
function LeftShift(n:integer; x:RealList):RealList;
{Pre:x[1..n] initialized.}
{Post:LeftShift[k] = x[k+1] for k=1..n-1 and LeftShift[n]=x[1]}
var
   k:integer;
begin
   for k:=1 to  n-1 do
      LeftShift[k] := x[k+1];
   LeftShift[n] := x[1];
end
```

To ensure adequate experience with both styles of implementation, we'll work with permutation functions in this section and take the procedural approach in §21.2 and §21.3.

The left shift operation is a *one pass* permutation. Each value in the array `x[1..n]` is *referenced* just once. Stated another way, if you were one of the values in `x[1..n]`, then you would "move"' just once from your "home" in `x` to your destination in the target array.

Now suppose that we want to apply the left shift to `x[1..n]` m times. The most obvious implementation is to apply `LeftShift` m times:

```
function MultipleLeftShift1(m,n:integer; x:RealList):RealList;
{Pre:x[1..n] initialized and 1<=m<=n}
{Post:MultipleLeftShift1[k]= x[k+m] for k=1..n-m, and x[k-n+m] for k=n-m+1..n.}
var
   i:integer;
begin
   for i:=1 to m do
      {Perform the i-th shift.}
      x := LeftShift(n,x);
   for i:=1 to n do
      MultipleLeftShift1[i]:= x[i];
end
```

This involves m passes. A better strategy is to move each entry to its final destination directly:

```
function MultipleLeftShift2(m,n:integer; x:RealList):RealList;
{Pre:x[1..n] initialized and 1<=m<=n}
{Post:MultipleLeftShift2[k]= x[k+m] for k=1..n-m, and x[k-n+m] for k=n-m+1..n.}
var
   k:integer;
begin
   for k:=1 to n-m do
      MultipleLeftShift2[k]:=x[k+m];
   for k:=n-m+1 to n do
      MultipleLeftShift2[k]:=x[k-n+m];
end
```

This implementation is more efficient because the amount of *data motion* is reduced by a factor of m. There is only a single pass. Because permutations involve no floating point arithmetic, the efficiency of a particular implementation depends essentially upon the amount of data motion. `Example21_1` benchmarks the two multiple left shift implementations and shows the superiority of `MultipleLeftShift2`.

```
program Example21_1;
{Benchmark multiple shifts.}
uses
   Chap21Codes;
const
   n = nmax; {Size of vector to be shifted.}
   prob_rep = 50; {Benchmark repetition factor.}
   mMax = 10; {Maximum shift factor.}
var
   StartTime,StopTime:real; {Clock snapshots}
   T1,T2:real; {Times for two methods}
   m:integer; {Size of shift.}
   x:RealList; {Array to be shifted.}
   k:integer;
   prob:integer;
begin
   ShowText;
   for k:=1 to n do
      x[k] := k;
   writeln(' m T1/T2');
   writeln('---------------');
   for m := 1 to mMax do
      {Benchmark two methods for performing left shift m times.}
      begin
         StartTime := tickcount;
         for prob := 1 to prob_rep do
            x := MultipleLeftShift1(m,n,x);
         StopTime := tickcount;
         T1 := StopTime - StartTime;
         StartTime := tickcount;
         for prob := 1 to prob_rep do
            x := MultipleLeftShift2(m,n,x);
         StopTime := tickcount;
         T2 := StopTime - StartTime;
         writeln(m:4,(T2/T1):8:3);
      end
end.
```

The functions `MultipleLeftShift1` and `MultipleLeftShift2`, the constant `nmax`, and the type `RealList` are declared in `Chap21Codes`. Sample output:

m	T1/T2
1	1.231
2	2.154
3	2.929
⋮	⋮
9	8.286
10	9.923

PROBLEM 21.1. Write a function, analogous to MultipleLeftShift2, that performs multiple right shifts.

PROBLEM 21.2. Write a function that reverses the order of values in an array. Thus,

2	1	5	8

should be transformed to

8	5	1	2

We next consider a permutation called the *perfect shuffle*. Suppose x[1..n] is initialized and that n is even. The perfect shuffle permutation involves shuffling the first and second halves of x[1..n]. Thus, if

x[1..6] =

10	20	30	40	50	60

then

y[1..6] =

10	40	20	50	30	60

is its perfect shuffle. The formation of the perfect shuffle involves mapping the first and second halves of x onto the odd and even-indexed portions of y:

```
y[1]:=x[1]; y[2]:=x[3+1];
y[3]:=x[2]; y[4]:=x[3+2];
y[5]:=x[3]; y[6]:=x[3+3]
```

In general we have

```
function PerfShuffle(n:integer; a:RealList):RealList;
{Pre:n is even and x[1..n] is initialized.  Set m = n/2}
{Post:For k=1..m, PerfShuffle[2k-1] = a[k] and PerfShuffle[2k] = a[k+m] }
var
   k,m:integer;
begin
   m:=n div 2;
   for k:=1 to m do
      begin
         PerfShuffle[2*k-1] := a[k];
         PerfShuffle[2*k]  := a[k+m];
      end
end
```

An interesting feature of the perfect shuffle is that if it is repeatedly applied, then the original array is retrieved. For example, if $n = 52$ and x[1..52] is initialized, then after 8 perfect shuffles, the original ordering is restored. See Example21_2. The program makes use of the output procedure

```
procedure ShowArray(i,j:integer; a:RealList; p:integer);
{Pre:a[i..j] initialized,p>0}
{Post:Prints a[i..j] on a single line with p:0 format.}
```

and prints the first and second halves of the "card deck" after each shuffle.

```
program Example21_2;
{Illustrate Perfect Shuffle}
uses
   Chap21Codes;
const
   n = 52; {Array length.}
   nPerms = 8; {Number of shuffles to perform.}
var
   m:integer; {n div 2}
   k:integer;
   a:RealList; {The array being shuffled.}
begin
   Showtext;
   m := n div 2;
   {Initialize and display a[1..n].}
   writeln('Initial array:  ');
   for k := 1 to n do
      a[k] := k;
   ShowArray(1,m,a,3);
   ShowArray(m+1,n,a,3);
   for k := 1 to nPerms do
      {Perform the k-th shuffle.}
      begin
         a := PerfShuffle(n,a);
         writeln;
         writeln('After ', k:2, ' permutations:');
         ShowArray(1,m,a,3);
         ShowArray(m+1,n,a,3);
      end
end.
```

The function **PerfShuffle**, the procedure **ShowArray**, and the type **RealList** are declared in **Chap21Codes**. Output:

```
Initial array:

01 02 03 04 05 06 07 08 09 10 11 12 13 14 15 16 17 18 19 20 21 22 23 24 25 26
27 28 29 30 31 32 33 34 35 36 37 38 39 40 41 42 43 44 45 46 47 48 49 50 51 52

After 1 shuffle:

01 27 02 28 03 29 04 30 05 31 06 32 07 33 08 34 09 35 10 36 11 37 12 38 13 39
14 40 15 41 16 42 17 43 18 44 19 45 20 46 21 47 22 48 23 49 24 50 25 51 26 52

  ⋮
```

PROBLEM 21.3. Write a program that prints the minimum number of times that the perfect shuffle must be applied to a length n array x[1..n] before the original order is restored. Report these integers for $n = 2, 4, \ldots, 100$.

PROBLEM 21.4. Complete the following function:

```
function EvenOdd(n:integer; a:RealList):RealList;
{Pre:n is even and x[1..n] is initialized.  Set m = n/2}
{Post:For k=1..m, EvenOdd[k] = a[2k-1] and EvenOdd[m+k] = a[2k] }
```

Thus, [10 20 30 40 50 60] becomes [10 30 50 20 40 60].

PROBLEM 21.5. The transpose of an n-by-n array A is obtained by swapping the values in A[i,j] and A[j,i] for all i and j that satisfy $1 \le i \le n$ and $1 \le j \le n$. Thus, the transpose of

20	10	30	50
70	30	10	40
80	20	30	90
50	40	60	10

is

20	70	80	50
10	30	20	40
30	10	30	60
50	40	90	10

Define an integer 2-dimensional array type Integer2DArray and complete the following procedure:

```
procedure Transpose(n:integer; var A:Integer2DArray);
{Pre:A[1..n,1..n] initialized}
{Post:A is overwritten with its transpose.}
```

PROBLEM 21.6. If A is an n-by-n array, then its left column shift is obtained by shifting each column one position to the left with column 1 becoming column n. Thus,

20	10	30	50
70	30	10	40
80	20	30	90

is transformed to

10	30	50	20
30	10	40	70
20	30	90	80

Define an integer 2-dimensional array type Integer2DArray and complete the following procedure:

```
procedure ColRotate(n:integer; var A:Integer2DArray);
{Pre:A[1..n,1..n] initialized}
{Post:Column j-1 is replaced by column j for j=2..n.  Column n is replaced by column 1.}
```

21.2 Sorting

We say that real array a[1..n] is *sorted* if the values satisfy

$$a[1] \le a[2] \le \cdots \le a[n]$$

In other words, the values range from smallest to largest. Our goal in this section is to develop procedures that permute the values in a given array a[1..n] so that they are sorted.

Consider the $n = 3$ case. Here is a fragment that sorts a[1..3] by swapping "out-of-order" pairs:

```
if a[1] > a[2] then
   begin
      t:=a[1]; a[1]:=a[2]; a[2]:=t;
   end
{a[1]<=a[2]}
if a[2] > a[3] then
   begin
      t:=a[2]; a[1]:=a[3]; a[3]:=t;
   end
{a[1]<=a[2], a[2]<=a[3]}
if a[1] > a[2] then
   begin
      t:=a[1]; a[2]:=a[3]; a[2]:=t;
   end
{a[1]<=a[2]<=a[3]}
```

The first two if's constitute a pass through the data and upon completion we know that the largest value has been "pushed" to the third position. At that stage, a[1..2] may not be sorted and that's why the third "compare-and-swap" is necessary.

There are two aspects of this fragment that are common to all array sorting procedures. Array entries are compared and array entries are moved. The sequence of the compare-and-swap operations differ from method to method.

Let's generalize from the $n = 3$ fragment, beginning with this sequence of compare and move operations that move the largest value to the end of the array:

```
for k:=1 to n-1 do
   if a[k]>a[k+1] then
      begin
         t:=a[k];
         a[k]:=a[k+1];
         a[k+1]:=t;
      end
```

To illustrate how this loop works, suppose $n = 6$ and

a[1..6] = | 30 | 50 | 10 | 60 | 40 | 20 |.

Five pairwise comparisons are made. Indicating the compared entries in boldface, here is what happens as the loop progresses:

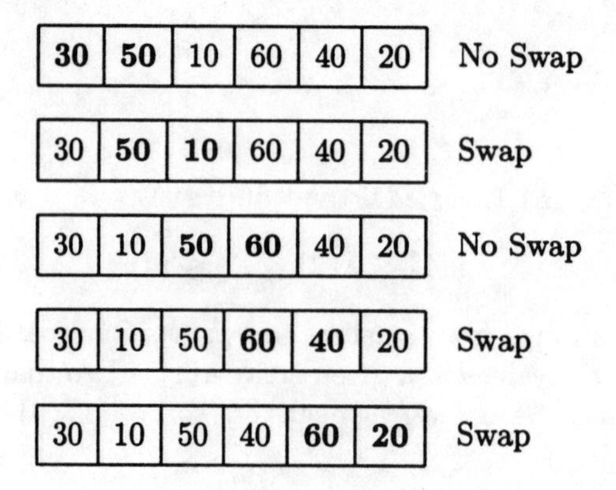

Upon completion the values are arranged as follows:

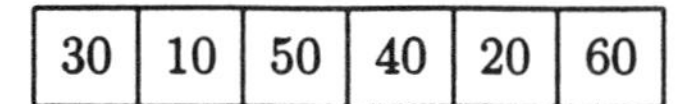

The array is not sorted, but with the reordering of its values, it is clearly *closer* to being sorted than at the start. In particular, the largest value is situated in a[6] and the remaining task is to sort a[1..5]. Thus, we repeat the execution of the loop with the value of n set to 5.

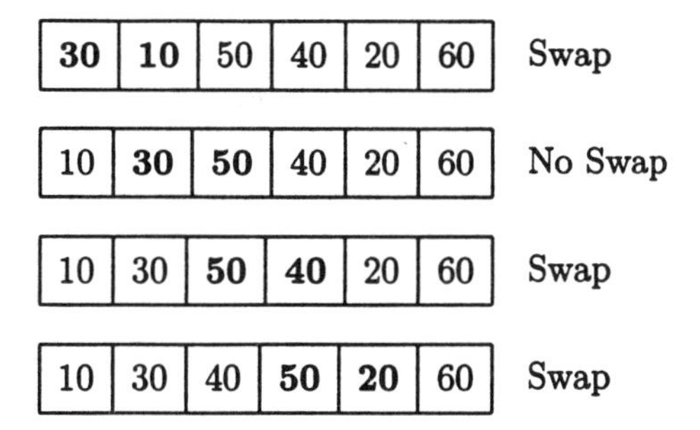

The last two entries are now are now "finalized":

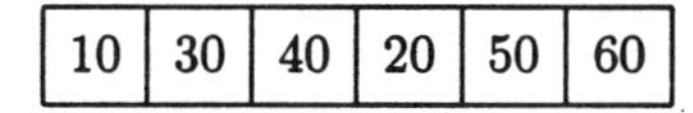

At this point, the pattern should be clear. We compare and swap our way across a[1..4],

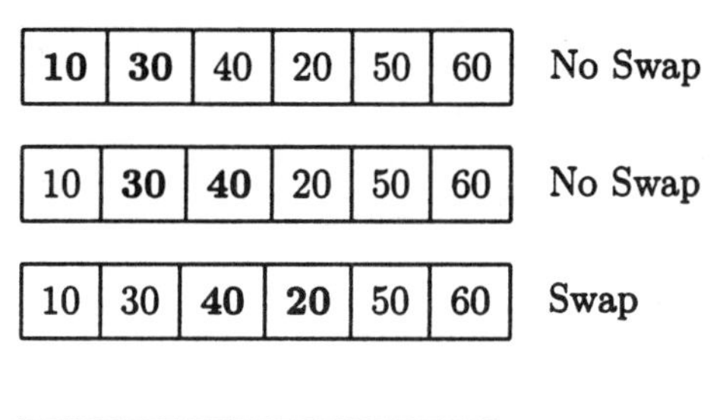

and get

10	30	20	40	50	60

This is followed by a reordering of a[1..3]:

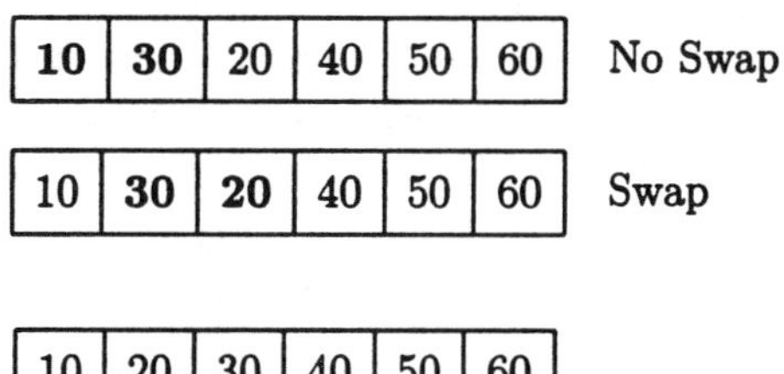

giving

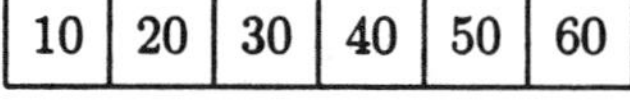

Lastly, we check a[1..2],

10	**20**	30	40	50	60

No Swap

and observe that the array is sorted at last:

10	20	30	40	50	60

We have illustrated the method of *bubble sort*. In general, after $n-1$ passes, the array is sorted. However, if we go through a pass without ever having to swap entries, then a[1..n] is sorted and the process terminates "early". This explains the while-loop in the following implementation[2]:

```
procedure bSort(n:integer; var a:RealList);
{Pre:a[1..n] initialized.}
{Post:a[1..n] is a permutation of a[1..n] and a1] <= a[2] <=...<= a[n].}
var
   pass:integer; {Number of passes through array}
   swaps:integer; {Number of swaps encountered during a pass}
   k:integer; {index}
   MoreToSort:boolean; {Remains true until a pass is completed with no swaps}
   t:real; {Temporary for swapping}
begin
   MoreToSort := true;
   pass := 1;
   while MoreToSort do
      begin
         swaps := 0;
         for k:=1 to n - pass do
            if a[k] > a[k+1] then
               {Swap required.}
               begin
                  t := a[k];
                  a[k] := a[k+1];
                  a[k+1] := t;
                  swaps := swaps + 1;
               end;
         MoreToSort := (swaps > 0);
         pass:=pass+1;
      end;
end
```

Note that the while-loop is guaranteed to terminate no later than after the $(n-1)$-st pass.

In bubble sort, a fair number of comparisons are made that do not result in any swapping. The method of *insertion sort* addresses this short coming and is based on the idea that if a[1..i] is sorted, then the sorting of a[1..i+1] is "easy." In particular, we need only "push" the value of a[i+1] left to its proper place in a[1..i]. Let's reconsider the example we used in the bubble sort discussion:

a[1..6] = | 30 | 50 | 10 | 60 | 40 | 20 |

Setting i=1 we have (obviously) that a[1..i] is sorted. It so happens that by comparing a[1] and a[2] that they too are sorted:

30	**50**	10	60	40	20	No Swap

Next we set i = 2 and push a[3] left with a sequence of swaps until a[1..3] is sorted. This

[2]The term "bubble" makes sense if you think of the large values in the array as bubbling up towards the high end of the array.

involves two comparisons and two swaps in our example:

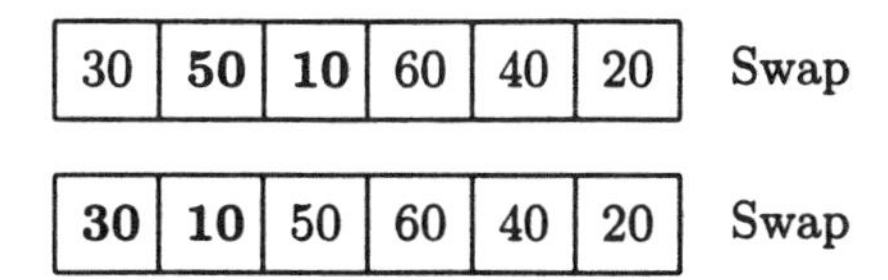

At this point a[1..3] is sorted:

10	30	50	60	40	20

The next task is to reposition the value in a[4]. But the comparison of a[3] and a[4] tells us that there is nothing to do:

10	30	**50**	**60**	40	20	No Swap

With a[1..4] sorted we move the value of a[5]:

10	30	50	**60**	**40**	20	Swap
10	30	**50**	**40**	60	20	Swap
10	**30**	**40**	50	60	20	No Swap

Lastly, we set i = 5 and move a[6]:

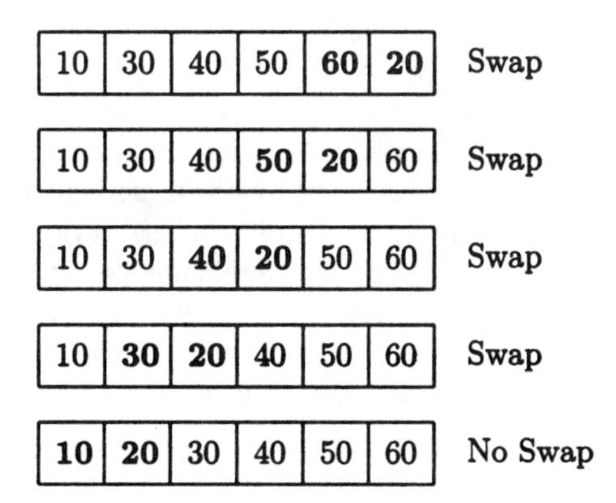

The general idea behind insertion sort should now be apparent:

```
for i:=1 to n-1 do
   Push the value in a[i+1] left through a sequence of swaps
   until a[1..i+1] is sorted.
```

It should be observed that every comparison in the algorithm results in a swap *except* when it is "discovered" that the value being "pushed" is in its proper location. Packaging the whole process in the form of a procedure we obtain the following:

```
procedure iSort(n:integer; var a:RealList);
{Pre:a[1..n] initialized.}
{Post:a[1..n] is a permutation of a[1..n] and a1] <= a[2] <=...<= a[n].}
var
   i,k:integer;
   t:real;
   MoreToSort:boolean;
begin
   for i:=1 to n - 1 do
      {Sort a[1..i+1] assuming that a[1..i] sorted.}
      begin
         k := i;
         MoreToSort := a[k+1] < a[k];
         while MoreToSort do
            begin
               t := a[k];
               a[k] := a[k+1];
               a[k+1] := t;
               k := k - 1;
               if k = 1 then
                  MoreToSort := false
               else
                  MoreToSort := a[k+1] < a[k];
            end;
      end;
end
```

How does this compare with bubble sort? There are several ways to approach this question. From the *worst case* point of view, both methods involve the same number of comparisons and swaps. Just consider the example when `a[1..n]` is sorted from large-to-small. Every comparison in bubble sort results in a swap and the same can be said of insertion sort. In a nutshell, the while-loops in both methods "go the distance." On the other hand, if `a[1..n-1]` is sorted and `a[n]` contains the smallest value in `a[1..n]`, then insertion sort requires about n comparisons and swaps to sort `a[1..n]`. This is an order of magnitude less work than bubble sort. However, it is not fair to portray bubble sort as an order of magnitude slower than insertion sort because this is just not the case on average. `Example21_3` supports this conclusion. The results show that bubble sort is about 70 percent slower than insertion sort when applied to the problem of sorting real arrays. The reason for this is that the latter method involves about half the number of comparisons as the former.

PROBLEM 21.7. Modify `Example21_3` so that instead of timings it displays the number of comparisons and swaps for each sorting method.

```
program Example21_3;
{Compare BubbleSort and InsertionSort}
uses
   DDCodes, Chap21Codes;
const n = 200; {Length of array to be sorted.}
   trialMax = 5; {Number of random trials.}
   seed = 0.329356; {To start the random sequence.}
var
   a,b:RealList; {For the initial and sorted array.}
   r:real; {Current random number}
   i,trial:integer;
   StartTime,StopTime:real; {For clock snapshots.}
   bTime,iTime:real; {BubbleSort and InsertionSort timings.}
begin
   ShowText;
   r := 0.329356;
   writeln('Trial Bubble/Insert Time');
   writeln('-------------------------------');
   for trial := 1 to trialMax do
      begin
         {Set up an array of random numbers.}
         for i := 1 to n do
            begin
               r := rand(r);
               b[i] := r
            end;
         {BubbleSort Benchmark}
         a:=b;
         StartTime := Tickcount;
         bSort(n, a);
         StopTime := Tickcount;
         bTime := (StopTime - StartTime)/60;
         {InsertionSort Benchmark}
         a:=b;
         StartTime := Tickcount;
         iSort(n,a);
         StopTime := Tickcount;
         iTime := (StopTime-StartTime)/60;
         writeln(trial:3, (bTime/iTime):17:3);
      end
end.
```

The functions bSort and iSort are declared in Chap21Codes. Sample output:

```
Trial     Bubble/Insert Time
--------------------------------
  1                    1.718
  2                    1.683
  3                    1.619
  4                    1.605
  5                    1.614
```

We now present the third and last of our sorting frameworks. It is based upon the repeated merging of ever-longer sorted arrays. Suppose we have two sorted arrays `a1[1..5]` and `a2[1..4]`, e.g.,

a1[1..4] = | 20 | 30 | 50 | 65 | a2[1..5] = | 10 | 30 | 45 | 75 | 80 |

Our goal is to *merge* these two arrays into a combined sorted array `b[1..9]` made up of their values:

b[1..9] = | 10 | 20 | 30 | 30 | 45 | 50 | 65 | 75 | 80 |

A step in the merge process involves comparing the smallest "remaining" value in **a1** with the smallest remaining value in **a2** and appropriately storing the lesser of the two compared values in **b**. At the start of the merge we have the arrays **a1**, **a2**, and **b**:

a1				a2					b								
20	30	50	65	**10**	30	45	75	80									

where the smallest remaining values in **a1** and **a2** are indicated by boldface. As a result of comparing `a1[1]` and `a2[1]` the value of the latter is copied into `b[1]`:

a1				a2					b								
20	30	50	65	**10**	30	45	75	80	10								

In this example, the process continues until all the values in **a2** have been selected:

a1				a2					b								
20	30	50	65	10	**30**	45	75	80	10	20							
20	**30**	50	65	10	**30**	45	75	80	10	20	30						
20	**30**	50	65	10	30	**45**	75	80	10	20	30	30					
20	30	**50**	65	10	30	**45**	75	80	10	20	30	30	45				
20	30	**50**	65	10	30	45	**75**	80	10	20	30	30	45	50			
20	30	50	**65**	10	30	45	**75**	80	10	20	30	30	45	50	65		

The remaining two values in **a2** are then copied into **b**:

a1				a2					b								
20	30	50	65	10	30	45	**75**	80	10	20	30	30	45	50	65	75	
20	30	50	65	10	30	45	75	**80**	10	20	30	30	45	50	65	75	80

There are several ways to encapsulate the process. Thinking ahead to how we intend to use the merge idea, we have chosen to work have a procedure that merges a pair of "adjacent" sorted subarrays. To illustrate with the above example, instead of the separate arrays `a1[1.4]` and `a2[1..5]`, the 9 input values are contained in a contiguous portion of a "master array", e.g., The input data might be housed as follows:

a[24..32] = | 20 | 30 | 50 | 65 | 10 | 30 | 45 | 75 | 80 |

Three indices are required to locate the subarrays to be sorted. One for indicating the start of the first subarray (`p = 24`), one for the end of the first subarray (`m = 27`), and one for the end of the second subarray (`q = 32`).

```
procedure Merge(p,m,q:integer; var a:RealList);
{Pre:p<=m<q and a[p..m] and a[m+1..q] are sorted.}
{Post:Permutes the values in a[p..q] so that they are sorted.}
var
   i1,i2,i:integer;
   b:RealList; {Work array}
begin
   i1 := p;
   i2 := m+1;
   i := p;
   while (i1 <= m) and (i2 <= q) do
      {Neither list is exhausted.}
      if a[i1] <= a[i2] then
         {Select value from a[p..m].}
         begin
            b[i] := a[i1];
            i := i + 1;
            i1 := i1 + 1
         end
      else
         {Select value from a[m+1..q].}
         begin
            b[i] := a[i2];
            i := i+1;
            i2 := i2+1
         end;
   while i1 <= m do
      {a[m+1..q] exhausted.  Copy the remaining portion of a[p..m].}
      begin
         b[i] := a[i1];
         i := i+1;
         i1 := i1+1
      end;
   while i2 <= q do
      {a[p..m] exhausted.  Copy the remaining portion of a[m+1..q].}
      begin
         b[i] := a[i2];
         i := i+1;
         i2 := i2+1
      end;
   for i := p to q  do
      a[i] := b[i];
end
```

The integers `i1`, `i2` and `i` key to understanding the dynamics behind this implementation. The index of the next value to be selected from the first subarray is `i1`. The index of the next value to be selected from the second subarray is `i2`. The index `i` indicates where the next merged value is to be placed in the work array `b`. Note that after the merge is completed, `b[p..q]` is copied into `a[p..q]`.

Repeated merging can be used to sort an entire array a[1..n]. Suppose we start with

a[1..8] =

80	40	30	50	70	10	60	20

Noting that a length-1 array is sorted, we merge a[1] with a[2], a[3] with a[4], a[5] with a[6], and a[7] with a[8]:

```
Merge(1,1,2,a);
Merge(3,3,4,a);
Merge(5,5,6,a);
Merge(7,7,8,a);
```

thereby obtaining

a =

40	80	30	50	10	70	20	60

Note that a is now made up of 4 length-2 sorted subarrays.

The next step is to merge a[1..2] with a[3..4] and a[5..6] with a[7..8]. Two calls to merge are required,

```
Merge(1,2,4,a);
Merge(5,6,8,a);
```

and we now have

a =

30	40	50	80	10	20	60	70

The sort is completed by merging a[1..4] and a[5..8]. After the call

```
Merge(1,4,8,a)
```

we obtain

a =

10	20	30	40	50	60	70	80

Picking up on the patterns we obtain the following implementation of *merge sort* if n is a power of two:

```
procedure mSort(n:integer; var a:RealList);
{Pre:n is a power of two and a[1..n] initialized.}
{Post:a[1..n] is a permutation of a[1..n] and a1] <= a[2] <=...<= a[n].}
var
   i:integer; {Index of the first element in the segment to be merged.}
   L:integer; {Length of subarrays being merged.}
   TwoL:integer; {Two times L.}
begin
   L := 1;
   while L < n do
      begin
         TwoL := 2 * L;
         for i := 1 to n div TwoL do
            Merge((i-1)*TwoL+1, i*TwoL,a);
         L := 2*L;
      end;
end;
```

Let's anticipate why merge sort is much more efficient than either bubble sort or insertion sort. Each time through the `while`-loop, a data item is moved once because it participates in exactly one merge. Thus, the number of passes required by the method equals the number of passes through the `while`-loop. If $n = 2^k$, then it is clear that $k = \log_2 n$ is this number. From what we said about the number of passes as a measure of efficiency, our intuition tells us that merge sort should be faster than our other sorting procedures by a factor of about $(n/\log_2(n))$. `Example21_4` points to this conclusion. The precise $n/\log_2 n$ ratio is not realized, but it should be obvious that as n grows, the merge sort advantage is decisive. The constraint that n be a power of two in `mSort` can be dropped without effectively changing performance. (See Problem 21.8.).

PROBLEM 21.8. Write a merge sort procedure that does not require n to be a power of two. Hint. After "Part A" of the k-th pass, `a[1..n]` should have $n div 2^k$ sorted subarrays each of length 2^k. In Part B, the last of these subarrays should be merged with whatever is left over. For example, if $n = 15$, then

Pass	Sorted Subarrays
1A	a[1..2], a[3..4], a[5..6], a[7..8], a[9..10], a[11..12], a[13..14],a[15..15]
1B	a[1..2], a[3..4], a[5..6], a[7..8], a[9..10], a[11..12], a[13..15]
2A	a[1..4], a[5..8], a[9..12], a[13..15]
2B	a[1..4], a[5..8], a[9..15]
3A	a[1..8],a[9..15]
3B	a[1..15]

In real applications there is more to solving a sorting problem than just picking the underlying algorithm, e.g., bubble sort, insertion sort, merge sort, etc. To stress this point we develop some examples that use the world data set introduced in Chapter 12. Assume the following types

```
country = record
      Name:string[25];
      population:longint;
      area:longint;
      capital:string[25];
      lat:real;
      long:real;
   end;
{If Nation is a country, then Nation.Name is its name, Nation.population}
{is its population, Nation.area is its area in square miles,}
{Nation.capital is the name of its capital, and Nation.lat and Nation.long}
{are the latitude and longitude of its capital in degrees.}
CountryList = array[1..200] of country;
```

and the availability of

```
procedure GetWorldData(var n:integer; var Nation:CountryList);
{Pre:Project must be opened in a folder that contains the file ''World''.}
{Post:World data represented in Nation[1..n].}
```

If `Nation` has type `CountryList`, then the call `GetWorldData(n,Nation)` sets up an array of records that represents the data in the file `World`. A typical entry looks like this

`Nation[32]` =

Canada	27400000	3851809	Ottawa	45.41	75.67

Consider the problem of printing a list of all the countries in order of decreasing population. The procedures of §21.2 are not applicable because they are designed to sort arrays with type

```
program Example21_4;
{Compare InsertionSort and MergeSort}
uses
   DDCodes, Chap21Codes;
const
   seed = 0.329356; {To start the random sequence.}
var
   a,b:RealList; {For the initial and sorted array.}
   r:real; {Current random number}
   i:integer;
   n:integer;
   StartTime,StopTime:real; {For clock snapshots.}
   iTime, mTime:  real; {InsertionSort and MergeSort timings.}
begin
   ShowText;
   r := 0.329356;
   writeln(' n Insert/Merge Time');
   writeln('------------------------------');
   n := 32;
   while n <= 1024 do
      begin
         {Set up an array of random numbers.}
         for i := 1 to n do
            begin
               r := rand(r);
               b[i] := r
            end;
         {InsertionSort Benchmark}
         a:=b;
         StartTime := Tickcount;
         iSort(n,a);
         StopTime := Tickcount;
         iTime := (StopTime - StartTime)/60;
         {MergeSort Benchmark}
         a:= b;
         StartTime := Tickcount;
         mSort(n,a);
         StopTime := Tickcount;
         mTime := (StopTime - StartTime)/60;
         writeln(n:4,(iTime/mTime):17:3);
         n := 2*n;
      end
end.
```

n	Insert/Merge Time
32	0.500
64	3.000
128	4.000
256	5.917
512	11.962
1024	21.411

RealList. However, we can tailor any of these sorting procedures to meet the requirements of this problem. Choosing bSort for simplicity, it can be revised to sort a CountryList array as follows:

```
procedure bSortSpecial(n:integer; var a:CountryList);
{Pre:a[1..n] initialized.}
{Post:The values in a[1..n] are permuted so that for i=1..n-1,}
{a[i].population>=a[i+1].population.}
var
   pass:integer; {Number of passes through array}
   swaps:integer; {Number of swaps encountered during a pass}
   k:integer; {index}
   MoreToSort:boolean; {Remains true until a pass is completed with no swaps}
   t:country; {Temporary for swapping}
begin
   MoreToSort := true;
   pass := 1;
   while MoreToSort do
      begin
         swaps := 0;
         for k:=1 to n - pass do
            if a[k+1].population > a[k].population  then
               {Swap required.}
               begin
                  t := a[k];
                  a[k] := a[k+1];
                  a[k+1] := t;
                  swaps := swaps + 1;
               end;
         MoreToSort := (swaps > 0);
         pass := pass + 1;
      end;
end
```

Since the sort is based upon population, the population field is referred to as the *key field.* Notice that we can adopt another sorting criteria merely by changing the comparison in the if statement. With

```
a[k+1].lat < a[k].lat
```

the countries in a[1..n] are sorted in order of increasing latitude. To arrange the countries from the most densely populated to the least densely populated we use

```
(a[k+1].population/a[k+1].area) > (a[k].population/a[k].area)
```

This suggests that we can impart a measure of generality to bSortSpecial by including a boolean-valued "comparison function" in the parameter list:

```
procedure bSortGeneral(n:integer; var a:CountryList;
                function InOrder(y,z:Country):boolean);
{Pre:a[1..n] initialized.}
{InOrder(y,z) is true if y should come before z in the sort.}
{Post:The values in a[1..n] are permuted so that for k=1..n-1,}
{InOrder(a[k],a[k+1]) is true.}
var
   pass:integer; {Number of passes through array}
   swaps:integer; {Number of swaps encountered during a pass}
   k:integer; {index}
   MoreToSort:boolean; {Remains true until a pass is completed with no swaps}
   t:country; {Temporary for swapping}
begin
   MoreToSort := true;
   pass := 1;
   while MoreToSort do
      begin
         swaps:= 0;
         for k:=1 to n - pass do
            if not InOrder(a[k],a[k+1]) then
               {Swap required.}
               begin
                  t := a[k];
                  a[k] := a[k+1];
                  a[k+1] := t;
                  swaps := swaps + 1;
               end;
         MoreToSort := (swaps > 0);
         pass := pass + 1;
      end;
end;
```

Thus, if

```
function BiggerPop(y,z:Country):boolean;
{Post:The population of y is bigger than the population of z.}
begin
   BiggerPop:=y.population>z.population;
end
```

then the call `bSortGeneral(n,a,BiggerPop)` is equivalent to the call `bSortSpecial(n,a)`.

A list of all the countries in order of decreasing population density would be obtained by passing in the function

```
function MoreDense(y,z:Country):boolean;
{Post:The population of y is bigger than the population of z.}
begin
   MoreDense:= (y.population/y.area) >= (z.population/z.area) ;
end
```

`Example21_5` further illustrates the use of `bSortGeneral`. It prints a list of the countries in order of increasing distance from the capital city to Ithaca, New York. It makes use of the following distance function:

```
function GCircleDist(Lat1,Long1,Lat2,Long2:real):real;
{Pre:(Lat1,Long1) and (Lat2,Long2) are the latitude/longitude coordinates (in degrees)}
{ of two points on Earth.}
{ -90<=Lat<=90,-180<=Long<=180}
{Post:The great circle distance (in miles) between them.}
```

PROBLEM 21.9. Write a program that prints a list of the 50 most populous countries and their population in the millions.

PROBLEM 21.10. Consider the problem of printing a list of the most densely populated countries that have more than 10 million people. The list should range from the most dense country to the least dense country. Implement and benchmark the following two approaches:

- Method 1. Form a list of all counties that have at least ten million. Sort and print this list in order of decreasing population density.
- Method 2. Sort the entire world data list so that the countries are arranged from most dense to least dense. Print only those countries in the list whose populations exceed ten million.

PROBLEM 21.11. Write a program that prints a list of the twenty most populous countries. The list should be in the order of increasing distance to Ithaca as measured from the capital city.

PROBLEM 21.12. Write a program that prints a list of all the countries. The list should be in the order of decreasing distance to the equator.

21.3 Representation

Our thinking about data motion needs to refined with the introduction of sorting programs that move whole records and not just single real numbers, e.g., **bSortGeneral**. The time t that it takes to move a record from one place to another in the computer can be effectively modeled by a simple function that is linear in the "length" of the record. i.e., $\alpha + \beta L$ where

- α is a "start-up" time associated with the initiation of any data movement.
- L measures the amount of data that is moved.
- β is the rate at which the data is moved.

An analogy may be made to airline travel. The time it takes to travel from your home to a destination airport is the time it takes to get to your home airport (e.g., $\alpha = 1$ hour) plus the length of the flight (e.g., $L = 3000$ miles) times the number of hours per mile (e.g., $\beta = 1/500$). Because records can be very long and involve a lot of data, the execution time for **bSortGeneral** depends strongly upon the size of the underlying record. For the world data that we have been using in the examples, an individual record contains about 10 to 20 times the amount of data as an individual real number.

One way to address this problem is to keep track of the re-orderings through an array of indices that indicate the "ranking" of each record. The only data that then moves during the sorting process are the integers in the "ranking" array. We present the idea through the same $n = 6$ example that we used in §21.2. Assume at the start that we have a real array

a[1..6] = | 30 | 50 | 10 | 60 | 40 | 20 |
|---|---|---|---|---|---|

```
program Example21_5;
{Distance to Ithaca}
uses
   DDCodes, Chap21Codes;
const
   IthLat = 42.45;
   IthLong = 76.5;
function CloserCap(y,z:Country):boolean;
{Post:The capital city of y is closer to Ithaca than the capital city of z.}
begin
   CloserCap := GCircleDist(y.Lat,y.Long,IthLat,IthLong) <=
                GCircleDist(z.Lat,z.Long,IthLat,IthLong)
end;
var
   n:integer; {Number of countries.}
   k:integer;
   Nation:  CountryList; {The world data.}
begin
   ShowText;
   GetWorldData(n,Nation);
   writeln('Capital City Distances to Ithaca:');
   writeln;
   bSortGeneral(n,Nation,InOrder);
   for k:=1 to n do
      Writeln(Nation[k].name:30,
         GCircleDist(Nation[k].Lat,Nation[k].Long,IthLat,IthLong):10:0);
end.
```

Output:

```
                       Canada    210
                United States    253
                      Bahamas   1190
                         Cuba   1384
                        Haiti   1692
                            :      :
                     Malaysia   9344
                    Mauritius   9413
                    Singapore   9469
                    Indonesia   9982
                    Australia  10020
```

and an integer "ranking" array that is initialized as follows:

i[1..6] =

1	2	3	4	5	6

The pairwise comparisons now begin. Since the first two values are in order, there is nothing to do and things stay the same:

30	50	10	60	40	20

1	2	3	4	5	6

The second and third values are out of order and so a swap is initiated. *However,* `i[2]` *and* `i[3]` *are swapped, not* `a[2]` *and* `a[3]`:

30	50	10	60	40	20

1	3	2	4	5	6

The next step sheds light on the role of the ranking array. Normally, at this step we would compare `a[3]` and `a[4]`, i.e., the 50 and the 60. This won't work because (in particular), the comparison must be made between `a[2]` and `a[4]`. The key is that `i[3]` and `i[4]` contain the relevant indices and so we compare `a[i[3]]` and `a[i[4]]`. It turns out that values are in order and there is nothing to do. But next we compare `a[i[4]]` and `a[i[5]]`. The values are out of order and we swap `i[4]` and `i[5]`:

30	50	10	60	40	20

1	3	2	5	4	6

Finally, we compare `a[i[5]]` and `a[i[6]]`. Since 60 is bigger than 20 a swap is performed:

30	50	10	60	40	20

1	3	2	5	6	4

This completes one pass.

Several observations are in order. First, the fragment

```
for k:=1 to n do
   writeln(a[i[k]])
```

prints the values 30, 10, 50, 40, 20, and 60. This re-ordering is *exactly* the same as the reordering obtained after one pass in `bSort`. The reordering now is represented by a pair of arrays: `a[1..n]` and `i[1..n]`. The former remains fixed and the latter represents the ranking. Upon completion,

`i[1..6]` =

3	6	1	5	2	4

and `i[k]` contains the index of the k-th smallest value in `a[1..6]`.

We stress again that the values in `a[1..n]` are compared but never moved. All motion is confined to the integer array `i[1..n]` and it is through this array that we access `a[1..n]`. Thus, instead of the loop

```
for k:=1 to n-pass do
   if a[k]>a[k+1] then
      begin
         t:=a[k];
         a[k]:=a[k+1];
         a[k+1]:=t;
      end
```

that we have in bSort, we have

```
for k:=1 to n-pass do
   if a[i[k]]>a[i[k+1]] then
      begin
         t:=i[k];
         i[k]:=i[k+1];
         i[k+1]:=t;
      end
```

Of course, t must be an integer. This leads to the following procedure that returns the ranking array:

```
procedure bSortI(n:integer; a:RealList); var i:IntegerList);
{Pre:a[1..n] initialized.}
{Post:For k=1..n-1, a[i[k]] <= a[i[k+1]]}
var
   i:integerList;
   pass:integer; {Number of passes through array}
   swaps:integer; {Number of swaps encountered during a pass}
   k:integer; {index}
   MoreToSort:  boolean; {Remains true until a pass is completed with no swaps}
   t:integer; {Temporary for swapping}
begin
   MoreToSort := true;
   pass := 1;
   for k:=1 to n do
      i[k] := k;
   while MoreToSort do
      begin
         swaps:=0;
         for k:=1 to n-pass do
            if a[i[k]] > a[i[k+1]] then
               {Swap required.}
               begin
                  t:=i[k];
                  i[k]:=i[k+1];
                  i[k+1]:=t;
                  swaps:=swaps+1;
               end;
         MoreToSort := (swaps > 0);
         pass:=pass+1;
      end;
   bSortI := i;
end;
```

As an application of this procedure, we reconsider Example21_5 which prints a list of countries in order of increasing distance of capital city to Ithaca. The comparison function used in the call to bSortGeneral requires two calls to GCDistance and thus, the overall sorting process involves $O(n^2)$ calls to this rather intensive distance function. This overhead can be reduced by an order of magnitude by computing once and for all, the distance of each capital city to Ithaca and storing

```
program Example21_7;
{Capital City Distance to Ithaca}
uses
   DDCodes,Chap21Codes;
const
   IthLat = 42.45;
   IthLong = 76.5;
var
   n:integer; {Number of countries.}
   k:integer;
   Nation:CountryList; {The world data.}
   a:RealList; {Array of capital distances to Ithaca.}
   i:IntegerList; {Ranking Array}
begin
   ShowText;
   GetWorldData(n,Nation);
   writeln('Capital City Distances to Ithaca:');
   writeln;
   {Compute the distance from each capital to Ithaca.}
   for k:=1 to n do
      a[k] := GCircleDist(Nation[k].Lat,Nation[k].Long,IthLat,IthLong);
   {Sort these values}
   bSortI(n,a,i);
   for k:=1 to n do
      {Print the k-th closest city.}
      Writeln(Nation[i[k]].name:30, a[i[k]]:10:0);
end.
```

RealList, IntegerList, CountryList, GCDistance, GetWorldData, and bSortI are declared in Chap21Codes. Output: See Example21_5.

the results in a real array a[1..n]. If we apply bSortI to this array then we obtain a ranking array whose values define the necessary ranking. See Example2_7.

PROBLEM 21.13. Complete the following procedure:

```
procedure bSortGeneralI(n:integer; a:CountryList;
      function InOrder(y,z:Country): boolean; var i:IntegerList);
{Pre:a[1..n] initialized. InOrder(y,z) is true if y comes before z in the sort.}
{Post:For k=1..n-1, InOrder(a[i[k]],a[i[k+1]]) is true.}
```

Use it to write a program that prints two columns of country names side-by-side. The first column should be the ten largest countries in area and the second column should be the ten smallest countries in area.

Chapter 22

Optimization

Optimization problems involve finding the "best" of "something. The search for the optimum ranges over a set called the *search space* and the notion of best is quantified through an *objective function*. One example encountered in §8.3 is to find the closest point on a line L to a given point P. Thus, L is the search space and Euclidean distance is the objective function. Using the calculus, a formula can be given that explicitly specifies the optimal point. This, however, is not typical. In practice, explicit recipes give way to algorithms and exact solutions give way to approximations. Suboptimal solutions are happily accepted if they are cheap to compute and "good enough."

To clarify these points we describe three different applications in this chapter. Our goal is to show how one goes about solving complicated optimization problems and build an appreciation for their role in computational science. In §22.1 we consider the traveling salesperson problem where the aim is to find the shortest roundtrip path that visits each of n given points exactly once. The search space is huge, consisting of $(n-1)!$ possible itineraries. A brute force search plan that considers every possibility is out of the question except for very small values of n. But with an appropriately chosen computational rule-of-thumb called a heuristic, we show that it is not necessary to scan the entire search space. A good, low-mileage itinerary can be produced relatively cheaply.

In §22.2 we use a small engineering design problem to discuss the important role that constraints play in optimization and how there is often more than one natural choice for an objective function. The problem is to build a 10-sprocket bicycle with a desirable range of gear ratios. As in the traveling salesperson problem, the number of possibilities to consider is huge, although finite. Constraints reduce the size of the search space and but extra care must be exercised to stay within the set of allowable solutions. The application is small as engineering design problems go, but rich enough in complexity to illustrate once again the key role of heuristics.

The last problem we consider is that of enclosing a given set of points with the smallest possible ellipse. In contrast to the previous two problems, this is a continuous optimization problem with a genuinely infinite search space. We set up a graphical environment that facilitates the search for the optimum ellipse.

22.1 Shortest Path

Suppose $P_1, \ldots, P_n$ are cities and that a traveling salesperson must visit each one exactly once during a round trip that starts and ends at one of the cities. To minimize expenses, the trip of minimal length is sought. This is called the *traveling salesperson problem* and is one of the most important and famous problems in computational science.

A path can be represented by a sequence of integers that specifies the order of visitation, e.g., $\{3,2,5,1,4\}$. With this notation we mean that city 3 is followed by city 2, city 2 is followed by city 5, etc. No assumption about the "starting city" is made and thus, $\{2,5,1,4,3\}$, $\{5,1,4,3,2\}$, $\{1,4,3,2,5\}$, and $\{4,3,2,5,1\}$ are equivalent paths. The two paths depicted In FIGURE 22.1 are $\{2,3,1,5,4\}$ and $\{2,5,3,1,4\}$. On the surface, this looks like just another search-for-the-min

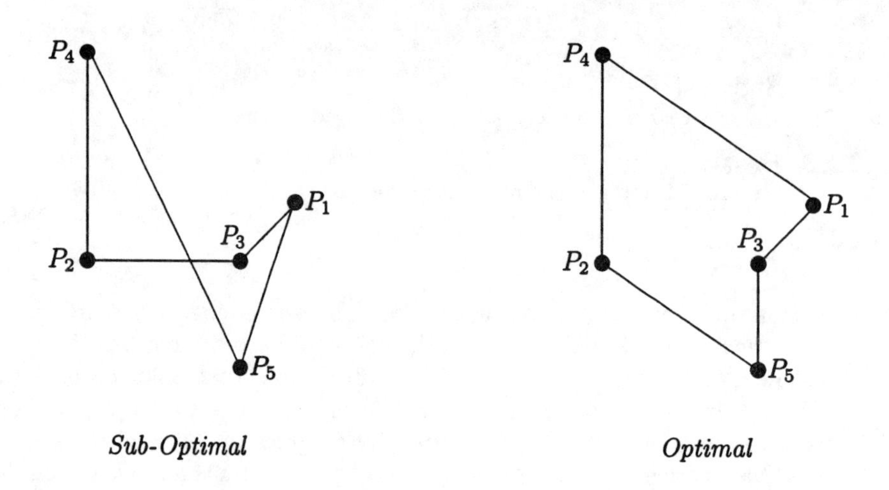

FIGURE 22.1 *Optimal and SubOptimal Paths*

problem. We merely compute all possible paths and pick the one having minimal length. For $n = 5$, there are $(n-1)! = 4! = 24$ possibilities:

$$
\begin{array}{cccccc}
\{1,2,3,4,5\} & \{1,2,3,5,4\} & \{1,2,4,3,5\} & \{1,2,4,5,3\} & \{1,2,5,3,4\} & \{1,2,5,4,3\} \\
\{1,3,2,4,5\} & \{1,3,2,5,4\} & \{1,3,4,2,5\} & \{1,3,4,5,2\} & \{1,3,5,2,4\} & \{1,3,5,4,2\} \\
\{1,4,2,3,5\} & \{1,4,2,5,3\} & \{1,4,3,2,5\} & \{1,4,3,5,2\} & \{1,4,5,2,3\} & \{1,4,5,3,2\} \\
\{1,5,2,3,4\} & \{1,5,2,4,3\} & \{1,5,3,2,4\} & \{1,5,3,4,2\} & \{1,5,4,2,3\} & \{1,5,4,3,2\}
\end{array}
$$

The check-all-itineraries approach is feasible for small n. But the factorial function grows too rapidly for this search plan to be a generally acceptable strategy. A computer that could process a billion itineraries per second would require about a century to solve a 100-city problem and 100 is not even large for the kind shortest path problems that arise in practice. It is interesting to note that although many improved search strategies have been proposed for the travelling salesperson problem, none are appreciably faster than the brute force method.

In order to solve practical problems it is necessary to be content with methods that merely produce "good" itineraries. We describe one such approach that generates a path by repeatedly visiting the nearest unvisited city. Here is the general idea assuming that s is the index of city where the journey begins:

```
Place all n cities on an unvisited city list.
Let the current city be city s and remove it from the unvisited cities list.
for k:=1 to n-1 do
   begin
      Determine the closest unvisited city and let it become the current city.
      Remove the current city from the unvisited cities list.
   end
Return to city s.
```

See FIGURE 22.2 where the next stop after P_3 is to be determined. P_2 is excluded from the search since it has already been visited.

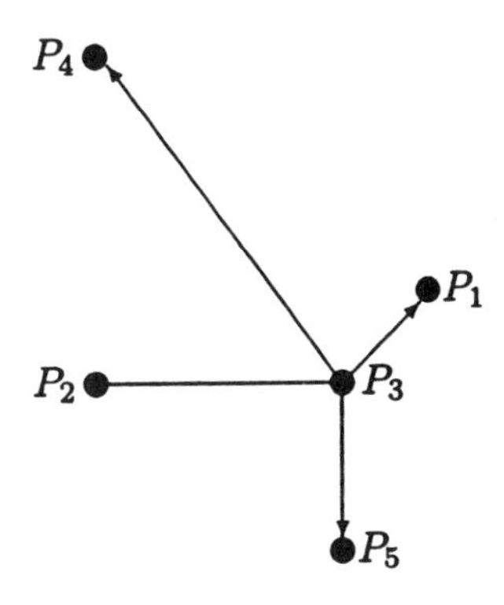

FIGURE 22.2 *Finding the Nearest Unvisited City*

The strategy of visiting the nearest unvisited city is an example of a *heuristic*. A heuristic is a computational rule-of-thumb that nudges us towards optimality, but without any guarantee that it will be achieved. Heuristics evolve from computational experience and intuition and to be effective, they must be relatively cheap to compute. Certainly, the idea of always moving on to the nearest unvisited city is a plausible plan. Let's see how the strategy holds up in practice.

We start with some declarations and from now for the sake of generality we'll talk about "points" instead of cities. To represent the list of points (cities), the list of unvisited points (cities), and the path (itinerary), we make use of the following types:

```
const
   MaxNumPoints = 200;
type
   RealPoint = record
         x,y:real
      end;
   RealPointList = array[1..MaxNumPoints] of RealPoint;
   IntegerList = array[1..MaxNumPoints] of Integer;
   BooleanList = array[1..MaxNumPoints] of Boolean;
```

We adopt the convention that if `UnVisited[k]` is `true`, then the k-th point has not been visited. If this component has the value `false`, then the k-th point has been visited.

The central computation is to find the nearest unvisited point. Assume that the distance function

```
function Distance(p,q:RealPoint):real;
{Post:The distance from p to q.}
```

is available. Recall from §14.2 that if p0 is an arbitrary point and p[1..n] is a list of given points, then

```
dmin:=Distance(p0,p[1]);
for j:=2 to n  do
   begin
      dj:=Distance(p0,p[j]);
      if dj < dmin then
         begin
            dmin:=dj;
            i:=j;
         end;
   end;
```

assigns to i the index of the nearest point to p0 and to dmin the distance from p0 to the nearest point. We *cannot* find the distance to the nearest unvisited point by simply changing the guard in the if to (dj<dmin)and UnVisited[j]. This is because p[1] may be the closest point and also a point that has been visited. One way around this problem is to assign some very large initial value to dmin:

```
dmin:=1000000;
for j:=1 to n do
   begin
      dj:=Distance(p0,p[j]);
      if (dj<dmin) and UnVisited[j] then
         begin
            dmin:=dj;
            i:=j;
         end;
   end
```

Notice that the loop now starts with $j = 1$. Although this style of initialization is simple, it is potentially flawed for there is always the chance that the closest unvisited point to $p0$ is greater than the large value initially assigned to dmin. A better approach is to use a while loop to determine the index of the first unvisited point:

```
first:=1;
while not UnVisited[first] do
   first := first + 1;
for j := first+1  to n  do
   begin
      dj := Distance(p0,p[j]);
      if (dj < dmin) and UnVisited[j] then
         begin
            dmin := dj;
            i := j;
         end;
   end;
```

The efficiency of this fragment can be improved by noting that there is no reason to compute the distance to the jth point unless UnVisited[j] is true. This suggests that we replace the body of the j-loop with

```
if UnVisited[j] then
   begin
      dj:=distance(p0,p[j]);
      if (dj<dmin) then
         begin
            dmin:=dj;
            i:=j;
         end;
   end
```

This is an example of *lazy evaluation*, a principle that says *never compute a quantity until it is absolutely required.* In this case, the distance function is quite cheap to compute. However, lazy evaluation is a good habit to develop because some day you may be computing with much more costly functions and you'll want to be conservative in their use. Encapsulating all of these ideas we obtain:

```
procedure NearesTotSubset(p0:RealPoint; n:integer; p:RealPointList;
            UnVisited:BooleanList; var i:integer; var dmin:real);
{Pre:n>=1 and UnVisited[1..n] and p[1..n] are initialized.  }
{UnVisited[1..n] contains at least one true value.}
{Post:dmin is the distance from p0 to the set of unvisited points among}
{p[1..n] and i is the index of the closest unvisited point.}
var
   first:integer;
   j:integer;
   dj:real; {Distance from p0 to the j-th point.}
begin
   {Compute the index of the first selected point}
   first:=1;
   while  not UnVisited[first] do
      first := first + 1;
   i:=first;
   dmin := Distance(p0,p[first]);
   for j:=first to n do
      begin
         if UnVisited[j] then
          {The j-th point is an UnVisited point.}
           begin
              dj := Distance(p0,p[j]);
              if dj < dmin then
                 begin
                    dmin := dj;
                    i := j
                 end
           end
      end
end;
```

Next, we turn our attention to the representation of a path and here we use an integer array `Path[1..n]`. The idea is simply to store the index of the k-th stop in `Path[k]`. Thus,

`Path[1..5]` = | 3 | 1 | 4 | 2 | 5 |

represents the itinerary Point 3 → Point 1 → Point 4 → Point 2 → Point 5 → Point 3 . As an illustration of this representation, we have the following graphics procedure that displays a given path:

```
procedure ShowPath(n:integer; p:RealPointList; Path:IntegerList);
{Pre:n>=1, p[1..n] and Path[1..n] are initialized.}
{Post:Shows the path whose k-th "stop" is p[k].}
var
   i:integer;
   start:integer; {Index of the point where drawing begins.}
   next:integer; {Index of the i-th point on the path.}
begin
   start := Path[1];
   cMoveToPoint(p[start]);
   cDrawBigDot(p[start]);
   for i:= 2 to n do
      begin
         next := Path[i];
         LineToPoint(p[next]);
         ShowPoint(p[next]);
      end;
   {Draw line back to the first point on the path.}
   LineToPoint(p[start])
end;
```

Notice that integers stored in `Path` are used as subscripts for the `x` and `y` arrays. This is an example of *indirect addressing*. See §21.3.

We now have all the tools necessary to generate the path. At the start of the trip, we make the following initializations assuming that `s` contains the index of the starting point:

```
path[1] := s;
current := s;
for k := 1 to n do
   UnVisited[k] := true;
UnVisited[s] := false;
```

This establishes the current location, the first stop, and the list of unvisited points. The fragment

```
NearestSub(p[current],n,p,UnVisited,next,dmin);
current:=next;
path[2]:=current;
UnVisited[current]:=false;
```

determines the second stop, records its index in `Path[2]`, and updates the `UnVisited` array so that it reflects the fact that the current point has been visited. Likewise, the fragment

```
NearestSub(p[current],n,p,UnVisited,next,dmin);
current:=next;
path[3]:=current;
UnVisited[current]:=false;
```

takes us on to the third stop. The pattern that is emerging can clearly be handled by a for-loop as can be seen in the following procedure:

```
procedure Trip(s,n:integer;p:RealPointList;var Path:IntegerList;
        var odometer:real);
{Pre:n>=1, 1<=start<=n, and p[1..n] is initialized.}
{Post:Path[1..n] is a permutation of 1..n and odometer is }
{the roundtrip distance of the path that visits points Path[1],..,Path[n].}
var
   k:integer;
   current:integer; {Index of the current point.  (The k-1st point.)}
   next:integer; {Index of the next point.  (The k-th point.)}
   dLeg:real; {Distance from current point to next point.}
   UnVisited:BooleanList; {UnVisited[k] is true if k-th point is unvisited.}
begin
   {Initialize list of unvisited points.}
   path[1] := s;
   here := s;
   for k := 1 to n do
      UnVisited[k] := true;
   UnVisited[s] := false;
   odometer := 0;
   for k := 2 to n do
      begin
         NearestSub(p[current],n,p,UnVisited,next,dLeg);
         path[k] := next;
         current := next;
         UnVisited[next] := false;
         odometer := odometer + dLeg;
      end;
   {Return to the first point.}
   odometer := odometer + Distance(p[current],p[s]);
end;
```

An arbitrary starting point is allowed. The itinerary and the total round trip distance are returned. Notice that **UnVisited** is a local array.

Example22_1 permits experimentation with **Trip**. All possible starting points are considered and the one that renders the shortest route is reported. Most examples show that **Trip** can produce different paths from different starting points. This is not surprising because the procedure relies on a heuristic and is not guaranteed to produce the absolutely shortest path. For example, in FIGURE 22.1 the suboptimal path results if P_2 is the starting point and the optimal path is obtained if P_1 is the starting point.

```
program Example22_1;
{Shortest Path}
uses
   DDCodes, Chap18Codes, Chap22Codes;
var
   n:integer; {The number of points.}
   p:RealPointList; {The points.}
   Path:IntegerList; {The itinerary.}
   odometer:real; {Length of the path.}
   Start:integer; {Index of the first point on the path.}
   BestStart:integer; {Index of the starting point of the shortest path.}
   BestLength:real; {Length of the shortest path examined.}
begin
   ShowDrawing;
   SetMap(0,0,10);
   GetManyPoints(n,p);
   start := 1;
   while (start<=n) and YesNo('Try another starting point?') do
      begin
         ClearScreen;
         Trip(start,n,p,Path,odometer);
         ShowPath(n,p,Path);
         MoveTo(10,40);
         writeDraw('Odometer = ',odometer:6:3,'.Index of start = ',start:2);
         if start = 1 then
            begin
               BestLength := odometer;
               BestStart := 1;
            end
         else if odometer < BestLength then
            begin
               BestLength := odometer;
               BestStart := start;
            end;
         start := start + 1;
      end;
   ClearScreen;
   Trip(BestStart,n,p,Path,odometer);
   ShowPath(n,x,y,Path);
   MoveTo(10,40);
   writeDraw('Odometer = ', odometer:6:3, '.  Starting index:', BestStart:2);
end.
```

SetMap and ClearScreen are declared in DDCodes. RealPointList and GetManyPoints are declared in Chap18Codes. IntegerList, Trip, and ShowPath are declared in Chap22Codes. For output see FIGURE 22.3.

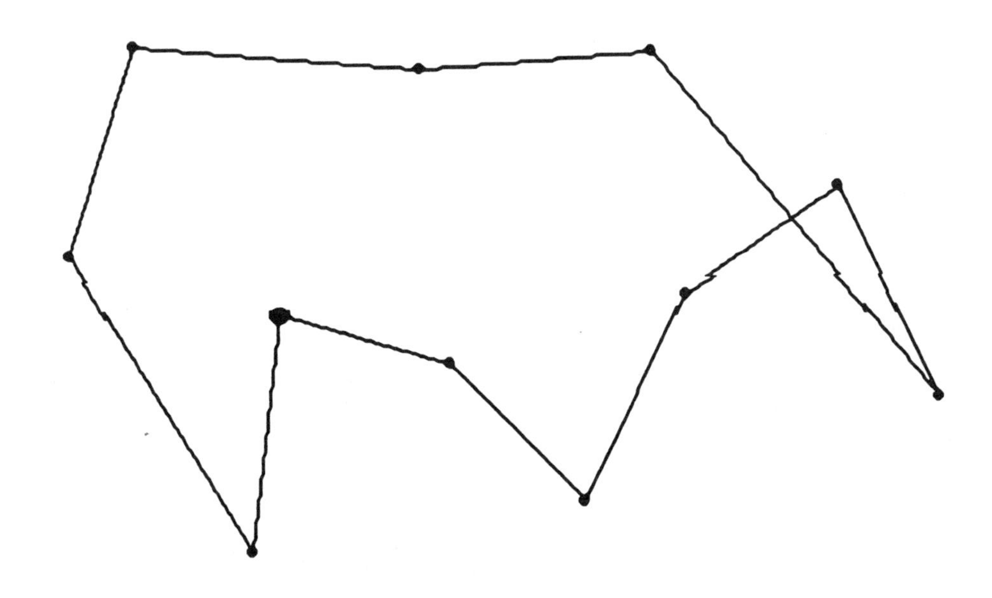

FIGURE 22.3 *Output from* `Example22_1`.

PROBLEM 22.1. Modify `Example22_1` so that it displays all possible paths for the case $n = 5$.

PROBLEM 22.2. What is the average length of the optimal path through n points that are randomly placed in the square $\{(x, y) : -r \le x \le r,\ -r \le y \le r\}$?

PROBLEM 22.3. Modify `Example22_1` so that it displays a low-mileage, one-way, route from the first to the last point.

PROBLEM 22.4. Modify `Trip` and `NearestSub` so that the distance function is passed as a parameter. Use these generalized procedures to compute a low-mileage, roundtrip path though the capitals of the 50 largest countries. Use great circle distance and obtain the necessary data using `GetWorldData`, available through the unit `Chap21Codes`.

22.2 Best Design

A bicycle has 3 *pedal sprockets* and 7 *wheel sprockets.* If the chain goes over a pedal sprocket with p teeth and a wheel sprocket with w teeth, then the *gear ratio* for that pair of sprockets is given by p/w. With three possible numerators and seven possible denominators, we see that there are 21 possible quotients. (The higher the ratio the harder it is to pedal and the further you travel per pedal revolution.) As an example, if pedal sprockets p_1, p_2, p_3 and wheel sprockets $w_1, \ldots, w_7$ are given by

$$[p_1\ p_2\ p_3\ /\ w_1\ w_2\ w_3\ w_4\ w_5\ w_6\ w_7\ ] \;=\; [48\ 45\ 22\ /\ 22\ 19\ 17\ 15\ 14\ 13\ 12\ ]\ ,$$

then the gear ratios are as follows:

	$w_1 = 22$	$w_2 = 19$	$w_3 = 17$	$w_4 = 15$	$w_5 = 14$	$w_6 = 13$	$w_{12} = 12$
$p_1 = 48$	2.18	2.53	2.82	3.20	3.43	3.69	4.00
$p_2 = 45$	2.05	2.37	2.65	3.00	3.21	3.46	3.75
$p_3 = 22$	1.00	1.16	1.29	1.47	1.57	1.69	1.83

Our initial goal is to choose the 10 *sprockets so that the* 21 *gear ratios are as equally spaced as possible across the interval* $[1, 4]$. With perfect spacing, the k-th smallest gear ratio would equal $1 + .15(k - 1)$. This is not quite the case for the above bicycle:

k	p	w	r	Ideal
1	22	22	1.00	1.00
2	22	19	1.16	1.15
3	22	17	1.29	1.30
4	22	15	1.47	1.45
5	22	14	1.57	1.60
6	22	13	1.69	1.75
7	22	12	1.83	1.90
8	45	22	2.05	2.05
9	48	22	2.18	2.20
10	45	19	2.37	2.35
11	48	19	2.53	2.50
12	45	17	2.65	2.65
13	48	17	2.82	2.80
14	45	15	3.00	2.95
15	48	15	3.20	3.10
16	45	14	3.21	3.25
17	48	14	3.43	3.40
18	45	13	3.46	3.55
19	48	13	3.69	3.70
20	45	12	3.75	3.85
21	48	12	4.00	4.00

FIGURE 22.4 *A Sample Bicycle*

Indeed, it is unlikely that we will be able to build the ideal bicycle simply because there are 21 conditions to satisfy but only 10 parameters to "play" with. Thus, we will need to formulate an objective function that rewards proximity to the ideal bicycle.

Before proceeding, we need to establish a type for bicycle representation. There are several plausible choices but we will use a 3-field record of arrays that encodes the pedal sprockets, the wheel sprockets, and the gears:

```
PedalList = array[1..3] of integer;
WheelList = array[1..7] of integer;
Gear = record
        i,j:integer;
        r:real
        {If g is a gear, then g.r is the ratio formed by the pairing}
        {of pedal sprocket i and wheel sprocket j.}
    end;
GearList = array[1..21] of gear;
Bike = record
        p:PedalList;
        w:WheelList;
        g:GearList;
        {If b is a bike, then b.p[1..3] and b.w[1..7] contain the pedal}
        {and wheel sprockets and b.g[1..21] are the gear ratios which we assume}
        {are sorted as follows:  b.g[1].r <= b.g[2].r <= ...  <= b.g[21].r.}
    end
```

A gear itself is a 3-field record that contains the ratio and the indices of the two sprockets that define it.

Suppose b is a variable with type Bike. To represent the bicycle described in FIGURE 22.3, we set

```
b.p[1..3]   =  | 48 | 45 | 22 |

b.w[1..7]   =  | 22 | 19 | 17 | 15 | 14 | 13 | 12 |

b.g[1..21]  =  | | 3 | 1 | 1.00 | | | 3 | 2 | 1.16 | | ··· | | 1 | 7 | 4.00 | |
```

Thus, b.p[2] = 45 because second pedal sprocket has 45 teeth and b.w[6] = 13 because the sixth wheel sprocket has 13 teeth. The second gear ratio is 1.16 because b.g[2].r has that value and is the quotient of b.p[b.g[2].i]] and b.w[b.g[2].j]]. The act of "building" a bike from ten chosen sprocket values involves setting up the three arrays b.p, b.w, and b.g. For this synthesis we assume the availability of the following function:

```
function MakeBike(p:PedalList; w:WheelList):bike;
{Post:The bike with pedal sprockets p[1..3] and wheel sprockets w[1..7]}
```

Note that the computation of the gear ratio list b.g requires the sorting of a length-21 real array. Bubble sort is used in our implementation.

As a measure of nearness to the ideal bicycle, we use the summation

$$|r_1 - 1.00| + |r_2 = 1.15| + \cdots + |r_{21} - 4.00|.$$

where r_k is the k-th smallest gear ratio. To that end we define

```
function f(b:bike):real;
{Post:The distance from b to the ideal bike.}
var
   s:real;
   k:integer;
begin
   s := 0;
   for k:=1 to 21 do
      s := s + abs(1+(k-1)*0.15-b.g[k].r);
   f:=s;
end
```

and set out to find the bicycle that minimizes its value. Other reasonable objective functions exist. See Problems 22.5, 22.6, and 22.7.

Turning our attention to the search, we start by assuming that the number of teeth on any sprocket is between 12 and 52 prompting us to specify the following two constraints:

$$C_1:\quad 12 \le p_i \le 52$$

$$C_2:\quad 12 \le w_i \le 52$$

If we follow the standard look-for-the-min strategy, then we are led to the following fragment[1]:

```
{Initializations}
for p[1]:=12 to 52 do
   for p[2]:=12 to 52 do
      for p[3]:=12 to 52 do
         for w[1]:=12 to 52 do
            for w[2]:=12 to 52 do
               for w[3]:=12 to 52 do
                  for w[4]:=12 to 52 do
                     for w[5]:=12 to 52 do
                        for w[6]:=12 to 52 do
                           for w[7]:=12 to 52 do
                              Check the bike with pedal sprockets
                              p[1..3] and wheel sprockets w[1..7].
```

With (52-12+1) = 41 possible sprocket sizes and 10 sprockets to choose, a fragment of this form would check $41^{10} \approx 10^{17}$ bicycles. Evaluating `f` for each of these and selecting the minimizer solves the problem, but it is not practical. For example, a computer that could evaluate a billion bicycles per second would require about 10 years to complete the task. Thus, we must take steps to constrict the size of the search space.

Observe that the above search strategy has a number of redundancies. It does not make sense to have repeats among either the pedal or wheel sprockets. From the standpoint of minimizing `f`, we can always "do better" with a bike that has distinct pedal and wheel sprockets. It should also be recognized that the value of `f` doesn't change if we reorder the sprockets. The set of 21 gear ratios stays the same and those numbers determine the value of the objective function. Thus, the bicycles [48 45 22 / 22 19 17 15 14 13 12] and [48 22 45 / 19 22 17 15 14 13 12] are equivalent.

[1] ThinkPascal does not allow array variables to be used as loop indices. So this fragment and the modifications of it that follow are not syntactically correct.

To avoid checking equivalent bikes and bikes with repeated sprockets, we impose the following constraints:

$$C_3: \quad 12 \le p_3 < p_2 < p_1 \le 52$$

$$C_4: \quad 12 \le w_7 < w_6 < w_5 < w_4 < w_3 < w_2 < w_1 \le 52$$

By tailoring the loop bounds in the checking fragment to reflect these inequalities we obtain

```
{Initializations}
for p[1]:=12 to 52 do
   for p[2]:=12 to p[1]-1 do
      for p[3]:=12 to p[2]-1 do
         for w[1]:=12 to 52 do
            for w[2]:=12 to w[1]-1 do
               for w[3]:=12 to w[2]-1 do
                  for w[4]:=12 to w[3]-1 do
                     for w[5]:=12 to w[4]-1 do
                        for w[6]:=12 to w[5]-1 do
                           for w[7]:=12 to w[6]-1 do
                              Check the bike with pedal sprockets
                              p[1..3] and wheel sprockets w[1..7].
```

With this fragment the number of checked bicycles is reduced by a factor of about 30,000. Our billion bike/second computer now requires about a day to produce the optimal bike. We mention that conditions C_3 and C_4 do not guarantee 21 distinct gear ratios. They merely keep us away from bikes that do not have 3 distinct pedal sprockets and 7 distinct wheel sprockets.

Sometimes economic constraints are imposed and these to have the effect of reducing the size of the search space. Let's assume that there are only six, economically feasible pedal sprocket sizes:

$$C_5: \quad p_1,\, p_2,\, p_3 \in \{\, 52,\ 48,\ 42,\ 39,\ 32,\ 28 \,\}$$

If we define the integer array

A[1..6] =

52	48	42	39	32	28

then the checking fragment can be implemented as follows:

```
{Initializations}
for i1:=1 to 6 do
   p[1]:= A[i1];
   for i2:=i1+1 to 6 do
      p[2]:=A[i2];
      for i3:=i2+1 to 6 do
         p[3]:=A[i3];
         for w[1]:=12 to 52 do
            for w[2]:=12 to w[1]-1 do
               for w[3]:=12 to w[2]-1 do
                  for w[4]:=12 to w[3]-1 do
                     for w[5]:=12 to w[4]-1 do
                        for w[6]:=12 to w[5]-1 do
                           for w[7]:=12 to w[6]-1 do
                              Check the bike with pedal sprockets
                              p[1..3] and wheel sprockets w[1..7].
```

Note that the first three loop indices "point" to sprocket values and are not sprocket values themselves, another example of *indirect addressing.* At this stage, the number of bicycles to check is about three billion and our computer requires a few seconds to find the best bike.

As a final step in the constriction of the search space, let's assume that a study points to a strong customer preference for bicycles whose lowest and highest gear ratios are exactly 1 and 4 respectively. The "marketing department" therefore prompts us to add the following constraints:

$$C_6:\ p_3/w_1 = 1$$

$$C_7:\ p_1/w_7 = 4$$

The first constraint follows because p_3/w_1 is the ratio of the smallest pedal sprocket to the largest wheel sprocket and so is the minimum gear ratio. Likewise, C_7 forces the largest pedal spocket p_1 to be four times bigger than the smallest wheel sprocket, w_7. Those two sprockets determine the highest gear ratio.

Before we rewrite the checking fragment to incorporate these changes, we make an observation about the possible values that p_1 can take on. Since p_1 is chosen from the set $\{52, 48, 42, 39, 32, 28\}$ and must be divisible by 4, we see that 52, 48, 32, and 28 are the only options. But setting $p_1 = 32$ or 28 implies $w_7 = 8$ or 7, and that is impossible because of the constraint C_4. Thus, we must have either $p_1 = 52$ or $p_1 = 48$. Leaving out the begin's and end's for clarity we obtain

```
{Initializations}
for i1:=1 to 2 do
   p[1]:= A[i1];
   w[7]:= p[1] div 7;
   for i2:=i1+1 to 6 do
      p[2]:=A[i2];
      for i3:=i2+1 to 6 do
         p[3]:=A[i3];
         w[1]:=p[3];
         for w[2]:=12 to w[1]-1 do
            for w[3]:=12 to w[2]-1 do
               for w[4]:=12 to w[3]-1 do
                  for w[5]:=12 to w[4]-1 do
                     for w[6]:=w[7]+1 to w[5]-1 do
                        Check the bike with pedal sprockets
                        p[1..3] and wheel sprockets w[1..7].
```

With this modification, the number of bicycles to be checked is about 3.1×10^5 and our billion bike/second computer requires about a millesecond to perform the search. In that kind of environment we would have to consider the problem as solved. This number of cheap f-evaluations is just not a major computational challenge.

However, it is instructive to proceed under the assumption that the search space is still too large, simply to see once again the role of heuristics. In particular, we'll incorporate some heuristics that reduce the number of possibilities when it comes to choosing $w_2, \ldots, w_6$. The other sprockets are more tightly constrained and do not need any further attention. (It is easy to show that are only 16 possible ways to choose p_1, p_2, and p_3. and once the pedal sprockets are selected, w_1 and w_7 are determined.) $w_2, \ldots, w_6$, because the remaining sprocket values are quite tightly constrained.

Suppose for $i = 2..6$ that we have an estimate of w_i based on intuition and experience. We can bracket our estimate of w_i with an "interval of interest" $[L_i, H_i]$ and confine our search

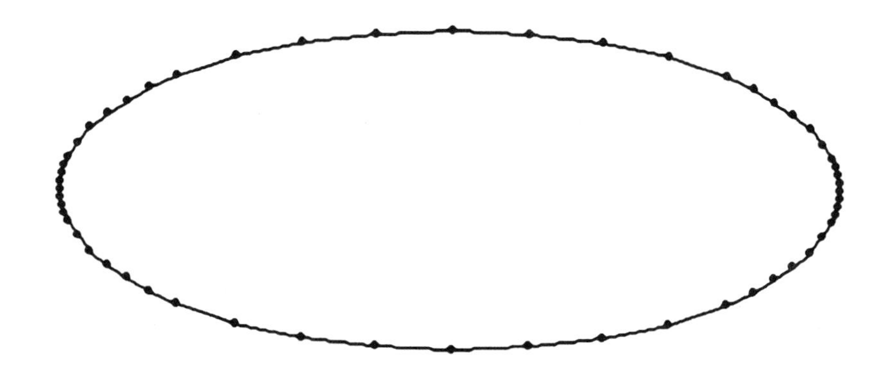

FIGURE 23.4 *Sample* Example23_5 *Output*

PROBLEM 23.11. Modify Example23_5 so that it prints the number of P-evaluations when DrawPathRecur is called.

PROBLEM 23.12. Recall the method of bisection that we developed in §10.x:

```
procedure Bisection(function f(x:real):real; L,R:real;
      eps:real; var root,froot:real);
{Pre:f(x) is continuous on [L,R], f(L)f(R)<0, and eps > 0}
{Post:root is is within eps of a true root and froot = f(root).}
```

Write an equivalent recursive version of this procedure. Take steps to avoid redundant function evaluations.

PROBLEM 23.13. We say that the rectangle with screen vertices at (hL,vT), (hL,vB), (hR,vT), (hR,vB) is *acceptable* if abs(hL-hR)<2, or abs(vT-vB)<2, or hR - hL = vB - vT. In otherwords, a rectangle is acceptable if it is either a square or is "small enough". If a rectangle is unacceptable, then it can be partitioned into a "largest possible" acceptable rectangle with a shared upper left hand corner and a non-acceptable rectangle:

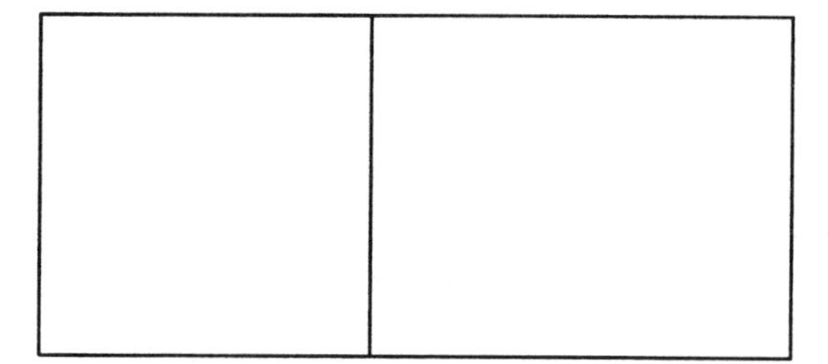

Clearly, the process can be repeated on the unacceptable portion and if we continue the recursion we end up with a partitioning of the original rectangle into acceptable rectangles:

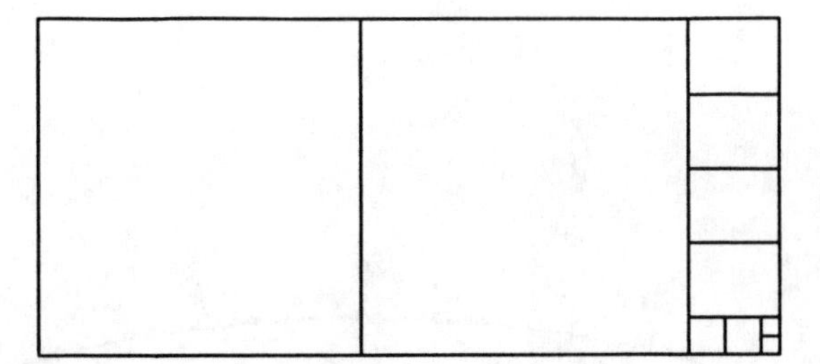

With that in mind, write a recursive procedure `DrawRectPartition(vT,hL,vB,hR:integer)` that displays the rectangle (vT,hL,vB,hR) and its partitioning into acceptable rectangles.

PROBLEM 23.14. Same as Problem 23.13, only center the largest acceptable rectangle inside the given rectangle:

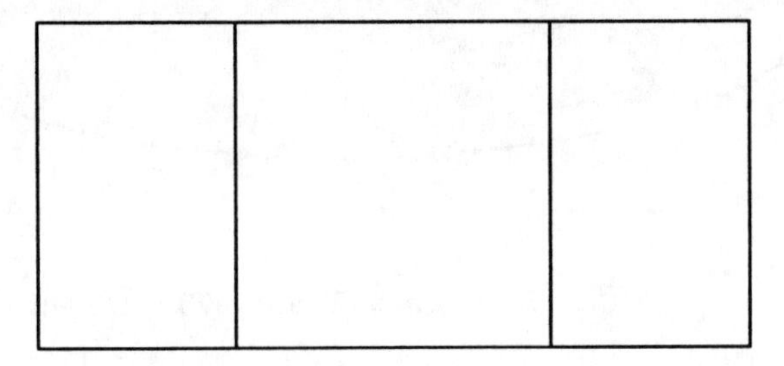

The two non-acceptable rectangles that result must each be partitioned.

PROBLEM 23.15. Let *tol* be a positive real number and suppose triangle T has sides α, β, and γ. We say that T is acceptable if α, β, and γ are all less than *tol*. If T is not acceptable, then the bisector of its largest vertex can be used to split it into two smaller triangles T_1 and T_2. By recurring on this idea we can partition T into the union of acceptable triangles. Write a recursive procedure `Split(T:GridTriangle;tol:real)` that does this. Define the type `GridTriangle` so that there are no calls to the distance function are required inside `Split`.

PROBLEM 23.16. Consider the problem of finding the largest equilateral triangle that fits inside a given triangle:

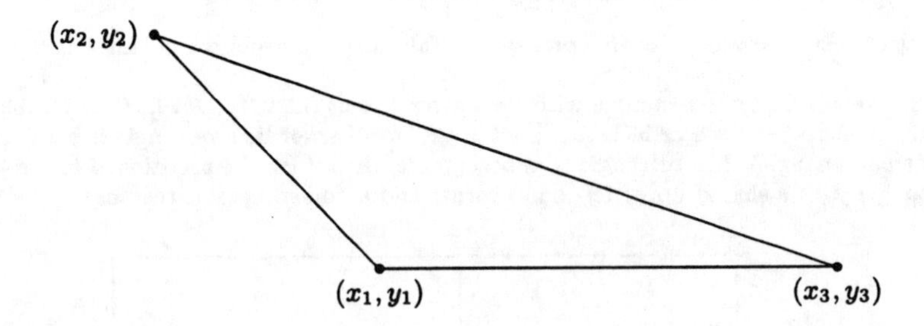

Assume that the side between (x_1, y_1) and (x_2, y_2) is the shortest side and that side between (x_2, y_2) and (x_3, y_3) is the longest side. Define the quantities

$$\alpha = (x_1 - x_3)^2 + (y_1 - y_3)^2$$

$$\beta = (x_2 - x_3)^2 + (y_2 - y_3)^2$$

$$\gamma = (x_1 - x_3)(x_2 - x_3) + (y_1 - y_3)(y_2 - y_3)$$

$$\lambda = 1 - 2\frac{(\alpha\beta - \gamma^2) - \gamma\sqrt{3(\alpha\beta - \gamma^2)}}{\alpha\beta - 4\gamma^2}$$

$$\mu = \alpha\frac{1+\lambda}{2\gamma}$$

It can be shown that if

$$\begin{aligned} u_1 &= = (1-\lambda)x_3 + \mu x_1 \\ v_1 &= = (1-\lambda)y_3 + \mu y_1 \\ u_2 &= = (1-\mu)x_3 + \mu x_2 \\ v_2 &= = (1-\mu)y_3 + \mu y_2 \end{aligned}$$

then the sought-after equilateral triangle has vertices (x_1, y_1), (u_1, v_1), and (u_2, v_2):

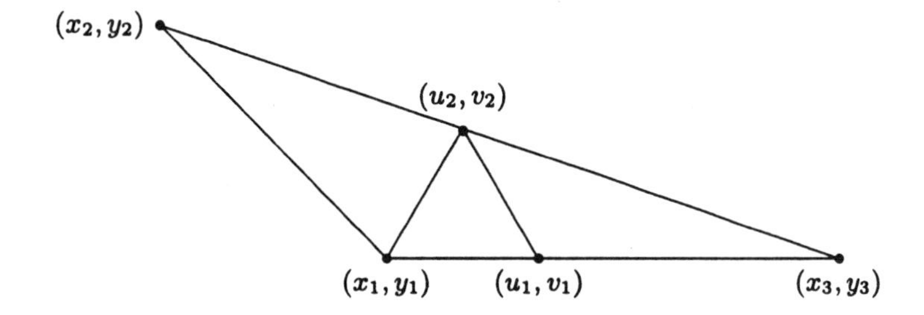

Write a procedure `Split(T:Triangle; var A,B,C:Triangle)` that performs this partition, assigning to `A` the equilateral triangle. After that, write a recursive procedure

```
procedure TriangleMesh(T:triangle;tol:real);
{Pre:tol>0}
{Post:Partitions T into equilateral triangles and very small triangles.}
{ A triangle is small if it has a side whose length is <=tol.}
```

PROBLEM 23.17.

```
procedure Partition(function f(x:real):real; L,fL,R,fR,tol:real;
      nfirst:integer; var n:integer; var x,y:RealList0);
{Pre:nFirst>=0, tol>0, fL=f(L), and fR = f(R).}
{Post:For k=nFirst..n-1, the line segment connecting (x[k],y[k]) and}
{ (x[k+1],y[k+1]) is an acceptable}
{approximation of f(x) on the interval [x[k],x[k+1]].}
```

Chapter 24

Models and Simulation

§**24.1** Prediction and Intuition
Nested loops, boolean expressions, arrays, functions and procedures.

§**24.2** The Effect of Dimension
1, 2, and 3-dimensional arrays.

§**24.3** Building Models from Data
Records of records, functions and procedures.

Scientists use models to express what they know. The level of precision and detail depends upon several factors including the mission of the model, the traditions of the parent science, and the mathematical expertise of the model-builder. When a model is implemented as a computer program and then run, a computer simulation results. This activity is at the heart of computational science and we have dealt with it many times before. In this closing chapter we focus more on the model/simulation "interface" shedding light on how simulations are used, what makes them computationally intensive, and how they are tied up with data acquisition.

Suppose a physicist builds a complicated model that explains what happens to a neutron stream when it bombards a lead shield. A simulation based upon this model could be used to answer a design question: How thick must the shield be in order to make it an effective barrier from the safety point of view? The simulation acts as a predictor. The computer makes it possible to see what the underlying mathematics "says." Alternatively, the physicist may just be interested in exploring how certain model parameters effect the simulation outcome. In this setting the simulation has a more qualitative, intuition-building role to play. The precise value of the numerical output is less important than the relationships that produce it. In §24.1 we examine these two roles that computer simulation can play using Monte Carlo, which we introduced in §6.3.

The time required to carry out a simulation on a grid usually depends strongly upon the grid's dimension. In §24.2 we build an appreciation for this by experimenting with a family of one, two, and three dimensional problems.

In the last section we discuss the role that data plays in model-building. The least squares fitting of a line to a set of points in the plane illustrates that a model's parameters can sometimes be specified as a solution to an optimization problem. A ray tracing application shows how a two-dimensional density model can be obtained by gathering lots of data from one-dimensional snapshots.

24.1 Prediction and Intuition

Scientists use models to describe physical, biological, and social phenomena. Often the model is set forth in the language of mathematics. Expressing the model in the form of a computer program and then running the program results in a *computer simulation.* This typically involves packaging the mathematics into functions and procedures and developing some kind of environment that supports experimentation. For example, we may choose to model the trajectory of a baseball leaving home plate with the equations

$$\begin{aligned} x(t) &= s_0 \cos(\theta)t \\ y(t) &= s_0 \sin(\theta)t - g * t^2/2 \end{aligned}$$

where s_0 is the initial speed, θ is the angle of departure, and g is $32.2 ft/sec^2$. The following procedure encapsulates the model:

```
procedure Baseball(t,s0,theta:real; var x,y:real);
{Pre:s0 = initial speed (feet/sec) and theta = angle of departure (degrees).}
{Post:(x,y) = baseball's position at time t.}
begin
   x := s0*t*cos(pi*theta/180);
   y := s0*t*sin(pi*theta/180) - 16.1*sqr(t);
end
```

To support experiments we implement a special plotting procedure:

```
procedure ShowTrajectory(s0,theta:real);
{Post:Plots the trajectory of a baseball that leaves homeplate with speed s0}
{and angle of departure theta degrees.}
```

See `Example24_1`. Running this program *simulates* the flight of a baseball. The environment is designed to permit easy variation of both the initial speed and the angle of departure, the two parameters that determine the flight of the baseball in this particular model. `ShowTrajectory` displays the position of the baseball every tenth of a second during its airborne flight. The position of home plate and an outfield fence are included to provide a frame of reference. By playing around with `Example24_1`, we discover that the ball travels the farthest when $\theta = 45^o$ and that the trajectory is parabolic. These facts could be confirmed analytically because of the simplicity of the model. However, in a more typical situation, the equations that define the underlying model are much more complex and mathematical intuition could only be acquired by running the simulation.

PROBLEM 24.1. Augment `Example24_1` so that it indicates whether or not the ball clears the fence.

PROBLEM 24.2. Let r_k and f_k be the densities of a rabbit and fox population at time $t_k = k\Delta$ where $\Delta > 0$ is a given time step. Assume that these densities evolve with time as follows:

$$\begin{aligned} r_k &= r_{k-1} + \Delta(2r_{k-1} - \alpha r_{k-1} f_{k-1}) \\ f_k &= r_{k-1} + \Delta(-f_{k-1} + \alpha r_{k-1} f_{k-1}) \end{aligned}$$

Here, α is an "interaction parameter" and it is assumed that the initial densities (r_0 and f_0) are known. This is a standard *predator-prey* model and it can be used to anticipate the future of the two populations. Write a program that simulates the dynamics of the rabbit-fox interaction. Use plots to show what happens. Try the values $\Delta = .05$, $r_0 = 300$, $f_0 = 150$, and $\alpha = 0.1$ and run the simulation for $n = 200$ time steps.

```
program Example24_1;
{Simulate the flight of a baseball.}
uses
   DDCodes;
var
   s0:real; {Initial Speed (feet per second)}
   theta:real; {Angle of Departure (degrees).}
procedure Baseball(t,s0,theta:real; var x,y:real);
      ⟨:⟩
procedure ShowTrajectory(s0,theta:real);
      ⟨:⟩
begin
   ShowDrawing;
   s0 := 100;
   theta := 60;
   while YesNo('Another Try?')  do
      begin
         s0 := ReviseReal('s0',s0,0,200,200);
         theta := ReviseReal('theta',theta,0,90,90);
         ClearScreen;
         ShowTrajectory(s0,theta);
      end;
end.
```

ClearScreen and YesNo are defined in DDCodes. Sample output:

Home Plate

Fence at 400 feet.

Many models are *stochastic* in nature, meaning that they have a probabilistic component. Consider a particle that is about to enter a row of m cells, e.g.,

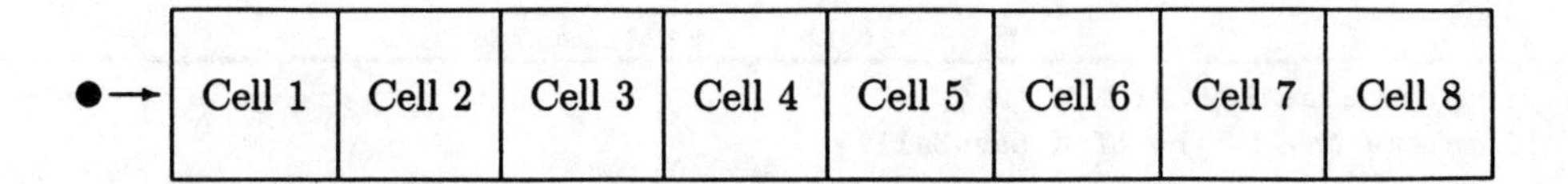

Assume that the particle moves from left to right and when it "visits" a cell, one of two things can happen. *Either* it is absorbed *or* it moves to the next cell with reduced velocity. The probability that it is absorbed increases as it slows down. We would like to know the probability that the particle is never absorbed and exits the last cell. Such a problem might arise if we wanted to design a protective shield that is "thick enough" to stop 99.999999 percent of all the particles.

In order to answer this kind of a question, we set up a very simple absorption model and implement a Monte Carlo simulation. Ignoring units, we assume that the particle enters Cell 1 with velocity $v = 1$. Thereafter, its fate is modeled by the following rules:

(a) If a particle enters a cell with velocity v and is not absorbed, then it moves on to the next cell with velocity $f * v$ where f is a given *absorption factor* that satisfies $0 < f < 1$.

(b) If a particle enters a cell with velocity v, then the probability of absorption is $1 - v^2$.

Let us write a fragment that traces the path of a single particle across the strip of cells subject to these rules. Assume that variables `r`, `f`, and `m` are initialized and respectively house a random number, the absorption factor, and the length of the cell strip. Assuming that $0 < r < 1$, $0 < f < 1$, and $m \geq 1$ we have

```
k:=1;
absorbed:=false;
v := 1.0;
while (k <= m) and (not absorbed) do
   {Particle enters cell k with velocity v.}
   begin
      r := rand(r);
      if r < 1-sqr(v) then
         {Particle is absorbed.}
         absorbed := true
      else
         {Particle is not absorbed.  Slows down and moves to next cell.}
         begin
            v := v*f;
            k := k + 1;
         end;
   end;
```

The `while`-loop body determines what happens to the particle in cell k. At the beginning of the loop, a random number is produced and compared to the value of `1-sqr(v)` where v is the entering velocity. Notice that when it enters the strip, v is 1 and so the particle can never be absorbed in the first cell. However, every time the particle moves to the next cell, its velocity decreases thereby increasing the likelihood that the comparison `r < 1-sqr(v)` is true. The larger the value of the absorption factor `f`, the more rapidly v is reduced. Upon completion of the fragment, the boolean variable **absorbed** indicates what happens to the particle. If it is true, then the particle is absorbed. Otherwise, it escapes the m-th cell. In the latter case the **while** loop is brought to a close because the value of **cell** exceeds m making the boolean expression

```
(cell < m) and (not absorbed)
```

false.

Of course, tracing what happens to a single particle does not tell us very much. To estimate the probability that the particle emerges from the strip we are required to "fire" a large number of particles into the strip and tabulate the number that make it out. To that end, we must set up another loop that oversees the successive firings. Assume that **nmax** houses the number of particles to be traced. The following fragment records in the integer array **count[1..nmax]**, a history of the **nmax** firings:

```
{Initialize count[1..nmax].}
for k:=1 to m do
   count[k] := 0;
for p:=1 to  nmax do
   {Process the p-th particle.}
   begin
      k:=1;
      absorbed:=false;
      v := 1.0;
      while (k <= m) and (not absorbed) do
         {Particle enters cell k with velocity v.}
         begin
            r := rand(r);
            if  r < 1 - sqr(v) then
               {Particle is absorbed.}
               begin
                  absorbed := true;
                  count[k] := count[k] + 1
               end
            else
               {Particle is not absorbed.  Slows down and moves to next cell.}
               begin
                  v:= v*f;
                  k:= k+1;
               end
         end
   end
```

In particular, **count[k]** contains the number of particles that manage to *leave* cell k. In a design situation, we may want to experiment with different absorption factors f. See **Example24_2**.

```
program Example24_2;
{Simulation of particle paths along a strip.}
uses
   DDcodes;
const
   seed = 0.123456;
   nmax = 1000; {Number of trials}
   m = 15; {Maximum number of tiles.}
type
   IntegerList = array[1..m] of integer;
var
   r:real; {For the random number sequence.}
   f:real; {Absorption factor.}
   i,p,k:  integer; {Indices.}
   v:real; {Particle speed.}
   absorbed:boolean; {Absorption flag.}
   count:IntegerList; {Count[k] = number of particles absorbed in k-th cell.}
   numberLeft:integer; {Number of particles that get through k cells.}
begin
   ShowText;
   r := seed;
   f := 1.0;
      ⟨:⟩
   for i:=1 to 10 do
      begin
         f := f - 0.01;
         {Process nmax particles.}
            ⟨:⟩
         writeln;
         write(f:3:2,' | ');
         NumberLeft := nmax;
         for k := 1 to m do
            begin
               NumberLeft := NumberLeft - count[k];
               write(Numberleft:5);
            end
      end
end.
```

Sample output: Entries are the number of particles that continue beyond the m-th cell for the specified f.

f	1	2	3	4	5	6	7	8	9	10	11	12	13	14	15
0.99	1000	981	938	898	828	762	676	588	486	387	310	252	201	140	105
0.98	1000	959	887	795	676	550	425	312	222	152	102	61	37	19	10
0.97	1000	934	836	679	535	401	279	192	107	60	32	10	3	1	0
0.96	1000	915	766	628	445	311	194	119	68	31	14	7	2	1	0
0.95	1000	913	741	546	360	208	114	53	24	8	4	1	0	0	0
0.94	1000	885	704	499	309	156	71	32	11	4	0	0	0	0	0
0.93	1000	847	633	414	226	111	51	15	5	1	0	0	0	0	0
0.92	1000	846	595	370	185	82	30	8	1	0	0	0	0	0	0
0.91	1000	829	569	315	165	72	22	13	3	1	0	0	0	0	0
0.90	1000	833	537	271	121	41	12	3	0	0	0	0	0	0	0

PROBLEM 24.3. This problem is about a very famous stochastic process called a *random walk*. In two dimensions, the idea is simply to take a walk with each step in one of four directions: north, east, south, or west. The chosen direction for each individual step is randomly selected with equal probability. To simulate this with a nice graphical trace of the evolving path, complete the following procedure:

```
procedure DrawRandWalk(hc,vc,s:integer; var steps:longint; var r:real);
{Pre:0<r<1.}
{Post:The value of steps is the number of steps required for a 2-dimensional walk}
{to leave the bounding square (vc-s,hc-s,vc+s,hc+s) after starting at (hc,vc).  The random}
{number stream is generated in r and its value is updated accordingly.}
```

Use the rand function in DDCodes as follows:

```
{(h,v) = current mouse position.}
r:=rand(r);
if r<=0.25 then
   h:=h+1
else if (0.25<r) and (r<=0.50) then
   h:=h-1
else if (0.50<r) and (r<=0.75) then
   v:=v+1
else if (0.75<r) then
   v:=v-1
LineTo(h,v)
```

Write a program that uses DrawRandWalk and sheds light on the average number of steps required before the walk crosses the bounding square. In particular, confirm that this average increases quadratically with s.

An interesting two-dimensional Monte Carlo simulation that physicists use to understand pole alignment in a magnetic substance involves the *Ising model.* The components of an Ising model are (a) an n-by-n array A of *cells* that have one of two states, (b) a probability p, and (c) a temperature T. We use an integer array type

```
IsingArray = array[0..nmax-1,0..nmax-1] of integer;
```

for representing the Ising array with the understanding that the values $+1$ and -1 are used to indicate state. At the start of the simulation, the cells in A[0..n-1,0..n-1] are set to $+1$ with probability p and -1 with probability $1-p$:

```
{0<r<1}
for i:=0 to n-1 do
   for j:=0 to n-1 do
      begin
         r:=rand(r);
         if r<=p then
            A[i,j]:=1
         else
            A[i,j]:=-1
      end
```

During the simulation, the states of the cells change in a probabilistic fashion. Whether or not a particular cell changes state depends upon the temperature and the states of the four neighbor cells. Here is an algorithmic specification of the state change rules:

Repeat:
Choose at random a cell from `A[0..n-1,0..n-1]`. This requires the random selection of two integers i and j from $\{0, 1, \ldots, n-1\}$.
Compute the potential E as follows:

```
North:=A[(n+i-1) mod n,j];
South:=A[(i+1) mod n,j];
West :=A[i,(n+j-1) mod n];
East :=A[i,(j-1) mod n];
E:=A[i,j]*(North+South+West+East);
```

if `E<=0` **then**
`A[i,j]` changes state, i.e., `A[i,j]:=-A[i,j]`.
else
`A[i,j]` changes state with probability `exp(-2*E/T)`.

A single pass through this loop is called a *Metropolis step* and we make three observations. (1) Neighbors are determined using modular arithmetic to handle the case when `A[i,j]` is on the "edge" of `A[0..n-1,0..n-1]`. Thus, the north neighbor of `A[0,j]` is `A[n-1,j]`. (2) `E` is not positive if the state of cell (i,j) disagrees with more than one neighbor state. (3) If `E` is positive, then state (i,j) agrees with at least three of its neighbor states and the chance that its state is flipped increases with temperature.

A number of procedures are available that facilitate the exploration of the Ising simulation. To permit the easy solicitation of the three model parameters we have

```
procedure SetUpIsing(var n:integer; var p,T,r:real; var A:IsingArray);
{Pre:n,p, and T are the current array size, probability factor and }
{temperature.  r contains the current random number.}
{Post:n,p, and T are revised interactively, A[0..n-1,0..n-1] is}
{initialized, and r is contains a new random number.}
```

This procedure also sets up the array of initial states. The graphical display of the array is handled by

```
procedure ShowState(i,j:integer; A:IsingArray);
{Pre:A[i,j] = 1 or -1}
{Post:If A[i,j] = 1, draws a gray square.  Otherwise a white square.}
```

`SetUpIsing` calls this n^2 times to establish the initial display. Thereafter, `ShowState` is called each time a cell state is changed. FIGURE 24.1 shows what the display typically looks like after initialization.

An important statistic that sums up the state of the entire array is the *field strength*, which is amply defined in the specification of the following function:

```
function FieldStrength(n:integer; A:IsingArray):integer;
{Pre:A[0..n-1,0..n-1] initialized.}
{Post:The sum of the values in A[0..n-1,0..n-1].}
```

In the Ising simulation, it is customary to note the field strength after every n^2 Metropolis steps. This is called a *Metropolis sweep* and we encapsulate it as follows:

```
procedure MetropolisSweep(n:integer; p,T:real; var r:real;
      var A:IsingArray; var FS:integer);
{Pre:r is the current random number and A is the current Ising array.}
{Post:r is the new random number, A the new Ising array, and }
{FS is its field strength.}
var
   i,j:integer; {(i,j) = randomly selected cell}
   k:integer; {Number of cells processed.}
   North,South,West,South:integer; {The states of A[i,j]'s neighbors.}
   EP:integer; {Potential of (i,j) cell.}
begin
   for k:=1 to sqr(n) do
      {Metropolis Step.}
      begin
         {Select an array entry at random.}
         r := rand(r); i := trunc(n*r);
         r := rand(r); j := trunc(n*r);
         {Compute Potential}
         North:=A[(n+i-1) mod n,j];
         South:=A[(i+1) mod n,j];
         West:=A[i,(n+j-1) mod n];
         East:=A[i,(j+1) mod n];
         EP := A[i,j]*(North + South + West + East);
         if EP <= 0 then
            {Negative potential, flip state.}
            begin
               A[i,j] := -A[i,j];
               ShowState(i,j,A)
            end
         else
            begin
               r := rand(r);
               if r <= exp(-2*EP/T) then
                  {With probability exp(-2*EP/T), we flip A[i,j]'s state.}
                  begin
                     A[i,j] := -A[i,j];
                     ShowState(i,j, A);
                  end;
            end;
      end;
   FS := FieldStrength(n,A);
end
```

The Ising model captures the idea that subject to random fluctuations, a cell will tend to have the same state as its neighbors, i.e., its magnetic polarity tends to be that of its neighbors. By running **Example24_3** you can witness the dynamics of the alignment process defined by this model.

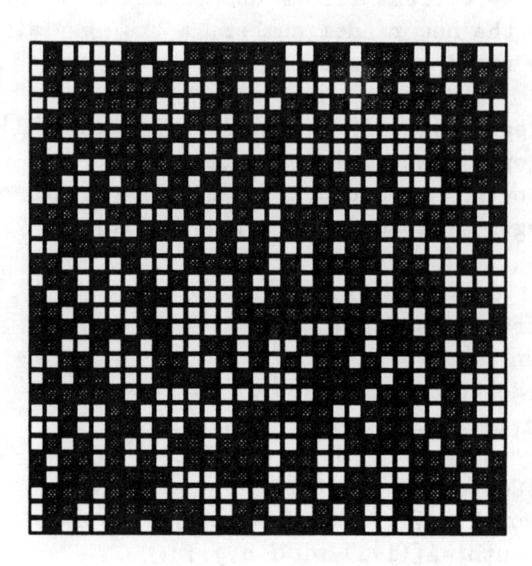

FIGURE 24.1 *Ising Array after Initialization* ($n = 30, p = .49, T = 1.0$).

FIGURE 24.2 *Ising Array after Many Updates* ($n = 30, p = .49, T = 1.0$).

```
program Example24_3;
{Ising Model}
uses
   DDcodes;
const
   nmax = 30; {Max array size.}
   seed = 0.123456;
type
   IsingArray = array[0..nmax, 0..nmax] of integer;
var
   r:real; {Random number.}
   A:IsingArray; {Particle array.}
   k:integer; {Metropolis step counter.}
   i,j:integer; {Indices.}
   n:integer; {Array size.}
   p:real; {A probability.}
   T:real; {Temperature.}
   EP:real; {Potential.}
   Sweep:integer; {Number of Metropolis sweeps.}

function FieldStrength (n:integer; A:IsingArray):integer;
      ⟨:⟩
procedure ShowState(i,j:integer; A:IsingArray);
      ⟨:⟩
procedure SetUpIsing(var n:integer; var p,T,r:real; var A:IsingArray);
      ⟨:⟩
begin
   ShowDrawing;
   {Initializations}
   T:=1.0; p:=0.49; n:=30; r:=seed;
   while YesNo('Another Simulation?')    do
      begin
         {Set parameters and display.}
         ClearScreen;
         SetUpIsing(n,p,T,r,A);
         sweep:=0;
         while not button do
            {Process a Metropolis Sweep.}
            begin
               MetropolisSweep(n,p,T,r,A,FS);
               {Update and display the sweep number and field strength.}
               sweep := sweep + 1;
               EraseRect(80,0,110,150);
               MoveTo(20,90);
               WriteDraw('Completed Sweeps = ', sweep:3);
               MoveTo(20,105);
               WriteDraw('Field Strength = ', FS:5);
            end
      end {Simulation}
end.
```

For sample output, see FIGURE 24.2.

PROBLEM 24.4. Modify `Example24_3` so that it is more efficient in two ways. (a) All the necessary calls to `exp` are precomputed before the loop in `MetropolisSweep`. (b) The field strength should be updated after every cell state change instead of "computed from scratch" at the end of `MetropolisSweep`.

PROBLEM 24.5. Modify `Example24_3` by extending the notion of neighbor so that it includes the "Northwest", "Northeast", "Southwest" and "Southeast" cells. Modify the state change rules accordingly and experiment.

24.2 The Effect of Dimension

If a simulation involves the repeated updating of every value in a d-dimensional array, then the intensity of the computation increases steeply with d. We already have something of an appreciation for this point if $d = 1$ and $d = 2$, having seen many times that memory and time are more problematic in two dimensions than in one. Three-dimensional simulations abound in computational science for the simple reason that the physical world has three dimensions. In this section we try to dramatize the computational intensity of such problems.

To focus on what happens as the dimension grows from 1 to 2 to 3, we have chosen an "averaging" operation that typifies many simulations that arise when a certain broad family of differential equations are solved through discretization. Each component of the array houses the value of some continuous function that is being approximated at a grid point. Simulating how this function varies with time amounts to a sequence of array updates.

We start with a 1-dimensional example and work with the following real array type:

```
Real1DList = array[-n..n] of real;
```

(In all the examples that follow, array size will equal problem size.) Components `A[-n]` and `A[n]` are *boundary values* and the components of `A[-n+1..n-1]` are *interior values*. A time step in the simulation consists of replacing each interior value with the average of itself and its two neighbors. For example, if $n = 3$ we have

Before: $A[-3..3]$ =	3	10	11	6	19	14	0
After: $A[-3..3]$ =	3	8	9	12	13	11	0

Think of the values as temperature and that the update is an attempt to simulate a cooling process. The "11" is replaced by (10+11+6)/3 = 9 because at that point in the grid, the temperature is warmer than in the surrounding neighborhood. The boundary values are left alone. Here is an encapsulation of this operation:

```
procedure Step1D(n:integer; var A:Real1DList);
{Pre:A[-n..n] initialized}
{Post:Each value in A[-n+1..n-1] is replaced by the average of itself and}
{its two neighbors.}
var
   i:integer; {Index}
   B:Real1DList; {Workspace.}
begin
   B:= A;
   for i:=-n+1 to n-1 do
      A[i] := (B[i] + B[i-1] + B[i+1])/3;
end;
```

Suppose we repeatedly apply this procedure to the array `A[-n..n]` with `A[i]` $= \exp(-5(i/n)^2)$. In other words, the array contains a discrete snapshot of the function $\exp(-5x^2)$ across the interval [-1,+1]. If $n = 5$, then here is a trace of the values obtained after the first few updates:

Step	A[-5]	A[-4]	A[-3]	A[-2]	A[-1]	A[0]	A[1]	A[2]	A[3]	A[4]	A[5]
0	.007	.041	.165	.449	.819	1.000	.819	.449	.165	.041	.007
1	.007	.071	.218	.478	.756	.879	.756	.478	.218	.071	.007
2	.007	.099	.256	.484	.704	.797	.704	.484	.256	.099	.007
3	.007	.120	.280	.481	.662	.735	.662	.481	.280	.120	.007
4	.007	.136	.294	.474	.626	.686	.626	.474	.294	.136	.007
5	.007	.145	.301	.465	.596	.646	.596	.465	.301	.145	.007
6	.007	.151	.304	.454	.569	.612	.569	.454	.304	.151	.007
7	.007	.154	.303	.442	.545	.583	.545	.442	.303	.154	.007
8	.007	.154	.300	.430	.524	.558	.524	.430	.300	.154	.007
9	.007	.154	.295	.418	.504	.535	.504	.418	.295	.154	.007

Clearly, the initial "bell shaped" temperature distribution is being damped out. See **Example24_4** where the updates continue until the temperature range `A[0]-A[n]` is reduced to one percent of its original value. The number of array updates required and the time per update are reported.

Next, we turn to the 2-dimensional analog of the above simulation and work with the type

```
Real2DList = array[-n..n,-n..n] of real;
```

Extending our terminology, we say that `A[i,j]` is a boundary value if either $|i| = n$ or $|j| = n$ and an interior value otherwise. The averaging process at an interior value now means replacing `A[i,j]` with the average of itself and its four neighbors `A[i-1,j]`, `A[i+1,j]`, `A[i,j-1]` and `A[i,j+1]`. Thus, if

$$A[-2..2,-2..2] = \begin{array}{|c|c|c|c|c|}\hline 1 & 7 & 2 & 5 & 4 \\ \hline 4 & 9 & 6 & 3 & 8 \\ \hline 3 & 4 & 0 & 3 & 1 \\ \hline 2 & 9 & 2 & 8 & 6 \\ \hline 1 & 3 & 6 & 1 & 9 \\ \hline \end{array},$$

then after an update it becomes

$$A[-2..2,-2..2] = \begin{array}{|c|c|c|c|c|}\hline 1 & 7 & 2 & 5 & 4 \\ \hline 4 & 6 & 4 & 8 & 8 \\ \hline 2 & 5 & 3 & 3 & 1 \\ \hline 2 & 4 & 5 & 4 & 6 \\ \hline 1 & 3 & 6 & 1 & 9 \\ \hline \end{array}.$$

```
program Example24_5;
{One-Dimensional Grid Simulation.}
const
   n = 5; {Problem and Array size.}
type
   Real1DList = array[-n..n] of real;
var
   Range:real; {Initial A[0]-A[n]}
   step:integer; {Number of simulation steps.}
   StartTime,StopTime:  real; {Clock Snapshots}
   A:Real1DList; {For carrying out the grid simulation.}
   i:integer; {Index.}

procedure Step1D(n:integer; var A:Real1DList);
      ⟨:⟩
begin
   ShowText;
   StartTime := TickCount;
   {Set up A, a discrete version of exp(-10x^2) on [-1,1].}
   for i:=-n to n do
      A[i] := exp(-10*sqr(i/n));
   Range := A[0] - A[n];
   step := 0;
   while A[0]-A[n] >= Range/100 do
      {Threshold not reached.  Another simulation step.}
      begin
         Step1D(n,A);
         step := step + 1;
      end;
   StopTime := TickCount;
   writeln(' Steps = ', step:2);
   writeln('Time per step = ', ((StopTime-StartTime)/(60*step)):6:4);
end.
```

Output:

```
        Steps  = 128
Time per step  = 0.0012
```

Here is a procedure that carries out this computation:

```
procedure Step2D(n:integer; var A:Real2DList);
{Pre:A[-n..n,-n..n] initialized}
{Post:A[i,j] is replaced by the average of itself and its 4 neighbors, }
{i=-n+1..n-1,j=-n+1..n-1}
var
   i,j:integer; {index}
   B:Real2DList; {Workspace.}
begin
   {Copy A into B.}
   B := A;
   for i:=-n+1 to n-1 do
      for j:=-n+1 to n-1 do
         A[i,j] := (B[i,j] + B[i-1,j] + B[i+1, j] + B[i,j-1] + B[i,j+1])/5;
end;
```

Notice that it uses a workspace. If Step2D is repeatedly applied, then the values tend to converge to a steady state equilibrium. In analogy with **Example24_4**, **Example24_5** starts with a discrete representation of the function $\exp(-5(x^2+y^2))$ and iterates until the range of values is reduced by a factor of 100.

PROBLEM 24.6. Make **Example24_5** more efficient by exploiting symmetry. If successful, then your modification should require one-fourth the storage, only $O(n)$ exp evaluations, and one-fourth the arithmetic per array update.

Lastly, we cover the 3-dimensional case. We have not yet encountered 3-dimensional arrays, but their typing is predictable:

```
Real3DList = array[-n..n,-n..n,-n..n] of real;
```

Three subscripts are required to specify a component, e.g., A[i,j,k]. Here is a fragment that stores in A[i,j,k] the value of $\exp(-5(x^2+y^2+z^2))$ at $(x_i, y_j, z_k) = (ih, jh, kh)$ where h is some positive real number:

```
for i:=-n to n do
   for j:=-n to n do
      for k:= -n to n do
         A[i,j,k] := exp(-5*(sqr(i*h) + sqr(j*h) + sqr(k*h)));
```

We say that A[i,j,k] is a boundary value if $|i| = 1$, or $|j| = 1$, or $|k| = 1$. Otherwise, it is an interior value. One way to think of a 3-dimensional array is as a stacking of 2-dimensional arrays. For a given k, we can regard A[-n..n,-n..n,k..k] as the k-th slice of the array. If A[i,j,k] is an interior value, then it has four neighbors in its slice: A[i-1,j,k], A[i+1,j,k], A[i,j-1,k], and A[i,j+1,k]. It also has a neighbor in each of the two adjacent slices: A[i,j,k-1] and A[i,j,k+1]. The 3-dimensional analog of the previous simulations involves replacing each interior value with the average of its six neighbors and itself, i.e.,

```
program Example24_5;
{Two-Dimensional Grid Simulation.}
const
   n = 5; {Array and Problem size.}
type
   Real2DList = array[-n..n,-n..n] of real;
var
   Range:real; {Initial A[0,0] - A[n,n].}
   step:integer; {Number of simulation steps.}
   StartTime,StopTime:real; {Clock Snapshots}
   A:Real2DList; {For the grid simulation.}
   i,j:integer; {Indices.}

procedure Step2D(n:integer; var A:Real2DList);
      ⟨:⟩
begin
   ShowText;
   StartTime := TickCount;
   {Set up A, a discrete version of exp(-5(x^2+y^2)) on [-1,1]x[-1,1].}
   for i:=-n to n do
      for j:= -n to n do
         A[i,j]:=exp(-5*(sqr(i/n) + sqr(j/n)));
   Range:= A[0,0] - A[n,n];
   step:= 0;
   while A[0,0]-A[n,n] >= Range/100 do
      {One percent value not reached.  One more time step.}
      begin
         Step2D(n,A);
         step := step + 1;
      end;
   Stoptime := tickcount;
   writeln(' Steps = ', step:3);
   writeln('Time per step = ', ((StopTime - StartTime)/(60*step)):6:4);
end.
```

Output:

```
                    Steps  = 110
            Time per step  = 0.0136
```

```
procedure Step3D(n:integer; var A:Real3DList);
{Pre:A[-n..n,-n..n,-n..n] initialized}
{Post:Each value in A[-n+1..n-1,-n+1..n-1,-n+1,1..n-1] is replaced by}
{the average of itself and its 6 neighbors.}
var
   i,j,k:integer; {indices}
   B:Real3DList;
begin
   B := A;
   for i:=-n+1 to n-1 do
      for j:=-n+1 to n-1 do
         for k:=-n+1 to n-1 do
            A[i,j,k] := (B[i,j,k] + B[i-1,j,k] + B[i+1,j,k] +
                        B[i,j-1,k] + B[i,j+1,k] + B[i,j,k-1] + B[i,j,k+1])/7;
end
```

This procedure is used in `Example24_6`. At the start of the simulation, the `A` array is a discrete snapshot of the function $f(x, y, z) = \exp(-5(x^2 + y^2 + z^2))$ on the cube $\{(x, y, z) : -1 \leq x \leq 1,\ -1 \leq y \leq 1,\ -1 \leq z \leq 1\}$.

Looking at the benchmarks produced by our 1, 2, and 3-dimensional simulations we see that the amount of work per array update is proportional to n^d, where d is the dimension:

d	Time/Update	Interior Values	Flops/Interior Value
1	.0012	$2n - 1$	3
2	.0136	$(2n - 1)^2$	5
3	.1590	$(2n - 1)^3$	7

In some areas of computational science, grid simulations are required that have dimension greater than three. The volume of computation in these situations is immense, even when the most sophisticated supercomputers are used.

PROBLEM 24.7. Make `Example24_6` more efficient by exploiting symmetry. If successful, then your modification should require one-eighth the storage, only $O(n)$ exp evaluations, and one-eighth the arithmetic per array update.

```
program Example24_6;
{Three-Dimensional Grid Simulation.}
const
   n = 5; {Problem and Array size.}
type
   Real3DList = array[-n..n,-n..n,-n..n] of real;
var
   Range:real; {Initial A[0,0,0]-A[n,n,n]}
   step:integer; {Number of simulation steps.}
   StartTime,StopTime:real; {Clock Snapshots}
   A:Real3DList; {For carrying out the grid simulation.}
   i,j,k:integer; {Indices.}

procedure Step3D(n:integer; var A:Real3DList);
    ⟨:⟩
begin
   ShowText;
   StartTime := TickCount;
   {Set up A, a discrete version of exp(-5(x^2 + y^2 + z^2)) }
   {on [-1,1]x[-1,1]x[-1,1].}
   for i:=-n to n do
      for j:=-n to n do
         for k:= -n to n do
            A[i,j,k] := exp(-5*(sqr(i/n) + sqr(j/n) + sqr(k/n)));
   Range := A[0,0,0] - A[n,n,n];
   step := 0;
   while A[0,0,0]-A[n,n,n] >= Range/100 do
      {One Percent-value not reached.  One more time step.}
      begin
         Step3D(n,A);
         step := step+1;
      end;
   StopTime := tickcount;
   writeln(' Steps = ', step:3);
   writeln('Time per step = ', ((StopTime - StartTime)/(60*step)):6:3);
end.
```

Output:

```
        Steps   = 89
Time per step   = 0.159
```

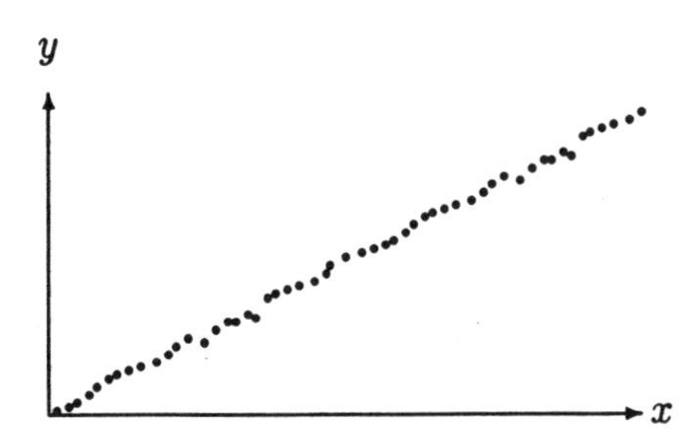

FIGURE 24.3 *Linearly Related Data*

24.3 Building Models from Data

Suppose we conjecture that the height of a plant is determined by the amount of sunlight it receives during its first month as a young seedling. To explore the possibilities, we plant m seeds. For $i = 1$ to m we expose the i-th seed to x_i hours of sunlight and measure the height y_i that is attained. Let us assume that a y-versus-x plot of the m experiments reveals something of a linear relationship between y and x as depicted in FIGURE 24.3 This suggests that we might be able to *model* the data with a linear function

$$y(x) = ax$$

The constant a is best thought of as a *model parameter.* In the *method of least squares*, this quantity is chosen so that the *least squares residual*

$$d_2(a) = \sqrt{\sum_{i=1}^{m} (ax_i - y_i)^2}$$

is minimized. Recognize that $|ax_i - y_i|$ is the distance between the data point (x_i, y_i) and the point predicted by the model (x_i, ax_i). From the equation $d_2'(a) = 0$ it is possible to show that $d_2(a)$ is minimized by setting

$$a = \frac{\sigma_{xy}}{\sigma_{xx}}$$

where

$$\sigma_{xy} = \frac{1}{m}\sum_{i=1}^{m} x_i y_i \qquad \sigma_{xx} = \frac{1}{m}\sum_{i=1}^{m} x_i^2.$$

An exploratory environment is set up in `Example24_7` for studying the least squares approach to model-building.

It uses the `RealList` data type and other features of the `Chap14Codes` unit. The functions `LS` and `LS_Residual` are straightforward and are respectively used to compute the optimum a and the quality of the fit as measured by the minimum residual.

Least squares fitting plays a huge role in computational science. We have considered the case when a 1-parameter linear model is sought. But nonlinear model building with many parameters can also be handled. Moreover, as the problems show, there are other criteria that can be used to measure the quality of the fit.

```
program Example24_7;
{Least square fitting of a line y = ax to data.}
uses
   DDCodes,Chap14Codes;
const
   x0 = 0; y0 = 0; {(x0,y0) = screen center}
   p = 20; {Pixels per xy unit.}
var
   m:integer; {Number of data points.}
   x,y:RealList; {The x[1..m] and y[1..m] house the data coordinates.}
   a:real; {The least squares parameters.}
   d:real; {Plot y=ax over the interval[-d,d].}
   r:real; {The minimum sum of squares.}
function LS(m:integer; x,y:RealList):real;
{Pre:x[1..m] and y[1..m] initialized.}
{Post:If a = LS(m,x,y), then y=ax is the best fit of the data}
{(x[i],y[i]), i=1..m, in the least squares sense.}
      ⟨:⟩
function LS_Residual(m:integer; x,y:RealList; a:real):real;
{Pre:x[1..m] and y[1..m] initialized.}
{Post:sqrt(|ax[1]-y[1]|^2+...+ |ax[m]-y[m]|^2)}
      ⟨:⟩
begin
   {Initialize the graphics coordinates.}
   ShowDrawing;
   SetMap(x0,y0,p);
   DrawAxes;
   {Get the data and determine the least squares fit.}
   GetPoints(m,x,y);
   ShowPoints(m,x,y);
   a := LS(m,x,y);
   {Plot the line y = ax.}
   d := ScreenWidth/p;
   cMoveTo(-d,-a*d);
   cLineTo(d, a*d);
   {Compute and display the minimum sum of squares.}
   r := LS_Residual(m,x,y,a);
   MoveTo(20,75);
   writedraw('Residual = ', r:10:4);
   MoveTo(20,90);
   WriteDraw('a =', a:10:4);
end.
```

SetMap and ScreenWidth are declared in DDCodes. DrawAxes, GetPoints, ShowPoints, cMoveTo, and cLineTo are declared in Chap114Codes. For sample output, see FIGURE 24.4.

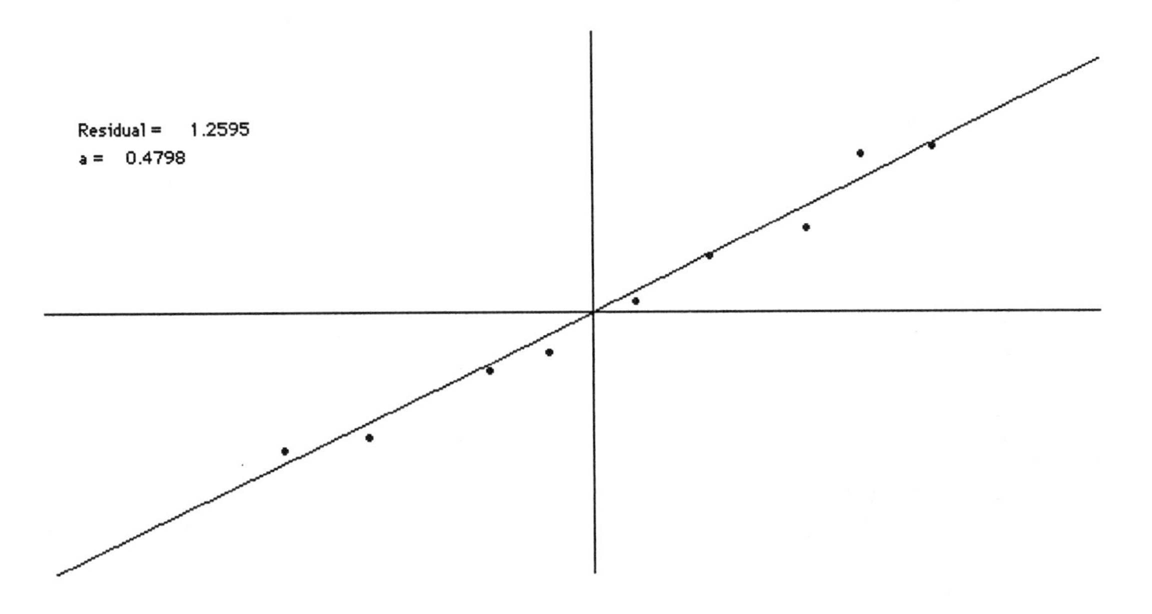

FIGURE 24.4 *Sample Output From* `Example24_7`

PROBLEM 24.8. The Least Square fit sums the squares of the vertical distances from each (x_i, y_i) to the fitting line. In the *total least squares* fit, the perpendicular distances are summed. The optimum a is given by

$$a = \frac{2\sigma_x y}{(\sigma_x x - \sigma_y y) + \sqrt{(\sigma_x x - \sigma_y y)^2 + 4\sigma_x y^2}}$$

where

$$\sigma_{xx} = \frac{1}{m}\sum_{i=1}^{m} x_i^2, \qquad \sigma_{xy} = \frac{1}{m}\sum_{i=1}^{m} x_i y_i, \text{ and } \quad \sigma_{yy} = \frac{1}{m}\sum_{i=1}^{m} x_i^2$$

Augment `Example24_7` so that it also displays the TLS fitting line and its residual.

PROBLEM 24.9. In what is called L_1 fitting, that model parameter a is chosen to minimize

$$d_1(a) = \sum_{i=1}^{m} |ax_i - y_i|.$$

Unlike the least squares case, there is no simple formula that prescribes the L_1 minimizer. In view of this, build an environment that supports an interactive search for the optimum a. Note that the fitting line can be specified by a single point that can be obtained by a mouse click.

PROBLEM 24.10. Same as the previous problem, only the minimizer of

$$d_\infty(a) = \max_{1 \le i \le m} |ax_i - y_i|$$

should be sought. This is called L_∞ fitting.

PROBLEM 24.11. The least square fitting of the line $y = ax + b$ to the data (x_i, y_i), $i = 1..m$ is only slightly more complicated than the case considered above. Indeed, it can be shown that the two model parameters a and b are given by

$$a = \frac{\sigma_{xy} - \sigma_x\sigma_y}{\sigma_{xx} - \sigma_x^2} \qquad b = \frac{\sigma_{xx}\sigma_y - \sigma_x\sigma_{xy}}{\sigma_{xx} - \sigma_x^2}$$

where σ_x, σ_{xy}, σ_y, and σ_{xx} are defined by

$$\sigma_x = \frac{1}{m}\sum_{i=1}^{m} x_i \qquad \sigma_{xy} = \frac{1}{m}\sum_{i=1}^{m} x_i y_i \qquad \sigma_y = \frac{1}{m}\sum_{i=1}^{m} y_i \qquad \sigma_{xx} = \frac{1}{m}\sum_{i=1}^{m} x_i^2$$

The idea behind an imaging system like a CAT scan is to obtain a characterization of a 3-dimensional object by assimilating a large number of 2-dimensional "snapshots." The act of assimilation is tantamount to building a model of the solid object. Our goal in this final part of the section is to show what some of the underlying model-building computations look like.

To do this economically we go down a notch in dimension and consider how we might estimate the density throughout a 2-dimensional object by bombarding it with a series of 1-dimensional "rays" and measuring their strength upon exit. Our intuition tells us that there should be a correlation between the strength of the emerging beam and the density of the material encountered along its "journey." The reconstruction process depends heavily upon the tracking of each ray through the medium and it is upon this portion of the computation that we focus.

Assume that $\rho(x, y)$ is the sought-after density function and that it is defined over the unit square

$$S = \{(x, y) : 0 \leq x \leq 1,\ 0 \leq y \leq 1\}.$$

Given a positive integer n_T, we discretize the problem by partitioning S into an an n_T-by-n_T array of square tiles, defining tile T_{ij} by

$$T_{ij} = \{(x, y) : (i-1)h \leq x \leq ih,\ (j-1)h \leq y \leq jh$$

where $h = 1/n_T$. The following picture should clarify the tile-naming convention:

T_{15}	T_{25}	T_{35}	T_{45}	T_{55}
T_{14}	T_{24}	T_{34}	T_{44}	T_{54}
T_{13}	T_{23}	T_{33}	T_{43}	T_{53}
T_{12}	T_{22}	T_{32}	T_{42}	T_{52}
T_{11}	T_{21}	T_{31}	T_{41}	T_{51}

$n_T = 5.$

Our goal is to approximate the continuous function $\rho(x, y)$ on T_{ij} with a single number ρ_{ij}. That is, we are going to assume that $\rho(x, y)$ is constant on each tile and that our job is to determine a good estimate for its value. The task of computing these n_T^2 numbers is the task of building a discrete model of $\rho(x, y)$.

Assume that n_R rays are shot through the square. Let d_{kij} be the length of the k-th ray's intersection with T_{ij} and assume that we have an instrument that can report the following *density measure*:

$$b_k = \sum_{i=1}^{n_T}\sum_{j=1}^{n_T} d_{ijk}\rho_{ij}.$$

If n_R is much bigger than n_T^2, then some kind of least squares fit could be used to estimate the best collection of ρ_{ij} assuming that we know the measurements $b_1, \ldots, b_{n_R}$ *and* the associated intersection distances d_{kij}. The details of the solution process are well beyond the scope of the text, but it is instructive to consider the computation of the d_{kij} and how they might be represented.

A three-dimensional array `D[1..nR,1..nT,1..nT]` would be the simplest way to package the d_{kij} information. But this is out of the question when you consider the enormity of the problem–n_T^2 might be a million and n_R might be several times that. Moreover, the 3D array route would be extremely wasteful because most of these numbers are zero. For example, if this is the path of the k-th ray,

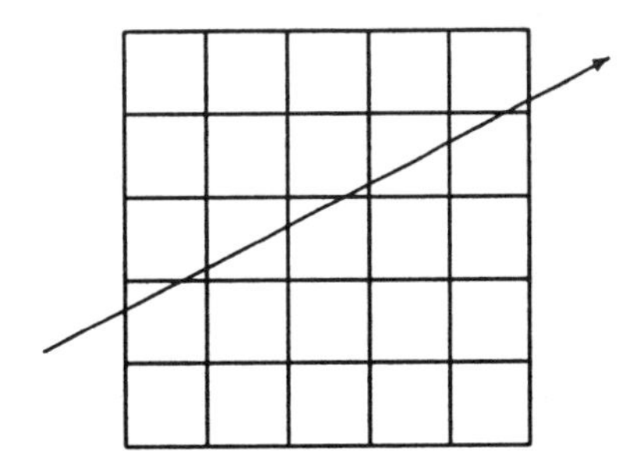

then only d_{k12}, d_{k13}, d_{k23}, d_{k24}, d_{k34}, d_{k44}, d_{k54} and d_{k55} are nonzero.

To solve this problem we establish a data structure that is economical with respect to memory but rich enough to support all the necessary, model-building computations. In essence, for a given ray we must tabulate the indices of the tiles that it crosses and the corresponding intersection distances. This is an example of a *ray tracing* problem and before we get into the supporting data structures, we need to look more carefully at the underlying point-slope computations.

Assume that the ray enters the square on its left edge and that it has slope m. In order to track its path we need to look at what happens when it crosses a particular tile. There are three possible entry edges: left (`'L'`), bottom (`'B'`), and top (`'T'`). The entry edge, a fraction $f_1 \in [0, 1]$, and some conventions completely specify the entry point:

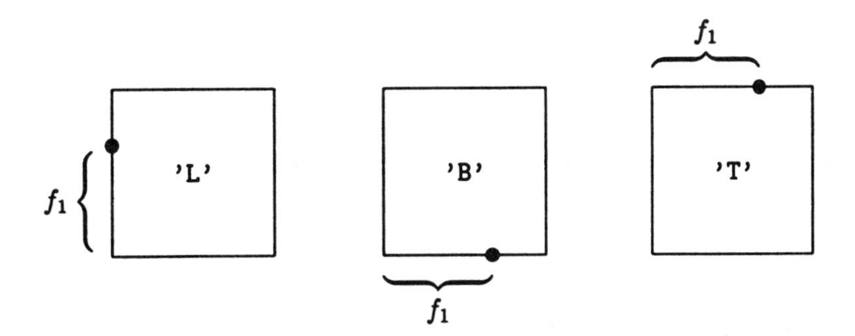

Thus, if (x_0, y_0) is the lower left corner of the tile, then the three points indicated above are given by $(x_0, y_0 + f_1 h)$, $(x_0 + f_1 h, y_0)$, and $(x_0 + f_1 h, y_0 + h)$ respectively.

Exit points may be specified similarly. If the ray enters from the left, then there are three possible exit sides:

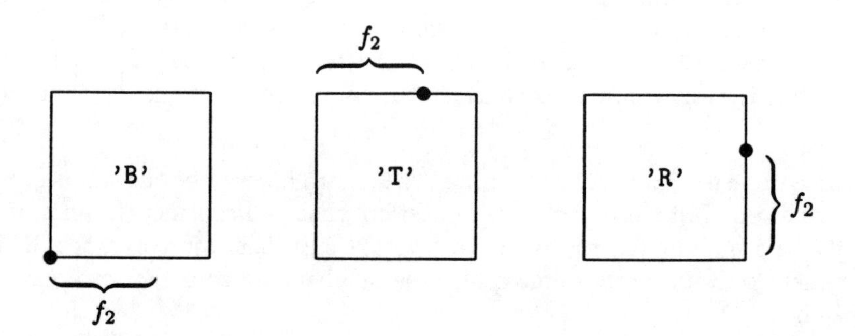

The quantity $f_2 \in [0,1]$ has the same role to play as f_1. Note that if the ray enters from the top, then it must exit from either the right or the bottom edges. Likewise, a ray entering from the bottom must exit from either the right or top edges. Here is a template for the overall, single tile crossing computation:

```
procedure CrossTile(EnterSide:char; f1,m,h:real; var ExitSide:char;
          var f2,d:real);
{Pre:m is the slope of the incoming ray, 0<=f1<=1, EnterSide = 'T' (for top),}
{'B' (for bottom), or 'L' (for left).}
{Let (x0,y0) = location of lower left corner and h = tile side.  Then}
{EnterSide = 'T' and the entry point = (x0+h*f1,y0+h)}
{EnterSide = 'B' and the entry point = (x0+h*f1,y0)}
{EnterSide = 'L' and the entry point = (x0,y0+h*f1)}
{Post:0<=f2<=1 and ExitPoint = 'T', 'B', or 'R'.}
{Exitside = 'T' and the exit point = (x0+h*f2,y0+h)}
{Exitside = 'B' and the exit point = (x0+h*f2,y0)}
{ExitSide = 'R' and the exit point = (x0+h,y0+h*f2)}
{d is the distance from the entry point to the exit point}
begin
  if EnterSide = 'L' then
     {Enter on left and can exit either from top, bottom, or right.}
          ⟨:⟩
  else if EnterSide = 'T' then
     {Enter on top and can exit either from bottom or right.}
          ⟨:⟩
  else if EnterSide = 'B' then
     {Enter on bottom and can exit either from top or right.}
          ⟨:⟩
end
```

Elementary point-slope computations are required to complete the procedure. Discussion of the first case where the ray enters from the left should be enough to see what is involved. The entry point is $(x_0, y_0 + f_1 h)$. A line from this point to the upper right corner $(x_0 + h, y_0 + h)$ has slope $1 - f_1$ while the line to the lower right corner has slope $-f_1$. Thus, simple comparisons with the ray's slope can be used to identify the three exit situations:

```
if m > (1 - f1) then
   {Exit on the top.}
         ⟨:⟩
else if m < -f1 then
   {Exit on the bottom.}
         ⟨:⟩
else
   {Exit on the right.}
         ⟨:⟩
```

Once the exit side has been determined, then the exit fraction f_2 and the intersection distance d can be easily determined. For example, if the exit point is on the top edge, then the following assignments are made:

```
ExitSide := 'T';
f2:=(1-f1)/m;
d:=h*sqrt(sqr(1-f1)+sqr(f2));
```

The computations are similar for the other seven entry/exit possibilities.

This brings us to the issue of how to store economically all the trajectory information. To that end we build upon the following record type that represents "what happens" when a ray crosses an individual tile:

```
TileCrossing = record
      i,j:integer;
      d:real;
   end
   {If t has type TileCrossing, then a path of length t.d }
   {was traversed across tile T(t.i,t.j) }
```

A list of TileCrossing variables is required to represent the tile crossing distance data for a single ray:

```
TileCrossingList = array[1..Lmax] of TileCrossing
```

Here, Lmax is an upper bound on the number of tiles that can be traversed by a single ray. For n_T-by-n_T tilings, it suffices to set this constant to $2n_T$. (Why?) Finally, a record of the form

```
RayPath = record
      L:integer;
      TC:TileCrossingList;
   end;
   {If R is a RayPath, then it crosses tiles (R.TC[p].i,R.TC[p].j), p=1..R.L.}
```

encapsulates everything about a particular ray path. Assignment to a RayPath variable is handled by the following function:

```
function MakeRayPath(m,y:real):RayPath;
{Pre:0<=y<=1}
{Post:If R = MakeRayPath(m,y), then R represents the trace }
{of a ray that enters the square at the point (0,y) with slope m.}
```

It begins by setting the stage for the first tile crossing:

```
h:=1/nT;
{Determine the entry tile and fraction.}
i := 1;
if y = 1 then
   j:=nT
else
   j:=trunc(y/h) + 1;
f1:= (y-(j-1)*h)/h;
EnterSide:='L';
```

This is followed by a while loop that oversees the tile-by-tile tracing. The indices of the tile that is about to be entered are i and j. If either of these indices fall outside the range $[1, n_T]$, then the iteration terminates because the ray has, in effect, exited the square S. Here is the implementation:

```
L:=0;
while (1<=j) and (j<=nTiles) and (i<=nTiles) do
   {Ray is about to cross Tile (i,j).}
   begin
      CrossTile(EnterSide,f1,m,h,ExitSide,f2,d);
      L:=L+1;
      MakeRayPath.TC[L]:=MakeTileCrossing(i,j,d);
      f1:=f2;
      if ExitSide = 'R' then
         {Moving into right neighbor tile.}
         begin
            i:=i+1;
            EnterSide := 'L'
         end
      else if ExitSide = 'T' then
         {Moving into top neighbor tile.}
         begin
            j:=j+1;
            EnterSide := 'B'
         end
      else if ExitSide = 'B' then
         {Moving into bottom neighbor tile.}
         begin
            j:=j-1;
            EnterSide:= 'T'
         end;
      end;
   MakeRayPath.L := L;
   end;
```

The function MakeTileCrossing has type TileCrossing and is used to package the intersection distance and the tile indices.

To illustrate the use of these functions, we have built a simple environment that shows how the ray tracing idea can used to locate a dense "mystery spot" in the unit square. A function

```
procedure GenRandGrid(var A:Real2DGrid; var i0,j0:integer);
{Post:i0 and j0 are random integers from [1,nTiles].}
{ A[i,j] = exp(-6((i-i0)^2+(j-j0)^2/nTiles).}
```

is used to initialize an array of density values, randomly locating a single "mountain" in the middle of tile (i_0, j_0). Here, `Real2DGrid = array[1..nT,1..nT] of real`.

A ray source location is obtained by clicking a "source bar". For each such click, a ray is directed toward each of twenty-five targets that are arranged in a line on the opposite side of the square. For each ray, the path density defined by the summation $\sum_k \sum_i \sum_j d_{kij}\rho_{kij}$ is reported by evaluating the following function:

```
function Density(R:RayPath; rho:Real2DGrid):real;
{Pre:rho[1..nT,1..nT] initialized.}
{Post:The ''density'' of the ray R.}
var
   sum:real;
   i,j:integer;
   k:integer;
begin
   sum := 0;
   for k:=1 to R.L do
      {Incorporate the contribution of the k-th tile touched,}
      {i.e., density x distance}
      begin
         i := R.TC[k].i;
         j := R.TC[k].j;
         {Tile (i,j) is the k-th tile crossed.}
         sum := sum + R.TC[k].d * rho[i, j];
      end;
   Density := sum;
end
```

See FIGURE 25.5. **Example24_8** puts it all together.

PROBLEM 24.12. Modify **Example24_8** so that the sources are automatically selected along the source bar with equal spacing. Display only the max ray from each scan.

PROBLEM 24.13. We say that a ray is *interior* to its scan in **Example24_8** if it is neither the first or last ray in its scan. Note that the mystery spot is located once two interior max rays are discovered. Modify **Example24_8** so that the **while**-loop terminates once this is the case.

PROBLEM 24.14. Modify **Example24_8** so that the density array has two mystery spots. Do this by calling **GenRandGrid** twice and adding the results. Through experimentation, develop a search strategy for locating both of these points.

```
program Example24_8;
{Ray Tracing Environment}
uses
   DDCodes, Chap24Codes;
var
   A:Real2DGrid; {The density array.}
   ScanNo:integer; {Index of the current scan.}
   k:integer; {Index of the ray being processed.}
   yL,yR:real; {The y coordinate of the source and target for the current ray.}
   R:RayPath; {The current ray.}
   slope:real; {The slope of the current ray.}
   d:real; {The path density of the current ray.}
   y:real; {(0,y) is the point of impact of the current ray.}
   i0,j0:integer; {(i0,j0) = location of the mystery spot.}
   kmax:integer; {The index of the ray with the largest path density.}
   dMax:real; {The largest path density of the scan.}
begin
   ShowDrawing;
   SetMap(0.5, 0.5, 250);
   GenRandGrid(A,i0,j0);
   DrawTarget;
   ScanNo := 0;
   while YesNo('AnotherScan?') do
      {Process another scan.}
      begin
         ScanNo:= ScanNo + 1;
         yL:=GetSource(ScanNo);
         kmax:=0;
         dmax:=0;
         for k:=1 to nTargets do
         {Process the k-th ray.}
               ⟨:⟩
         {Highlight the densest path in scan.}
            ⟨:⟩
      end;
   {Show the mystery point.}
   cDrawBigDot((i0-0.5)/nT,(j0-0.5)/nT);
end.
```

SetMap, YesNo, and cDrawBigDot are declared in DDCodes. GenRandGrid, DrawTarget, GetSource, and nTargets are declared in Chap24Codes. For sample output, see FIGURE 24.5.

Probability = 0.49
Temperature = 1.00

Probability = 0.49
Temperature = 1.00
Completed Sweeps = 28
Field Strength = 102

<table>
<tr><td>1</td><td>7</td><td>2</td><td>5</td><td>[illegible]</td></tr>
<tr><td>4</td><td>6</td><td>4</td><td>8</td><td>8</td></tr>
<tr><td>2</td><td>5</td><td>3</td><td>3</td><td>1</td></tr>
<tr><td>2</td><td>4</td><td>5</td><td>4</td><td>6</td></tr>
<tr><td>1</td><td>3</td><td>6</td><td>1</td><td>9</td></tr>
</table>

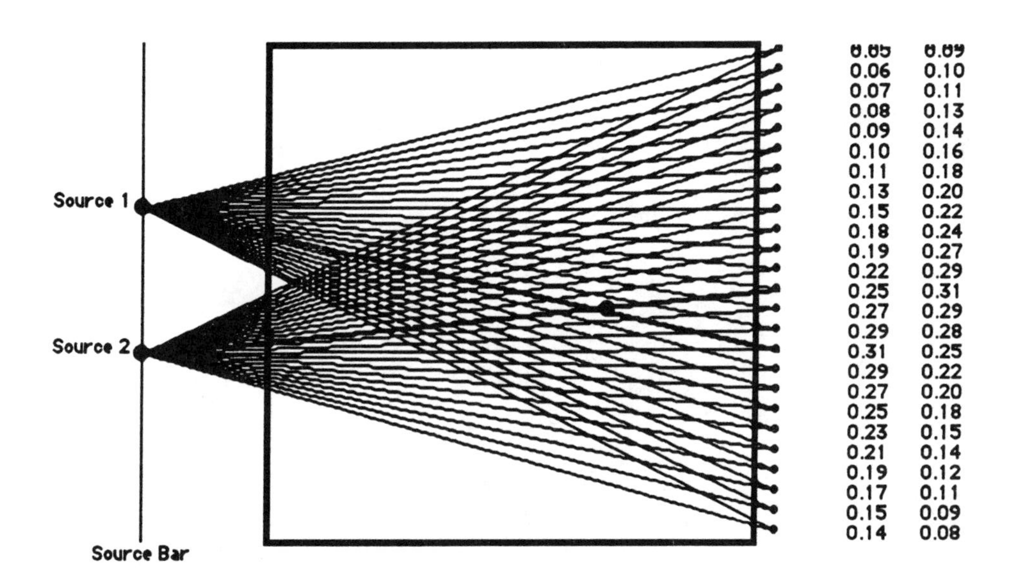

FIGURE 24.5 *A Ray Tracing Environment*

Appendix I
Synopsis of the Think Pascal Units

DDCodes

```
unit DDcodes;

interface

const
   h_center = 300;
   v_center = 170; {(h_center,v_center) = screen center.}
   ScreenWidth = 700;
   ScreenHeight = 340;

function rand(x:real):real;
{Pre:0<=x<=1
{Post:A random value in [0,1] (uniform distribution.}

{************************ Graphical I/O **********************}

procedure GetPosition (var h, v:  integer);
{Post:(h,v) is the location of the next mouseclick.}

procedure Wait;
{Post:Returns as soon as the mouse is clicked.}

procedure ClearScreen;
{Post:Clears the entire DrawWindow}

function YesNo (Message:  string):  boolean;
{Post:Displays two boxes at the top of the screen.}
{True if the displayed ``yes box'' is clicked.}
{False if the displayed ``no box '' is clicked.}
{In the latter case, both boxes are erased.}

function ReviseInteger(s:string; OldValue,L,R,Num:integer):integer;
{Pre:L <= OldValue <= R and Num <> 0.}
{Post:An interactively determined value between L and R is returned.}
{The string s should name the parameter to be varied.}
{The new value is chosen from a finite subset of values in [L,R].}
{The spacing between possible values is (R-L) div Num.}

function ReviseReal(s:string; OldValue,L,R:real; Num:integer):real;
{Pre:L <= OldValue <= R and Num <> 0.}
{Post:An interactively determined value between L and R is returned.}
{The string s should name the parameter to be varied.}
{The new value is chosen from a finite subset of values in [L,R].}
{The spacing between possible values is (R-L)/Num.}
```

```
{********************* x/y Graphics Tools ***********************************}

var
   x_0, y_0:real; {Global variables.  (x_0,y_0) maps to (h_center,v_center)}
   p_scale:real; {Global variable.  Number of pixels per xy unit.}

procedure SetMap(x0,y0,p:real);
{Post:Sets the global ''mapping'' variables x_0,y_0,p_scale.}

procedure cMoveTo(x,y:real);
{Pre:(x_0,y_0) has screen location (h_center,v_center).  p_scale = pixels per xy unit.}
{Post:Moves mouse to a point on the screen that corresponds to (x,y).}

procedure cLineTo(x,y:real);
{Pre:(x_0,y_0) has screen location (h_center,v_center).  p_scale = pixels per xy unit.}
{Post:Draws a line from the current mouse position to the point on the }
{screen that corresponds to (x,y).}

procedure cGetPosition(var x,y:real);
{Pre:(x_0,y_0) has screen location (h_center,v_center).  p_scale = pixels per xy unit.}
{Post:The xy coordinates of the next mouseclick.}

procedure cDrawDot(x,y:real);
{Pre:(x_0,y_0) has screen location (h_center,v_center).  p_scale = pixels per xy unit.}
{Post:Draws a dot at the screen location corresponding to (x,y).}

procedure cDrawBigDot(x,y:real);
{Pre:(x_0,y_0) has screen location (h_center,v_center).  p_scale = pixels per xy unit.}
{Post:Draws a big dot at the screen location corresponding to (x,y).}

procedure cDrawCircle(xc,yc,r:real);
{Pre:(x_0,y_0) has screen location (h_center,v_center).  p_scale = pixels per xy unit.}
{Post:Draws a circle with center (xc,yc) and radius r.}

procedure cFrameRect(yT,xL,yB,xR:real);
{Pre:(x_0,y_0) has screen location (h_center,v_center).  p_scale = pixels per xy unit.}
{Post:Draws the rectangle with vertices (xL,yB), (xR,yB), (xL,yT), and (xR,yT).}

procedure DrawAxes;
{Pre:(x_0,y_0) has screen location (h_center,v_center).  p_scale = pixels per xy unit.}
{Post:Draws the x and y axes.}
```

Chap7Codes

```
unit Chap7Codes;

interface

function Factorial(n:integer):longint;
{Pre:0<=n<=12}
{Post:n!}
```

```
function Stirling(n:integer):real;
{Pre:  n>=0}
{Post:Stirling approximation to n!}

function BinCoeff(n,k:integer):longint;
{Pre:0<=k<=n}
{Post:Number of ways to select k objects from a set of n objects}
```

Chap8Codes

```
unit Chap8Codes;

interface

procedure DrawGrid(n,s,hL,vT:integer);
{Pre:n>0,s>0.}
{Post:Draws an n-by-n grid with s-by-s tiles.  The upper left corner}
   { of the upper left tile has screen coordinate (hL,vT).}

procedure PaintETriangle(h0,v0:integer; r:real; Orient,shade:char);
{Pre:r>=0, Orient = 'T', 'B', 'L', or 'R', shade = 'w', 'b', or 'g.'}
{Post Draws an equil.  triangle with one vertex at (h0,v0) and sides = r.}
{If Orient = 'T' then the side opposite (h0,v0) is horizontal and below.}
{If Orient = 'B' then the side opposite (h0,v0) is horizontal and above.}
{If Orient = 'L' then the side opposite (h0,v0) is vertical and to the right.}
{If Orient = 'R' then the side opposite (h0,v0) is vertical and to the left.}
{If shade = 'w' then the triangle is white.}
{If shade = 'b' then the triangle is black.}
{If shade = 'g' then the triangle is gray.}

procedure PaintDiamond (h0,v0:integer; r:real; Orient,Shade:char);
{Pre:r>=0, Orient = 'T', 'L', 'B', 'R'}
{Post:Draws a diamond with one vertex at (h0,v0).  This vertex is }
{is either at the top (Orient = 'T'), the left (Orient = 'L'), the bottom}
{(Orient = 'B', or the right (Orient = 'R'). }
{Diamond is gray if shade = 'g' and black otherwise.}

procedure FramePoly(n,h0,v0:integer; r,theta:real);
{Pre:n>=3}
{Post:Draws a polygon whose vertices are rounded versions of}
{(h0+rcos(theta +2pi*k/n),v0-r*sin(theta+2pi*k/n)), k=0,..,n-1.}

procedure PaintHex(h0,v0:integer; r:real);
{Pre:r>=0}
{Post:Draws a shaded hexagon with vertices that are rounded versions}
{of ( h0 + r * cos ( k * pi / 3 ) , v0 - r * sin ( k * pi / 3 ) ) for k=0,1,..,5}
```

Chap9Codes

```
unit Chap9Codes;

interface

function Distance(x1,y1,x2,y2:real):real;
{Post:The distance from (x1,y1) to (x2,y2).}

function PointToLine(x0,y0,x1,y1,x2,y2:real):real;
{Pre:(x1,y1) and (x2,y2) are distinct points.}
{Post:Distance from (x0,y0) to line through (x1,y1) to (x2,y2).}

function PointToLineSegment(x0,y0,x1,y1,x2,y2:real):real;
{Pre:(x1,y1) and (x2,y2) are distinct points.}
{Post:Distance from (x0,y0) to line segment from (x1,y1) to (x2,y2).}

function ChordIntersect(a,b,c,d:real):boolean;
{Pre:0<=a<=b<=pi and 0<=c<=d<=pi}
{Post:The line segment from (cos(a),sin(a)) to (cos(b),sin(b)) intersects}
{the line segment from (cos(c),sin(c)) to (cos(d),sin(d)).}

function IsColinear(x1,y1,x2,y2,x3,y3:real):boolean;
{Post:(x1,y1),(x2,y2), and (x3,y3) are approximately colinear.}
```

Chap10Codes

```
unit Chap10Codes;

interface

procedure QuadRoots(a,b,c:real; var r1,r2:real);
{Pre:a<>0, b*b - 4ac >= 0.}
{Post:r1 and r2 are the roots of the quadratic}
{equation ax*x + bx + c = 0.}

procedure GenQuadRoots(a,b,c:real; var r1,r2:real; var complex:boolean);
{Pre:a<>0.}
{Post:If both roots of ax^2 + bx + c = 0 are real, then }
{ complex = false, and r1 and r2 are the roots.}
{ If the roots are complex, then complex = true, and r1 and r2}
{ return the real and imaginary parts of the roots, i.e., the two }
{ roots are r1 + i*r2 and r1 - i*r2 where i = sqrt(-1).}

procedure Bisection(function f(x:real): real; L,R:real; eps:real;
          var root,froot:real);
{Pre:f(x) is continuous on [L,R], f(L)f(R)<0, and eps > 0}
{Post:root is is within eps of a true root and froot = f(root).}
```

```
procedure Newton(function f(x:real):real; function fp(x:real):real;
      InitialGuess:real; ItMax:integer; eps:real; L,R:real; var root,froot:real;
      var its:integer);
{Pre:ItMax > 0.  L <= R. Functions f and fp are defined on [L,R]. eps > 0.}
{ InitialGuess is in [L,R] and is an estimate of a root.}
{Post:froot = f(root).  Either|froot| <= eps or:}
{ Its = ItMax indicating that too many iterations required.}
{ Its < ItMax indicating that another Newton step would jumps out of [L,R].}
```

Chap11Codes

```
unit Chap11Codes;

interface

function TilesInDisk(h:real):longint;
{Pre:h>0}
{Post:The number of h-by-h tiles that fit entirely within a circle}
{of radius 1 centered at the origin.  The base tile has vertices (0,h),}
{(h,h), (h,0), and (0,0).}

function Distance(x1,y1,x2,y2:real):real;
{Post:The distance from (x1,y1) to (x2,y2).}

function T(x1,y1,x2,y2,x3,y3:real):real;
{Post:The area of a triangle with vertices (x1,y1), (x2,y2), (x3,y3).}

function Q (x1, y1,x2,y2,x3,y3,x4,y4:real):real;
{Pre:The line segment that connects (x1,y1) and (x3,y3)}
{intersects the line segment that connects (x2,y2) and (x4,y4).}
{Post:The area of the quadrilateral ABCD obtained by connecting the points}
{A= (x1,y1), B = (x2,y2), C = (x3,y3), and D = (x4,y4) in order.}

function Q1(x1,y1,x2,y2,x3,y3,x4,y4:real):real;
{Pre:The line segment that connects (x1,y1) and (x3,y3)}
{intersects the line segment that connects (x2,y2) and (x4,y4).}
{Post:The area of the quadrilateral ABCD obtained by connecting the points}
{A = (x1,y1), B = (x2,y2), C = (x3,y3), and D = (x4,y4) in order.}

function RectangleRule(function f(x:real):real; a,b:real; n:longint):real;
{Pre:f(x) is defined on [a,b] and n > 0.}
{Post:The integral of f from a to b using the rectangle rule with n}
{approximating rectangles.}
```

Chap12Codes

```
unit Chap12Codes;

interface

procedure TakeApart(s:string; var d,m,y:string);
{Pre:s is a string of the form m/d/y where:}
{ m is one of the strings '1' , '2' , ...,'12' and indicates month.}
{ d is one of the strings '1' , '2' , ...,'31' and indicates day.}
{ y is one of the strings '00' , '2' , ...,'99' and indicates year.}
{Post:d, m, and y are the day, month and year substrings.}

function Char2Int(c:char):integer;
{Pre:c is one of the ten digits.}
{Post:The value of c as an integer}

procedure CountryParts(s:string; var Name,Pop,Area,Capital,Lat,Long:string);
{Pre:s has the form s1#s2#s3#s4#s5#s6 where s1..s6 are substrings not containing #.}
{Post:These substrings are assigned to Name, Pop, Area, Capital, Lat, and Long resp.}

function RemoveCommas(s:string):string;
{Post:The string obtained by deleting all commas from s.}

function Str2Int(s:string):longint;
{Pre:s is a string encoding of an integer x that saisfies 0<=x<=2**31 - 1.}
{Post:x}

procedure MantExp(x:real; var m:real; var e:integer);
{Post:if x<>0, then x = m*10^e where 0<m<1.}
{Otherwise, m=0 and e=0}

function Digit2Char(n:integer):char;
{Pre:0<=n<=9}
{Post:The character equivalent of n.}

function Int2Str(n:longint):string;
{Post:The string 'equivalent' of n.}
{Thus, 31 becomes '31', -578 becomes '-578', etc.}

function SciNot(x:real):string;
{Post:x in scientific notation}

function Int2Bin(n:longint):string;
{Pre:n>=0}
{Post:The base-2 representation of n.}
```

Chap13Codes

```
unit Chap13Codes;

interface

uses
   DDCodes;

procedure PaintDisk(h,v,r:integer);
{Post:Draws a bordered gray disk with radius r and center (h,v)}

procedure DrawTarget(s:string; vT,hL,vB,hR,r:integer);
{Post:Draws a border of width r around a white square defined by}
{hL<=h<=hR and vT<=v<=vB. The string s is displayed.}

procedure DrawSpiral(n,turn,hc,vc:integer; d:real);
{Post:Draws a polygonal line with n "legs".  The k-th leg has length k*d}
{and turns left turn degrees from leg k-1.  The first leg starts at (hc,vc) and ends}
{at (hc+d*cos(turn),vc-d*sin(turn)).}

procedure SimplePlot(a,b:real; function f(x:real):real);
{Pre:(x_0,y_0) has screen location (h_center,v_center).  p_scale = pixels per xy unit.}
{ a<b, f defined on [a,b]}
{Post:Plots f(x) on [a,b] with axes though (x_0,y_0).}
```

Chap14Codes

```
unit Chap14Codes;

interface

uses
   DDCodes;
const
   nmax = 500;
type
   RealList = array[1..nmax] of real;
   IntegerList = array[1..nmax] of integer;
   BooleanList = array[1..nmax] of boolean;

procedure GetPoints (var n:integer; var x,y:RealList);
{Post:n points are clicked and x[1..n] and y[1..n] contain the associated}
{xy values.}

procedure GetRandomPoints(n:integer; xL,xR,yB,yT: real; var rn:real;
            var x,y:RealList);
{Pre:0<rn<1, xL<=xR, yB<=yT}
{Post:x[1..n] houses random real numbers selected from the interval [xL,xR].}
{y[1..n] houses random real numbers selected from the interval [yB,yT].}
{rn returns with a value obtained by executing 2n times the statement rn:=rand(rn).}
```

```
procedure ShowPoints(n:integer; x,y:RealList);
{Pre:x[1..n] and y[1..n] initialized.}
{Post:Draws the points (x[1],y[1]),...,(x[n],y[n]).}

procedure ConnectPoints(n:integer; x,y:RealList);
{Pre:x[1..n] and y[1..n] are initialized}
{Post:Draws the polygon defined by the points (x[1],y[1]),...,(x[n],y[n]).}

function Distance(x1,y1,x2,y2:real):real;
{Post:The distance from (x1,y1) to (x2,y2)}

procedure Centroid (n:integer; x,y:RealList; var xbar,ybar:real);
{Pre:n>=1 and x[1..n] and y[1..n] are initialized}
{Post:(xbar,ybar) is the centroid of (x[k],y[k]), k=1,...,n}

function MaxInList(n:integer; x:RealList):real;
{Pre:n>=1 and x[1..n] is initialized}
{Post:The largest value in x[1..n].}

function MinInList(n:integer; x:RealList):real;
{Pre:n>=1 and x[1..n] is initialized}
{Post:The smallest value in x[1..n].}

procedure Nearest(x0,y0:real; n:integer; x,y:RealList; var p:integer; var dmin:real);
{Pre:n>=1 x[1..n] and y[1..n] are initialized.}
{Post:dmin is the minimum distance from (x0,y0) to the set}
{(x[1],y[1]),...,(x[n],y[n]) and dmin is the distance from (x0,y0) to (x[p],y[p])}

procedure MinSep(n:integer; x,y:RealList; var p,q:integer; var dmin:real);
{Pre:n>=1 x[1..n] and y[1..n] are initialized.}
{Post:Among the the points (x[1],y[1]),...,(x[n],y[n]) , point p and point q}
{have the minimum separation and the distance between them is dmin.}

procedure ShowPath(n:integer; x,y:RealList; Path:IntegerList);
{Pre:n>=1, x[1..n],y[1..n] and Path[1..n] are initialized.}
{Post:Shows the path whose i-th "stop" is (x[Path[i]],x[Path[i]]).}

procedure NearestSub(x0,y0:real; n:integer; x,y:RealList; select:BooleanList;
      var p:integer; var dmin:real);
{Pre:n>=1 x[1..n] and y[1..n] are initialized.  select[1..n] contains at least
            one true value.}
{Post:dmin is the distance from (x0,y0) to the set of selected points among}
{(x[1],y[1]),...,(x[n],y[n]) and p is the index of the closest selected point.}

procedure Trip (start,n:integer; x,y:RealList; var Path:IntegerList;
         var odometer:real);
{Pre:n>=1, 1<=start<=n, and x[1..n] and y[1..n] are initialized.}
{Post:Path[1..n] is a permutation of 1..n and odometer is the roundtrip distance of }
{the path that visits points Path[1],..,Path[n].}
```

Chap15Codes

```
unit Chap15Codes;

interface

const
   fEvalMax = 200;
var
   h_Left:integer; {Left edge of the graph window.}
   h_Right:integer; {Right edge of the graph window.}
   v_Top:integer; {Top edge of the graph window.}
   v_Bottom:integer; {Bottom edge of the graph window.}

type
   RealList0 = array[0..fEvalMax] of real;

function SetUp(n:integer; xval:  RealList0; function f(x:real):real):RealList0;
{Pre:0<=n<=fEvalMax.  f(x) defined at x = xval[0],...,xval[n].}
{Post:SetUp[k] = f(x[k]), k=0,...,n.}

function xEqual(n:  integer; L, R: real):  RealList0;
{Pre:0<= n<=fEvalMax}
{Post:xEqual[k] = L + k(R-L)/n, k=0,...,n.}

procedure StandardPlot;
{Post:Sets up a standard plot window.}

function MaxInList0(n:integer; x:RealList0):real;
{Pre:0<=n<=fEvalMax.}
{Post:The largest value in x[0..n].}

function MinInList0 (n:integer; x:RealList0):  real;
{Pre:0<=n<=fEvalMax.}
{Post:The smallest value in x[0..n].}

procedure Plot(n:integer; x,y:RealList0; xmin,xmax,ymin,ymax:real);
{Pre:0<=n<=fEvalMax}
{Post:Plots y[0..n] vs x[0..n] with window (v_Top,h_Left,v_Bottom,h_Right) }
{corresponding to (ymax,xmin,ymin,xmax)}

procedure AutoPlot(n:integer; x,y:RealList0);
{Pre:0<=n<=fEvalMax}
{Post:Plots y[0..n] vs x[0..n] with auto scaling.}

procedure SinCos1(n:integer; var s,c:RealList0);
{Pre:n>0}
{Post:s[k] = sin(kR/n), c[k] = cos(kR/n), k=0,...,n.}

procedure SinCos2(n:integer; var s,c:RealList0);
{Pre:n>0, n mod 4 = 0}
{Post:s[k] = sin(k*pi/n), c[k] = cos(k*pi/n), k=0,...,n.}
```

```
function Add(n:integer; a,b:RealList0):RealList0;
{Pre:0<=n<=fEvalMax}
{Post:Add[k] = a[k]+b[k], k=0,...,n.}

function Sub(n:integer; a,b:RealList0):RealList0;
{Pre:0<=n<=fEvalMax.}
{Post:Sub[k] = a[k]-b[k], k=0,...,n.}

function Mult(n:integer; a,b:RealList0):RealList0;
{Pre:0<=n<=fEvalMax.}
{Post:Mult[k] = a[k]*b[k], k=0,...,n.}

function Divide(n:integer; a,b:RealList0):RealList0;
{Pre:0<=n<=fEvalMax, b[k] <>0, k=0,...,n}
{Post:Divide[k] = a[k]+b[k], k=0,...,n.}

function Scale(n:integer; a:real; b:RealList0):RealList0;
{Pre:0<=n<=fEvalMax.}
{Post:Scale[k] = a*b[k], k=0,...,n.}

function Translate(n:integer; a:real; b:RealList0):RealList0;
{Pre:  0<=n<=fEvalMax.}
{Post:  Translate[k] = a+b[k], k=0,...,n.}

function Constant (n:integer; a:real):RealList0;
{Pre:0<=n<=fEvalMax.}
{Post:Constant[k] = a, k=0,...,n.}

function CopyV (n:integer; a:RealList0):RealList0;
{Pre:0<=n<=fEvalMax.}
{Post:CopyV[k] = a[k], k=0,...,n.}

function AbsV(n:integer; a:RealList0):RealList0;
{Pre:  0<=n<=fEvalMax.}
{Post:AbsV[k] = |a[k]|, k=0,...,n.}

function LookUpE (z:real; n:integer; L,R:real; y:RealList0):real;
{Pre:L<=z<=R and for k=0,...,n y[k], is the value of a function at L + kh }
{where h = (R-L)/n.}
{Post:The interpolated value of the function at xval.}

function LinearSearch (z:  real; n:integer; xval:RealList0):integer;
{Pre:n>=1, xval[0]<xval[1]<...<xval[[n] and xval[0]<=z<=xval[n]}
{Post:If k = LinearSearch(z,n,xval), then 0<=k<n and xval[k]<=z<=xval[k+1].}

function BinarySearch(z:real; n:integer; xval:RealList0):integer;
{Pre:n>=1, xval[0]<xval[1]<...<xval[[n] and xval[0]<=z<=xval[n]}
{Post:If k = BinarySearch(z,n,xval), then 0<=k<n and xval[k]<=z<=xval[k+1].}

function LookUpULin(z:real; n:integer; xval,yval:  RealList0):real;
{Pre:xval[0]<...xval[n] and xval[0]<=z<=xval[n].  For k=0,...,n }
{yval[k], is the value of a function at xval[k].  }
{Post:The interpolated value of the function at z}
```

```
function LookUpUBin(z:real; n:integer; xval,yval:RealList0):real;
{Pre:xval[0]<...xval[n] and xval[0]<=z<=xval[n].  For k=0,...,n }
{yval[k], is the value of a function at xval[k].  }
{Post:The interpolated value of the function at z}
```

Chap16Codes

```
unit Chap16Codes;

interface

const
   pmax = 1000;
type
   LongIntList = array[1..pmax] of longint;

procedure DivisorList(x:longint; var n:integer;var d:LongIntList);
{Pre:x>=1}
{Post:n is the number of divisors of x and they are contained }
{in d [1..  n], ranging from smallest to largest.}

function IsPrime(p:longint):boolean;
{Pre:p>=1}
{Post:p is a prime}

procedure PrimeList(x:longint; var n:  integer; var p:  LongIntList);
{Pre:x>=1}
{Post:n is the number of prime numbers <=x and they are contained}
{in p[ 1..n], ranging from smallest to largest.}

const
   nStates = 51;
type
   IntegerList = array[1..nStates] of longInt;
   RealList = array[1..nStates] of real;
   StringList = array[1..nStates] of string[25];

procedure GetPopulationData(year:string; var state:stringlist; var p:IntegerList);
{Pre:The relevant census year.  Either '1980' or '1990'}
{Post:Define Washington DC as the 51st state.}
{The names of the 51 states are in state[1..51] and their populations in p[1..51].}

function MaxIndex(n:integer; x:RealList):integer;
{Pre:x[1..n] initialized.}
{Post:x[k] >= every component of x[1..n].}
```

```
procedure Apportion (C,n:  integer; p:IntegerList; function f(pop:longint;
      di:integer):  real; var d:  IntegerList; var PV: RealList);
{Pre:C = number of districts, n = number of states, p[1..n] their populations,
{f is the priority function.}
{Post:For i=1..n, d[i] is the number of districts awarded to state i and}
{ PV[i] = f(p[i],d[i]) }

function SmallestDivisors(pop:longint; di:integer):real;

function MajorFractions(pop:longint; di:integer):real;

function EqualProportions(pop:  longint; di:integer):real;

function HarmonicMean(pop:longint; di:integer):real;

function GreatestDivisors(pop:longint; di:integer):real;
```

Chap17Codes

```
unit Chap17Codes;

interface

uses
   DDCodes, Chap14Codes, Chap15Codes;
const
   rowmax = 30;
   colmax = 20;
type
   Real2DArray = array[1..rowmax, 1..colmax] of real;
   Real2DArray0 = array[0..rowmax, 0..colmax] of real;
   Int2DArray = array[1..rowmax, 1..colmax] of integer;
   LongInt2DArray = array[1..rowmax, 1..colmax] of longint;

procedure ShowElement(hL,vT,p,size:integer; x:real);
{Pre:0<=p<=3}
{Post:Prints the value of x to p decimal places in a size-by-size square with}
{upper left corner at (hL,vT).}

procedure Show2DArray(hL,vT,p,size:integer; A:Real2DArray; m,n:integer);
{Pre:0<=p<=3, A[1..m,1..n] initialized.}
{Post:Displays the array A[1..m,1..n], with upper left corner at (hL,vT). }
{The value of each entry is printed to p decimal places in a }
{size-by-size square.}

procedure HighLightSubArray(hL,vT,size,i1,i2,j1,j2:integer);
{Post:Draws a thick line rectangle defined by hL+(j1-1)size <= h <= hL+(j2-1)size }
{and vT+(i1-1)size <= v <= vT+(i2-1)size}

function SubArray(A:Real2DArray; i1,i2,j1,j2:integer):Real2DArray;
{Pre:1<=i1<=i2, 1<=j1<=j2 and A[i1..i2,j1..j2] initialized.}
{Post:The array A[i1..i2,j1..j2]}
```

```
function SetUp2D(m,n:integer; xval,yval:RealList0;
         function f(x,y:real):real):Real2DArray0;
{Pre:xval[0..m] and yval[0..n] and f defined at all (xval[i],yval[j]).}
{Post:SetUp2D[i,j] = f(xval[i],yval[j), i=0..m, j=0..n}

function RandM(m,n:integer; L,R:real; var rn:real):Real2DArray;
{Pre:0<rn<1, L < R}
{Post:An m-by-n array with entries chosen randomly from the interval [L,R].}
{The value returned in rn is the value obtained by executing m*n times the}
{assignment rn:=rand(rn).}

function Pascal(n:integer):LongInt2DArray;
{Post:The n-by-n Pascal array.}

function Normalize(m,n:integer; A:Real2DArray):Real2DArray;
{Pre:A[1..m,1..n] initialized}
{Post:Each column of Normalize[1..m,1..n] is the normalization of the}
{corresponding column of A.}

procedure MaxTriangle(n:integer; x,y:RealList; var imax,jmax,kmax:integer;
      var Pmax:real);
{Pre:x[1..n] and y[1..n] contain the xy coordinates of n points and that n>=3.}
{Post:Of all the triangles defined by selecting three distinct points, the one with}
{vertices (x[imax],y[imax]), (x[jmax],y[jmax]), and (x[kmax],y[kmax])}
{has the longest perimeter which is given by Pmax.}

{**************************** Production Cost *************************************}

const
   sizeH = 35;
   sizeV = 24; {Array entries displayed in boxes of size sizeH-by-sizeV}
   hCost = 160;
   vCost = 30; {(hCost,vCost) = upper left corner of displayed cost array.}
   hPO = 160;
   vPO = 164; {(hPO,vPO) = upper left corner of displayed purchase order array.}
   hInv = 160;
   vInv = 222; {(hInv,vInv) = upper left corner of displayed inventory array.}
   nfact = 4; {Number of factories.}
   nprod = 8; {Number of products.}

type
   PurchaseOrder = array[1..nprod] of integer;
   IntTable = array[1..nfact, 1..nprod] of integer;

var
   PO:PurchaseOrder;
   Cost,Inventory:IntTable;

function iCost(i,n:integer; Cost:IntTable; w:PurchaseOrder):integer;
{Pre:Cost[i..i,1..n] and w[1..n] initialized.}
{Post:Cost[i,1]w[1] +...+ Cost[i,n]w[n]}
```

```
procedure Cheapest(m,n:integer; Cost:IntTable; w:PurchaseOrder;
        var imin,MinCost:integer);
{Pre:Cost[1..m,1..n] and w[1..n] initialized}
{Post:i=imin minimizes Cost[i,1]w[1]+...+ Cost[i,n]w[n] and MinCost is the}
        { correponding value.}

function iCanDo(i,n:integer; Inv:IntTable; w:PurchaseOrder):boolean;
{Pre:Inv[i,1],...,Inv[i,n] initialized}
{Post:Inv[i,j]>=w[j] for j=1..n}

procedure ShowMessage(s:string);

procedure DisplayEntry(hL,vT,x:integer);
{Post:Displays x inside the rectangle (vT,hL,vT+sizeV,hL+sizeH)}

procedure HighLight(vT,hL,vB,hR:integer);
{Post:Draws rectangle (vT,hL,vB,hR) with thick lines.}

procedure InitializeTables;
{Post:Sets up menus, initializes and displays Cost[1..m,1..n], Inventory[1..m,1..n],}
{and PO[1..n].}

procedure ShowStatus;
{Pre:Cost[1..m,1..n], Inv[1..m,1..n], and w[1..n] initialized.}
{Post:Displays production costs for each factory and whether or not there is}
          {sufficient inventory.}

procedure UpDateTable(hL,vT,m,n,step:integer; s:string; var A:IntTable);
{Post:Updates a clicked entry}

procedure UpDatePurchaseOrder(n,step:integer; s:string; var w:PurchaseOrder);

procedure Deplete(i,n:integer; PO:PurchaseOrder; var Inv:IntTable);
{Pre:Inv[i..i,1..n] and PO[1..n] are initialized.}
{Post:Reduces Inv[i,j] by PO[j], j=1..n and updates the displayed array accordingly.}

{******************************* Bit Map ***************************************}

type

   Boolean2DArray = array[1..50, 1..50] of boolean;
   DotMatrix = array[1..7, 1..5] of boolean;
   ListOfDotMatrix = array[1..20] of DotMatrix;
   DotMatrixDigits = array[0..9] of DotMatrix;

function Build7x5(hL,vT,d:integer):DotMatrix;
{Pre:Interactively set's up a 7-by-5 boolean array.}
{The displayed 7-by-5 array is made up of d-by-d tiles}
{and the upper left corner is at (hL,vT).}

procedure Show7x5(hL,vT,d:integer; B:DotMatrix);
{Post:Displays B as a 7-by-5 array of d-by-d tiles.}
{The upper left corner of the upper left tile is at (hL,vT). }
{The (i,j) tile is ''inked'' if B[i,j] is true.}
```

```
function Digit7by5(n:integer):DotMatrix;
{Pre:0<=n<=9}
{Post:The 7-by-5 dot matrix encoding of n.}
```

```
procedure DrawBigDigit(k,hL,vB,p:integer; D:DotMatrixDigits);
{Pre:0<=k<=9, D houses the 7-by-5 dot matrix representations.}
{Post:Draws the digit k as a 7-by-5 dot array with radius p dots.  The lower}
{left corner of the digit area is at (hL,vB).}
```

Chap18Codes

```
unit Chap18Codes;

interface

uses
   DDCodes;

{**************************** Single Point tools ******************************}

type
   RealPoint = record
         x,y:real;
      end;
   {If p is a RealPoint, then it represents the point (p.x,p.y).}

function MakePoint(x,y:real):RealPoint;
{Post:The point (x,y).}

procedure ShowPoint(p:RealPoint);
{Post:Draws a dot at the screen location corresponding to p.}

procedure ShowBigPoint(p:RealPoint);
{Post:Draws a big dot at the screen location corresponding to p.}

function GetPoint(Message:string):RealPoint;
{Post:Displays a message and returns the next clicked point.}

procedure MoveToPoint(p:RealPoint);
{Post:Moves mouse to the screen cooordinate corresponding to p.}

procedure LineToPoint(p:RealPoint);
{Post:Draws a line from the current mouse position to the}
{screen coordinate corresponding to p.}

function MidPoint(P1,P2:RealPoint):RealPoint;
{Post:The midpoint of the line segment extending from P1 to P2.}
```

```
function Distance(p1,p2:RealPoint):real;
{Post:The distance from p1 to p2.}

{*************************** Point Set tools ******************************}

const
   MaxNumPoints = 500;
type
   RealPointList = array[1..MaxNumPoints] of RealPoint;

procedure GetManyPoints(var n:integer; var p:RealPointList);
{Post:n points are clicked and p[1..n] contains the associated points.  }

procedure ShowManyPoints(n:integer; p:RealPointList);
{Pre:p[1..n] initialized,}
{Post:Draws the points p[1]..p[n].}

procedure ConnectPoints(n:integer; p:RealPointList);
{Post:Connects the points p[1],...,p[n],p[1] in order.}

function Centroid (n:integer; p:RealPointList):RealPoint;
{Pre:p[1..n] initialized}
{Post:  The centroid.}

{**************************** Line Segment tools ******************************}

type
   LineSegment = record
         p,q:RealPoint;
      end;

{If L is a LineSegment, then it represents the line segment from L.p to L.q.}

function MakeLineSegment(p,q:RealPoint):LineSegment;
{Post:The line segment from p to q.}

procedure ShowLineSegment (L:LineSegment);
{Post:Draws L.}

function PointToLineSegment(p0:RealPoint; L:LineSegment):RealPoint;
{Post:The closest point to p0 on L}

function SameSide(p,q:RealPoint; L:LineSegment): boolean;
{Post:p and q are on the same side of the line defined by L.}

{*************************** Triangle tools ******************************}

type

   Triangle = record
         p,q,r:RealPoint
      end;
   {If T is a triangle, then it represents the triangle with vertices T.p, T.q,}
   {and T.r.}
```

```
function MakeTriangle(p,q,r:RealPoint):Triangle;
{Post:The triangle with vertices p, q, and r.}

procedure ShowTriangle(T:Triangle);
{Post:Draws T.}


{**************************** Rectangle tools ******************************}

type
   Rectangle = record
         x0,x1:real;
         y0,y1:real
   end;
   {If R is a rectangle, then it represents the set of all points (x,y)}
   {that satisfy R.x0 <= x <= R.x1 and R.y0 <= y <= R.y1.}

function MakeRectangle(p,q:RealPoint):Rectangle;
{Post:The rectangle that has the line segment from p to q as a diagonal.}

procedure ShowRectangle (R:Rectangle);
{Post:Draws rectangle R.}

{**************************** Polygon tools ******************************}

type
   Polygon = record
         n:integer;
         v:RealPointList
      end;
   {If P is a polygon, then it obtained by connecting the points }
   { P.v[1] , P.v[2],.., P.v[ P.n], P.v[1] in turn .  }

function MakePolygon(n:integer; v:RealPointList):Polygon;
{Post:The polygon obtained by connecting vertices v[1],..,v[n].}

procedure ShowPolygon (P: polygon);
{Post:Draws P.}
```

Chap19Codes

```
unit Chap19Codes;

interface

const
   nmax = 200; {The representation can handle integers with <= nmax+1 digits.}

type
   digit = 0..9;
```

```
VeryLongInt = record
      sign:boolean;
      L:integer;
      a:array[0..nmax] of digit;
   end;
{If x has type VeryLongInt, then x represents an integer i where:}
{ The sign of i is positive if x.sign is true and negative if x.sign is false.}
{ The absolute value of i is x.a[0] + x.a[1]*10 + ...  + x.a[q]*(10^q) }
{ with q = x.L.}

rational = record
      num:longint;
      denom:longint;
   end;
{If x is rational, then x.num/x.denom is its value.}

complex = record
      r:  real;
      i:  real;
   end;
{If z is complex then z.r + i*z.i is its value with i = sqrt(-1).}
```

```
{************ Very Long Integer Functions ****************}

function Int2VLInt(n:longint):VeryLongInt;
{Post:The representation of n as a VeryLongInt.}

function AddPosVLInt(x,y:VeryLongInt):VeryLongInt;
{Post:The sum of x and y|}

function VLInt2Str(x:VeryLongInt; w:integer):  string;
{Post:Converts x to equivalent string representation.  }
{If fewer than w characters are needed, then enough blanks are added to}
{to the front so that the returned string has length w.}

{************ Rational Functions ****************}

function gcd(p,q:longint):longint;
{Pre:p and q positive integers.}
{Post:the greatest common divisor of p and q.}

function reduce(r:rational):rational;
{Post:r reduced to lowest terms.}

function AddRational(a,b:rational):rational;
{Post:The sum of a and b reduced to lowest terms.}

function SubRational(a,b:  rational):rational;
{Post:The diference a - b reduced to lowest terms.}

function MultRational(a,b:  rational):rational;
{Post:The product of a and b reduced to lowest terms.}
```

```
function DivRational (a,b:rational):rational;
{Pre:b <> 0}
{Post:The quotient a/b reduced to lowest terms.}

{*********** Complex Functions ***************}

function AddComplex(u,v:complex):complex;
{Post:u+v}

function SubComplex(u,v:complex):complex;
{Post:u-v}

function MultComplex(u,v:complex):complex;
{Post:u*v}

function DivComplex(u,v:complex):complex;
{Pre:v<> 0}
{Post:u/v}

function ScaleComplex(u:complex; s:real):complex;
{Post:s*u}

function Cabs(z:complex):real;
{Post:The absolute value of z.}

function ExpComplex(z:complex):complex;
{Post:Define S(n) = 1+z + z^2/2!  +...+ z^n/n!  and let m be the smallest}
{value positive integer so that |z^m/m!| <= .00001*|S(-1)|.  If m<= 30, then }
{S(m) returned.  Otherwise, S(30) is returned.}

function Newton(z:complex):complex;
{Post:  (3z^4+1)/(4z^3)}

function Circle(t:real):complex;
{Post:  cos(2pi*t) + i*sin(2pi*t)}

procedure DisplayOrbits(function f(z:complex):complex; function g(t:real):complex;
      n,m:integer; a,b,p real);
{Pre:  (a+bi) corresponds to the center of the screen.}
{p pixels per unit distance along both the real and imaginary axes.}
{f(z) is defined if z = g(t), 0<=t<=1.}
{Post:Plots the length-m orbits the points g(k/n), k=0..n under the action of f.}
```

Chap20Codes

```
unit Chap20Codes;

interface

uses
   DDCodes;
const
   DegreeMax = 20;
type
   CoeffList = array[0..DegreeMax] of real;
   polynomial = record
         deg:integer;
         a:CoeffList
      end;
   RatFunc = record
         num:polynomial;
         denom:polynomial;
      end;

function DerPoly(p:polynomial):polynomial;
{Post:The derivative of p.}

function AddPoly(alfa,beta:real; p,q:polynomial):  polynomial;
{Post:The polynomial alfa * p and beta * q.}

function LinearTimesPoly(b,c:real; p:polynomial):polynomial;
{Pre:1 + p.deg <= degmax.}
{Post:The polynomial (b+cx)p(x).}

function Chebychev(n:integer):polynomial;
{Pre:n>=0}
{Post:The n-th Chebychev polynomial}

function MultPoly(p,q:polynomial):polynomial;
{Pre:(p.deg)*(q.deg) <=MaxDegree}
{Post:The product of p and q.}

function EvalPoly(p:polynomial; t:real):real;
{Post:The value of p at t.}

function SetUpPoly(n:integer; p:polynomial; xval:RealList0):RealList0;
{Pre:0<=n<=nmax.  p(x) defined at x = xval[0],...,xval[n].}
{Post:SetUpPoly[k] = p(x[k]), k=0,...,n.}

function Pade(n,m:integer):RatFunc;
{Pre:n,m>=0}
{Post:The (n,m) Pade approximate of exp(x).}

function SetUpRatFunc(n:integer; r:RatFunc; xval:RealList0):RealList0;
{Pre:0<=n<=nmax.  p(x) defined at x = xval[0],...,xval[n].}
{Post:SetUpPoly[k] = p(x[k]), k=0,...,n.}
```

Chap21Codes

```
unit Chap21Codes;

interface

const
   nmax = 1024;
   inmax = 250;

type
   RealList = array[1..nmax] of real;
   IntegerList = array[1..inmax] of integer;
   country = record
         Name:string[25];
         population:longint;
         area:longint;
         capital:string[25];
         lat:real;
         long:real
      end;
   CountryList = array[1..200] of country;

function LeftShift (n:integer; x:RealList):RealList;
{Pre:x[1..n] initialized.}
{Post:LeftShift[k] = x[k+1] for k=1..n-1.  and LeftShift[n]=x[1]}

function MultipleLeftShift1(m,n:integer; x:RealList):RealList;
{Pre:x[1..n] initialized and 1<=m<=n}
{Post:MultipleLeftShift1[k]= x[k+m] for k=1..n-m, and x[k-n+m] for k=n-m+1..n.}

function MultipleLeftShift2(m,n:integer;x:RealList):RealList;
{Pre:x[1..n] initialized and 1<=m<=n}
{Post:MultipleLeftShift2[k]= x[k+m] for k=1..n-m, and x[k-n+m] for k=n-m+1..n.}

function PerfShuffle(n:integer; a:RealList):RealList;
{Pre:n is even and x[0..n-1] is initialized.}
{Post:PerfShuffle[k] = a[k div 2] if k is even and a[(n + k ) div 2] if k is odd.}

procedure bSort(n:integer; var a:RealList);
{Pre:a[1..n] initialized.}
{Post:a[1..n] is a permutation of a[1..n] and a[1] <=a[2]<=...<=a[n].}

procedure iSort(n:integer; var a:RealList);
{Pre:a[1..n] initialized.}
{Post:a[1..n] is a permutation of a[1..n] and a[1] <=a[2]<=...<=a[n].}

procedure Merge(p,m,q:integer; var a,b:RealList);
{Pre:p<=m<q, a[p]<=...<=a[ m] ,a[m+1]<=...<=a[q].}
{Post:Permutes the values in a[p..q] so that a[p]<=...<=a[q]}

procedure mSort(n:integer; var a,b:RealList);
{Pre:n a power of two and a[1..n] initialized.}
{Post:a[1..n] is a permutation of a[1..n] and a[1] <=a[2]<=...<=a[n].}
```

```
procedure ShowArray(i,j:integer; a:RealList; p:integer);
{Pre:a[i..j] initialized,p>0}
{Post:Prints a[i..j] on a single line with p:0 format.}

procedure GetWorldData(var n:integer; var Nation:CountryList);
{Pre:Project must be opened in the Chapter 21 folder.}
{Post:World data represented in Nation[1..n].}

function GCircleDist(Lat1, Long1, Lat2, Long2:  real):  real;
{Pre:(Lat1,Long1) and (Lat2,Long2) are the latitude/longitude coordinates}
{ (in degrees) of two points on Earth.}
{ -90<=Lat<=90,-180<=Long<=180}
{Post:The great circle distance (in miles) between them.}

procedure bSortGeneral(n:integer; var a:CountryList;
          function InOrder(y,z:Country):boolean);
{Pre:a[1..n] initialized.  InOrder(y,z) is true if y comes before z in the sort.}
{Post:The values in a[1..n] are permuted so that for k=1..n-1,}
{InOrder(a[k],a[k+1]) is true.}

procedure bSortI(n:integer; a:RealList; var i:IntegerList);
{Pre:a[1..n] initialized.}
{Post:For k=1..n-1, a[i[k]] <= a[i[k+1]] is true}
```

Chap22Codes

```
unit Chap22Codes;

interface

uses
   DDCodes, Chap18Codes;
type
   IntegerList = array[1..MaxNumPoints] of integer;
   BooleanList = array[1..MaxNumPoints] of boolean;

procedure ShowPath(n:integer; p:RealPointList; Path:IntegerList);
{Pre:n>=1, p[1..n] and Path[1..n] are initialized.}
{Post:Shows the path whose i-th "stop" is p(Path[i]).}

procedure NearestToSubset(p0:RealPoint; n:integer; p:RealPointList;
         UnVisited:BooleanList;
      var i:integer; var dmin:real);
{Pre:n>=1 , p[1..n] are initialized.  UnVisited[1..n] contains at least one true value.}
{Post:dmin is the distance from p0 to the set of unvisited points among}
{(x[1],y[1]),...,(x[n],y[n]) and i is the index of the closest unvisited point.}

procedure Trip(s,n:integer; p:RealPointList; var Path:IntegerList; var odometer:real);
{Pre:n>=1, 1<=start<=n, and p[1..n] is initialized.}
{Post:Path[1..n] is a permutation of 1..n and odometer is the roundtrip distance of}
{the path that visits points Path[1],..,Path[n].}
```

```
type
   Range = record
         L,H:array[2..6] of integer;
      end;
      {If R has type Range then:  }
      { R.L[2] < R.L[3] < R.L[4] < R.L[5] < R.L[6]}
      { R.H[2] < R.H[3] < R.H[4] < R.H[5] < R.H[6]}
      { R.L[k] <= R.H[k], k=2,..,6.}
   PedalList = array[1..3] of integer;
   WheelList = array[1..7] of integer;
   Gear = record
         i,j:integer;
         r:real
      end;
      {If g is a gear, then g.r is the ratio formed by the pairing}
      {of pedal sprocket i and wheel sprocket j.}
   GearList = array[1..21] of gear;
   Bike = record
         p:PedalList;
         w:WheelList;
         g:GearList;
      end;
      {If b is a bike, then b.p[1..3] and b.w[1..7] contain the pedal and wheel }
      {sprockets and b.g[1..21] are the gears, sorted from small to large.}

procedure PrintBike(hL,vT:integer; b:bike; fb:real);
{Post:Displays the values of b.p and b.w and b.val with upperleft corner of the}
{ 'information box' at (hL,vT). fb is the value of the bike.}

function SetUpRange (OldRange:  range):  range;
{Post:Displays the old range table, solicits updates, displays updates, }
{and returns new range values.}

function MakeBike(p:PedalList; w:WheelList):bike;
{Post:The bike with pedal sprockets p[1..3] and wheel sprockets w[1..7]}

function TheBestBike(R:range; function f(b:bike):real):bike;
{Post:The bike b that minimizes f subject to the constraints that}
{ (1) b.p[3] < b.p[2] < b.p[1]}
{ (2) b.w[7] < b.w[6] < b.w[5] < b.w[4] < b.w[3] < b.w[2] < b.w[1] }
{ (3) b.p[1], b.p[2], b.p[3] have one of the values 52, 48, 42, 39, 32, 28}
{ (4) b.p[1]/b.w[7] = 4}
{ (5) b.p[3] = b.w[1]}
{ (6) R[k].low <= b.w[k] <= R[k].high , k=2..6}
```

```
type
   Ellipse = record
         F1,F2,P,Center:RealPoint;
         d:real;
         rx,ry:real;
         c,s:real;
         Area:real;
      end;
   {If E is an ellipse, then E.F1 and E.F2 are its focii and E.P is a perimeter.}
   {E.Center is its center, E.rx and E.ry are the semiaxes, and E.d is the string}
   {distance, E.c and E.s are the cosine and sine of its tilt and E.Area is its area.}

procedure DefineEllipse(var F1,F2,P:RealPoint);
{Post:F1,F2,P are the next three clicked points.  Displayed as large black dots.}

function MakeEllipse(F1,F2,P:RealPoint):Ellipse;
{Post:The ellipse E with focii F1 and F2 with the property that P is on E.}

procedure DrawEandP(E:ellipse; n:integer; PointSet:RealPointList);
{Post:Draws E and PointSet.}

function Contain(n:integer; PointSet:RealPointList; E:Ellipse):boolean;
{Pre:PointSet[1..n] initialized.}
{Post:Every point in PointSet is inside E.}
```

Chap24Codes

```
unit Chap24Codes;

interface

uses
   DDCodes;

const
   seed = 0.123456; {For starting the random number sequence.}
   xL = -0.25; {The rays originate from a source bar along the vertical x=xL.}
   xR = 1.05; {The targets are located along the vertical x = xR.}
   nTargets = 25; {The number of targets.}
   nT = 25; {Size of Grid}
   LMax = 100; {Max number of tiles traversed by ray.}

type
   Real2DGrid = array[1..nT, 1..nT] of real;
```

```
   TileCrossing = record
         i,j:integer;
         d:real;
      end;
      {If t has type TileCrossing, then a path of length d }
      {was traversed across tile T(t.i,t.j) }
   TileCrossingList = array[1..Lmax] of TileCrossing;
   RayPath = record
         L:integer;
         TC:TileCrossingList;
      end;
      {If R is a RayPath, then it crosses tiles (R.TC[p].i,R.TC[p].j), p=1..R.L}

procedure DrawTarget;
{Pre:The constant nTargets specifies the number of targets.}
{Post:Displays Square, Source Bar, and Targets}


function GetSource(ScanNo:integer):real;
{Post:The y-coordinate of the source.  Also displays the source and its index.}

procedure CrossTile(EnterSide:char; f1,m,h:real; var ExitSide:char; var f2,d:real);
{Pre:m is the slope of the incoming ray, 0<=f1<=1, EnterSide = 'T' (for top),}
{'B' (for bottom), or 'L' (for left).}
{Let (x0,y0) = location of lower left corner and h = tile side.  Then}
{EnterSide = 'T' and the entry point = (x0+h*f1,y0+h)}
{EnterSide = 'B' and the entry point = (x0+h*f1,y0)}
{EnterSide = 'L' and the entry point = (x0,y0+h*f1)}
{Post:0<=f2<=1 and ExitPoint = 'T', 'B', or 'R' (for right).}
{Exitside = 'T' and the exit point = (x0+h*f2,y0+h)}
{Exitside = 'B' and the exit point = (x0+h*f2,y0)}
{ExitSide = 'R' and the exit point = (x0+h,y0+h*f2)}
{d is the distance from the entry point to the exit point}

function MakeTileCrossing(i,j:integer; d:real):TileCrossing;

function MakeRayPath(m,y:real):RayPath;
{Pre:0<=y<=1}

procedure GenRandGrid(var A:Real2DGrid; var i0,j0:integer);
{Post:i0 and j0 are random integers from [1,n].  }
      {A[i,j] = exp(-6((i-i0)^2+(j-j0)^2/nTiles).}

function Density(R:RayPath; A:Real2DGrid):real;
{Pre:A[1..nTiles,1..nTiles] initialized.}
{Post:The ''density'' of the ray R.}
```

Index